VEGETABLE SCIENCE

VEGETABLE SCIENCE

Editor

Hartwig Rosenberger

scitus
academics

Vegetable Science

Edited by **Hartwig Rosenberger**

ISBN: 9781681179414
Library of Congress Control Number: 2017911248

SCITUS Academics LLC,
616, Corporate Way, Suite 2, 4766,
Valley Cottage, NY 10989

www.scitusacademics.com

Printed in United States of America.

Preface

Vegetable growing is a substantial branch of horticulture. But it is no longer confined to growing in a small garden as vegetables are now cultivated on a large scale commercially. The study of vegetable culture is called as 'Olericulture'. Vegetables are rich and comparatively cheaper source of vitamins. Consumption of these items provides taste, palatability, increases appetite and provides fiber for digestion and to prevent constipation. They also play key role in neutralizing the acids produced during digestion of pretentious and fatty foods and also provide valuable roughages which help in movement of food in intestine. It is the most important factor influencing crop yield therefore for getting proper yields; it is essential to know the climatic requirement of the vegetable crops. Almost all the vegetables have optimum range of climatic components and beyond those limits vegetables cannot perform up to the mark. The perishable nature of vegetables demand comprehensive planning for movement, Storage, processing and distribution of vegetable products. The growth of vegetable industry as a commercial proposition largely depends on mainly allied enterprises like storage, processing marketing and maintenance and service enterprises to encourage vegetable growing.

This volume covers original researches in the field of breeding, physiology, ecology, production, management, seed technology, genomics molecular breeding, plant genetic resources, disease and pest management, marketing, post-harvest etc. innovative articles on all aspects of vegetable production, including growth regulation, pest management, sustainable production, harvesting, handling, storage, shipping, and final consumption. While the primary focus of this volume is on the production of improved, indigenous, and wild vegetables, consideration of their nutritional, medicinal and industrial importance and genetic conservation are also outlined to provide a comprehensive overview of their contributions to the concepts of stability and sustainability of tropical agriculture. The contributed chapters are written by researchers, practitioners, and academics present current findings on new crops and protected culture as well as traditional crops, examine marketing trends in the commercial vegetable industry, and address vital issues of concern to breeders, production managers, and processors working in all continents where vegetables are grown.

Table of Contents

CHAPTER 1

New Approaches to Irrigation Scheduling of Vegetables

Michael D. Cahn[1,*] and Lee F. Johnson[2]

[1]*University of California, Cooperative Extension, Monterey County, 1432 Abbott St., Salinas, CA 93901, USA*

[2]*NASA ARC-CREST/CSUMB, MS 232-21, Moffett Field, CA 94035, USA*

ABSTRACT

Using evapotranspiration (ET) data for scheduling irrigations on vegetable farms is challenging due to imprecise crop coefficients, time consuming computations, and the need to simultaneously manage many fields. Meanwhile, the adoption of soil moisture monitoring in vegetables has historically been limited by sensor accuracy and cost, as well as labor required for installation, removal, and collection of readings. With recent improvements in sensor technology, public weather-station networks, satellite and aerial imaging, wireless communications, and cloud computing, many of the difficulties in using ET data and soil moisture sensors for irrigation scheduling of vegetables can now be addressed. Web and smartphone applications have been developed that automate many of the calculations involved in ET-based irrigation scheduling. Soil moisture sensor data can be collected through wireless networks and accessed using web browser or smartphone apps. Energy balance methods of crop ET estimation, such as eddy covariance and Bowen ratio, provide research options for further developing and evaluating crop coefficient guidelines of vegetables, while recent advancements in surface renewal instrumentation have led to a relatively low-cost tool for monitoring crop water requirement in commercial farms. Remote sensing of crops using satellite, manned aircraft, and UAV platforms may also provide useful tools for vegetable growers to evaluate crop development, plant stress, water consumption, and irrigation system performance.

Keywords: evapotranspiration; crop coefficients; soil moisture sensors; NDVI; web application; decision support tools; UAV; remote sensing

1. INTRODUCTION

1.1. Water Scarcity and Commercial Vegetable Production

Most economically important vegetable production regions of the world have Mediterranean, semi-arid, or desert climates in which supplemental irrigation is required to maximize yield and quality. Efficient irrigation management has become a major concern for vegetable farmers in these areas as water supplies have become more restricted and environmental impairments to ground and surface water from agricultural run-off and drainage has received more attention from regulatory agencies. Agricultural water resources have become increasingly stressed in the vegetable producing regions of Australia, Southern Europe, Chile, the Middle East, North Africa, China, and the western United States. California has been under severe drought conditions for four years since 2012 [1,2], and ground water supplies have been depleted to historically low levels on the central coast, where most of the salad vegetables are grown for the US and export markets. Lowering of the ground water table below sea-level through agricultural pumping has caused salt water to intrude into coastal aquifers and threatened the sustainability of thousands of hectares of prime farmland used for vegetable production [3]. Australia also experienced a severe decade-long drought during the early 2000's that impacted vegetable farmers in the Murray-Darling Basin in New South Wales and Queensland [4]. The vegetable production regions of Chile [5] and Spain [6] have also experienced recurring multiyear droughts.

Since many vegetables require high rates of nitrogen fertilizer and have shallow root systems, leaching of nitrate during the irrigation season and during periods of heavy precipitation has resulted in nitrate contamination of aquifers in many key vegetable production regions [7,8,9]. Irrigation run-off from vegetable fields also contaminates surface water bodies such as rivers, creeks, estuaries, and lakes with nutrients and pesticides [10,11,12]. Growers in California and Europe must report on the amount of fertilizer nitrogen that they apply to their crops to comply with environmental regulations [13,14]. They are also required to implement best practices to minimize nitrate losses, which will require improving irrigation management.

1.2. Challenges for Irrigation Scheduling in Modern Vegetable Operations

Vegetable production poses several unique challenges in managing irrigation water efficiently. One of the major challenges for growers is the number of fields that must be concurrently managed in a medium to large size vegetable operation. Fields sizes for vegetable crops tend to be small (<5 ha) relative to agronomic crops, especially fields planted with leafy green, crucifer, and other salad vegetables. Large vegetable growing operations in the Salinas valley of California manage 1–2 thousand ha of vegetables, with an average field size of just 4 ha. Small fields permit intensive management, allowing plantings to be staggered so that a steady supply of vegetables can be delivered to buyers and shippers. However, smaller field sizes means that farming decisions must be coordinated for many fields that may vary with respect to maturity, soil texture, cultural practices, microclimate, or other site-specific conditions.

In addition to field size, the diversity of vegetables and number of crop rotations per season increases management complexity. Most diversified operations will produce more than 30 types of vegetables, each with unique nutrient and water requirements. Many vegetables are grown over short cropping cycles. Leafy green salad mixes like leaf and crisphead lettuce, for instance, reach maturity over just 30–65 day intervals during the summer. As a result, farmers in Mediterranean climates often grow two to three rotations of vegetable crops per year. Fields often are intensively tilled between crops to remove compacted zones caused by traffic from harvest machinery and to prepare raised beds for seeding or transplanting the next crop. These activities require that crews retrieve drip tape or sprinkler pipe after each crop is harvested.

Another irrigation scheduling challenge is the number of field operations that must be coordinated during a crop cycle. Fields are typically sprayed to protect against insect and disease pests several times per crop cycle, and must be periodically cultivated to control weeds. Fertilizer applications are made several times during the growth cycle to satisfy the nutrient requirements of vegetables. Irrigations must be timed to accommodate access for tractor and hoeing crews so that activities can be completed on schedule. There may be several extended periods when a vegetable crop cannot be irrigated to permit the soil to dry sufficiently to allow tractor access. Irrigation equipment, such as sprinkler and mainline pipes, may need to be moved before each pass of a tractor through a field.

Irrigation scheduling is also complicated by the numerous methods that vegetable growers use to supply water to their crops. Depending on the crop type, plant density, development stage, water source, and field and soil characteristics, growers may choose to irrigate using overhead sprinklers, drip, furrow, or a combination of methods. In California, leafy green vegetables harvested for salad mixes are seeded at high densities (8.5 million plants per hectare) [15] on 2 m wide raised beds and are irrigated almost exclusively with sprinklers. Small seeded vegetables such as crisphead and romaine lettuce are typically germinated with sprinklers, and after several weeks, irrigated with surface drip until harvest. Celery transplants may be established using overhead sprinklers, then drip irrigated, and occasionally flood irrigated to rewet the shoulders of raised beds.

Since vegetable production is frequently characterized by short crop cycles, intensive crop rotations, and numerous small fields at different stages of maturity, farm managers are challenged to schedule and coordinate all the field activities while taking time to carefully schedule irrigations to optimize water use. Consequently, many farm managers may follow predetermined irrigation schedules to simplify water management and make small adjustments during the cropping season depending on their observations of the crop, soil, and weather conditions. Usually the amount of water applied to high value vegetables exceeds crop evapotranspiration (ET) to avoid water stress. In the Salinas Valley of California, for example, applied water amounts on broccoli, cauliflower, and cabbage averaged >200% of the estimated crop evapotranspiration (ETc) requirement [16]. While these applications rates may avoid water stress so that yield and quality are maximized, over-irrigating may strain water supplies and result in the leaching of nitrate and/or generate run-off that degrades the quality of receiving surface water bodies.

Considering the numerous challenges and limitations of managing vegetables, growers, farm managers, and irrigators need convenient methods to schedule irrigations so that they can better determine how much water to apply to match the requirements of their crops. During the past few decades, significant improvements and lower costs for wireless communications, computing power, sensor technology, and aerial and satellite imagery have increased the potential to develop accurate and intuitive approaches for scheduling irrigations in vegetables.

2. ADVANCES IN SOIL MOISTURE SENSOR TECHNOLOGY

2.1. Recent Developments in Soil Moisture Sensing

The monitoring of soil moisture has long been a standard way to determine when crops need to be irrigated. Growers and farm managers typically evaluate soil moisture by probing with a shovel or auger or monitoring with sensors. Sensors for volumetric moisture content and soil water tension have been commercially available for more than 40 years but were originally used more in research than for commercial crop production. Volumetric soil moisture sensors provide readings in units of $m^3 \cdot m^{-3}$, and tension-based sensors readings are typically in units of kPa, where a greater absolute value corresponds to drier soil conditions [17]. In the past, the primary barriers to more widespread use of soil moisture sensors in irrigation management have included both cost as well as labor required for the installation, removal, and collection of readings [18]. In recent years, there has been a proliferation of commercially available soil moisture monitoring systems for agriculture. Many sensors interface with dataloggers and wireless communication systems to provide near real-time status of soil moisture from several depths and locations within a field. Data are automatically uploaded by radio or cell phone communications to cloud-based computer servers and are accessible through apps on smartphones and tablet computers. These communication advancements greatly improve the convenience of accessing data and can be configured to provide timely alerts when crops require irrigation. Many of these wireless communication systems for soil moisture sensors also support on-farm weather stations, digital flow meters, and control valves, which facilitates the monitoring of irrigation system operations.

Soil moisture sensors have also evolved during the last few decades in terms of size, cost, and accuracy. Electromagnetic (EM) soil moisture sensors, used to determine volumetric soil moisture content, include time domain reflectometry (TDR) and capacitance sensors. These were once bulky and expensive instruments. With improvements in electronic manufacturing and better designs, current EM sensors (Figure 1A,B) are generally much smaller, cheaper, and more accurate than earlier models. Some versions are integrated with soil temperature and salinity sensors (e.g., Model 5TE, Decagon Devices Inc., Pullman, WA, USA). Others are integrated with dataloggers and radio communications (e.g., Model gStake, gThrive Inc., Santa Clara, CA, USA) to facilitate field installation and removal.

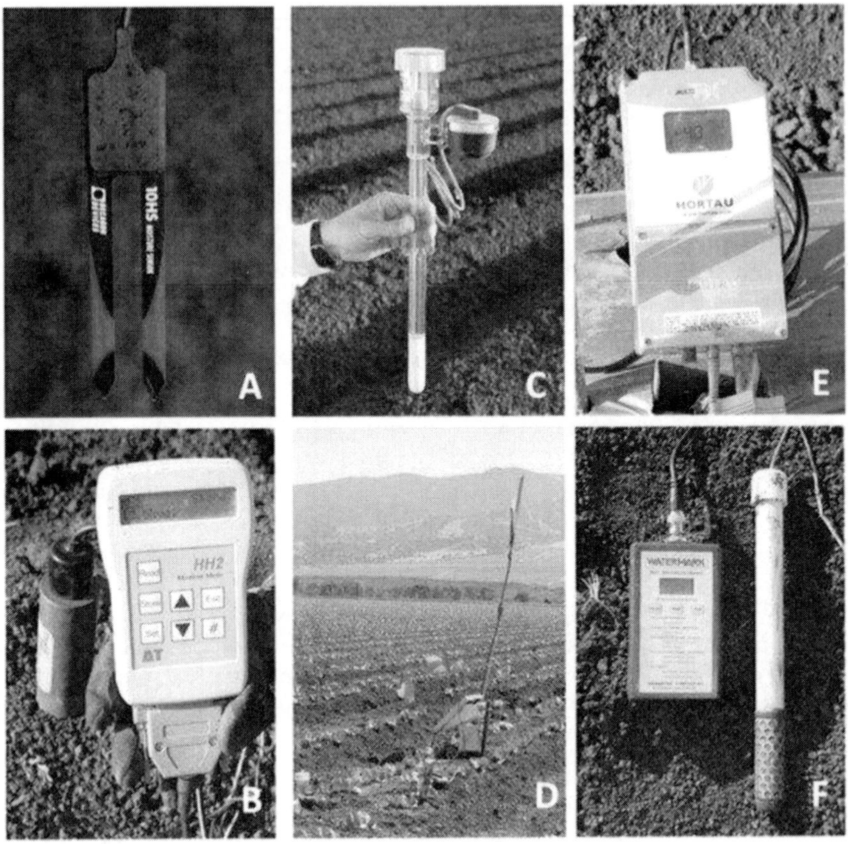

Figure 1. Examples of various soil moisture sensors: (**A**,**B**) capacitance sensors; (**C**) tensiometer with an electronic gauge; (**D**) tensiometers with electronic gauges installed in a lettuce field and interfaced with a datalogger and radio communications; (**E**) tensiometer integrated with pressure transducer, datalogger, and radio communications; (**F**) granular matrix sensor and reader.

Tension-based soil moisture sensors have been considered the best method to determine if a crop needs water since most vegetables experience reductions in yield and quality under prolonged periods of high soil water tensions. Tension thresholds that optimize production have been determined for many vegetable species [17,19], including broccoli [20], cabbage [21], cauliflower [22], lettuce [23,24], potato [25], and tomato [26]. An advantage of tension thresholds is that they are less influenced by soil texture than volumetric moisture thresholds. Tensiometers can accurately measure soil water tension in a range of 0–

85 kPa using a mechanical vacuum gauge attached to a water filled tube with a porous ceramic cup [17] (Figure 1C). Tensiometers must be installed without air gaps between the ceramic cup and soil to function properly. To reduce labor for collecting readings, several recent models have sensitive electronic pressure transducers that measure vacuum pressure and can be interfaced to dataloggers and wireless communication services (Figure 1D,E) so that data may be monitored remotely (Figure 2) (e.g., Hortau Inc., Lévis, QC, Canada). Though tensiometers do not need calibration, they usually require periodic maintenance to assure that they are functioning properly. Entrapped air that develops at high tensions must be replaced with de-aired water. Adding a weak solution of algaecide or bleach can prevent biological growth inside the tensiometer tube, which may potentially clog the ceramic cup [26].

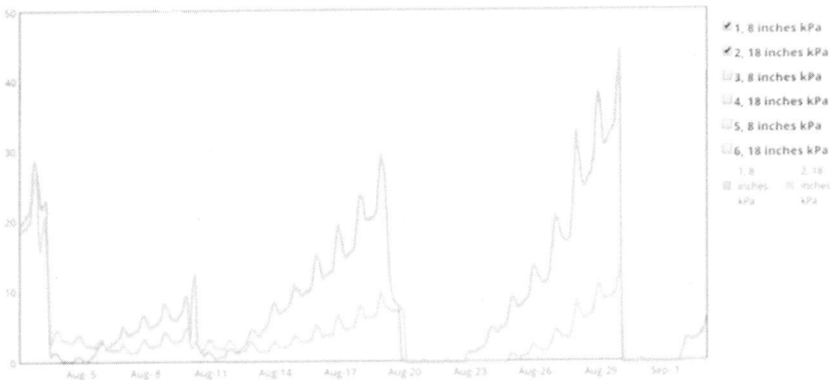

Figure 2. Fifteen minute readings (kPa) from tensiometers installed in a lettuce field at 8 (20 cm) and 18 inch (46 cm) depths are displayed in an online irrigation scheduling application.

Granular matrix sensors (GMS) (Figure 1F) indirectly measure soil water tension using electrical resistance and are often used as an alternative to tensiometers (e.g., Model watermark 200SS, Irrometer Company Inc., Riverside, CA, USA). Most GMS can be interfaced to dataloggers and wireless communications [26]. An advantage of GMS compared to tensiometers is that they do not require regular servicing. These sensors retain sensitivity at higher soil moisture tensions (up to 200 kPa) but are less accurate at low soil moisture tensions (0–15 kPa) than tensiometers [17,26]. GMS readings are also affected by fluctuations in soil temperature [17]. Another limitation of GMS is that the response time to wetting and drying cycles is slower compared to tensiometers [26,27].

A number of studies have evaluated automated irrigation scheduling based on soil moisture readings for improving the water use efficiency of vegetables irrigated with drip tape. Automatic controllers used in these studies irrigated crops for short durations several times per day when soils dried below a predetermined threshold [28,29]. Some control systems have relied on tension based sensors [26,28], while others have used volumetric soil moisture sensors [28,30,31,32,33,34] to determine when to irrigate. The use of TDR soil moisture sensors for triggering irrigations in sub-surface drip irrigated sweet corn resulted in an 11% savings in water use with similar yields compared to a standard sprinkler irrigated treatment [30]. Small plot trials have demonstrated water savings, reduced nitrate leaching, and improvements in yield in drip irrigated fresh market tomatoes, bell peppers, and zucchini squash using capacitance soil moisture sensors to trigger irrigations [31,32,33,34,35]. Using tensiometers, Mũnoz-Carpena et al. [26] were able to reduce water use in fresh market tomato by 67% compared to the grower standard practice without significant reductions in marketable yield. Most of the water savings were during the early crop stages when evapotranspiration rates were low. Capacitance soil moisture sensors were generally found to be more reliable for automated irrigation scheduling than tensiometers and GMS. As discussed earlier, tensiometers require regular maintenance, and the GMS response to changes in soil moisture was too slow to use these sensors for triggering irrigations. While these studies have demonstrated the potential for water savings in small plots, automated irrigation scheduling may have several limitations in commercial vegetable fields. The optimal volumetric soil moisture threshold for triggering an irrigation would need to be empirically determined for different soil types. Pressurized water would need to be continuously available to accommodate frequent short irrigations. In large fields, short irrigation cycles could lead to significant drainage at the lower end of a field when drip lines depressurize.

2.2. Limitations of Soil Moisture Sensors for Irrigation Scheduling in Vegetables

While much progress has been made in improving the accuracy and utility of soil moisture sensors, several factors still limit their use for irrigation scheduling of vegetables. Though costs for individual sensors may be less than in the past, the addition of dataloggers, cell phone modems, and radio communications, which facilitate real-time monitoring, has added to the total costs. Labor for installation and removal of sensors remains a significant cost, especially in vegetable crops with short production cycles. Considering labor and capital costs, many growers elect to use soil moisture monitoring equipment in a small percentage of their vegetable fields.

Though soil moisture sensors are useful for determining when to irrigate vegetables, they are less useful for estimating how much water to apply. Soil moisture tension readings must be converted to volumetric moisture to estimate soil water depletion since a previous irrigation or rain event. Volumetric soil moisture sensors may also need calibration. Most capacitance sensors use the manufacturer's calibration equations to convert readings to volumetric water content. Clay content, organic matter, salinity, bulk density, and temperature can affect the accuracy of capacitance sensors [36,37,38,39,40]. The effect of these factors on water content readings can vary in different soil types. For example, Kargas and Soulis [38] evaluated the accuracy of the 10HS capacitance sensor, and found that temperature had a larger effect on soil water content readings in clay than course textured soils. They concluded that calibration for specific soil types was needed to achieve accurate readings.

Even with accurate sensors, spatial variability can limit the reliability of soil moisture estimates if readings are collected from only a few locations, especially if soil hydraulic properties vary within a field or the irrigation system applies water unevenly. Soil maps can be useful for guiding the placement of soil moisture sensors in locations that represent the dominant soil properties of fields. In drip-irrigated fields, determining the optimal location to accurately monitor soil moisture depletion can be particularly challenging. Soil moisture is typically higher under drip tape than adjacent to the plants, where root activity is concentrated. Placing sensors too close to a drip line may lead to under-irrigating a crop, and placing sensors too far away may result in over-irrigating [41]. Through computer modeling experiments, Soulis et al. [42] concluded that the optimal positioning of soil moisture sensors in drip-irrigated crops was influenced by soil hydraulic properties, crop evapotranspiration rate, and the configuration of the irrigation system. An additional consideration is the ideal depth at which to place sensors relative to the crop root zone. Broccoli roots can reach depths greater than 1.2 m at maturity [16], while roots of many leafy greens such as spinach may reach less than 0.5 m by harvest [43].

3. ET-BASED APPROACHES TO SCHEDULING IRRIGATIONS IN VEGETABLES

Water requirements of vegetable crops can also be determined from estimates of crop evapotranspiration (ETc). Using Penman-Monteith

[44,45], or similar equations [46], evapotranspiration (ET_0) of a well-watered reference crop, such as grass, or alfalfa, is commonly estimated from measurements of air temperature, relative humidity, wind speed, and solar radiation [47]. The evapotranspiration of a well-watered vegetable crop can be estimated relative to a reference crop by multiplying ET_0 by a crop coefficient (Kc). Crop coefficients for most major vegetable crops were summarized by Allen et al. [48], Guerra et al. [49], and Grattan et al. [50].

Networks of public weather stations to monitor reference ET have been established in many vegetable production regions of the world where irrigation is commonly used. The California Irrigation and Information System (CIMIS) is operated by the Department of Water Resources in California and consists of more than 145 weather stations sited on reference crops throughout the state. ET_0 and other meteorological data, including precipitation, air and soil temperature, wind speed, solar radiation, and relative humidity, are available for users to download from the state operated website. Similar networks have been developed for other states including Arizona (AZMET), Colorado (CoAgMet), Florida (FAWN), Nevada (NICENET), Oklahoma (MesoNet), Oregon (AgriMet), and Washington (AgWeatherNet). Most European countries also have weather station networks that provide daily ET_0 data through public websites (e.g., Spain [51], Italy [52]). In addition, CIMIS offers a satellite based product that estimates ET_0 at a 2 km resolution based on data from the Geostationary Operational Environmental Satellite [53].

Many commercial companies also offer affordable weather stations (e.g., ET107, Campbell Sci. Inc., Logan, UT, USA; WatchDog 2900ET Weather Station, Spectrum Technologies, Aurora, IL, USA; HOBO U30-NRC Weather Station, Onset Computer Corp. Bourne, MA, USA) that can be used to estimate ET_0 on a farm. These systems offer a way to monitor ET_0 in regions not covered by a public network, or for farms that are not located reasonably near an established station. However, in practice, few growers site or maintain stations in accordance with World Meteorological Organization standards for reference evapotranspiration measurements. Many growers site weather stations near trees, buildings, or parking lots, which limits the fetch and can influence the micro-climate, and potentially bias estimates of ET_0. Stations are infrequently sited on a suitable well-watered reference crop. Most farmers do not allocate time to regularly maintain and calibrate meteorological instrumentation on their stations. Commercially available atmometers (ETgage Co., Loveland, CO, USA) can be used to monitor ET_0 but must also be sited over a reference crop.

Although ET_0 data have been useful for water agencies to estimate the seasonal water use of crops at a regional scale, growers have generally considered this approach to be impractical for scheduling irrigations in vegetable systems. A major difficulty is that the Kc value can change daily as vegetables grow and leaf area increases. During the establishment phase, when crop foliage covers a small percentage of the soil surface, ET is mostly from soil evaporation rather than from crop transpiration. For many vegetable crops, such as leafy greens and brassicas, the canopy cover is less than 10% until halfway through the crop cycle. The Kc during this early stage will depend on factors influencing evaporative losses from the soil, such as method and frequency of irrigations, and soil physical properties [48]. Later, in the rapid growth phase, Kc values increase daily as the canopy cover develops and covers the soil.

Solutions for calculating a daily Kc for lettuce were proposed by Gallardo et al. [54] in which soil evaporation from unshaded soil (E) is estimated separately from the water lost by transpiration and soil evaporation under the crop canopy (T). Using this approach, Kc is proportional to the fractional cover (Fc) of the leaf canopy shading the soil, and E is related to the hydraulic properties of the soil and the time since the crop was irrigated. The general approach is consistent with Allen et al. [48,55], who also proposed separating the Kc of crops that have a significant period of minimal canopy cover into dual components consisting of a basal crop coefficient (Kcb) that represents water loss by transpiration and an evaporation coefficient (Ke), representing loss from soil evaporation. Direct relationships between Fc and Kcb have been reported for several crops such as broccoli [56], wheat [57], cotton [58], sugar beets [58], and grapes [59].

Both Gallardo et al. [54] and Allen et al. [48] describe calculations for estimating soil evaporation, which divide Ke into stage 1 and stage 2 periods. Stage 1 occurs immediately following an irrigation or rain event that saturates the soil to field capacity. During stage 1, water loss is limited by the energy available for evaporation. Stage 2 occurs after the soil is visibly dry and water loss is limited by soil hydraulic properties. The Stage 2 rate of evaporation diminishes with time as the soil water content declines.

4. SOFTWARE FOR ET BASED SCHEDULING OF VEGETABLES

4.1. Overview of Sofware Tools

Even with publicly available ET_0 data and accurate Kc models, the implementation of ET-based irrigation scheduling at the scale of a commercial vegetable farm may be difficult for growers. Daily ET_0 values need to be retrieved from a location representative of the field of interest. An estimate of canopy cover is needed to determine the average Kc for each irrigation event. Soil evaporation calculations require knowledge of the soil properties of the field, irrigation method, and interval between irrigations or rainfall events. To convert crop ET into an irrigation recommendation, the application rate and distribution uniformity of the irrigation system, and in some cases the leaching fraction, also needs to be integrated into the calculations.

Recognizing that most growers have limited time to dedicate to making decisions on water management, several universities and public institutions have developed computer programs to facilitate ET based scheduling. Spreadsheet and Windows-based irrigation scheduling programs, such as CROPWAT [60], KanSched 2.0, Basic Irrigation Scheduling (BIS) [61], and consumptive use program (CUP) [62] automate the irrigation scheduling calculations but require users to retrieve and enter daily ET_0 values. Since separate spreadsheet files maybe needed for each crop, these programs may be difficult to implement in large growing operations that manage many fields. Washington Irrigation Scheduling Expert (WISE) [63,64] is a downloadable JavaScript application that runs on personal computers. More recently developed web-based applications such as Irrigation Scheduler mobile (Washington State University) [65], Irrigation Management Online (Oregon State University) [66], Wateright (Fresno State University) [67,68], and CropManage (University of California, Division of Agriculture and Natural Resources) [69,70] were developed to automatically retrieve ET_0 data from weather station networks such as AgWeatherNet and CIMIS and support irrigation scheduling of multiple fields. These applications are accessed through web browsers, some of which automatically resize the user interface to smartphone screens. Colorado State University WISE Irrigation Scheduler [71] and SmartIrrigation (University of Florida, University of Georgia) [72,73,74] have similar capabilities but can operate exclusively on smartphones through downloadable applications. Outside of the United States, several public agencies have developed online irrigation scheduling services including IRRINET in Italy [75], ISS-ITAP in Spain [76], IRRISA in France [77], and IrriSatSMS [78] in Australia. In addition, a growing number of ET-based software tools are commercially available. Two examples are Probe Schedule (IRRinet LLC, Dalles, OR, USA) and Irrigation Advisor (PowWow Energy Inc., San Mateo, CA, USA). Further details of these commercial products are not provided here since they are proprietary and usually not documented in the scientific literature.

Most online irrigation scheduling tools have cloud-based databases that retain information associated with each planting, such as soil type, planting and harvest dates, weather station name, and irrigation system application rate, so that the user does not need to reenter critical information for each irrigation event. After entering the information required to initiate a new planting, users can quickly look up how long to irrigate their crops. Some scheduling software provides weekly or daily summaries of how much water to apply (e.g., IrrigationScheduler, Wateright, SmartIrrigation, etc.), while others such as CropManage enable the user to input specific dates for each irrigation [70]. The online format facilitates automated retrieval of current and forecasted weather data in advance of planned irrigation events, so that growers can be alerted when a field will need water. In addition to linking with weather station networks, decision support tools can also incorporate other online services such as UC Davis SoilWeb for identifying the soil physical properties of a field using the United States Department of Agriculture (USDA) Soil Survey Geographic (SSURGO) database [79] or Google Maps, which can be customized for determining the longitude and latitude of a field or viewing locations of nearby weather stations. CropManage can also be customized to automatically retrieve flowmeter and soil moisture data from internet-accessible dataloggers.

While many of these irrigation-scheduling programs will provide recommendations for multiple commodities, few have been developed or tested specifically for vegetables. Most irrigation scheduling applications use single Kc values for the four crop growth stages described by Doorenbos and Pruitt [80] to simplify calculations of ETc. CropManage was initially developed for vegetable irrigation and employs a dual crop coefficient approach for estimating crop ET similar to Gallardo et al. (1996), as described in Johnson et al. [81] and Smith et al. [16]. The vegetable crops currently supported include broccoli, cabbage, cauliflower, lettuce, and spinach. Since the user enters the date of each irrigation event, CropManage adjusts watering recommendations for the frequency and method of irrigation. Empirical models of fractional cover are included for each supported vegetable crop so that the user can customize the Kcb curves for a specific season, bed width, and planting configuration. Replicated field trials demonstrated that the CropManage irrigation recommendations using the ET based algorithm with a dual crop coefficient optimized water use and yield in crisp head lettuce and broccoli [81].

4.2. Achieving Widescale Adoption of Irrigation Software

Despite significant progress in developing ET-based irrigation scheduling software for commercial vegetable farms, many challenges exist that will need to be addressed to achieve large-scale adoption by growers.

Although public ET weather station networks have been established in many vegetable producing regions of the US and Europe, they are less common in other regions such as India, China, and Latin America. In areas where ET_0 data are available, the density or siting of weather stations may not provide sufficient resolution to fully capture regional variation in microclimates. In these instances, growers may choose to install and operate their own weather stations, as described above.

Another challenge is to deliver accurate irrigation recommendations without requiring users to provide excessive details about their crop, irrigation system, and field conditions. Morrison [82] found that time constraints and concerns about data entry errors were factors that deterred grower adoption of the irrigation scheduling software. Irrigation scheduling apps also need to be intuitive for irrigators and farm managers to use in the field. The user-interface displayed on smartphones and tablet computers must be easy to read and to understand, while providing sufficient detail for the user to verify that the values for ET_0, Kc, and other variables used in the scheduling calculations are accurate. Also, since the principal language of many irrigators in the western US is Spanish, irrigation scheduling applications need to support multiple languages.

The cost of computer programming services is also a major barrier to developing and maintaining irrigation scheduling software. As software has become more sophisticated, and needs to be compatible with personal and tablet computers and smartphones, the complexity and consequently costs of development have increased. Software products will incur annual costs associated with updating to new technology, introducing new features, and troubleshooting identified errors. Most publicly available irrigation scheduling applications were initially developed under research grants, and need either continued grant funding or income generated from user subscriptions to support updating and maintenance costs. Several online irrigation scheduling services that were initially funded by the European Union became inactive after grants ended [83]. It may be cost-effective for institutions to collaborate on the development of software products that are sufficiently flexible to be customized for different growing regions and commodities. SmartIrrigation is an example of such collaboration between the University of Florida and University of Georgia. There also may be opportunities for public agencies, commodity boards, and commercial companies to partner on the development of commercial irrigation scheduling software. For instance, SureHarvest Inc. (Santa Cruz, CA, USA) recently developed an online irrigation-scheduling tool for almonds in collaboration with the University of California and the Almond Board of California.

5. FIELD MEASUREMENTS OF CROP ET

Field measurements of crop ET are needed to develop and verify crop coefficients as well as to investigate the interaction of water stress on crop yield and quality. Several reviews detail the advantages and disadvantages of various methods of measuring crop ET [84,85]. Weighing lysimeters have often been used to measure crop water use [85,86]. While considered the most accurate ET measurement method, the expense of these instruments limits their use to research stations and constrains the range of sites, crops, and management practices that can reasonably be evaluated. Energy balance methods such as Bowen ratio [85], eddy covariance [85], and surface renewal [87] are more affordable alternatives to lysimetry for evaluating crop ET and can be deployed in commercial fields for evaluating crop water use under a wide range of growing conditions. Instrumentation costs for these methods can exceed $10,000 USD per station, which has limited their use primarily to research studies. Energy balance methods involve monitoring of net radiation, soil heat flux, and sensible heat flux to estimate the latent heat flux (Wm^{-2}), which can be converted to water flux or evapotranspiration rate through the latent heat of vaporization. Generally, the sensors used for energy balance measurements have become more reliable, cheaper, and more accurate in recent years. The capabilities of dataloggers used for operating the instrumentation have also improved, with more memory and faster computation speeds, while post-processing software modules have become more accessible. Energy balance methods have been used for evaluating the water use of vegetable crops in commercial fields, including artichokes, broccoli, beans, lettuce, onions, and processing tomatoes [50,88,89,90,91,92]. These methods have been useful for reevaluating existing crop coefficients, as production practices and varieties have improved during recent years [93]. However, considering the diversity of vegetables and production methods, relatively few ET studies have been published using these techniques during the last decade. Surface renewal, which uses thermocouples to measure rapid changes in air temperature, has greatly reduced the costs of estimating sensible heat flux and may be well suited to vegetable fields due to a smaller fetch requirement compared to eddy covariance. Advances in surface renewal methodology [94] have culminated in a commercial service that provides growers with ET estimates and irrigation recommendations via a web application (Tule Technology Inc., Oakland, CA, USA). However, this tool has primarily been used in perennial crops such as trees, grapes, and strawberries.

6. SATELLITE-BASED CROP ET DETERMINATION

6.1. Energy Balance

Researchers have developed satellite-based energy balance models to estimate ET at various spatial and temporal scales (e.g., reviews of Courault et al. [95]; Kalma et al. [96]; Gonzalez-Dugo et al. [97]). The Surface Energy Balance Algorithm for Land (SEBAL) and the Mapping EvapoTranspiration at high Resolution with Internalized Calibration (METRIC) are two widely used approaches [98,99]. Optical and thermal band data from Landsat or other satellites is used to help parameterize the energy balance equation [100]. At the moment of satellite overpass, latent heat flux (LE) is retrieved through the calculation of surface energy balance that is based on radiative, aerodynamic, and energy balance physics. LE is converted to ET for the corresponding hour and subsequently to reference ET fraction (ETrF) by comparison with the measured reference ET for the hour. Finally, daily (2) ET is derived as the product of ETrF and reference ET for the day, under the assumption that ETrF approximates the average daylight evaporative fraction. The energy balance approach accounts for the effects of ET reduction in water-stressed crops, due to stomatal regulation and increased ET due to bare-soil evaporation. Linear or spline interpolation can be used to estimate ETrF (hence daily ET) between satellite overpasses, which occur every eight days in the case of Landsat, assuming clear-sky conditions. Manual intervention is used to calibrate the sensible heat flux computation by identifying portions of the image (hot and cold 'anchor' pixels) that represent extreme ET conditions, where ET can be estimated and assigned a priori. In a typical agricultural situation, cold pixels are associated with well-irrigated agricultural fields with high Fc and hot pixels with bare dry fields. Overall the ET retrieval error for the satellite-based energy balance is typically 5–15%, when estimates are produced by an experienced operator, and rises to 30–40% when operated by non-specialists or novices to the fields of hydrologic science, environmental physics, remote sensing, or agricultural systems [85]. An automated calibration method has been recently developed to facilitate and improve operation by non-specialists [101].

6.2. Vegetation Index

Hybrid approaches estimate fractional cover (Fc) from remote sensing and use ground-based ET_0 data. Fc is a good indicator of light interception, which is a strong driver of ETc [48]. Weighing lysimeter observations by Bryla et al. [102] revealed a strong relationship between

Fc and ETc for vegetable crops in California's San Joaquin Valley, and the potential for using satellite-based Fc to estimate ETc in vegetable crops was demonstrated by Johnson and Trout [103]. Fc can be estimated from various spectral visible-region bands, where chlorophyll absorption dominates, combined with near-infrared (NIR), where vegetation is highly reflective. A common formulation is the normalized difference vegetation index, or NDVI, which is derived as (NIR – red)/(NIR + red) [104]. Trout et al. [105] and Johnson and Trout [103] found a strong relationship between NDVI and Fc for vegetables and other crop types in California's San Joaquin Valley. NDVI began to lose sensitivity above 80 percent cover, a point generally regarded as effective full cover for water management applications [106]. As above, interpolation can be applied to estimate NDVI, and hence ETc, for days between satellite overpasses as needed.

7. SATELLITE BASED IRRIGATION MANAGEMENT SERVICES

7.1. Prototype Systems

By furnishing field scale estimates of Fc, a key factor for estimating Kcb and basal ETc (ETcb), optical remote sensing may potentially improve the accuracy of ET estimates in vegetables. Several satellite-based models have been developed for irrigation management. An early proof-of-capability satellite demonstration on wheat and corn was undertaken by the DEMETER project in Europe, which involved timely delivery of Landsat-based crop coefficients to agricultural end-users, with available online visualization and analysis capabilities [107].

More recently, the fully-automated Satellite Irrigation Management Support (SIMS) [108] uses atmospherically corrected Landsat data to map NDVI, Fc, and Kcb for multiple crop types, including vegetables across about eight million acres of California irrigated farmland [109]. These variables are updated every eight days at 30 m spatial resolution (cloud cover and data availability permitting) from 2010 to the present. The SIMS uses daily 2 km ET_0 statewide grids produced by the California Irrigation Management Information System [53] to generate basal crop ET. Web data services allow users to display annual time-series graphs and download data for any given location (Figure 3). An application programming interface (API) enables on-demand transfer of SIMS data products to support external irrigation advisory services such as CropManage and also allows the user and other software tools to specify the crop type via the API and retrieve crop-specific Kcb data from SIMS.

Where information is available on applied water, an FAO56 based soil water balance model can be used to adjust for soil evaporation and crop stress and to retrospectively derive agricultural water use fractions at the field scale for the evaluation of irrigation management and system performance [110].

Figure 3. SIMS map of fractional cover of vegetable crops near Salinas, California as of mid-July 2016.

IrriSAT [111] is a weather based irrigation management and benchmarking technology that uses satellite remote sensing to generate crop water management information in Australia, primarily for grape and cotton growers at present [112,113]. As with SIMS, Landsat is used to estimate Kc at a 30 m resolution. Unlike SIMS, Kc is derived directly from a single linear relationship with satellite NDVI. Daily crop water use is determined as the product of Kc and ET_0 observations from nearby weather stations, and a seven-day ET_0 forecast is also produced. The delivery platform, which is built on the Google App Engine helps irrigators track soil moisture and better manage irrigation schedules. In addition, IrriSAT facilitates the calculation of seasonal agronomic performance metrics, including the irrigation water use index and gross production water use index.

Additional satellite-based services have been implemented Europe and Australia [114]. These include IRRISAT-Italy (as distinct from IrriSAT-Australia described above), EO4Water (Austria), and IrriEye (southern Australia). These services generally involve the use of optical DEIMOS 20 m satellite imagery to generate and deliver leaf area index with derived Kc and related map overlays within a browsing and querying toolbox [115]. Personalized irrigation guidance regarding crop water requirement is dispensed through a secure website, SMS, and email. End-users are able to provide evaluative feedback. A variety of crops are being tested, including vegetables, sugar beet, corn, alfalfa, and orchards.

Earth Engine Evapotranspiration Flux (EEFlux) [116] is a METRIC implementation on the Google Earth Engine cloud computing platform [117]. Landsat scenes from 1984 to the present are combined with gridded weather data to allow the analysis of most land areas worldwide. The products include maps of ET, associated energy balance flux components, and land cover. A time-series can be constructed by processing several scenes and interpolating between image dates. Automated image calibration for anchor pixels is offered in a Level 1 version to accommodate operation by non-experts, with some sacrifice in accuracy. A Level 2 version, which allows custom calibration by the operator, requires an annual license fee. EEFlux is a general tool that operates on both agricultural and non-agricultural land cover types.

7.2. Satellite Remote Sensing Considerations for Irrigation Scheduling

Satellite observation is recognized as a useful tool that will continue to support crop coefficient development and ET monitoring for new types and varieties of crops, as facilitated by the wider availability of gridded weather data and continued advancement in convenient and user-friendly mapping technologies [118]. The use of public domain satellites with open-data policies, i.e., the data source is available at no charge, is an advantage of this approach, as is the capability to collect data for many fields simultaneously. Vegetable crops, however, do present some unique challenges for satellite-based evaluation. The frequency of satellite overpasses, presence of cloud cover, and added time needed to perform custom atmospheric correction (e.g., Vermote and Saleous [119]) may limit the use of remote sensing for real time irrigation scheduling of fast-growing vegetable crops. As more satellite platforms become available, such as the European Space Agency Sentinel-2 system, the observation interval between images may soon decrease to four to five days for NDVI-based ET models. Also, upon further testing, it may be shown that use of a simplified atmospheric correction approach (e.g., Tasumi et al. [120]) may be adequate for application in some or most regions. Where real-time irrigation scheduling operations are unsupported, remote sensing may still prove useful for the retrospective evaluation of irrigation system performance or compilation of water use metrics. Another challenge is the typically small field size for vegetables as noted above (4–5 ha), which is near the minimum land area recommended for evaluation by Landsat or other public Earth observation satellites. An additional challenge is posed by leaf color, especially for crops such as red lettuce that depart widely from the green-leaf norm. A recent study has also shown that while NDVI was strongly related to Fc in both leafy greens ($r^2 = 0.88$) and cole crops ($r^2 = 0.93$), the relationships were different [121]. Thus, the development of customized relationships between NDVI and Fc by crop category is recommended to increase the accuracy of Fc estimates.

8. REMOTE SENSING USING MANNED AIRCRAFT AND UNMANNED AERIAL VEHICLES (UAV)

The use of manned aircraft and UAVs for monitoring vegetables has become a viable alternative or complement to satellite observation. Image resolution is generally much finer than that of Landsat or other open-data satellites, and passes over fields can be scheduled on a more frequent and flexible basis. However, data calibration to assure image consistency in support of time-series observations may be lacking. Several commercial companies offer NDVI and thermal spectrum images at <1 m resolution taken at weekly intervals using small planes in California and elsewhere (e.g., CERESimaging Inc., Oakland, CA, USA; TerrAvion Inc., San Leandro, CA, USA). Images can be accessed from the web generally within 24 h of collection. The high resolution of these scenes can provide enhanced information about within-field spatial variability in crop growth and water stress and for determining Fc used to parameterize ET equations. Patterns can show when a crop is under-irrigated such as in Figure 4, where the NDVI values for a romaine lettuce crop are lowest midway between sprinkler lines, presumably caused by poor irrigation distribution uniformity.

The acquisition costs of UAVs and the training needed for mission planning, flight operation, and image processing have diminished significantly during recent years. A number of studies describe the use of UAVs for monitoring water stress [122,123] and vegetative cover [124,125] of crops, though few have specifically focused on vegetables. Due to the ease of operation, arranging flights at specific times and locations with UAVs may be simpler than with manned aircraft, which usually need to image a minimal number of fields to justify flight costs. This flexibility may be useful in vegetables when timely field scouting is needed to make decisions or diagnose a problem with a fast-growing crop. UAVs can fly at very low altitudes and carry sensor packages that can provide high-resolution imagery in the order of a few centimeters. A disadvantage of UAV operation, in the case of an owned system, would be a higher up-front cost than the engagement of a manned aircraft provider and additional time required for data retrieval and image processing.

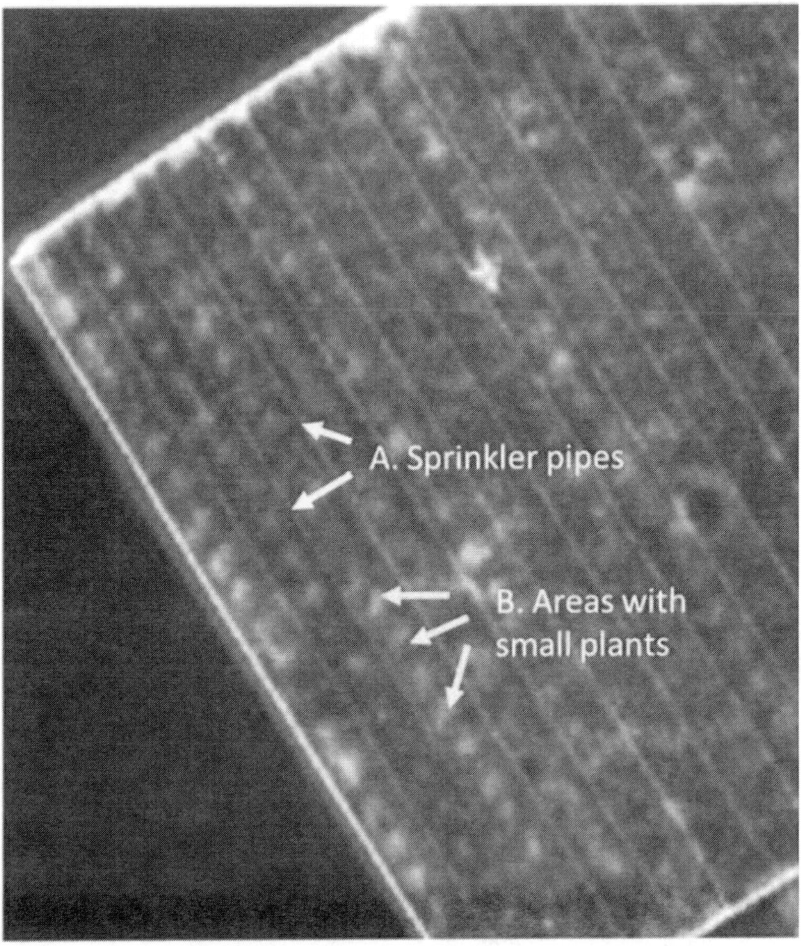

Figure 4. NDVI image of a romaine lettuce field irrigated with sprinklers on lateral pipes (**A**) spaced 12.2 m apart; Plants were frequently smaller in areas midway between pipes (**B**) where the NDVI values were low. Photo credit: CERESimaging Inc., Oakland, CA, USA.

9. CONCLUSIONS

Water scarcity will most likely continue to be a significant problem in many of the important vegetable production regions of the world [126]. Improving water use efficiency through more accurate scheduling of irrigation can conserve water and address water quality impacts associated with commercial vegetable operations. However, irrigation scheduling in vegetables presents some unique challenges due to the diversity of crop types, intensive rotations, and number of fields that must be managed, as well as competing cultural operations that are involved in growing marketable crops. Advances in soil moisture sensors, wireless communications, ET measurements, remote sensing, computer technology, and cloud computing offer many potential opportunities to develop robust irrigation advisory tools to help farmers accurately determine and meet crop water needs. While much progress has already been made in soil moisture sensing and ET-based irrigation scheduling, achieving wide-scale adoption in the vegetable industry will require continued innovation. Continued collaboration between public research institutions, universities, commodity boards, and commercial firms will likely be needed to develop simple-to-use tools that will be broadly accepted by vegetable growers.

ACKNOWLEDGMENTS

The authors wish to thank Forrest Melton, Richard Smith, and Rick Snyder for their careful review and edits in the preparation of this manuscript.

ABBREVIATIONS

The following abbreviations are used in this manuscript:

API	application programming interface
BIS	basic irrigation scheduling
CIMIS	California irrigation management and information system
CUP	consumptive use program
EEFlux	earth engine evapotranspiration flux
EM	electromagnetic

ET	evapotranspiration
ET_0	reference evapotranspiration
ETc	crop evapotranspiration
Fc	fractional cover
GMS	granular matrix sensor
Kc	crop coefficient
LE	latent heat flux
METRIC	mapping evapotranspiration at high resolution with internalized calibration
NDVI	normalized difference vegetation index
NIR	near infra-red
SEBAL	surface energy balance algorithm for land
SIMS	satellite irrigation management system
SSURGO	soil survey geographic database
UAV	unmanned aerial vehicle
UC	University of California
USDA	United States Department of Agriculture
WISE	Washington irrigation scheduling expert

REFERENCES

1. Mount, J.; Chappelle, C.; Gray, B.; Hanak, E.; Howitt, R.; Lund, J.; Frank, R.; Gartrell, G.; Grantham, T.; Medellín-Azuara, J.; et al. California's Water: Managing Droughts. Public Policy Institute of California, Water Policy Center, 2016; p. 4. Available online: http://www.ppic.org/content/pubs/report/R_1016JM2R.pdf (accessed on 18 February 2017).
2. Hanak, E.; Mount, J.; Chappelle, C.; Lund, J.; Medellín-Azuara, J.; Moyle, P.; Thompson, B.; Viers, J.; Seavy, N. What If California's Drought Continues? Public Policy Institute of California, Water Policy Center, 2015; p. 20. Available online: http: //www.ppic.org/content/pubs/report/R_815EHR.pdf (accessed on 18 February 2017).
3. Pedrero, F.; Alarcon, J.J.; Nicolas, E.; Asano, T. Monterey Regional Water Pollution Control Agency (MRWPCA): Transforming wastewater into a profitable resource. *Agric. Vergel Frutic. Hortic. Floric. Citric. Vid Arroz* **2009**, *28*, 464–469.
4. Khan, S. Managing climate risks in Australia: Options for water policy and irrigation management. *Aust. J. Exp. Agric.* **2008**, *48*, 265–273.

5. Arumí, J.L.; Rivera, D.; Holzapfel, E.; Muñoz, E. Effect of drought on groundwater in a Chilean irrigated valley. *Proc. Inst. Civ. Eng. Water Manag.* **2013**, *166*, 231–241.
6. Ortega-Reig, M.; Palau-Salvador, G.; Cascant i Sempere, M.J.; Benitez-Buelga, J.; Badiella, D.; Trawick, P. The integrated use of surface, ground and recycled waste water in adapting to drought in the traditional irrigation system of Valencia. *Agric. Water Manag.* **2014**, *133*, 55–64.
7. Harter, T.; Lund, J.R. *Addressing Nitrate in California's Drinking Water with a Focus on Tulare Lake Basin and Salinas Valley Groundwater. Report for the State Water Resources Control Board Report to the Legislature*; Center for Watershed Sciences, University of California: Davis, CA, USA, 2012; p. 78.
8. Hallberg, G.R. *Nitrate in Ground Water in the United States*; Follett, R.F., Ed.; Elsevier Science Publishers: Amsterdam, The Netherlands, 1989.
9. Zhang, L.; Ju, X.; Liu, C.; Kou, C. A study on nitrate contamination of ground water sources in areas of protected vegetables-growing fields—A case study in Huimin County, Shandong Province. *Sci. Agric. Sin.* **2010**, *43*, 4427–4436.
10. Howarth, R.W.; Sharpley, A.; Walker, D. Sources of nutrient pollution to coastal waters in the United States: Implications for achieving coastal water quality goals. *Estuaries* **2002**, *25*, 656–676.
11. Ng, C.M.; Weston, D.P.; Lydy, M.J. Pyrethroid insecticide transport into Monterey Bay through riverine suspended solids. *Arch. Environ. Contam. Toxicol.* **2012**, *63*, 461–470.
12. Anderson, B.S.; Hunt, J.W.; Phillips, B.M.; Nicely, P.A.; Gilbert, K.D.; Vlaming, V.D.; Connor, V.; Richard, N.; Tjeerdema, R.S. Ecotoxicologic impacts of agricultural drain water in the Salinas River, California, USA. *Environm. Toxicol. Chem.* **2003**, *22*, 2375–2384.
13. Agricultural Order No. R3-2012-0011 Conditional Waiver of Waste Discharge Requirements for Discharges from Irrigated Lands. 2012; p. 94. Available online: http://www.waterboards.ca.gov/centralcoast/water_issues/programs/ag_waivers/docs/ag_order/agorder_final_011014.pdf (accessed on 20 January 2017).
14. Council Directive 91/676/EEC of 12 December 1991 Concerning the Protection of Waters against Pollution Caused by Nitrates from Agricultural Sources. 1991. Available online: http://eur-lex.europa.eu/legal-content/EN/TXT/?uri=CELEX%3A31991L0676 (accessed on 20 January 2017).
15. Koike, S.; Cahn, M.D.; Cantwell, M.; Fennimore, S.; LeStrange, M.; Natwick, E.; Smith, R.; Takale, E. Spinach production in California. In *Vegetable Production Series*; Vol. ANR Publication 7212; UCANR: Richmond, CA, USA, 2011.
16. Smith, R.; Cahn, M.; Hartz, T.; Love, P.; Farrara, B. Nitrogen dynamics of cole crop production: Implications for fetility management and environmental protection. *HortScience* **2016**, *51*, 1586–1591.
17. Shock, C.C.; Wang, F. Soil water tension, a powerful measurement for productivity and stewardship. *HortScience***2011**, *46*, 178–185.
18. Pardossi, A.; Incrocci, L. Traditional and New Approaches to Irrigation Scheduling in Vegetable Crops. *HortTechnology* **2011**, *21*, 309–313.
19. Shock, C.; Pereira, A.; Hanson, B.; Cahn, M. Vegetable irrigation. In *Irrigation of Agricultural Crops*, 2nd ed.; Lanscano, R., Sojka, R., Eds.; ASA-CSSA-SSSA (American Society of Agronomy-Crop Science Society of America-Soil Science Society of America): Madison, WI, USA, 2007; pp. 535–606.
20. Thompson, T.L.; Doerge, T.A.; Godin, R.E. Subsurface drip irrigation and fertigation of broccoli: I. Yield, quality, and nitrogen uptake. *Soil Sci. Soc. Am. J.* **2002**, *66*, 186–192.
21. Smittle, D.A.; Dickens, W.L.; Stansell, J.R. Irrigation regimes affect cabbage water use and yield. *J. Am. Soc. Hortic. Sci.***1994**, *119*, 20–23.
22. Thompson, T.L.; Doerge, T.A.; Godin, R.E. Nitrogen and water interactions in subsurface drip-irrigated cauliflower: I. Plant response. *Soil Sci. Soc. Am. J.* **2000**, *64*, 406–411.

23. Gallardo, M.; Jackson, L.E.; Schulbach, K.; Snyder, R.L.; Thompson, R.B.; Wyland, L.J. Production and water use in lettuces under variable water supply. *Irrig. Sci.* **1996**, *16*, 125–137.
24. Thompson, T.L.; Doerge, T.A. Nitrogen and water rates for subsurface trickle-irrigated romaine lettuce. *HortScience* **1995**, *30*, 1233–1237.
25. Wang, F.-X.; Wu, X.-X.; Shock, C.C.; Chu, L.-Y.; Gu, X.-X.; Xue, X. Effects of drip irrigation regimes on potato tuber yield and quality under plastic mulch in arid Northwestern China. *Field Crops Res.* **2011**, *122*, 78–84.
26. Munoz-Carpena, R.; Dukes, M.D.; Li, Y.C.; Klassen, W. Field comparison of tensiometer and granular matrix sensor automatic drip irrigation on tomato. *HortTechnology* **2005**, *15*, 584–590.
27. Markovic, M.P.; Josipovic, M.A.; Sostaric, J.I.; Zebec, V.B.; Rapcan, I.A. Effectiveness of granular matrix sensors in different irrigation treatments and installation depths. *J. Agric. Sci. Belgrad.* **2016**, *61*, 257–269.
28. Munoz-Carpena, R.; Dukes, M.D.; Li, Y.; Klassen, W. Design and Field Evaluation of a New Controller for Soil-Water Based Irrigation. *Appl. Eng. Agric.* **2008**, *24*, 183–191.
29. Phene, C.J.; Howell, T.A. Soil sensor control of high-frequency irrigation systems. *Trans. Am. Soc. Agric. Eng.* **1984**, *27*, 392–396.
30. Dukes, M.D.; Scholberg, J.M. Soil moisture controlled subsurface drip irrigation on sandy soils. *Appl. Eng. Agric.* **2005**, *21*, 89–101.
31. Zotarelli, L.; Dukes, M.D.; Barreto, T.P. Use of Soil Moisture Sensor-based Irrigation on Vegetable Crops. *Proc. Fla. State Hort. Soc.* **2009**, *122*, 218–220.
32. Zotarelli, L.; Dukes, M.D.; Scholberg, J.M.S.; Femminella, K.; Munoz-Carpena, R. Irrigation scheduling for green bell peppers using capacitance soil moisture sensors. *J. Irrig. Drain. Eng.* **2011**, *137*, 73–81.
33. Zotarelli, L.; Dukes, M.D.; Scholberg, J.M.S.; Munoz-Carpena, R.; Icerman, J. Tomato nitrogen accumulation and fertilizer use efficiency on a sandy soil, as affected by nitrogen rate and irrigation scheduling. *Agric. Water Manag.* **2009**, *96*, 1247–1258.
34. Zotarelli, L.; Scholberg, J.M.; Dukes, M.D.; Munoz-Carpena, R.; Icerman, J. Tomato yield, biomass accumulation, root distribution and irrigation water use efficiency on a sandy soil, as affected by nitrogen rate and irrigation scheduling. *Agric. Water Manag.* **2009**, *96*, 23–34.
35. Zotarelli, L.; Dukes, M.D.; Scholberg, J.M.; Hanselman, T.; Le Femminella, K.; Muñoz-Carpena, R. Nitrogen and water use efficiency of zucchini squash for a plastic mulch bed system on a sandy soil. *Sci. Hortic.* **2008**, *116*, 8–16.
36. Thompson, R.B.; Gallardo, M.; FerncLndez, M.D.; Valdez, L.C.; Martcnez-GaitcLn, C. Salinity Effects on Soil Moisture Measurement Made with a Capacitance Sensor. *Soil Sci. Soc. Am. J.* **2007**, *71*, 1647–1657.
37. Visconti, F.; de Paz, J.M.; Martcnez, D.; Molina, M.J. Laboratory and field assessment of the capacitance sensors Decagon 10HS and 5TE for estimating the water content of irrigated soils. *Agric. Water Manag.* **2014**, *132*, 111–119.
38. Kargas, G.; Soulis, K.X. Performance analysis and calibration of a new low-cost capacitance soil moisture sensor. *J. Irrig. Drain. Eng.* **2012**, *138*, 632–641.
39. Kelleners, T.J.; Soppe, R.W.O.; Ayars, J.E.; Skaggs, T.H. Calibration of capacitance probe sensors in a saline silty clay soil. *Soil Sci. Soc. Am. J.* **2004**, *68*, 770–778.
40. Parvin, N.; Degré, A. Soil-specific calibration of capacitance sensors considering clay content and bulk density. *Soil Res.* **2016**, *54*, 111–119.
41. Soulis, K.X.; Elmaloglou, S. Optimum soil water content sensors placement in drip irrigation scheduling systems: Concept of time stable representative positions. *J. Irrig. Drain. Eng.* **2016**, *142*, 04016054.
42. Soulis, K.X.; Elmaloglou, S.; Dercas, N. Investigating the effects of soil moisture sensors positioning and accuracy on soil moisture based drip irrigation scheduling systems. *Agric. Water Manag.* **2015**, *148*, 258–268.
43. Heinrich, A.; Smith, R.; Cahn, M. Nutrient and water use of fresh market spinach. *HortTechnology* **2013**, *23*, 325–333.

44. Penman, H.L. Natural evaporation from open water, bare soil and grass. *Proc. R. Soc.* **1948**, 120–145.
45. Monteith, J.L. Evaporation and the environment. *Symp. Soc. for Exp. Biol.* **1965**, *19*, 205–234.
46. Allen, R.G.; Walter, I.A.; Elliot, R.; Itenfisu, D.; Brown, P.; Jensen, M.E.; Mecham, B.; Howell, T.A.; Snyder, R.; Eching, S.; et al. The ASCE standardized reference evapotranspiration equation. In *Water for a Sustainable World, Limited Supplies and Expanding Demand*, Proceedings of the Second International Conference on Irrigation and Drainage, Phoenix, AZ, USA, 12–15 May 2003.
47. Allen, R.G.; Pruitt, W.O.; Wright, J.L.; Howell, T.A.; Ventura, F.; Snyder, R.; Itenfisu, D.; Steduto, P.; Berengena, J.; Baselga Yrisarry, J.; et al. A recommendation on standardized surface resistance for hourly calculation of reference ETo by the FAO56 Penman-Monteith method. *Agric. Water Manag.* **2006**, *81*, 1–22.
48. Allen, R.G.; Pereira, L.S.; Raes, D.; Smith, M. Crop evapotranspiration: Guidelines for computing crop water requirements. In *FAO Irrigation and Drainage Paper*; Food and Agriculture Organization (FAO): Rome, Italy, 1998.
49. Guerra, E.; Ventura, F.; Snyder, R.L. Crop coefficients: A literature review. *J. Irrig. Drain. Eng.* **2016**, *142*, 06015006.
50. Grattan, S.R.; Bowers, W.; Dong, A.; Snyder, R.L.; Carroll, J.J.; George, W. New crop coefficients estimate water use of vegetables, row crops. *Calif. Agric.* **1998**, *52*, 16–21.
51. Estaciones Agroclimáticas, Consejería de Agricultura, Pesca y Desarrollo Rural. Available online: http://www.juntadeandalucia.es/agriculturaypesca/ ifapa/ria /servlet/FrontController?action=Init (accessed on 27 January 2017).
52. Rete Agrometeorologica Nazionale, Ministero delle Politiche Agricole Alimentari e Forestali. Available online: https://www.politicheagricole.it/ flex/FixedPages /Common/miepfy200_reteAgrometeorologica.php/L/IT (accessed on 27 January 2017).
53. Hart, Q.; Brugnach, M.; Temesgen, B.; Rueda, C.; Ustin, S.; Frame, K. Daily reference evapotranspiration for California using satellite imagery and weather station measurement interpolation. *Civ. Eng. Environ. Syst.* **2009**, *26*, 19–33.
54. Gallardo, M.; Snyder, R.L.; Schulbach, K.; Jackson, L.E. Crop growth and water use model for lettuce. *J. Irrig. Drain. Eng.* **1996**, *122*, 354–359.
55. Allen, R.G.; Pereira, L.S.; Smith, M.; Raes, D.; Wright, J.L. FAO-56 dual crop coefficient method for estimating evaporation from soil and application extensions. (Special Issue: Evapotranspiration prediction in irrigated agriculture). *J. Irrig. Drain. Eng.* **2005**, *131*, 2–13.
56. El-Shikha, D.W.P.; Hunsaker, D.; Clarke, T.; Barnes, E. Ground-based remote sensing for assessing water and nitrogen status of broccoli. *Agric. Water Manag.* **2007**, *92*, 183–193.
57. Hunsaker, D.F.G.; French, A.; Clarke, T.; Ottman, M.; Pinter, P. Wheat irrigation management using multispectral crop coefficients. *Trans. ASABE* **2007**, *50*, 2017–2033.
58. Gonzalez-Dugo, M.M.L. Spectral vegetation indices for benchmarking water productivity of irrigated cotton and sugarbeet crops. *Agric. Water Manag.* **2008**, *95*, 48–58.
59. Campos, I.N.C.; Calera, A.; Balbontin, C.; Gonzalez-Piqueras, J. Assessing satellite-based basal crop coefficients for irrigated grapes. *Agric. Water Manag.* **2010**, *98*, 45–54.
60. Smith, M. CROPWAT. A computer program for irrigation planning and management. In *FAO Irrigation and Drainage Paper*; FAO Land and Water Development Division: Rome, Italy, 1992.
61. Snyder, R.L.; Orang, M.N.; Bali, K.M.; Eching, S. Basic Irrigation Scheduling (BIS), 2000. p. 10. Available online: https://www.researchgate.net/publication/ 255580842_BASIC_IRRIGATION_SCHEDULING_BIS (accessed on 20 January 2017).

62. Orang, M.N.; Matyac, J.S.; Snyder, R.L. Consumptive use program (CUP) model. *Acta Hortic.* **2004**, *664*, 461–468.
63. Lieb, B. Washington Irrigation Scheduling Expert. Available online: https://sourceforge.net/projects/wsuwise/(accessed on 20 January 2017)
64. Leib, B.G.; Elliott, T.V. Washington Irrigation Scheduling Expert (WISE) Software. In *National Irrigation Symposium*, Proceedings of the 4th Decennial Symposium, Phoenix, AZ, USA, 14–16 November 2000; pp. 540–548.
65. Washington State University IrrigationScheduler Mobile. Available online: http://weather.wsu.edu/is/ (accessed on 20 January 2017).
66. Hillyer, C.; English, M. Irrigation Management Online. Available online: http://oiso.bioe.orst.edu/RealtimeIrrigationSchedule/index.aspx (accessed on 20 January 2017).
67. Wateright. Available online: http://www.wateright.org (accessed on 20 January 2017).
68. Thompson, W. Irrigation scheduling made simple(r): Wateright website does the hard work for you. *Calif. Grow. Avocados Citrus Subtrop.* **1998**, *22*, 22–23.
69. CropManage. Available online: https://cropmanage.ucanr.edu (accessed on 20 January 2017).
70. Cahn, M.; Smith, R.; Farrara, B.; Hartz, T.; Johnson, L.; Melton, F.; Post, K. Irrigation and nitrogen management decision support tool for vegetables and berries. In Proceedings of the U.S Committee on Irrigation and Drainage Conference: Groundwater Issues and Water Management—Strategies Addressing the Challenges of Sustainability USCID, Sacramento, CA, USA, 4–7 March 2014; pp. 53–64.
71. Bartlett, A.C.; Andales, A.A.; Arabi, M.; Bauder, T.A. A smartphone app to extend use of a cloud-based irrigation scheduling tool. *Comput. Electron. Agric.* **2015**, *111*, 127–130.
72. SmartIrrigation Apps. Available online: http://smartirrigationapps.org/ (accessed on 20 January 2017).
73. Migliaccio, K.W.; Morgan, K.T.; Vellidis, G.; Zotarelli, L.; Fraisse, C.; Zurweller, B.A.; Andreis, J.H.; Crane, J.H.; Rowland, D.L. Smartphone apps for irrigation scheduling. *Trans. ASABE* **2016**, *59*, 291–301.
74. Vellidis, G.; Liakos, V.; Andreis, J.H.; Perry, C.D.; Porter, W.M.; Barnes, E.M.; Morgan, K.T.; Fraisse, C.; Migliaccio, K.W. Development and assessment of a smartphone application for irrigation scheduling in cotton. *Comput. Electron. Agric.* **2016**, *127*, 249–259.
75. Mannini, P.; Genovesi, R.; Letterio, T. IRRINET: Large Scale DSS Application for On-farm Irrigation Scheduling. *Procedia Environ. Sci.* **2013**, *19*, 823–829.
76. Montoro, A.; Lopez-Fuster, P.; Fereres, E. Improving on-farm water management through an irrigation scheduling service. *Irrig. Sci.* **2011**, *29*, 311–319.
77. Deumier, J.M.; Leroy, P.; Peyremorte, P. Tools for improving management of irrigated agricultural crop systems. *Water Rep.* **1996**, *8*, 39–49.
78. Car, N.J.; Christen, E.W.; Hornbuckle, J.W.; Moore, G.A. Using a mobile phone Short Messaging Service (SMS) for irrigation scheduling in Australia Farmers participation and utility evaluation. *Comput. Electron. Agric.* **2012**, *84*, 132–143.
79. USDA SSURGO. Available online: https://sdmdataaccess.nrcs.usda.gov/ (accessed on 20 January 2017).
80. Doorenbos, J.; Pruitt, W.O. *Guidelines for Predicting Crop-Water Requirements*, 2nd ed.; Food and Agriculture Organization of the United Nations: Rome, Italy, 1977; p. 156.
81. Johnson, L.F.; Cahn, M.; Martin, F.; Melton, F.; Benzen, S.; Farrara, B.; Post, K. Evapotranspiration-based irrigation scheduling of head lettuce and broccoli. *HortScience* **2016**, *51*, 935–940.
82. Morrison, M. Encouraging The Adoption of Decision Support Systems by Irrigators. *Rural Soc.* **2009**, *19*, 17–31.

83. Giannakis, E.; Bruggeman, A.; Djuma, H.; Kozyra, J.; Hammer, J. Water pricing and irrigation across Europe: Opportunities and constraints for adopting irrigation scheduling decision support systems. *Water Sci. Technol. Water Supply* **2016**, *16*, 245–252.
84. Rana, G.; Katerji, N. Measurement and estimation of actual evapotranspiration in the field under Mediterranean climate: A review. *Eur. J. Agron.* **2000**, *13*, 125–153.
85. Allen, R.G.; Pereira, L.S.; Howell, T.A.; Jensen, M.E. Evapotranspiration information reporting: I. Factors governing measurement accuracy. *Agric. Water Manag.* **2011**, *98*, 899–920.
86. Allen, R.G. *Lysimeters for Evapotranspiration and Environmental Measurements*; American Society of Civil Engineers: New York, NY, USA, 1991.
87. Snyder, R.L.; Spano, D. Using surface renewal measurements to estimate crop evapotranspiration. In Proceedings of the Third Congress of the European Society for Agronomy, Abano Padova, Italy, 18–22 September 1994.
88. Hanson, B.R.; May, D.M. Crop coefficients for drip-irrigated processing tomato. *Agric. Water Manag.* **2006**, *81*, 381–399.
89. Cahn, M.; Snyder, R.; Hanson, B. Estimating evapotranspiration in processing tomato. *Acta Hortic.* **1999**, *487*, 493–497.
90. Hanson, B.R.; May, D.M. Crop evapotranspiration of processing tomato in the San Joaquin Valley of California, USA. *Irrig. Sci.* **2006**, *24*, 211–221.
91. Rosa, R.; Dicken, U.; Tanny, J. Estimating evapotranspiration from processing tomato using the surface renewal technique. (Special Issue: Sensing technologies for sustainable agriculture.). *Biosyst. Eng.* **2013**, *114*, 406–413.
92. Lopez Avendano, J.E.; Diaz Valdes, T.; Watts Thorp, C.; Rodriguez, J.C.; Castellanos Villegas, A.E.; Partida Ruvalcaba, L.; Velazquez Alcaraz, T.D.J. Evapotranspiration and crop coefficient of bell pepper in Culiacan Valley, Mexico. *Terra Latinoam.* **2015**, *33*, 209–219.
93. Hanson, B.R.; May, D.M. New crop coefficients developed for high-yield processing tomatoes. *Calif. Agric.* **2006**, *60*, 95–99.
94. Shapland, T.M.; McElrone, A.J.; Khaw Tha Paw, U.; Snyder, R.L. A turnkey data logger program for field-scale energy flux density measurements using eddy covariance and surface renewal. *Ital. J. Agrometeorol.* **2013**, *18*, 5–16.
95. Courault, D.; Seguin, B.; Olioso, A. Review on estimation of evapotranspiration from remote sensing data: From empirical to numerical modeling approaches. *Irrig. Drain. Syst.* **2005**, *19*, 223–249.
96. Kalma, J.; McVicar, T.; McCabe, M. Estimating land surface evaporation: A review of methods using remotely sensed surface temperature data. *Surv. Geophys.* **2008**, *29*, 421–469.
97. Gonzalez-Dugo, M.; Neale, C.; Mateos, L.; Kustas, W.; Prueger, J.; Anderson, M.; Li, F. A comparison of operational remote sensing-based methods for estimating crop evapotranspiration. *Agric. For. Meteorol.* **2009**, *149*, 1843–1853.
98. Bastiaanssen, W.; Meneti, M.; Feddes, R.; Holtslag, A. A remote sensing surface energy balance algorithm for land (SEBAL). *J. Hydrol.* **1998**, *212*, 198–212.
99. Allen, R.G.; Tasumi, M.; Trezza, R. Satellite-based energy balance for mapping evapotranspiration with internalized calibration (METRIC) Model. *J. Irrig. Drain. Eng.* **2007**, *133*, 380–394.
100. Teixeira, A.; Bastiaanssen, W.; Ahmad, M.; Bos, M. Reviewing SEBAL input parameters for assessing evapotranspiration and water productivity for the Lower-Middle Sao Francisco river basin, Brazil, Part A: Calibration and validation. *Agric. For. Meteorol.* **2008**, *149*, 462–476.
101. Allen, R.; Burnett, B.; Kramber, W.; Huntington, J.; Kjaersgaard, J.; Kilic, A.; Kelly, C.; Trezza, R. Automated calibration of the METRIC-Landsat evapotranspiration process. *J. Am. Water Resour. Assoc.* **2013**, *49*, 563–576.
102. Bryla, D.R.; Trout, T.J.; Ayars, J.E. Weighing Lysimeters for Developing Crop Coefficients and Efficient Irrigation Practices for Vegetable Crops. *HortScience* **2010**, *45*, 1597–1604.

103. Johnson, L.F.; Trout, T. Satellite NDVI assisted monitoring of vegetable crop evapotranspiration in California's San Joaquin Valley. *Remote Sens.* **2012**, *4*, 439–455.
104. Tucker, C. Red and photographic infrared linear combinations for monitoring vegetation. *Remote Sens. Environ.* **1979**, *8*, 127–150.
105. Trout, T.; Johnson, L.; Gartung, J. Remote sensing of canopy cover in horticultural crops. *HortScience* **2008**, *43*, 333–337.
106. Neale, C.; Jayanthi, H.; Wright, J. Irrigation water management using high resolution airborne remote sensing. *Irrig. Drain. Syst.* **2005**, *19*, 321–326.
107. Calera-Belmonte, A.; Jochum, A. Operational Tools for Irrigation Water Management Based on Earth Observation: The DEMETER Project. In *Remote Sensing for Agriculture, Ecosystems, and Hydrology VIII*; SPIE: Stockholm, Sweden, 2006.
108. Melton, F.; Johnson, L.F. Satellite Irrigation Management System. Available online: https://ecocast.arc.nasa.gov/dgw/sims (accessed on 27 January 2017).
109. Melton, F.; Johnson, L.; Lund, C.; Pierce, L.; Michaelis, A.; Hiatt, S.; Guzman, A.; Adhikari, D.; Purdy, A.; Rosevelt, C.; et al. Satellite Irrigation Management Support with the Terrestrial Observation and Prediction System. *IEEE J. Sel. Top. Appl. Earth Obs. Remote Sens.* **2012**, *5*, 1709–1721.
110. Johnson, L.; Cassel-Sharma, F.; Goorahoo, D.; Melton, F. Landsat-based calculation of agricultural water use fractions in California. In Proceedings of the 19th William T. Pecora Memorial Remote Sensing Symposium, Denver, CO, USA, 17–20 November 2014.
111. IrriSat. Available online: https://irrisat-cloud.appspot.com/ (accessed on 27 January 2017).
112. Montgomery, J.; Hornbuckle, J.; Hume, I.; Vleeshouwer, J. IrriSAT—Weather based scheduling and benchmarking technology. In Proceedings of the 17th ASA Conference, Hobart, Australia, 20–24 September 2015.
113. Hornbuckle, J.; Vleeshouwer, J.; Ballester, C.; Montgomery, J.; Hoogers, R.; Bridgart, R. *IrriSAT Technical Reference*; Deakin University: Victoria, Australia, 2016.
114. Vuolo, F.; D'Urso, G.; De Michele, C.; Bianchi, B.; Cutting, M. Satellite-based irrigation advisory services: A common tool for different experiences from Europe to Australia. *Agric. Water Manag.* **2015**, *147*, 82–95.
115. D'Urso, G. Current status and perspectives for the estimation of crop water requirements from earth observation. *Ital. J. Agron.* **2010**, *5*, 107–120.
116. EEflux. Available online: http://eeflux-level1.appspot.com/ (accessed on 27 January 2017).
117. Allen, R.; Morton, C.; Kamble, B.; Kilic, A.; Huntington, J.; Thau, D.; Ratcliffe, I. EEFlux: A Landsat-based evapotranspiration mapping tool on the Google Earth Engine. In Proceedings of the ASABE/IA Irrigation Symposium, Long Beach, CA, USA, 10–12 November 2015.
118. Pereira, L.; Allen, R. Crop evapotranspiration estimation with FAO56: Past and future. *Agric. Water Manag.* **2015**, *147*, 4–20.
119. Vermote, E.; Saleous, N. *LEDAPS Surface Reflectance Product Description*; University of Maryland: College Park, MD, USA, 2007.
120. Tasumi, M.; Allen, R.; Trezza, R. At-surface albedo from Landsat and MODIS satellites for use in energy balance studies of evapotranspiration. *J. Hydrol. Eng.* **2008**, *13*, 51–63.
121. Johnson, L.; Cahn, M.; Rosevelt, C.; Guzman, A.; Melton, F.; Lockhart, T.; Farrara, B.; Melton, F. *Satellite Estimation of Fractional Cover in Several California Specialty Crops*; American Geophysical Union: San Francisco, CA, USA, 2016.
122. Gago, J.; Douthe, C.; Coopman, R.E.; Gallego, P.P.; Ribas-Carbo, M.; Flexas, J.; Escalona, J.; Medrano, H. UAVs challenge to assess water stress for sustainable agriculture. *Agric. Water Manag.* **2015**, *153*, 9–19.

123. Gonzalez-Dugo, V.; Zarco-Tejada, P.; Nicolas, E.; Nortes, P.A.; Alarcon, J.J.; Intrigliolo, D.S.; Fereres, E. Using high resolution UAV thermal imagery to assess the variability in the water status of five fruit tree species within a commercial orchard. *Precis. Agric.* **2013**, *14*, 660–678.

124. Torres-Sanchez, J.; Pena, J.M.; de Castro, A.I.; Lopez-Granados, F. Multi-temporal mapping of the vegetation fraction in early-season wheat fields using images from UAV. *Comput. Electron. Agric.* **2014**, *103*, 104–113.

125. Torres-Sanchez, J.; Lopez-Granados, F.; Pena, J.M. An automatic object-based method for optimal thresholding in UAV images: Application for vegetation detection in herbaceous crops. *Comput. Electron. Agric.* **2015**, *114*, 43–52.

126. Food and Agriculture Organization of the United Nations; The World Water Council. Toward a Water and Food Secure Future: Critical Perspectives for Policy-Makers. 2015, p. 61. Available online: http://www.fao.org/nr/water/docs/FAO_WWC_white_paper_web.pdf (accessed on 20 January 2017).

CHAPTER 2

Organic Plant Breeding: A Key to Improved Vegetable Yield and Safe Food

RodelMaghirang *, Maria Emblem Grulla, Gloria Rodulfo, Ivy Jane Madrid and Maria Cielo Paola Bartolome

Institute of Plant Breeding, College of Agriculture, University of the Philippines Los Baños, Laguna 4031, Philippines

ABSTRACT

Most often, organic farming focuses on the improvement of management practices such as nutrient application and pest control, and very seldom deals with variety improvement or breeding. Because it has been dependent on commercially-available varieties developed under conventional high-input methods, traits are expressed resulting in low yields that are commonly attributed to organic farming practices rather than to the adaptability of the cultivar to the system. A research program in the Philippines involving several regions and institutions has pioneered in the evaluation and improvement of varieties through breeding under low-input organic conditions. After making several crosses, pedigree selection, replicated yield and on-farm trials, promising and potential varieties were developed and identified in squash, cucumber, lettuce and yardlong bean. The most promising yield advantages over the respective check varieties ranged up to 47% in squash, 31% in yardlong bean, 42% in lettuce, and 43% in cucumber. Pest and disease resistance were also considered during the selection process, and top performers were moderately to highly resistant. General acceptability in appearance, taste and marketability provided additional selection criteria for considering the top performers and potential varieties. Commercial varieties developed and performing well under conventional high-input methods were mostly not suitable under organic low-input conditions. Hence, breeding under organic low-input conditions is a must to achieve high yield in organic farming systems.

Keywords: low input; squash; cucumber; lettuce; yardlong bean; Philippines

1. INTRODUCTION

In the last 14 years, the total organic agricultural land in the world has increased by almost 300%. India has the largest number of organic producers (650,000), Australia has the largest area (17.2 million hectares), the USA has the largest market (24,347 million Euros), and Switzerland has the highest per capita consumption as reported by the International Federation of Organic Agriculture Movements (IFOAM) [1]. It is common practice to approach organic production by replacing application of synthetic chemicals with biologically- or naturally-derived alternatives. Numerous products and methods of cultivation, soil inputs, weed control, nutrient recycling and application, and non-chemical pest control have been developed. Recently, especially in the Asia and Pacific region, efforts have been initiated in breeding varieties for organic production systems. This is an approach not yet been widely exploited by researchers.

Organic varieties or seeds are required in organic production. However, limitations in the availability of organic varieties triggered policy makers to adjust and consider this gap. Conventional F1 varieties from private seed companies are allowed in organic production provided that seeds are produced for at least one generation under an organic system [2,3]. However, previous variety trials show that conventional varieties do not always perform well under organic conditions. This may be because conventional varieties are intentionally developed under optimum conditions but are not adaptable to low input conditions and an organic environment.

Basically, the concept of organic breeding is the same as conventional breeding, with Phenotype (P) = Genotype (G) + Environment (E) + Genotype × Environment (GE), wherein the performance of the variety (P) is dependent on the genetic trait (G), effect(s) of the environment (E), and the interaction between the variety and the environment (GE). However, varieties from conventional breeding may have traits that may be unsuitable for organic production systems, and some important traits for organic farming systems may not be found in the conventional varieties [4]. Breeding under organic conditions may result in improved levels of stress tolerance and disease resistance in the resulting varieties. Hence, the evaluation and improvement of varieties through breeding under organic conditions at minimum levels of input application are essential for the development of the organic sector and for the quality of organic products.

As IFOAM [5] mentioned in a position paper, "organic plant breeding is inevitable." Thus, this research has the objectives of selecting, recommending or developing organic vegetable cultivars such as yardlong bean (Vignaunguiculatasubsp. sesquipedalis), squash (Cucurbitamoschata), lettuce (Lactuca sativa) and cucumber (Cucumissativus) for organic production and consequently to increase vegetable production in organic systems.

2. EXPERIMENTAL SECTION

2.1. Variety Development

From 1992 to 2002, selected open-pollinated lines and segregating populations of yardlong bean, squash, lettuce and cucumber were subjected to primary evaluation under the DA-BAR funded project "Varietal Evaluation of Selected Vegetables under Organic Conditions" [6]. From these lines and populations, entries with good quality and performance under organic conditions were selected and crosses were made among the selections. The crosses made were evaluated in two sites at the University of the Philippines Los Baños (UPLB), one at the Institute of Plant Breeding (IPB) main compound and another at the PAMANA station along Pili Drive. Pedigree selection was employed under low-input organic conditions.

At the experimental sites, raised beds with dimensions of 1 m × 5 m were prepared for the evaluation of the lines and crosses. Each raised bed was lined with two rows having 20 hills per row and a distance of 50 cm between hills. The evaluation sites were calibrated to be able to select for genotypes with tolerance to weed populations, resistance or tolerance to pests and diseases, and efficiency in nutrient utilization and related stress(es), following Philippine National Standards (PNS) for Organic Agriculture. Organic fertilizers (vermicompost and other compost materials) were applied at a minimal level of about 50%–75% of the nutrition required by the plant, specifically at 2–3 $t \cdot ha^{-1}$. Fermented plant juice and fermented fruit juice were also applied at specific stages of the plants. Minimal weeding and pest control measures were also practiced. Plastic mulch was used to control weeds, while aromatic pest-repellent plants such as lemon grass and marigold were planted as a means of pest control. Furthermore, flowering plants such as Mexican sunflower and

Zinnia were planted to attract pollinators and other beneficial insects, while barrier plants were established around the experimental areas to prevent contamination from neighboring farms. These farm practices were employed for every season of evaluation of the selected lines.

For every generation, individual plants were selected based on general vigor, fruit characteristics, resistance to pests and diseases, and tolerance to environmental stress. Selfing was performed for each individual plant selection, and the selfed seeds obtained were used for the following generation. At the F6 generation, the selected lines were evaluated in an observational trial to determine the top lines to be included in the yield trials. The observational trials were conducted at IPB-UPLB, Palawan Agricultural Center (PAC), DA-Palawan Agricultural Experiment Station (DA-PAES), and the Bureau of Plant Industry—Los Baños National Crop Research and Development Center (Los Baños, Laguna, Philippines).

2.2. Replicated Yield Trials

Replicated yield trials were conducted at PAMANA, Pili Drive, UPLB and DA-PAES. The entries included for each yield trial were the top selected lines and a commercial check variety for each crop. Each trial was carried out using a Randomized Complete Block Design (RCBD) with three replications. The replicated yield trials also followed the aforementioned farm practices employed during the preliminary evaluation of the lines. Data on horticultural and fruit characteristics were gathered from five plant samples per replication. Total yield was obtained and fruits were graded as marketable and non-marketable.

2.3. On-Farm Trials (OFT)

The top performing lines for each crop were further evaluated through on-farm trials in order to assess their acceptability and performance in different locations. The OFT sites were located in some Philippine provinces of Region 4, namely, Laguna, Quezon, Cavite, and Palawan and were all under organic farming conditions. All entries were planted in plots of commercial sizes with at least 20 m^2 per entry. A commercial hybrid variety which served as the check was also included for each trial. Cultural management followed the practices of the farmer cooperators. At the end of each trial, the cooperators were asked to rank the lines based on their preferences and to identify the lines they preferred the most, as well as the basis of their selection. Notes on the performance of the entries on-farm were taken. Seed production was also performed by the cooperators in order to use their selected lines in the next season.

2.4. Breeder Seed Production

Breeder's seeds of the potential varieties were produced on-station at PAMANA to ensure purity of supply for the following trials and also for variety registration. All top-performing entries across crops were produced following the cultural management practices for organic production. Fruits were allowed to mature prior to extraction of seeds, after which seeds were air dried for 2–3 days and then dried under the sun for another 2–3 days. Completely dried seeds were kept in tightly sealed containers.

2.5. Data Analysis

All data were analyzed by ANOVA using the Statistical Analysis System (SAS) Software (SAS Institute Inc., Cary, NC, USA). Means were compared using Fisher's Least Significant Difference (LSD) test at $p = 0.05$.

3. RESULTS AND DISCUSSION

The study resulted in the recommendation of four selections per crop that were vigorous, prolific, early, high yielding and highly acceptable to farmers and the market (Table 1).

3.1. Yardlong Bean

In the development of varieties of yardlong bean, average yields of 22.29 $t{\cdot}ha^{-1}$ (selection 0801-5-1-1-0), 22.13 $t{\cdot}ha^{-1}$(selection 1096-1-1-0-0), 20.29 t/ha (selection 10116-1-1-0-0) and 21.25 $t{\cdot}ha^{-1}$ (selection 10421-0-0) were achieved. These selections had average yield advantages of 31%, 30%, 25% and 19%, respectively, over the check variety. The selections were highly acceptable in the market due to their medium to long pods. Green podded 1096-1-1-0-0 had an average pod length range of 58cm, light green podded 10421-0-0 was 57 cm, dark green podded 10116-1-1-0-0 was 54 cm, and green podded 0801-5-1-1-0 was 52 cm (Figure 1). Selection 1096-1-1-0-0 had the heaviest pod weight ranging of 22.9–32.7 g, 0801-5-1-1-0-0 ranged from 23–24.4 g, 10116-1-1-0-0 ranged from 23.3–25.7 g, and 10421-0-0 ranged from 23.9–26.3 g.

Table 1. Yield and percent yield advantage above the check yield of potential varieties developed for organic low-input conditions.

Crop/Entry	Yield (t/ha)	Yield Advantage (%)
	Yardlong bean	
0801-5-1-1-0	22.29	31.0
1096-1-1-0-0	22.13	30.1
10421-0-0	21.25	24.9
10116-1-1-0-0	20.29	19.3
Check	17.01	
	Cucumber	
11622	26.03	43.0
11621	18.87	3.7
11617	18.54	1.8
11624	17.97	−1.3
Check	18.21	
	Lettuce	
Le 1103(Looseleaf)	10.96	44.0
Le 0701 (Looseleaf)	6.38	−16.1
Check (Looseleaf)	7.61	
Le 0702 (Cos)	7.74	0.9
Le 1104 (Cos)	6.93	−9.7
Check (Cos)	7.67	
	Squash	
10128-1-1	61.03	47.5
1058-1-1	48.03	16.1
1056-1-1	42.83	3.5
10127-1-2	49.67	20.1
Check	41.37	

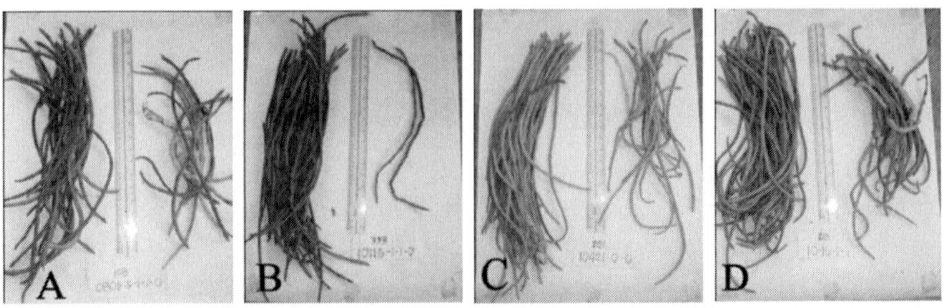

Figure 1. Yardlong bean top yielders: (**A**) 0801-5-1-1-0; (**B**) 10116-1-1-0-0; (**C**) 10421-0-0 and (**D**) 1096-1-1-0-0.

In Alfonso, Cavite, green-podded yardlong bean, 0801-5-1-1-0 and 10116-1-1-0-0 were sold in the local market. Selection 0801-5-1-1-0 was also chosen for its resistance to bean rust. In Majayjay, Laguna, 10116-1-1-0-0, 10421-0-0 and 0801-5-1-1-0 were preferred for their characteristics, while 1096-1-1-0-0 was not chosen because it did not survive the intermittent rains and waterlogging in the area. The Majayjay cooperator preferred 0801-5-1-1-0 because of its high yield capacity, for good pod qualities such as wider, longer pods, and for its dark green color. In Quezon, the light-green pods of 10421-0-0 were highly preferred. Generally, all cooperators chose all entries because of their moderate to high resistance to common bean diseases and pests, long pods that were stringless and snappy, high yield, and market acceptability.

3.2. Cucumber

For cucumber variety selections, average yields observed were 11622 with 26.03 $t{\cdot}ha^{-1}$, 11,621 with 18.87 $t{\cdot}ha^{-1}$, 11,617 with 18.51 $t{\cdot}ha^{-1}$ and 11,624 with 17.97 $t{\cdot}ha^{-1}$. Entry 11622 had an average yield advantage over the check variety of 43%, 11621 was 4% more, and aa617 was 1.8% more. Cucumber selection 11,617, which has a light green or bicolor (green with white streaks) fruit skin, had an average fruit weight of 241.6 g and an average length of 14.7 cm; bicolor fruit skin 11621 had a 277.7 g average fruit weight and 16.6 cm average length; light green or bicolor skin 11622 had a 234.4 g average weight and 13.5 cm average length; and, 11,624, which is a bicolored, had a 202.8 g average fruit weight and 11.4 cm average length (Figure 2). All selections flowered on the 26th day after planting. All were thin skinned and sweet.

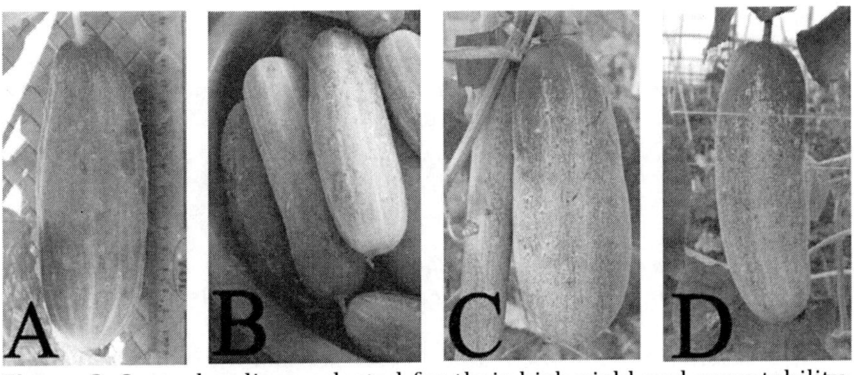

Figure 2. Cucumber lines selected for their high yield and acceptability: (**A**) 11622; (**B**) 11621; (**C**) 11617 and (**D**) 11624.

Lines 11,621 and 11,624 were the top selections of the farmer-cooperator in Alfonso for their juicy flesh, green and thin skin, and high yield capacity. In its preliminary OFT, line 11,621 was also selected for its vigor and earliness. These cucumber selections were highly accepted by the market around Alfonso and nearby municipalities. In Lucban, Quezon, the cooperator selected 11,621 for its prolific fruiting, thin skin and juicy flesh.

3.3. Lettuce

There are two types of lettuce recommended from the study, the loose-leaf and the Cos or Romaine type (Figure 3). Two selections were identified per lettuce type. Loose-leaf type Le 1103 had the highest average yield of 10.96 $t{\cdot}ha^{-1}$ with average yield advantage over the check of 44%, and Le 0701 yielded 6.38 $t{\cdot}ha^{-1}$. While Le 0701 had a lower average yield, it was highly acceptable. Le 0702, on the other hand, was the top yielder of the Cos type with a 7.74 $t{\cdot}ha^{-1}$ and a yield advantage of 0.9%. Loose-leaf Le 1103 has light green color with a purple tinge, is curly and spreading, while Le 0701 is light green, curly and also has spreading leaves. The Cos-type Le 0702 has green, closed-heads with slightly curly leaf blades, and is upright, while Le 1103 is green, has a flat-head and is semi-upright. These selections were also late bolting, highly uniform and vigorous.

Consumers in the markets in Alfonso, and Tagaytay, Cavite, preferred loose-leaf types—Le 07-01 and Le 11-03 and Cos-type Le 07-02. The cooperator-farmers chose these selections because of their late-bolting trait, too. In Lucban, Quezon, Le 0701, a loose-leaf type, was selected for its late-bolting, high yield, and vigor, while Le 1104 was preferred for its resistance or tolerance to stem rot that was observed in the area. Le 1104 was also high yielding and late bolting. The cooperator in Silang, Cavite, chose Le 0701 for loose-leaf and Le 0702 for a Cos-type for their high yield. In Majayjay, Laguna, the cooperator chose loose-leaf Le 0701 and Cos type Le 0702 due to their high yield, acceptability based on their taste and appearance, and their high marketability. They were also late bolters and vigorous when grown on regular plots or in vertical gardens.

3.4. Squash

Squash selection 10128-1-1 was the top yielder from two preliminary yield trials in DA-PAES. Average total yield of this selection was 61.03 $t{\cdot}ha^{-1}$ with a yield advantage of 47.5% above the check variety, followed by line 1058-1-1 with 48.03 $t{\cdot}ha^{-1}$ (a 16% yield advantage) and line 1056-1-1 with 42.83 $t{\cdot}ha^{-1}$ (a 3.5% yield advantage). These three top yielders are the round types with varying sizes. The last squash selection was 10127-1-2 which is a long-necked or cacao-type squash that yielded an average 49.67 t·ha^{-1} (a yield advantage of 20%). Line 10128-1-1 has yellow orange flesh with average length and diameter of 14 cm and 15 cm, respectively, while 1058-1-1 has yellow flesh with an average fruit length and diameter of 10 cm and 13 cm, respectively, and 1056-1-1 also has yellow orange flesh with average length of 9.5 cm and average fruit diameter of 13 cm. Lastly, cacao type 10127-1-2 has a 33.5 cm average fruit length and a 10 cm average diameter. The flesh color is also yellow orange (Figure 4).

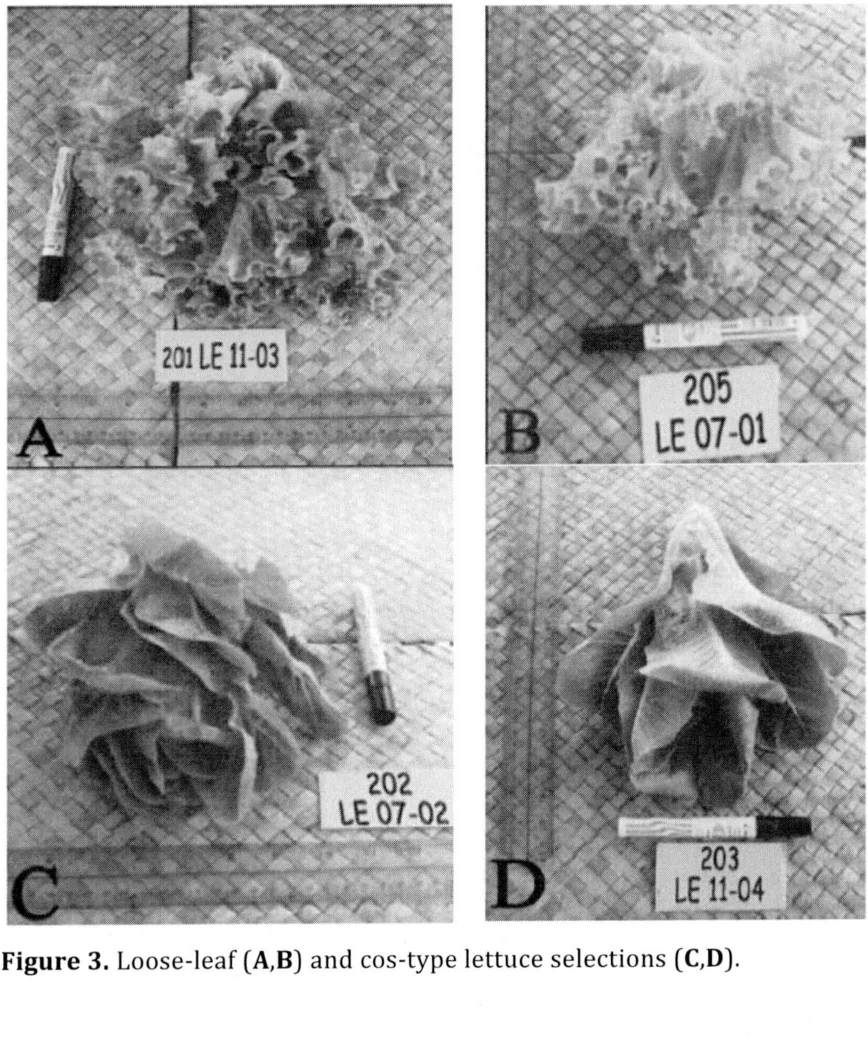

Figure 3. Loose-leaf (**A**,**B**) and cos-type lettuce selections (**C**,**D**).

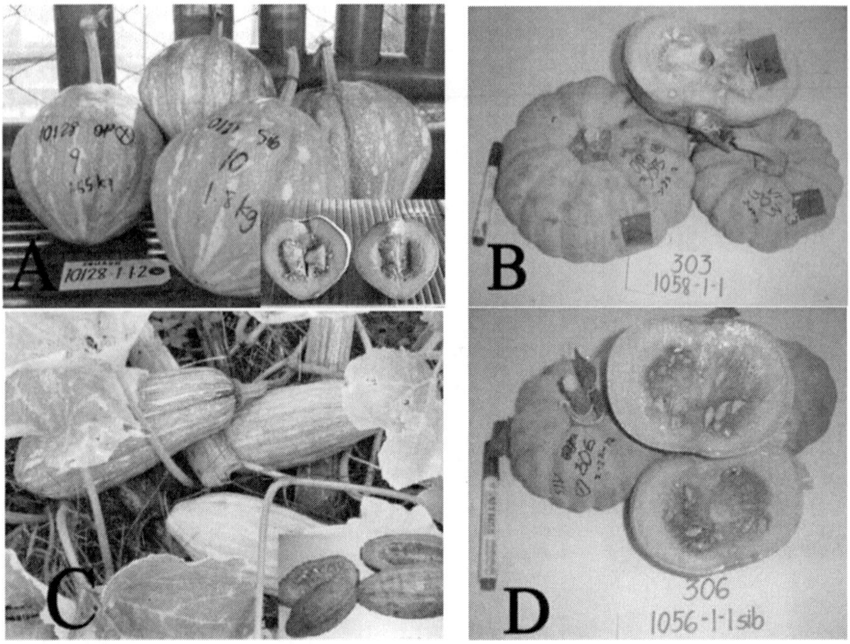

Figure 4. Round-type squash (**A**) 10128-1-1; (**B**) 1058-1-1; (**D**) 1056-1-1; and cacao-type squash (**C**) 10127-1-2 selections.

In Alfonso, Cavite, the farmer-cooperator selected the solo type 10128-1-1 for its size, shape and fruit yield capacity. Organic market-goers in Lucban, Quezon, preferred line 10128-1-1 because of its size which is suitable for a one-serving dish. It was also selected by the cooperator-producer because of its heart-shaped fruits and its low number of seeds. In Tayabas, Quezon, lines 10127-1-2 and 1056-1-1 were preferred for their prolific yield. Another cooperator located in Brgy. Isugod, Quezon, Palawan, selected line 10128-1-1 for its highly acceptable fruit.

4. CONCLUSIONS

Commercial varieties from conventional production systems do not always perform best under organic systems. Therefore, it is inevitable that breeding vegetable varieties adaptable for organic low-input conditions will be needed to achieve a high yield in organic farming. From the present studies, varieties developed under organic conditions were observed to be at par or yield more than commercial varieties commonly grown for the market. These lines have also shown resilience under adverse conditions such as high disease incidence and environmental aberrations along with high consumer acceptability. Thus, these varieties developed under organic agricultural systems show promise for answering the need for diversity, sustainability and efficiency in crop production.

ACKNOWLEDGMENTS

This research was conducted in collaboration with Department of Agriculture-Palawan Agriculture Experiment Station (DA-PAES) and Bureau of Plant Industry- LBNCRDC, and supported and funded by the Philippine Council for Agriculture, Aquatic and Natural Resources Research and Development (PCAARRD). The authors would like to deeply recognize and express gratitude to the research team for their dedication and commitment.

REFERENCES

1. IFOAM. *Consolidated Annual Report of IFOAM—Organics International*; IFOAM-Organics International Head Office: Bonn, Germany, 2015.
2. Ratanawaraha, C.; Ellis, W.; Panyakul, V. *Organic Agri-Business: A Status Quo Report for Thailand 2007*; Thai-German Programme for Enterprise Competitiveness, Sustainable Agriculture Foundation and GreenNet Foundation: Bangkok, Thailand, 2007.
3. Lammerts van Bueren, E.T. Organic Plant Breeding and Propagation: Concepts and Strategies. Ph.D. Thesis, Wageningen University, Wageningen, The Netherlands, 2002.
4. Legzina, L.; Skrabule, I. *Plant Breeding for Organic Farming: Current Status and Problems in Europe*; Compendium ENVIRFOOD: Talsi, Latvia, 2005.
5. IFOAM. *IFOAM Position on the Use of Organic Seed and Plant Propagation Material in Organic Agriculture*; IFOAM Head Office: Bonn, Germany, 2011.
6. Maghirang, R.G.; Rodulfo, G.S.; Enicola, E.E.; Candelaria, R. Organic Breeding and Seed Production in Selected Vegetables. In Proceedings of the Organic Seed Preconference of the 17th IFOAM Organic World Congress, Namyangju, Korea, 26–27 September 2011.

CHAPTER 3

Climate Change Impacts on Water Use in Horticulture

Richard L. Snyder

Department of Land, Air and Water Resources, University of California, Davis, CA 95616, USA

ABSTRACT

The evidence for anthropogenic global climate change is strong, and the projected climate changes could greatly impact horticultural production. For horticulture, two of the biggest concerns are related to the scarcity of water for crop production and the potential for increased evapotranspiration (ET). While ET is known to increase with air temperature, it is also known to decrease with increasing humidity and atmospheric CO_2 concentration. Considering all of these factors and a plausible climate projection, this paper demonstrates that ET may increase or decrease depending on the magnitude of atmospheric changes including wind speed. On the other hand, the evidence is still strong that water resources will become less reliable in many regions where horticultural crops are grown.

Keywords: evapotranspiration; environmental conditions; CO_2 concentrations; water requirements

1. INTRODUCTION

Most scientists agree that global climate change is occurring at an alarming rate and that these changes are likely to impact water use in horticulture, agronomy, and natural ecosystems. In some locations, climate change can potentially increase agricultural production, but it is generally believed that widespread detrimental impacts on agricultural production are more likely in much of the world. Currently, the global climate change trend is for increasing air temperature, mainly at night and during winter, and more near the poles than in lower latitudes. Because of global warming, energy storage in water has increased dramatically (much more than in the air), and higher water temperatures has led to rising sea level due to water expansion and additional heat storage mainly in the oceans. In addition, increasing air and water temperature enhances evaporation, and higher air temperature increases the saturation water vapor pressure, i.e., the amount of water vapor that is held in the air at saturation. These atmospheric changes can impact plant growth, production, and water usage, and this chapter presents ideas on the possible impact of higher temperature, humidity, and CO_2 concentration on the evapotranspiration (ET) of horticultural crops.

At this time, the main cause for global climate change is the increasing concentration of CO_2. According to the "Keeling curve," the CO_2 concentration recorded at the Mauna Loa Observatory in Hawaii (USA) increased from about 315 to 395 ppm from 1958 to 2013, and recently passed 400 ppm. This increase has serious implications for global climate change and its impact on nature and horticulture.

The greenhouse gases (GHGs) that contribute most to global climate change from the worst to least worse are as follows: (1) H_2O—water vapor; (2) CO_2—carbon dioxide; (3) NH_4—methane; (4) N_2O—nitrous oxide; (5) O_3—ozone. While H_2O is actually a more effective GHG than CO_2, the atmospheric concentration is spatially and temporally variable and it is not increasing rapidly at this time. However, as the oceans warm, higher concentrations of atmospheric H_2O are likely, and the higher levels can greatly contribute to warming. Carbon dioxide is a less effective GHG than H_2O, but the more evenly distributed global concentration is increasing steadily, and it is currently the GHG causing the most rapid global temperature rise. Methane is a concern for the future because large amounts of methane are stored in the ocean and in permafrost. Scientists believe that thawing the permafrost and warming the oceans might release this stored NH_4 and cause a rapid temperature increase. Nitrous oxide is used as an aerosol propellant, anesthetic, and an oxidizer in rockets and engine fuel. Ozone is not evenly distributed over the Earth's surface, but man-made O_3 does contribute to atmospheric warming in urban areas.

GHGs mostly reside in the troposphere, i.e., the lowest level of the atmosphere, where they intercept long waveband radiation (LWR) and raise atmospheric temperatures. The stratosphere, which is a stable layer of air above the troposphere, has considerably less greenhouse gas than the troposphere, and, interestingly, this upper atmospheric layer is cooling. If the global warming were caused by increasing solar output or reduced reflection of solar radiation, the troposphere and stratosphere would both likely have increasing temperatures. Consequently, the simultaneous troposphere warming and stratosphere cooling is a good indicator that the warming is caused by anthropogenic increases in GHG additions (mostly CO_2) to the atmosphere. The melting of polar ice packs is a strong indication of global warming, and it will likely compound the warming because less solar radiation will be reflected back to space as the ice melts.

Climate change is likely to cause sea level rise and damage plant and animal life in the oceans and on land. Of course, changes in temperature and humidity could also lead to changes in general circulation of the atmosphere, greater frequency of storms and floods, and changes in the length and severity of drought. In general, climate scientists are projecting more frequent and severe storms and drought due to changes in atmospheric circulation [1]. In terms of horticulture, the authors of [2] and [3] discuss several potential effects of climate change on horticulture. In some cases, increasing CO_2, temperature, humidity, and other greenhouse gases might be beneficial in regions where crop production is limited by cold temperatures due to (1) lower potential for frost damage, (2) faster growth, and (3) lengthened growing seasons. In addition, there is some benefit coming from CO_2 fertilization, which can enhance photosynthesis. On the other hand, climate change could negatively impact agriculture in regions where climate conditions are currently good for production. For example, climate change might (1) decrease chilling and inhibit bloom and fruit set in horticultural crops, (2) lead to high temperature and wind during bloom or ripening that could negatively impact fruit set or fruit quality, (3) increase ET rates that could lead to water deficits, and (4) increase problems with heat stress.

Some possible climate change impacts on agriculture include (1) droughts, (2) floods, (3) faster phenological development, (4) inadequate chilling requirements, (5) pollination affected by rainfall and other extreme events, (6) frost and chill damage, (7) the spread of new insects and diseases, and (8) lower or higher yield and quality due to warming and water relations during summer [2,3]. One example of warming impacts is decreased winter chilling, which leads to bad pollination, staggered bloom, reduced fruit set, and poor fruit quality. In low chill years, apricots and cherries can drop flower buds or not produce a crop when chilling is inadequate.

In some regions, winter fog is an important factor in achieving adequate chilling for some crops. Rainfall during bloom can inhibit bee pollination, and precipitation around harvest time can increase fruit and nut diseases, cause fruit cracking, and destroy a crop. In deciduous orchards, late spring and summer rainfall has a negative effect on fruit and nut quality and production. Clearly, there are a multitude of climate factors that can change to wreak havoc on the production of horticultural crops. For example, climate scientists are projecting more bimodal precipitation in California, with more precipitation in the spring and fall and less in the winter. However, California depends on water storage in the mountain snowpack, and higher snow lines in the mountains, especially in the north where the mountains are shorter, will have less snowpack storage and will result in less water delivery to agricultural land in the summer [4]. Snowpack storage will affect the collection and distribution of irrigation water, but precipitation timing will also affect rainfall damage to crops in the spring and fall as well as changes in fog formation due to winters with higher temperatures and less rainfall.

While all of the aforementioned changes and impacts are important to consider, it is difficult to project changes in general circulation, storms, and drought with a high degree of accuracy in any particular region on Earth. However, the general impact of widespread higher temperatures, humidity, and increased CO_2 on factors affecting crop production is possible. Following a short discussion on how rising CO_2 concentrations cause global warming, this paper will present some ideas on the possible impact of higher temperature, humidity, and CO_2 on ET.

2. INCREASING CO_2 AND GLOBAL WARMING

The impact of atmosphere CO_2 concentration on the greenhouse effect and the Earth's surface temperature was first described by Svante August Arrhenius [5]. Although climate science has advanced considerably since Arrhenius first made his estimates of increasing CO_2 concentration, its impact on global temperature is still quite accurate. Equations (1)–(3) are modified versions of the original equations from [5]. Equation (1) provides an estimate of Earth's emission temperature (T_e), which is the mean surface temperature it would have if there were no atmosphere:

$$T_e = \left[\frac{\pi r^2 (1-\alpha_p) R_{sc}}{4\pi r^2 \sigma}\right]^{\frac{1}{4}} = \left[\frac{(1-\alpha_p) R_{sc}}{4\sigma}\right]^{\frac{1}{4}} \approx 255 \text{ K} \qquad (1)$$

where R_{SC} = 1631 W·m^{-2} is the solar constant, $\alpha_p \approx$ 0.30 is the albedo (reflection of solar radiation from the surface), $r \approx 6371 \times 10^3$ m is the mean Earth radius, π = 3.1415927, and σ = 5.67 × 10^{-8} J·s^{-1}·m^{-2}·K^{-4} is the Stefan–Boltzmann constant. Based on Equation (1), if we had no atmosphere, the Earth's mean surface temperature would be approximately T_e = 255 K = −18 °C.

Adding gases to the atmosphere has a minimal impact on the surface solar radiation balance, but greenhouse gases in the atmosphere are known to intercept upward and downward fluxes of LWR, which raises the atmospheric temperature. Assuming a single layer atmosphere that is opaque to LWR, it can be shown that Equation (2) provides an estimate of the maximum Earth surface temperature (T_s) assuming that all of the LWR is absorbed by the atmosphere.

$$T_s = 2^{0.25} \cdot T_e \approx 1.19 \cdot T_e = 303 \text{ K}. \quad (2)$$

Thus, if the GHG in the atmosphere was 100% efficient at intercepting LWR, the maximum surface temperature would be T_s = 303 K = 30 °C. Therefore, with current solar radiation, short waveband radiation reflection, and GHGs, the actual surface temperature of Earth falls somewhere between −18 °C and 30 °C.

Assuming a single layer leaky atmosphere, which allows for some loss of LWR back to space, it can be shown that Equation (3) provides an estimate of the Earth surface temperature (T_s) as a function of T_e and the fraction of upward LWR absorbed by the atmosphere (ϵ). Until the GHG concentration began to increase, about 78% of the LWR was absorbed by GHGs in the atmosphere. Therefore, using Equation (3) and $\epsilon = 0.78$ shows that GHG increased the Earth's surface temperature from −18 °C to 15 °C (288 K).

$$T_s = \left(\frac{2}{2-\epsilon}\right)^{\frac{1}{4}} T_e = \left(\frac{2}{2-\epsilon}\right)^{\frac{1}{4}} (255) \approx 288 \text{ K}. \quad (3)$$

Equation (3) illustrates that T_s is affected only by T_e and ϵ, and Equation (1) shows that T_e is affected only by the amount of incoming and reflected solar radiation. Thus, only the incoming and reflected short waveband radiation and GHG concentration determine the Earth's global surface temperature. Since 1958, changes in R_{SC} and α_p were insignificant; however, the anthropogenic releases of CO_2 and other GHGs led to a rise of more than 80 ppm in atmospheric CO_2concentration. The rising CO_2 concentration increased ϵ and it is the most likely cause for most of the corresponding 1 °C increase in global temperature [6]. Based on [1], considerably more global warming is projected as CO_2 levels continue to rise. While global temperature is rising, other climate factors, e.g., humidity, precipitation, wind speed, cloudiness, and precipitation, are also changing, and the assessment of climate change impact on weather and horticultural crops should consider all factors.

3. IMPACT ON EVAPOTRANSPIRATION

Climate change is likely to increase temperature, humidity, and stomatal resistance of plants, and all of those parameters affect ET. It is common for people to associate higher ET with higher temperature, because evaporation rates do increase with higher temperature. However, evaporation is a physical process, whereas ET is both physical and biological. Increasing temperature will affect ET, but radiation, humidity, wind, and CO_2 concentration also affect ET. All of these factors are needed to properly assess the impact of climate change on plant water usage.

4. ESTIMATING CROP EVAPOTRANSPIRATION

For many decades, scientists, engineers, and irrigation managers have used ET and the water balance method to determine agricultural and urban water demand for water resource planning and delivery and for on-farm and urban irrigation management. Because ET is affected by soil, plant, and atmospheric factors, spatial ET variation is common on different scales. In some climates, one can estimate ET using weather data of large areas, e.g., 50–100 km, but in other locations, microclimates can limit weather-based estimates of ET to small areas, e.g., less than 5 km. In general, the most common practice is to estimate "potential" or "energy-limited" crop evapotranspiration (ET_c) as the product of reference evapotranspiration (ET_{ref}) and a crop coefficient (K_c). Reference evapotranspiration is the energy-limited ET rate from a broad expanse of a well-watered vegetated surface, e.g., grass or alfalfa.

There are several methods to determine ET_{ref}, but the most common is to monitor weather over a large grass surface and use an equation to estimate the water use for a selected reference surface. Recently, the authors of [7,8] recommended fixed coefficients to estimate the canopy and aerodynamic resistances as inputs to the Penman–Monteith equation [9] to estimate ET_{ref} for 0.12-meter-tall and 0.50-meter-tall vegetated surfaces and assigned the symbol ET_o for the short canopy and ET_r for the tall canopy. The equations were called "standardized reference evapotranspiration" because the procedures to compute the ET_o and ET_r were standardized. Strictly speaking, the ET_o and ET_r equations provide estimates of ET from a virtual surface with input coefficients to estimate the appropriate canopy and aerodynamic resistances. However, in practice, the ET_o and ET_r approximate the ET of 0.12-meter-tall cool-season grass and 0.50-meter-tall alfalfa, respectively. Once the ET_{ref} is known, the ET_c is estimated as either $ET_c = ET_o \times$

K_c or $ET_c = ET_r \times K_{cr}$, where K_{cr} is specific to ET_r. Because ET_o is more widely used than ET_r, the remaining discussion will only use the ET_o method. Crop coefficients are generally determined as $K_c = ET_c/ET_o$ by calculating the K_c ratio using measured daily ET_c and ET_o from weather data. Global climate change could affect either the ET_o or the K_c, so it is important to assess the impact of climate change on both factors.

5. REFERENCE EVAPOTRANSPIRATION

To assess the possible impact of climate change on ET_o, Snyder et al. [10] used the standardized reference evapotranspiration (ET_o) equation for short canopies [7,8] that uses daily radiation, temperature, humidity, wind speed, and canopy resistance data to calculate ET_o. The following version of the ET_o equation [11] was used for the analysis:

$$ET_o = \frac{0.408\Delta(R_n - G) + \frac{900\gamma(e_s - e_a)u_2}{T+273}}{\Delta + \gamma(1 + 0.34u_2)} \quad \left(\mathrm{mm{\cdot}d^{-1}}\right). \qquad (4)$$

In Equation (4), Δ ($\mathrm{kPa{\cdot}K^{-1}}$) is the slope of the saturation vapor pressure curve at the mean daily air temperature, R_n($\mathrm{MJ{\cdot}m^{-2}{\cdot}d^{-1}}$) is the net radiation over well-watered grass, G ($\mathrm{MJ{\cdot}m^{-2}{\cdot}d^{-1}}$) is the soil heat flux density, γ ($\mathrm{kPa{\cdot}K^{-1}}$) is the psychrometric constant, T (°C) is the mean daily air temperature at a 1.5–2.0 m height, u_2 ($\mathrm{m{\cdot}s^{-1}}$) is the wind speed measured at 2 m above the ground, and e_s and e_a (kPa) are the saturation and actual vapor pressures of the air measured at a 1.5 m height. Information on how to compute R_n, G, Δ, etc. is provided in [7].

The aerodynamic resistance to sensible and latent heat transfer (r_a) occurs indirectly in two locations in Equation (4). The 0.34 in the denominator comes from the following:

$$\frac{r_c}{r_a} = \frac{70}{208/u_2} \approx 0.34u_2. \qquad (5)$$

In Equation (4), the right-hand side of the numerator could be written as

$$\frac{900\gamma(e_s - e_a)u_2}{T+273} = \frac{187200[\gamma(e_s - e_a)/(T+273)]}{208/u_2} = \frac{187200[\gamma(e_s - e_a)/(T+273)]}{r_a}. \qquad (6)$$

Therefore, the 208 coefficient is also included in the numerator of Equation (4) (within the 900). The $r_c = 70\ \mathrm{s{\cdot}m^{-1}}$ was estimated from the typical stomatal resistance $r_s = 100\ \mathrm{s{\cdot}m^{-1}}$ (corresponding to stomatal conductance g_s = 0.010 m·s^{-1} = 10 mm·s^{-1}) for the actively transpiring C_3 grass leaf surface, which was estimated as half of the LAI = 2.88. Therefore, the canopy resistance for 0.12-meter-tall C_3 species grass r_c was calculated as

$$r_c = \frac{r_s}{0.5 \cdot LAI} = \frac{100 \text{ s} \cdot \text{m}^{-1}}{0.5 \times 2.88} = 69 \approx 70 \text{ s} \cdot \text{m}^{-1}. \quad (7)$$

Long et al. [12] studied the effect of CO_2 concentration on stomatal conductance of C_3 species plants, and the relationships they reported were used to estimate the impact of increasing CO_2 concentration on canopy resistance of the standardized reference surface. Assuming that the $r_c \approx 70$ s·m^{-1} applies to a 2004 CO_2 concentration of about 372 ppm, estimating a new r_c value for higher CO_2 concentration provides a method to estimate possible impacts of higher CO_2 on ET_o.

Based on eight climate models, the mean air temperature projection for California is about 2.2 °C by 2050 and 4.0 °C by 2100 [13]. The global CO_2 concentration is projected to reach about 550 ppm by 2050 and more than 700 ppm by 2100 [14]. Long et al. [12] reported that stomatal conductance for many C_3 species plants decreased by about 20% when the CO_2concentration was increased from 372 to about 550 ppm from about 200 independent measurements. Assuming this is true for the stomatal conductance of 0.12-meter-tall C_3 species grass with a stomatal resistance of 100 s·m^{-1}, the stomatal conductance for C_3 grass should decrease from about 10 mm·s^{-1} to 8 mm·s^{-1}, which corresponds to r_s = 125 s·m^{-1}. Using the same approach used to calculate r_c in the ET_o equation [7], the r_c for 550 ppm is calculated as

$$r_c = \frac{r_s}{0.5 \cdot LAI} = \frac{125 \text{ s} \cdot \text{m}^{-1}}{0.5 \cdot 2.88} \approx 87 \text{ s} \cdot \text{m}^{-1}. \quad (8)$$

Thus, increasing CO_2 concentration from 372 to 550 ppm should increase canopy resistance of 0.12-meter-tall C_3 grass from 70 to 87 s·m^{-1}.

Roderick and Farquhar [15] observed that the global mean daily maximum temperature (T_x) increased by approximately 0.1 °C per decade and that the daily minimum temperature (T_n) increased by about 0.2 °C per decade, but there was no change in the vapor pressure deficit ($VPD = e_s - e_a$) in the few preceding decades. Cayan et al. [13] projected a 3 °C mean temperature increase by 2050 as a worst-case scenario for California. Therefore, Snyder et al. [10] evaluated the possible impact of climate change on ET_o by assuming that the mean air temperature would increase by 3.0 °C, with T_n and $T_x$increasing by 4.0 °C and 2.0 °C, respectively. Globally, the temperatures were rising, with no increase in relative humidity, which implies that the vapor pressure was also rising.

In much of California, the mean daily dew point temperature (T_d) is often nearly equal to T_n, so it was assumed that the T_d like T_n would increase by 4 °C in 2050. It was also assumed that the CO_2 concentration would increase from 372 to 550 ppm by volume by 2050, which corresponds to increasing the r_c from 70 to 87 s·m^{-1}. It was assumed that aerodynamic resistance, which is dependent on the atmospheric stability, wind speed and plant canopy roughness, would not change.

Using 18 California weather stations from a wide range of climates, daily data from 2003 were used to calculate how the annual ET_o might change with the aforementioned climate projections. The results (Figure 1) indicate that the climate change scenario would have little impact on annual ET_o. The annual ET_o increased slightly where there were mean wind speeds less than 1.7 m·s^{-1}, and it decreased for wind speeds greater than 1.7 m·s^{-1}. However, the absolute magnitude of variation from the current annual ET_o was small for all weather stations. Based on the regression lines in Figure 1, the 4 °C rise in T_d and the 17 s·m^{-1} increase in r_c counteracted the impact of the 3 °C mean temperature increase.

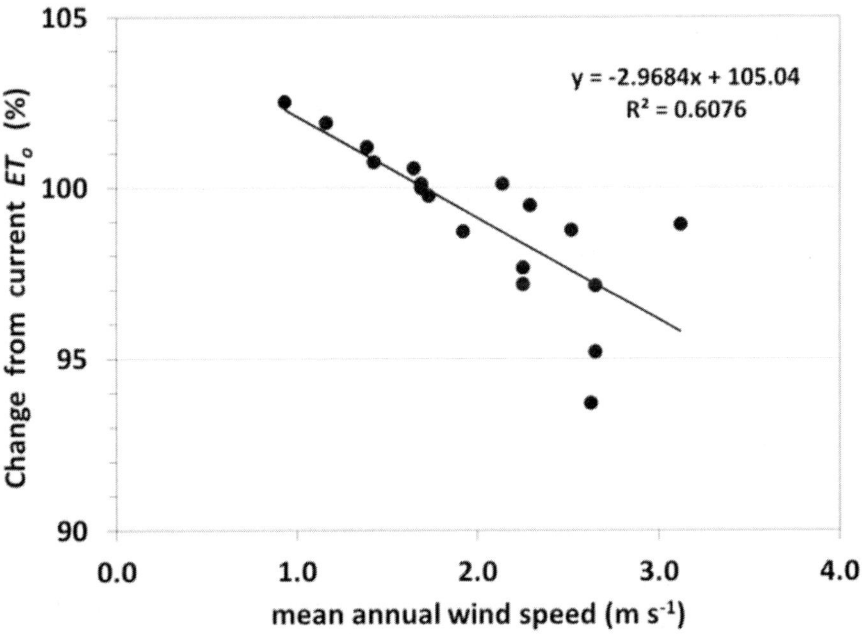

Figure 1. A plot of the change from current annual ET_o (mm·y^{-1}) assuming T_x increases by 2 °C, T_n increases by 4 °C, T_d increases by 4 °C, and r_c increases from 70 to 87 s·m^{-1} versus the mean annual wind speed. Figure 1 is from Snyder et al. [10].

6. CROP COEFFICIENTS

Allen et al. [7,8,11] reported a daily time step, standardized reference evapotranspiration equation for short canopies (Equation (7)). While the generated ET_o is actually for a virtual surface, i.e., having characteristics that determine the coefficients in Equation (4), the ET_o approximates the ET_c for a broad expanse of well-watered, cool-season grass. Allen et al. [7,8] also reported a daily time step equation for tall canopies (ET_r) that is expressed as

$$ET_r = \frac{0.408\Delta(R_n - G) + \frac{1600\gamma(e_s - e_a)u_2}{T+273}}{\Delta + \gamma(1 + 0.38u_2)} \quad \left(\text{mm}\cdot\text{d}^{-1}\right). \quad (9)$$

Note that Allen et al. [7] used $r_c = 45\ \text{s}\cdot\text{m}^{-1}$ and $r_a = 118/u_2$ to obtain the 0.38 = 45/118 in Equation (9). The canopy resistance for about 372 ppm CO_2 was derived using Equation (10) with $r_s = 66.7\ \text{s}\cdot\text{m}^{-1}$ and LAI = 2.96.

$$r_c = \frac{r_s}{0.5 \cdot LAI} = \frac{66.7\ \text{s}\cdot\text{m}^{-1}}{0.5 \times 2.96} = 45.1 \approx 45\ \text{s}\cdot\text{m}^{-1}. \quad (10)$$

Assuming a 20% reduction in alfalfa stomatal conductance when changing from 372 to 550 ppm CO_2 [12], the stomatal conductance changes from 0.015 $\text{m}\cdot\text{s}^{-1}$ to 0.012 $\text{m}\cdot\text{s}^{-1}$, the stomatal resistance changes from 66.7 to 83.3 $\text{s}\cdot\text{m}^{-1}$, and the canopy conductance at 550 ppm is estimated as

$$r_c = \frac{r_s}{0.5 \cdot LAI} = \frac{83.3\ \text{s}\cdot\text{m}^{-1}}{0.5 \times 2.96} = 56.3 \approx 56\ \text{s}\cdot\text{m}^{-1}. \quad (11)$$

Strictly speaking, the ET_r equation estimates the ET_c for a virtual surface with characteristics represented by coefficients used in Equation (9), but the ET_r rates are approximately equal to the ET_c of a broad expanse of well-watered, 0.5-meter-tall alfalfa. Derivations and explanations of the ET_o and ET_r equations are addressed in [7,8]. To accurately estimate ET_o and ET_r, the input weather data for Equations (4)–(9) are collected over a broad expanse of well-watered grass. Derivation of the net radiation (R_n) and ground heat flux (G) in $\text{MJ}\cdot\text{m}^{-2}\cdot\text{d}^{-1}$ are presented in [7]. The T (°C) is the mean air temperature measured at a height between 1.5 and 2.0 m, u_2 ($\text{m}\cdot\text{s}^{-1}$) is the mean daily wind speed monitored at a 2 m height, the saturation vapor pressure (e_s) in kPa is calculated from T, and the actual vapor pressure (e_a) in kPa is measured at the same height as T. The slope of the saturation vapor pressure (Δ) in $\text{kPa}\cdot°\text{C}^{-1}$ and the psychrometric constant γ in $\text{kPa}\cdot°\text{C}^{-1}$ are computed as described in [7].

While little change in ET_o rates is expected due to global climate change, crop coefficient (K_c) values might be affected depending on how climate factors change in the future. To evaluate the effect of changing weather variables on K_c values, an analysis was done using data from 49 California Irrigation Management Information System (CIMIS) weather stations [16]. Equations (4) and (9) were used to compute ET_o and ET_r using canopy resistances corresponding to both 372 and 550 ppm CO_2. The stomatal conductances g_s = 0.010 m·s^{-1} (grass) and g_s = 0.015 m·s^{-1} (alfalfa) were reduced by 20% to 0.008 m·s^{-1} (grass) and 0.012 m·s^{-1} (alfalfa) at 550 ppm CO_2 based on [12]. The K_c for alfalfa was calculated for 372 ppm CO_2 for all 49 stations using the mean climate data from July 2003. For 550 ppm CO_2, the K_c for alfalfa was calculated using the mean climate data from July 2003, but with the monthly mean daily maximum, minimum, and dew point temperatures increased by 2 °C, 4 °C, and 4 °C, respectively. The wind speed and equations for aerodynamic resistance were not changed.

A plot of the 550 ppm versus 372 ppm K_c values (Figure 2) indicates that the crop coefficients are likely to decrease slightly for a wide range of climates. The biggest K_c decrease was about 0.03 when $K_c \approx 1.10$, and the smallest decrease was about 0.01 for $K_c \approx 1.40$. The differences were approximately 0.01 when K_c values were high. The response of K_c to the projected climate change is most likely related to the alfalfa canopy being more coupled with the environment than the grass canopy, which has a higher aerodynamic resistance. Plant canopies that are more coupled to the environment are more likely to exhibit a reduction in transpiration rate than a canopy that is more controlled by the aerodynamic resistance, e.g., grass. This analysis provides some evidence that coupling with the environment might lead to reductions in crop coefficients due to global climate change.

While changes in the CO_2 concentration can have an effect on the canopy resistance, this analysis showed that the same percentage decrease stomatal conductance can lead to a bigger reduction in transpiration if the aerodynamic resistance is lower and the canopy is more coupled with the environment. There is a lack of information on how crop coefficients of orchard and vine crops might respond to climate change, but the alfalfa example in this paper provides some insight. Since taller rougher canopies, e.g., orchards and vineyards, have considerably lower aerodynamic resistance than alfalfa, it is likely that the K_c values for orchard and vine crops might decrease even more.

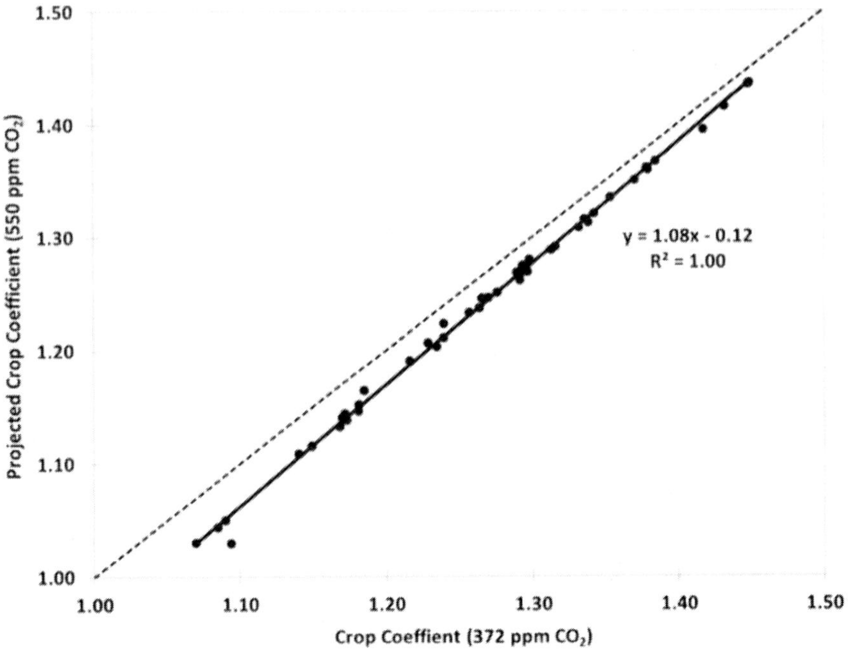

Figure 2. A plot of K_c values for alfalfa calculated from July 2003 mean daily weather data from 49 CIMIS weather stations in California for a climate with 550 ppm versus 372 ppm CO_2. For the projected 550 ppm CO_2 climate, the original daily maximum temperature data were increased by 2 °C, and the minimum and dew point temperatures were increased by 4 °C relative to July 2003 data. The wind speed and solar radiation were not changed from the original data. The alfalfa ET_c was computed using the ET_r equation (Equation (9)), and ET_o was computed using Equation (4).

For both C_4 and C_3 species, stomatal conductance is reduced by the increasing CO_2 concentration external to leaves [17]. However, small differences in C_4 and C_3 stomatal conductance responses to CO_2 concentration have been reported for grasses with the conductance differences decreasing at higher CO_2 concentrations [18]. On the other hand, grasses are more decoupled from the environment than taller, rougher plant canopies, so the K_c response of grass species due to a projected climate is likely to be smaller than the K_c response of taller rougher plant canopies with lower aerodynamic resistance.

7. CONCLUSIONS

In summary, the evidence for anthropogenic global climate change due to the excessive release of CO_2 into the atmosphere is strong. While there are many possible impacts of climate change on horticultural crops, the effect of changes on water use of horticultural crops is particularly important. The fact that global climate change is dependent only on global receipt and reflection of solar radiation and GHGs provides strong evidence that anthropogenic global climate change is real and concerning. The global change can impact many weather factors in addition to affecting physical and biological factors, and these weather factors can affect plant growth and agricultural production. The FACE studies showed that increasing atmospheric CO_2 will decrease stomatal conductance, and this will increase canopy resistance of C_3 species plants, which decreases plant transpiration. Additionally, climate change projections indicate that water vapor content of the air will increase as temperature rises, and increased atmospheric H_2O also decreases transpiration. Using the standardized reference evapotranspiration equation for short canopies to calculate ET_o, the impact of projected increases in atmospheric temperature and CO_2 and H_2O concentrations were evaluated, and a large effect on ET_o rates is unlikely. There was some evidence that ET_o would increase slightly at low wind speeds and it would decrease as wind speeds increased. The calculation of alfalfa K_c values at 372 and 550 ppm CO_2 with an increase of 2 °C, 4 °C, and 4 °C for maximum, minimum, and dew point temperatures in the higher CO_2 environment, showed that K_c values will probably slightly decrease. This decrease is likely due to the higher coupling of canopy to the environment for the alfalfa canopy. There is little information available about how K_c values might change for tree and vine crops, but trees and vines are even more coupled to the environment, so an even bigger decrease in K_c seems plausible for the taller rougher canopies. A similar K_c response to climate change is expected for both C_3 and C_4 plants. While the evapotranspiration responses to global change seem small, the projected changes in precipitation and water storage in snowpack are large and could have devastating impacts on horticulture in some regions.

REFERENCES

1. IPCC. *Climate Change 2013. The Physical Science Basis Working Group I Contribution to the Fifth Assessment Report of the Intergovernmental Panel on Climate Change*; Cambridge University Press: New York, NY, USA, 2014; pp. 1–1535.

2. Dixon, G.R.; Collier, R.H.; Bhattacharya, I. An assessment of the effects of climate change on horticulture. In *Horticulture: Plants for People and Places*; Dixon, G.R., Aldous, D.E., Eds.; Springer: Dordrecht, The Netherlands, 2014; Volume 2, pp. 817–857.
3. Glenn, D.M.; Kim, S.H.; Ramirez-Villegas, J.; Läderach, P. Response of Perennial Horticultural Crops to Climate Change. In *Horticultural Reviews*, 1st ed.; Janick, J., Ed.; John Wiley & Sons, Inc.: Hoboken, NJ, USA, 2013; Volume 41, pp. 47–129.
4. Anderson, J.; Chung, F.; Anderson, M.; Brekke, L.; Easton, D.; Ejeta, M.; Peterson, R.; Snyder, R.L. Progress on incorporating climate change into management of California's water resources. *Clim. Chang.* **2008**, *87*, S91–S108.
5. Arrhenius, S.A. On the Influence of Carbonic Acid in the Air Upon the Temperature of the Ground. *Philos. Mag. J. Sci.* **1896**, *41*, 237–276.
6. NOAA Layers of the Atmosphere. Available online: http://www.srh.noaa.gov/jetstream/atmos/layers.html (accessed on 18 November 2016).
7. Allen, R.G.; Walter, I.A.; Elliott, R.L.; Howell, T.A.; Itenfisu, D.; Jensen, M.E.; Snyder, R.L. *The ASCE Standardized Reference Evapotranspiration Equation*; American Society of Civil Engineers: Reston, VA, USA, 2005; pp. 1–173.
8. Allen, R.G.; Pruitt, W.O.; Wright, J.L.; Howell, T.A.; Ventura, F.; Snyder, R.L.; Itenfisu, D.; Steduto, P.; Berengena, J.; Baselga Yrisarry, J.; et al. A recommendation on standardized surface resistance for hourly calculation of reference ETo by the FAO56 Penman-Monteith method. *Agric. Water Manag.* **2006**, *81*, 1–22.
9. Monteith, J.L. Evaporation and Environment. *Symp. Soc. Exp. Biol.* **1965**, *19*, 205–234.
10. Snyder, R.L.; Moratiel, R.; Zhenwei, S.; Swelam, A.; Jomaa, I.; Shapland, T. Evapotranspiration Response to Climate Change. *Acta Hortic.* **2011**, *922*, 91–98.
11. Allen, R.G.; Pereira, L.S.; Raes, D.; Smith, M. *Crop Evapotranspiration: Guidelines for Computing Crop Water Requirements*; Irrigation and Drainage Paper No. 56; FAO of United Nations: Rome, Italy, 1998; pp. 1–300.
12. Long, S.P.; Ainsworth, E.A.; Rogers, A.; Ort, D.R. Rising atmospheric carbon dioxide: plants FACE the future. *Ann. Rev. Plant Biol.* **2004**, *55*, 591–628.
13. Cayan, D.; Luers, A.L.; Hanemann, M.; Franco, G. *Scenarios of Climate Change in California: An Overview*; CEC-500-2005-186-SF; California Energy Commission: Sacramento, CA, USA, 2006.
14. Prentice, I.C.; Farquhar, G.D.; Fasham, M.J.R.; Goulden, M.L.; Heimann, M.; Jaramillo, V.J.; Kheshgi, H.S.; Le Quéré, C.; Scholes, R.J.; Wallace, D.W.R. The Carbon Cycle and Atmospheric Carbon Dioxide. In *Climate Change 2001: The Scientific Basis. Contribution of Working Group I to the Third Assessment Report of the Intergovernmental Panel on Climate Change*; Cambridge University: Cambridge, UK; New York, NY, USA, 2002; pp. 183–238.
15. Roderick, M.L.; Farquhar, G.D. Changes in New Zealand Pan Evaporation since the 1970s. *Int. J. Climatol.* **2005**, *25*, 2031–2039.
16. Snyder, R.L.; Pruitt, W.O. *Evapotranspiration Data Management in California*; American Society of Civil Engineers: New York, NY, USA, 1992; pp. 128–133.
17. Morison, J.I.L.; Gifford, R.M. Plant growth and water use with limited water supply in high CO_2 concentrations. I. Leaf area, water use and transpiration. *Aust. J. Plant Physiol.* **1984**, *11*, 361–374.
18. Hager, H.A.; Ryan, G.D.; Kovacs, H.M.; Newman, J.A. Effects of elevated CO_2 on photosynthetic traits of native and invasive C_3 and C_4 grasses. *BMC Ecol.* **2016**, *16*, 28.

CHAPTER 4

Vegetable Grafting: The Implications of a Growing Agronomic Imperative for Vegetable Fruit Quality and Nutritive Value

Marios C. Kyriacou[1], Youssef Rouphael[2], Giuseppe Colla[3], Rita Zrenner[4] and Dietmar Schwarz[4*]

[1]*Department of Vegetable Crops, Agricultural Research Institute, Nicosia, Cyprus*

[2]*Department of Agricultural Sciences, University of Naples Federico II, Naples, Italy*

[3]*Department of Agricultural and Forestry Sciences, University of Tuscia, Viterbo, Italy*

[4]*Leibniz Institute of Vegetable and Ornamental Crops, Großbeeren, Germany*

ABSTRACT

Grafting has become an imperative for intensive vegetable production since chlorofluorocarbon-based soil fumigants were banned from use on grounds of environmental protection. Compelled by this development, research into rootstock–scion interaction has broadened the potential applications of grafting in the vegetable industry beyond aspects of soil phytopathology. Grafting has been increasingly tapped for cultivation under adverse environs posing abiotic and biotic stresses to vegetable crops, thus enabling expansion of commercial production onto otherwise under-exploited land. Vigorous rootstocks have been employed not only in the open field but also under protected cultivation where increase in productivity improves distribution of infrastructural and energy costs. Applications of grafting have expanded mainly in two families: the Cucurbitaceae and the Solanaceae, both of which comprise major vegetable crops. As the main drives behind the expansion of vegetable grafting have been the resistance to soilborne pathogens, tolerance to abiotic stresses and increase in yields, rootstock selection and breeding have accordingly conformed to the prevailing demand for improving productivity, arguably at the expense of fruit quality. It is, however, compelling to assess the qualitative implications of this growing agronomic practice for human nutrition. Problems of impaired vegetable fruit quality have not infrequently been associated with the practice of grafting. Accordingly, the aim of the current review is to reassess how the practice of grafting and the prevalence of particular types of commercial rootstocks influence vegetable fruit quality and, partly, storability. Physical, sensorial and bioactive aspects of quality are examined with respect to grafting for watermelon, melon, cucumber, tomato, eggplant, and pepper. The physiological mechanisms at play which mediate rootstock effects on scion performance are discussed in interpreting the implications of grafting for the configuration of vegetable fruit physicochemical quality and nutritive value.

Keywords: carotenoids, Cucurbitaceae, flavor, functional compounds, physiological mechanism, mRNA transport, rootstock, Solanaceae

INTRODUCTION

Retail cost for fresh horticultural products reflects capital investment in developing suitable plant stock, in fostering its cultivation, and in product storage and handling along the food supply chain. Although a multifaceted concept drawing on various implicated stakeholders, quality is ultimately what captures the expectations to be met at the retail customer's end of the agroindustry spectrum. The perception of quality is dependent on intrinsic traits of horticultural commodities, shaped by genotypic, cultural and ecophysiological effects, and on extrinsic traits formulated by the socio-economic and marketing environment (Schreiner et al., 2013). Multidisciplinary studies have highlighted that quality is more important to consumers than price when the latter varies within the anticipated range (Harker et al., 2003). Although consonance of quality with the cost of purchase influences consumer behavior, quality is that which largely commands recurring customers. In regulatory context, the issue of quality is addressed chiefly by crop-specific class standards based on limited key visual and organoleptic characteristics (Commission Implementing Regulation (EU) No 543/2011, 2011). Quality standards thus tend to define class criteria for minimum acceptability and provide practical, effective, mostly non-destructive means for standardization procedures. They fail, however, to address complex compositional aspects of quality pertaining to flavor, particularly the volatile aroma fraction, or to nutritional and bioactive value which consumers are becoming increasingly conscious of Schreiner et al. (2013).

Plant breeding on the other hand, has aimed preeminently at improving yield, endowing plant stock with disease resistance, at providing resilience to mechanical injury and improving overall postharvest performance, and to a lesser extent at improving sensory quality traits (Bai and Lindhout, 2007). However, the configuration of important sensory traits, such as volatile aroma components, seems mediated by ethylene-dependent biosynthetic pathways linked also to shelf-life performance (Pech et al., 2008), and to textural changes associated with cell wall matrix solubilization events (Dos-Santos et al., 2013). Hence breeding for shelf-life may elicit adverse pleiotropic effects on desirable sensory attributes (Causse et al., 2002). This is particularly critical in fruits characterized by autocatalytic climacteric ripening, as aptly exemplified by the distinct sensory profile of odorous climacteric vs. inodorous non-climacteric melons (Verzera et al., 2011). Collecting desirable traits while avoiding undesirable combinatorial effects complicates breeding efforts. In this respect, grafting may provide expedient means of selecting independently for rootstock and scion traits, provided the compatibility of the graft combination.

Driven initially by its efficiency as an alternative to the banned use of chlorofluorocarbon-based soil fumigants, the grafting of annual fruit crops has grown across crops and beyond applications restricted to addressing soilborne disease problems (Rouphael et al., 2010). Grafting has been increasingly tapped for cultivation under adverse environments posing abiotic and biotic stresses to vegetable crops, thus enabling expansion of commercial production onto otherwise under-exploited land. The ability of taping wild genetic resources for exploiting traits of root physiological tolerance to stress independently to scion characteristics has facilitated the application of grafting for the cultivation of annual fruit crops under marginal conditions of salinity, nutrient stress, water stress, organic pollutants, and alkalinity (Savvas et al., 2010; Schwarz et al., 2010; Borgognone et al., 2013). Moreover, the economic implications of the significant yield increase imparted by select vigorous commercial rootstocks has encouraged their use under protected cultivation where increase in productivity improves distribution of infrastructural and energy costs (Colla et al., 2011). Provided the anatomical and physiological compatibility of the graft combinations, rootstock effects on plant performance under soil biotic and abiotic stress conditions clearly outweigh those of the scion. Moreover, there is evidence of rootstock mediation in the configuration of scion fruit quality characteristics, widely reported in a range of crops but confounded with frequent rootstock–scion interaction which cannot always be explained in the context of narrow rootstock–scion specificity (Rouphael et al., 2010).

As the main drives behind the expansion of vegetable grafting have been the resistance to soilborne pathogens (Louws et al., 2010), tolerance to abiotic stresses (Schwarz et al., 2010; Kumar et al., 2015; Rouphael et al., 2016) and increase in yields (Lee et al., 2010), rootstock selection and breeding have accordingly conformed to the prevailing demand for improving productivity, arguably at the expense of fruit quality. Thus, grafting has not been employed as a method for improving vegetable fruit quality. Quite the contrary, often yield and quality are contradictory traits (Klee and Tieman, 2013). Considering the rapid expansion of the vegetable grafting industry, understanding the implications of grafting for fruit quality is imperative, and equally pressing is the unraveling of the mechanisms involved. Possible factors engaged in rootstock mediation of quality include changes in water and nutrient uptake efficiency, indirect effects on ripening behavior resulting from altered crop load and source–sink balance, and even an epigenetic component to the grafting process involving transfer of genetic material from rootstock to scion (Savvas et al., 2010; Soteriou et al., 2014; Avramidou et al., 2015). Previous reviews on the grafting of annual fruit crops that have covered aspects relating to fruit quality were published by Davis et al. (2008a, only Cucurbitaceae), Flores et al. (2010 only tomato), and Rouphael et al. (2010). The current review aims at providing an updated critical review of scientific advances addressing grafting effects on the fruit quality of annual crops; moreover, it discusses methodological postulates and mechanisms possibly mediating these effects. Current knowledge has been compiled in a crop specific approach where fruit quality attributes and rootstocks employed are discussed in a uniform and integrated context.

THE CONFIGURATION OF FRUIT QUALITY IN GRAFTED VEGETABLES

Cucurbitaceae

Watermelon [Citrullus lanatus (Thunb.) Matsum. and Nakai]

The adoption of grafting as a means to secure watermelon crop stand and productivity, mainly against conditions of biotic stress, by far exceeds that of any other open cultivated annual fruit crop (FAO, 2012). The use of rootstocks resistant to soilborne diseases has become a prerequisite for watermelon production, especially in areas where intensive cultivation is practiced and scarcity of arable land precludes the application of broad rotation schemes. While most studies assessing rootstock–scion interaction had initially laid emphasis on aspects of disease resistance and agronomic performance, a plethora of works has been produced that examine the implications of grafting for watermelon fruit quality, involving mainly inter-specific hybrids [*Cucurbita maxima* (Duchesne) × *C. moschata* (Duchesne ex Poir)] and gourd [*Lagenaria siceraria* (Molina) Standl.] rootstocks.

Morphometric characteristics

Although a trait prominently delineated by genotype, watermelon fruit weight might be influenced by environmental conditions and cultural practices, including grafting, that affect overall field performance (Alexopoulos et al., 2007; Cushman and Huan, 2008; Proietti et al., 2008; Soteriou and Kyriacou, 2014). Vigorous interspecific and *L. siceraria*rootstocks can improve yields significantly, which in genotypically large-fruited scions usually translates into a tendency for higher unit fruit weight, while in small-fruited cultivars it tends to increase the number of fruits per plant (Colla et al., 2006a; Alexopoulos et al., 2007; Cushman and Huan, 2008; Proietti et al., 2008; Soteriou and Kyriacou, 2014). Decrease in fruit weight against non-grafted control is usually an indicator of rootstock–scion incompatibility, while in compatible grafts maximum reported fruit weight increase approximates 55% (Yetisir and Sari, 2003; Yetisir et al., 2003; Huitrón et al., 2007; Cushman and Huan, 2008; Soteriou and Kyriacou, 2014).

Secondary morphological characteristics of watermelon fruit that may appeal to consumers' perception of quality include shape, expressed as the ratio of longitudinal to equatorial diameter, and rind thickness. Fruit shape constitutes a trait predominantly governed by scion genotype and little affected by environmental or cultural factors; hence the effect of grafting thereupon has been circumstantial and mostly non-significant or minimal (Colla et al., 2006a; Alan et al., 2007; Rouphael et al., 2008; Soteriou and Kyriacou, 2014; Fredes et al., 2017). On the other hand, rind thickness is a morphological trait more responsive to grafting, and to cultural practice at large, as it relates to watermelon harvest maturity (Soteriou et al., 2014; Kyriacou et al., 2016). On commercial *C. maxima* × *C. moschata* and *L. siceraria*rootstocks, especially on landraces of the latter, thickening of watermelon rind is often observed (Yetisir et al., 2003; Alexopoulos et al., 2007; Proietti et al., 2008; Kyriacou and Soteriou, 2015). However, this has not been a ubiquitous effect across the above rootstocks, or with less common rootstocks such as *C. moschata, Sicyos angulatus* L., *C. lanatus*var. *citroides* (L. H. Bailey) Mansf. and *C. pepo* L. which were only sporadically effective in this respect (Davis and Perkins-Veazie, 2005; Alan et al., 2007; Huitrón et al., 2008; Soteriou and Kyriacou, 2014; Fredes et al., 2017). Rootstock effect on watermelon rind thickness is in general limited and studies involving multiple rootstock–scion combinations have demonstrated the predominance of the relative effect of the scion cultivar on this attribute (Kyriacou and Soteriou, 2015). Thinning of the rind is known to characterize watermelon maturation, but also its postharvest life (Corey and Schlimme, 1988); therefore, potential rootstock effect should be examined under conditions that account carefully for the effect of harvest maturity (Soteriou et al., 2014). In any case, thickening of the rind can improve the postharvest performance of watermelon fruit and may also provide a tool for increasing the source of important bioactive compounds concentrated in the rind, such as citrulline, which constitute potential by-products of the fresh-cut industry (Tarazona-Díaz et al., 2011).

Colourimetric attributes

Among the physical characteristics of watermelon fruit that strongly influence consumer preference, is the intensity of red coloration of the pulp. Change in the intensity of red hue, expressed as increase in colourimetric CIELAB component a*, marks the development of watermelon pulp color during ripening; moreover, a widening of hue angle (h°), signifying transition from red to orange–yellow is characteristic of watermelon over-ripening and senescence (Brown and Summers, 1985; López–Galarza et al., 2004; Soteriou et al., 2014). Watermelon pulp color is directly dependent upon lycopene synthesis and its accumulation in chromoplasts, while cultivar differences in pulp color correlate highly with differences in lycopene content (Perkins-Veazie and Collins, 2006; Kyriacou and Soteriou, 2015). Grafting therefore may affect pulp color to the extent it affects lycopene content (Davis and Perkins-Veazie, 2005). Watermelon scions grafted on interspecific *Cucurbita* hybrid rootstocks may incur delayed pulp color development, compared to non-grafted control, expressed as a delayed peak in colourimetric component a* synchronous to the peak in lycopene content (Soteriou et al., 2014).

Textural characteristics

Pulp firmness constitutes one of the most important sensory traits of watermelon fruit subject to wide genotypic variation, with pronounced firmness observed usually in seedless, triploid cultivars (Leskovar et al., 2004; Soteriou and Kyriacou, 2015). Notwithstanding the genotypic effect of the scion, rootstock effects on watermelon pulp firmness can be significant hence the choice of rootstock instrumental for improving fruit quality and postharvest life (Yetisir et al., 2003; Cushman and Huan, 2008; Bruton et al., 2009; Kyriacou and Soteriou, 2015). Interspecific *Cucurbita* hybrid rootstocks most consistently increase watermelon pulp firmness in both diploid and triploid scions (Bruton et al., 2009; Huitrón et al., 2009; Soteriou et al., 2014; Soteriou and Kyriacou, 2015). The effect of grafting, however, might render the pulp of certain cultivars, especially mini triploids that are genotypically inclined to outstanding firmness, undesirably hard (Soteriou and Kyriacou, 2015). Among less commonly used rootstocks, the parents of interspecific hybrids *C. maxima* and *C. moschata*, *C. ficifolia* Bouché, and citron melon (*C. lanatus* var. *citroides*) have been reported to elicit firmer watermelon pulp (Cushman and Huan, 2008; Bruton et al., 2009), whereas cushaw squash (*C. argyrosperma* C. Huber) pumpkin had the opposite effect (Davis and Perkins-Veazie, 2005). Gourd rootstocks *L. siceraria* usually have no effect on pulp firmness although erratic cultivar-specific effects, both positive and negative, have been reported (Yetisir et al., 2003; Cushman and Huan, 2008; Bruton et al., 2009; Özdemir et al., 2016). Morphological abnormalities scarcely associated with watermelon grafting include yellow bands in the pulp bordering the rind, hollow heart, excessively hard and discolored pith, and overall poor texture (Lee, 1994; Yamasaki et al., 1994; Davis et al., 2008b; Soteriou and Kyriacou, 2014). However, most reports on commercially available *C. maxima* × *C. moschata* and *L. siceraria* rootstocks do not make reference to such defects which may reflect rootstock–scion incompatibility and adverse environmental conditions or cultural practices.

Sweetness and acidity

The most valued singular quality trait of watermelon is undoubtedly sweetness, sensorially triggered mostly but not entirely by soluble mono- and di-saccharides, since other juice solutes including organic acids, soluble pectins and amino acids, phenolic compounds and minerals influence sweet sensation (Kader, 2008; Magwaza and Opara, 2015). The soluble solids content (SSC) – containing sugars and acids, together with small amounts of dissolved vitamins, fructans, proteins, pigments, phenolics, and minerals – is the most important quality measure used to indicate sweetness of watermelon as well as other fruits (Magwaza and Opara, 2015). It is in general not highly compromised by grafting on most commercial *C. maxima* × *C. moschata* rootstocks (Colla et al., 2006a; Proietti et al., 2008; Huitrón et al., 2009; Soteriou and Kyriacou, 2014; Kyriacou et al., 2016). Scion response to *L. siceraria* rootstocks appears more erratic and rootstock-specific with most graft combinations not demonstrating a significant effect on SSC but exceptions of SSC reduction, especially on landraces, or SSC increase are not infrequent (Yetisir and Sari, 2003; Alan et al., 2007; Alexopoulos et al., 2007; Cushman and Huan, 2008; Çandır et al., 2013). Effects on watermelon sweetness have occasionally been demonstrated by more marginal or experimental rootstocks, such as reduction of SSC by *C. argyrosperma* and *C. pepo* (Davis and Perkins-Veazie, 2005), and increase by *C. lanatus* var. *citroides* (Fredes et al., 2017).

Sweetness depends mostly on the total concentration of soluble carbohydrates, which in most fruits constitutes the largest fraction of the SSC, but also on the relative proportions of the three main sugars, glucose, fructose, and sucrose, which contribute differentially to sweetness and combine to yield what is termed sweetness index (Elmstrom and Davis, 1981; Brown and Summers, 1985; Kader, 2008). Among cucurbit genotypes, variation in sugar content and sweetness index has been associated mostly with their ability to accumulate sucrose at the expense of fructose and glucose during ripening, owing to the activity of sucrose phosphate synthase and sucrose synthase and the decline in activity of soluble acid invertase (Stepansky et al., 1999; Yativ et al., 2010). In watermelon, fructose and glucose are the main sugars supplying the demands of the ovary during initial fruit development, due to the high activities of neutral and acid invertases (Lanchun et al., 2010). Sucrose is the main soluble carbohydrate accumulating in watermelon fruit during ripening at the expense of reducing sugars (Brown and Summers, 1985; Chisholm and Picha, 1986; López–Galarza et al., 2004; Soteriou et al., 2014), although less common genotypes accumulating reducing-sugars throughout ripening have been reported (Yativ et al., 2010). Lower accumulation of hexoses at the onset of fruit development and reduced sucrose accumulation during ripening have been implicated in moderate reduction of watermelon total sugar content in response to the use of *C. maxima* × *C. moschata* and *L. siceraria* rootstocks (Miguel et al., 2004; Liu et al., 2006; Kyriacou and Soteriou, 2015; Fredes et al., 2017). However, other studies involving the same rootstock types have revealed no significant effects on glucose, fructose, sucrose, or total sugars content (Colla et al., 2006a; Proietti et al., 2008; Soteriou et al., 2014). Disparity of results regarding the effect of grafting on non-structural carbohydrates in many cases reflects differential ripening events, notwithstanding the possible effects of cultural practice and climatic conditions particularly on flowering and fruit setting. Grafting may affect the earliness of flowering and thereby affect the time to commercial maturity (Satoh, 1996; Sakata et al., 2007), however, the delay in maturation relates mainly to retarded post-anthesis ripening events as a result of increased crop load on grafted plants (Soteriou et al., 2014; Kyriacou and Soteriou, 2015; Soteriou and Kyriacou, 2015).

Acidity balances sweetness in the taste profile of most fruits, although effectively the ratio between SSC and titratable acidity (TA) is considered crucial in terms of consumer acceptability mostly for sour fruits. Acidity in watermelon fruit is very low, with a pH range of 5.5–5.8 and acid concentration in its juice 0.7–1.2 g/l predominantly in malate form (Kyriacou and Soteriou, 2012; Çandır et al., 2013; Soteriou et al., 2014; Fredes et al., 2017). Grafting on *C. maxima* × *C. moschata* has been found to increase the TA and reduce the pH of the pulp (Colla et al., 2006a; Proietti et al., 2008; Soteriou et al., 2014). Increase in watermelon acidity has been elicited by grafting not only on hybrid rootstocks but also on *C. lanatus* var. *citroides* and on certain *L. siceraria* rootstocks, expressed mostly in higher malic acid levels in the juice (Çandır et al., 2013; Fredes et al., 2017). The TA of watermelon pulp declines linearly with ripening; grafting, however, sustains higher TA throughout the ripening period in comparison to non-grafted plants, which verifies that this is less mediated by maturity than the effect of grafting on sugars (Soteriou et al., 2014), moreover it predisposes the fruit for improved postharvest performance (Kyriacou and Soteriou, 2015).

Aroma profile

Alcohol and aldehyde characteristics of the Cucurbitaceae family constitute the main aroma volatiles in watermelon fruit, with the former usually in higher concentrations (Beaulieu and Lea, 2006; Saftner et al., 2007). The most abundant alcohols identified in the aroma profile of mini watermelons include (*Z*)-3-Nonen-1-ol (fresh melon), (*Z,Z*)-3,6-Nonadien-1-ol (pumpkin, cucumber), hexanol (flower, green), nonanol (herbaceous) and (*Z*)-6-Nonen-1-ol (pumpkin-like, green melon) (Yajima et al., 1985; Dima et al., 2014). Among identified aldehydes most abundant were (*Z*)-2-nonenal (honeydew melon, fruity), hexanal (green), (*E,Z*)-2,6-nonadienal (cucumber, green), nonanal (melon, orange peel), (*Z*)-6-nonenal (honeydew melon, fruity), 6-methyl-5-hepten-2-one (flower) and (*E*)-6-nonenal (earthy) (Dima et al., 2014). Although significant rootstock-specific effects on watermelon volatile profile have been identified, the effect of grafting on watermelon aroma profile remains at large a scarcely charted territory (Petropoulos et al., 2014; Fredes et al., 2017). Grafting midi-watermelon cultivars (≈6 kg) on *C. maxima* × *C. moschata* and *L. siceraria* rootstocks was found to increase fruit content in several aroma volatiles, including (*E*)-2-nonenal (fat, cucumber) and (*Z,Z*)-3,6-nonadien-1-ol (green, cucumber) (Petropoulos et al., 2014). Fredes et al. (2017) identified differential effects among *C. maxima* × *C. moschata* rootstocks in the levels of (*Z*)-6-nonenal and (*E,Z*)-2,6-nonadienal, associated with melon-like and cucumber-like aromas, respectively. A critical and consistent finding across *C. maxima* × *C. moschata* rootstocks, but not on *C. lanatus* var. *citroides*, is the increased level of (*Z*)-6-nonen-1-ol, which confers undesirable pumpkin-like odor in fruits from grafted plants. However, the identification of higher levels of lycopene degradation products, such as 6-methy-5-heten-2-one and geranylacetone in the volatile profile of fruit from non-grafted plants, characterized by earlier peak in lycopene content (Soteriou et al., 2014), suggests differential harvest maturity between treatments may be implicated in these findings (Lewinsohn et al., 2005). Available work is far from providing conclusive evidence on the effect of grafting on watermelon aroma profile. Future work needs to take carefully into consideration the evolution of aroma profile during ripening so that the potential effects of grafting are discerned from those of harvest maturity. Analysis of volatiles performed using a GC–MS-olfactory approach combined with extensively trained sensory panels would provide a more resilient basis for further investigation into rootstock-mediated effects on watermelon aroma profile (Saftner et al., 2007).

Functional compounds

Notwithstanding wide genotypic variation, watermelon is a lycopene-rich food source with higher lycopene concentration in its pulp than that of tomato (Perkins-Veazie et al., 2001; Fish and Davis, 2003; Soteriou et al., 2014). Grafting, particularly on *C. maxima* × *C. moschata* rootstocks, has been reported to raise lycopene levels significantly in watermelon fruit (Perkins-Veazie et al., 2007; Proietti et al., 2008; Soteriou et al., 2014; Kyriacou and Soteriou, 2015). Increase was also reported on selected *L. siceraria* genotypes (Çandır et al., 2013) and on *C. argyrosperma* and *C. pepo*but limited to seedless scions (Davis and Perkins-Veazie, 2005). Decrease in lycopene levels associated with certain rootstock–scion combinations involving *L. siceraria* and *C. argyrosperma* (Davis and Perkins-Veazie, 2005; Çandır et al., 2013), or absence of effect (Bruton et al., 2009; Soteriou and Kyriacou, 2014) have been more infrequently reported. Conflicting reports may be explained in the light of recent work demonstrating that lycopene content is affected more by maturity and less by grafting, as the peak in lycopene content appears about 1 week earlier in fruit from non-grafted than from grafted plants (Soteriou et al., 2014). Ripening-dependent accumulation of lycopene may derive from the inhibition of β-carotene synthesis or from an alternative ripening-specific pathway, such as the 1-deoxy-D-xylulose-5-phosphate (DOXP) pathway (Bramley, 2002; Schofield et al., 2008). It is also not known whether the progressive transition in pulp color from red to orange–yellow, which signifies over-ripening, derives from the conversion of accumulated lycopene to β-carotene, or from a senescence-related degradation of lycopene (Ronen et al., 2000; Schofield et al., 2008). The implications of grafting for both of the above processes remain uninvestigated. In addition, lycopene synthesis events are carried over to the postharvest period where they appear temperature-controlled and linked to changes in pulp color (Perkins-Veazie and Collins, 2006). Lycopene content peaked 7 days postharvest at 25°C and was further increased by grafting on *C. maxima* × *C. moschata* rootstocks (Kyriacou and Soteriou, 2015). Depending upon maturity at the time of harvest, postharvest lycopene synthesis may appear as a continuation of the ripening-dependent pattern observed preharvest.

A non-essential amino acid found in abundance in watermelon and other cucurbits is citrulline (Rimando and Perkins-Veazie, 2005). It is a metabolic intermediate in the nitric oxide cycle, active in biological functions such as vasodilation and muscle relaxation which derive from the dissipation of NO during conversion of citrulline to arginine (Nissinen et al., 2003). Earlier indications that grafting could increase amino acid content of watermelon fruit, particularly citrulline (Davis et al., 2008c) have been confirmed by more recent work. Grafting onto *C. maxima* × *C. moschata* rootstock resulted in higher citrulline content in the pulp throughout fruit ripening (Soteriou et al., 2014). Grafting improves the performance of watermelon under deficit irrigation (Proietti et al., 2008), while the accumulation of citrulline in watermelon vegetative tissues under drought conditions has been proposed to contribute to oxidative stress tolerance based on its novel hydroxyl radical scavenging activity (Akashi et al., 2001). Citrulline accumulation in watermelon rind and pulp, possibly relates to an osmotic role during cell expansion as it constitutes a potentially significant fraction of the non-carbohydrate soluble solids in the fruit (Curis et al., 2005; Davis et al., 2011; Tarazona-Díaz et al., 2011; Soteriou et al., 2014).

Melon (Cucumis melo L.)

Melon constitutes an annual fruit species of complex quality configuration owing to the diverse ripening patterns and associated aroma profiles of its botanical varieties. These are discerned into two major groups: the climacteric short shelf-life odorous varieties *cantalupensis* and *reticulates* (e.g., charentais and muskmelon) characterized by intense aroma, as opposed to the non-climacteric, long shelf-life, non-aromatic *inodorus* varieties, such as honeydew and canary melons (Pech et al., 2008; Allwood et al., 2014). Cantaloupes are among the most widely produced melon varieties but the range of specialty melon types cultivated commercially includes many others, such as Galia, Ananas, Persian, Honeydew, Piel de Sapo, Casaba, Crenshaw, Canary, and Asian melons (Strang et al., 2007). Melon grafting as a phytoprotective measure targets Fusarium and Monosporascus wilts by exploiting mainly resistant same-species (*C. melo*) genotypes, interspecific (*C. maxima* × *C. moschata*) pumpkin hybrids and white gourd [*Benincasa hispida* (Thunb.) Cogn.]; whereas grafting on resistant *Cucumis metuliferus* E. Mey. ex Naudin and *C. melo* subsp. *Agrestis* (Naudin) Pangalo rootstocks emerges as a growing practice against Meloidogyne root knot nematodes (Trionfetti-Nisini et al., 2002; Fita et al., 2007; Davis et al., 2008a; Louws et al., 2010; Guan et al., 2014). Graft incompatibility and deterioration in the fruit quality of grafted plants are common problems, particularly with *Cucurbita* hybrid rootstocks, further complicated by pronounced rootstock interaction with the wide range of melon scion genotypes (Traka-Mavrona et al., 2000; Rouphael et al., 2010; Soteriou et al., 2016).

Morphometric characteristics

Whereas compatible *C. melo* and *Cucurbita* hybrid rootstocks generally tend to have no effect on melon fruit weight, there is also widespread rootstock–scion interaction in the responses of different melon types to grafting. For instance, the fruit weight of muskmelon (cv. Proteo) was not influenced by either *C. melo* (cvs. Energia and Sting) or *Cucurbita* hybrid rootstocks (cvs. Polifemo, AS10, RS841, P360, and Elsi) (Condurso et al., 2012). However, the same scion (cv. Proteo) grown hydroponically on other *C. melo* (cvs. Dinero and Jador) and hybrid rootstocks (cvs. P360 and PS1313), incurred a limited mean increase of 6.8% in fruit weight (Colla et al., 2010a). In the case of inodorus honeydew melon (cv. Incas), fruit weight was not influenced by grafting onto *C. melo* (cvs. Belimo, Energia, Griffin, Sting, and ES liscio) and *Cucurbita*hybrid rootstocks (cvs. AS10, P360, ES99-13, and Elsi), although it was increased moderately when grafted onto hybrids 'RS841' and 'Polifemo' (Crinò et al., 2007; Verzera et al., 2014). Commercial hybrid rootstocks 'TZ148,' 'N101,' 'Carnivor,' and '30900' also had no effect on the fruit weight of an Ananas type (cv. Raymond) and two Galia type (cvs. Elario and Polynica) melons (Soteriou et al., 2016), as was also the case with cantaloupe (cv. Athena) grafted on interspecific hybrids 'Strong Tosa' and 'Tetsukabuto' (Zhao et al., 2011). By contrast, Schultheis et al. (2015) identified a general trend for reduction of fruit weight as a result of grafting in field trials of muskmelon, honeydew and specialty melons, tested, however, only on hybrid rootstock cv. Carnivor. Traka-Mavrona et al. (2000) found grafting had no effect on fruit weight of three inodorous melons (cvs. Thraki, Peplo, and Lefko Amynteou), and a cantaloupe (cv. Kokkini Banana) using two hybrids ('TZ-148' and 'Mamouth') and one pumpkin (*C. maxima*) landrace ('Kalkabaki') as rootstocks under protected and open field cultivation. Similarly, grafting galia (cv. Arava) and honeydew (cv. Honey Yellow) melons onto nematode-resistant *C. metuliferus* had no effect on fruit weight under organic or conventional production systems (Guan et al., 2014). Exceptional increase (29%) in fruit weight was reported for cantaloupe (cv. Cyrano) when grafted on hybrid 'P360' and grown under greenhouse salinity treatments (Colla et al., 2006b). Decrease in fruit weight of muskmelon cv. Proteo resulted from grafting onto *B. hispida*, whereas *C. metuliferus*, *C. zeyheri* Sond., *C.*

moschata, C. maxima, and *C. maxima× C. moschata* hybrids had no such effect on muskmelon cultivars Proteo and Supermarket (Trionfetti-Nisini et al., 2002). Finally, Park et al. (2013) examined four *C. melo* accessions and a Shintoza hybrid as rootstocks and found that none of these had an effect on the fruit weight of muskmelon ('Earl's elite') and honeydew ('Homerunstar') except a *C. melo* accession ('K134069') which produced fruits of greater weight and size than the non-grafted control.

Other morphological traits of relevance to melon quality include fruit shape, exocarp and pulp thickness. Reports on melon grafting do not present significant rootstock effects on these variables which rather seem strongly delineated by the scion genotype. In the case of honeydew, muskmelon, cantaloupe and Piel de Sapo melons these traits were not affected by grafting on *Cucurbita* hybrid, *C. melo, C. maxima* and *C. melo* subsp. *agrestis* rootstocks (Traka-Mavrona et al., 2000; Fita et al., 2007; Colla et al., 2010a; Verzera et al., 2014). Similar results on fruit shape were obtained with Ananas and Galia type melons (Soteriou et al., 2016), and with cantaloupe grafted on hybrid rootstocks (Colla et al., 2006b), notwithstanding a limited increase in rind thickness.

Textural characteristics

Fruit texture is an essential characteristic for the organoleptic assessment of melon fruit and one often reported to deteriorate as a result of grafting (Rouphael et al., 2010). Grafting honeydew melon onto *Cucurbita* hybrids and *C. melo*rootstocks had no effect on fruit dry matter content and pulp firmness (Crinò et al., 2007). On the contrary, Colla et al. (2006b) found that pulp firmness values recorded for cantaloupe grafted on hybrid rootstock were significantly higher (19–32%) than those observed for non-grafted plants, and the same effect was observed with muskmelon grafted either on *C. melo* or *Cucurbita* hybrid rootstocks (Colla et al., 2010a), despite that grafting resulted in lower pulp dry matter content in both these studies. By contrast, flesh firmness of Galia melon was consistently reduced by grafting on four different hybrid rootstocks and a similar tendency was evidenced with Ananas melon; however, these results were obtained from graft combinations that demonstrated incompatibility problems, variably causing plant decline, with loss of pulp firmness being one of the quality indices proposed for prognostication of incompatibility (Soteriou et al., 2016). Galia melon grafted onto interspecific hybrid rootstocks and onto *C. metuliferus* incurred reduced overall sensory rating but not reduced flesh firmness compared to non-grafted controls (Guan et al., 2015), although the same scion grafted on *C. metuliferus* and grown organically in nematode infested soil incurred a reduction in pulp firmness (Guan et al., 2014).

Grafting interacted with scion cultivar in respect to firmness but a general trend for loss of firmness was identified in two annual field trials involving numerous cultivars of muskmelon, honeydew and specialty melon scions, although assessment of firmness across the various melon types was hampered by the difficulty of harvesting melons of the same maturity (Schultheis et al., 2015). Grafting cantaloupe on two interspecific hybrid rootstocks reduced flesh firmness compared to non-grafted and self-grafted control, with differences minimized after prolonged postharvest storage (Zhao et al., 2011). Finally, no effect on flesh firmness was reported when Piel de Sapo-type melon was grafted on either Monosporascus-resistant *C. melo* subsp. *agrestis* or the widely used but less compatible interspecific hybrid RS841 (Fita et al., 2007). From the findings above it is evident that, unlike the case of grafted watermelon (Kyriacou and Soteriou, 2015; Kyriacou et al., 2016), melon grafting, whether on same or different species rootstocks, unequivocally does not increase flesh firmness; it either has no effect or it results in loss of textural quality, depending largely on rootstock–scion compatibility and to a lesser extent on cultural conditions.

Sweetness and acidity

Fruit sweetness is a major sensory feature of melon quality (Yamaguchi et al., 1977; Liu et al., 2010), which stems mainly from soluble carbohydrates but is commonly quantitated on the basis of the SSC derived from the temperature-compensated refractive index of the fruit juice. Understanding rootstock-mediated effects on melon sweetness is critical for safeguarding sensorial acceptability of melon fruit produced on grafted plants. Available reports describing these effects manifest widespread rootstock–scion interaction attesting the importance of selecting appropriate graft combinations. For instance, grafting cantaloupe (cv. Cyrano) on hybrid 'P360' under greenhouse conditions reduced the fruit SSC by an absolute 1.6% when compared to the non-grafted control; however, the SSC (10.9%) remained highly acceptable (Colla et al., 2006b). Similar findings were reported for muskmelon (cv. Proteo) grafted on hybrid rootstock ('P360') in open field cultivation, though both grafted (8.3%) and non-grafted (7.5%) plants exhibited quite low fruit SSC (Colla et al., 2010b). However, evaluating the same scion ('Proteo') on hybrid rootstocks 'P360' and 'PGM96-05,' revealed no effect on SSC though grafting both 'Proteo' and 'Supermarket' muskmelons on *B. hispida* and *C. metuliferus* did reduce their SSC (Trionfetti-Nisini et al., 2002). Crinò et al. (2007) found no effect on the SSC of honeydew melon (cv. Incas) by grafting on four hybrids and four *C. melo* rootstocks, as also verified by Verzera et al. (2014) using the same scion on three hybrids and two *C. melo* rootstocks, but not with hybrid rootstock 'AS10' which increased the SSC from 15.5 to 16.3%. Galia melon 'Arava' grafted onto hybrids 'Strong Tosa' and 'Carnivor' incurred reduced overall acceptability, flavor rating and SSC compared to non-grafted controls, but not when grafted onto *C. metuliferus* (Guan et al., 2014, 2015). Moreover, honeydew scion ('Honey Yellow') on the same rootstocks did not differ in sensory properties and SSC in comparison with either non-grafted or self-grafted controls (Guan et al., 2014, 2015). The SSC of Piel de Sapo melon was slightly reduced when grafted on Monosporascus-resistant rootstock *C. melo* subsp. *agrestis* ('Pat 81') and also on hybrid 'RS 841,' but not to an extent that might significantly affect marketability (Fita et al., 2007). Although decrease in the SSC is not an infrequent response to grafting, with potential implications on, the increment of decrease in none of the reported studies seemed decisive for overall quality and marketability.

Besides the case of hybrid 'AS10' above, few are the cases of rootstocks reported of causing increase in melon SSC. One such is the absolute increase of 1.2% obtained in the mean SSC (10.4%) of greenhouse grown Galia melon grafted onto hybrid rootstocks 'TZ148,' 'N101,' 'Carnivor,' and '30900' (Soteriou et al., 2016). Most often, grafting on compatible rootstocks has no effect on the fruit SSC. Out of four hybrid rootstocks onto which Ananas melon was grafted only one ('30900') had an effect on the SSC causing an absolute reduction by 1.12% relative to the non-grafted control (Soteriou et al., 2016). Field trials of muskmelon, honeydew and specialty melon types (Persian, Tuscan, Canary, Galia, Piel de Sapo) grafted on hybrid rootstock 'Carnivor' vs. self-grafted and non-grafted controls generally showed no effect on the SSC although limited grafting × scion interaction was evident (Schultheis et al., 2015). Traka-Mavrona et al. (2000) also found grafting had no effect on the fruit SSC of three inodorous melon and one cantaloupe cultivars tested on two hybrids and one pumpkin landrace rootstocks under protected and open field cultivation. Finally, in a study closely observing fruit harvest maturity based on the date of fruit setting, Park et al. (2013) reported that grafting both muskmelon ('Earl's elite') and honeydew melon ('Homerunstar') onto four *C. melo* accessions and one Shintoza type *Cucurbita* hybrid had no effect on the scions' fruit SSC, and they repudiated categorically claims of reduced fruit quality as a result of grafting, provided compatible rootstocks.

Further to the effect of grafting on SSC, the concentrations of soluble sugars are also critical as they dictate their relative contribution to the sweetness index (SI) of fruits (Elmstrom and Davis, 1981). According to Liu et al. (2010), accumulation patterns for hexoses, sucrose and oligosaccharides were similar during muskmelon ripening from non-grafted and grafted plants, despite differences in sugar levels between rootstocks. Moreover, during the period of fast sugar accumulation (32–48 days after anthesis) muskmelons from grafted plants maintained higher starch content than the non-grafted control, and the starch fraction was higher in the lower sugar content rootstock and lower in the non-grafted control. It was further postulated that the marked increase in mesocarp starch content may derive from competition by the vigorous rootstocks for the soluble sugars translocated to sink fruit which the rate of sucrose decomposition was unable to satisfy. Differential rootstock-mediated patterns of soluble sugars' accumulation were depicted in a recent study by Soteriou et al. (2016), wherein total soluble sugars in the pulp of both Galia and Ananas melons were not differentiated between four hybrid rootstocks and the non-grafted control; however, grafting Galia generally increased fruit sucrose levels at the expense of fructose and glucose whereas the opposite was observed with Ananas melon (Soteriou et al., 2016).

Like most cucurbits, melon is a fruit of very low acidity, usually below 0.2% in citrate equivalents, which nevertheless affects the sweet-to-sour balance in sensory perception (Crinò et al., 2007; Colla et al., 2010b; Verzera et al., 2014; Guan et al., 2015). Grafting honeydew melon and cantaloupe on hybrid rootstocks had a minimal effect on fruit TA which was inconsequential to fruit sensory quality (Colla et al., 2006b; Verzera et al., 2014). Similarly, no effect was found on the TA of muskmelon and Galia melon by grafting on either *Cucurbita* hybrid or *C. melo* rootstock (Crinò et al., 2007; Colla et al., 2010b; Zhao et al., 2011; Guan et al., 2015).

Aroma profile

The production of volatile compounds in melon is associated with ethylene-dependent pathways (Obando-Ulloa et al., 2008; Pech et al., 2008) and with textural changes related to cell wall matrix solubilization events (Dos-Santos et al., 2013). Hence, the climacteric (*cantalupensis* and *reticulates*) and the non-climacteric (*inodorus*) types demonstrate distinct volatile profiles, with C9 aliphatic aldehydes being the key aroma and flavor descriptors for inodorous honeydew melons (Verzera et al., 2014), as opposed to the mainly ester-based (ethyl butanoate, methyl 2-methylbutanoate and ethyl 2-methylpropanoate) descriptors for fruity and sweet aroma notes of *cantalupensis* and *reticulates* muskmelon cultivars (Kourkoutas et al., 2006; Beaulieu and Lea, 2007). Grafting seems to affect the aroma profile of both muskmelon and honeydew type melons. Grafting muskmelon on interspecific *Cucurbita* hybrids and on *C. melo* rootstocks generally elicited higher levels of non-key alcohol and aldehyde volatile compounds responsible for green and fresh notes, such as flower-green (1-hexanol), fruity (2-methyl-1-butanol), fatty-green (1-octanol), ethereal (ethanol), green [(*E*)-2-butenal], and fresh-lemon-green (octanal) aromas (Condurso et al., 2012). Ester-based aromas characteristic of muskmelon were generally higher in non-grafted control, such as cantaloupe-like, green fruity, melon (ethyl 2-methylbutanoate) and sweet-fruit (ethyl butanoate) aromas (Chuan-qiang et al., 2011; Condurso et al., 2012). However, significant exceptions to this motif were found among both *Cucurbita* spp. and *C. melo* rootstocks, rendering screening for optimum rootstock–scion combinations essential. In fact, some commercial *Cucurbita* hybrids (e.g., 'RS-841') and *C. melo* (e.g., 'Energia') rootstocks can be successfully used for controlling soilborne pathogens without any significant effect on the fruit aroma (Condurso et al., 2012). Similarly, Verzera et al. (2014) examined the effect of four inter-specific hybrids and two melon genotypes on the fruit aroma and sensory quality of honeydew melon cv. Incas (*C. melo* L. subsp. *melo* var. *inodorus* H. Jacq.). Prevalent volatiles in both grafted and non-grafted *inodorus* melon were mainly aldehydes and alcohols such as nonanal (melon, orange peel), (*Z*)-6-nonenal and (E)-2-nonenal (honeydew melon fruity), (*E,Z*)-2,6-nonadienal, 1-nonanol (herbaceous), (*Z*)-3-nonen-1-ol (melon, green, floral) and (*Z,Z*)-3,6-nonadien-1-ol (pumpkin, cucumber). Fruits from plants grafted on three of the interspecific hybrids (cvs. RS-841, P-360, Polifemo) and one *C. melo* rootstock ('Energia') had similar aroma profiles to the control, however, particular rootstocks from either type (e.g., 'AS10' and 'Sting') were found to decrease the amounts of key aroma compounds. Grafting was generally found to reduce the intensity of honeydew melon and herbaceous aroma descriptors and increase those related to fruity aroma and flavor. It is important to emphasize that selection is possible of resistant interspecific hybrid rootstocks (e.g., 'RS-841') that increase yield and fruit weight of both honeydew cv. Incas and muskmelon cv. Proteo scions, without having a detrimental effect on sensory characteristics, including the aroma profile (Condurso et al., 2012; Verzera et al., 2014).

Functional compounds

Melon is a rich source of α-, ζ-, and especially β-carotene but also of lutein, cryptoxanthin, phytoene, and the violaxanthin cycle carotenoids, however, little is known on the effect of grafting on these components (Laur and Tian, 2011). The fruit carotenoid profile of odorous melon, was either non-differentiated, or highly improved particularly with regards to the α- and β-carotene components in response to grafting on *C. maxima* × *C. moschata* hybrid rootstocks vis-à-vis the non-grafted control; whereas grafting on *C. melo* rootstocks resulted in significantly reduced β-carotene levels, which inadvertently emphasized ζ-carotene content, while lutein was increased with grafting on both types of rootstocks (Condurso et al., 2012). Carotenoid content is largely responsible for melon pulp color; hence the effects of grafting on these traits are expectedly associated. Colla et al. (2006b) reported that grafting cantaloupe (cv. Cyrano) on hybrid rootstock 'P360' influenced pulp colourimetric values positively, resulting in brighter (higher L^*) and more intense orange hue (higher a^*/b^* ratio), probably reflecting higher α- and β-carotene concentrations in the pulp (Condurso et al., 2012). Intriguingly, increased levels of both chlorophylls and β-carotene were obtained in the leaves of Galia type cvs. Arava and Resisto grafted on interspecific rootstocks 'Shintoza,' 'Kamel,' and particularly on 'RS841' (Romero et al., 1997).

Cucumber (Cucumis sativus L.)

Cucumber constitutes an annual vegetable species mostly grown under protected cultivation. The use of rootstocks resistant or tolerant to soilborne diseases, foliar pathogens, arthropods, and weeds has become instrumental for cucumber production, especially under intensive farming practices with limited crop rotations (Lee et al., 2010; Louws et al., 2010). Several rootstocks (*C. maxima* × *C. moschata*, *C. ficifolia*, *C. moschata*, *C. argyrosperma*, *L. siceraria*, *B. hispida*, *Luffa cylindrica* (L.) M. Roem., *Momordica charantia* L., *S. angulatus*, *Citrullus* spp.) have been used for cucurbit grafting; most enhance scion growth and productivity under unfavorable soil and environmental conditions, but some lack tolerance to specific stresses and others can have a detrimental effect on vegetable fruit quality (Rouphael et al., 2010, 2012). The most popular rootstocks for cucumbers belong to the genus *Cucurbita*. In particular, the interspecific cross *C. maxima* × *C. moschata* has been exploited as a favorable source of rootstocks, currently the most common commercial rootstocks for cucumber (Lee et al., 2010). Less frequent is the use of single non-hybrid *Cucurbita* species as rootstocks, such as accessions of *C. argyrosperma*, *C. ficifolia*, *C. maxima*, *C. moschata*, and *C. pepo*. Fruit quality deterioration in grafted plants, reported chiefly as decrease in sweetness and acidity, is a common problem particularly with *Cucurbita* hybrids which are frequently implicated in scion × rootstock interactions, further compounded by crop management practices (Davis et al., 2008b; Rouphael et al., 2010).

Morphometric characteristics

It is well established that vigorous *Cucurbita* interspecific hybrids can improve cucumber yields significantly (Davis et al., 2008b). More frequently, the effect on yield is related to the variation in fruit size, as grafted plants are characterized by a vigorous root system (high root length and density) able to enhance photosynthetic rate as well as water and nutrient uptake efficiency (particularly N, P, Ca, and Mg) and, consequently, crop productivity (Rouphael et al., 2010). Several authors have demonstrated a significant increase in fruit weight when cucumber plants were grafted onto *Cucurbita*interspecific hybrids ('RS841,' 'Strong Tosa,' 'PS1313,' and 'P360') and *Cucumis pustulatus* Naudin ex Hook.f. compared to non-grafted control (Colla et al., 2012, 2013; Goreta Ban et al., 2014; Liu et al., 2015). However, in some cases increased cucumber yield has been attained mainly by an increase in the number of fruits per plant rather than an increase in mean fruit size (Huang et al., 2009).

Other morphological traits that constitute primary criteria for making purchasing decisions are the fruit shape index and the colouration of the skin (Rouphael et al., 2010). Reports on cucumber grafting demonstrated that the effect of rootstocks on fruit shape has been mostly non-significant or minimal (Lee et al., 1999; Colla et al., 2013). Regarding color, Colla et al. (2012) reported that lightest colouration, expressed as an increase in colourimetric CIELAB component L*, was observed on the skin of cucumber cv. Akito grafted onto the commercial rootstock 'PS1313' (*C. maxima* × *C. moschata*) compared to fruit from plants grown on their own roots.

Textural characteristics

Fruit firmness constitutes also an important physical property influencing consumer acceptability (Rouphael et al., 2010). Hwang et al. (1992) demonstrated that cucumber from plants grafted onto *S. angulatus* 'Andong' rootstock tended to be firmer than those grafted onto figleaf gourd (*C. ficifolia* 'Heukjong'). On the contrary, Morishita (2001) reported that 'Kema' and 'Kifujin New Type' cucumber plants grafted onto the bloomless rootstock 'Big Ben Kitora' carried fruits of softer flesh than those grown on their own roots. Nevertheless, more popular rootstocks such as *C. moschata* or *C. maxima* × *C. moschata* had no effect on fruit firmness when compared to non-grafted plants (Sakata et al., 2007; Colla et al., 2013). The variation in fruit firmness induced by rootstocks may be attributed to several mechanisms such as the uptake and translocation of calcium, modulated water relations and nutritional status, increased synthesis of endogenous hormones as well as variation in cell morphology and turgor (Rouphael et al., 2010).

Sweetness and acidity

It has been reported that changes in grafted vegetable aroma and taste appear to be not only scion but also rootstock-dependent attesting the importance of selecting appropriate graft combinations (Rouphael et al., 2010). For instance, grafting cucumber onto 'Heukjong' figleaf gourd (*C. ficifolia*) reduced the fruit SSC and fructose concentration when compared to the non-grafted control; whereas the SSC remained high when 'Andong' (*S. angulatus*) was used as rootstock (Lee et al., 1999). Moreover, Huang et al. (2009) observed a lower accumulation of SSC in plants grafted onto figleaf gourd and grown under unstressed conditions compared to self-grafted control. However, under saline conditions of 60 mM NaCl the SSC and TA incurred significant increase in both grafted and self-grafted plants. Similarly, the SSC of 'Akito' cucumber increased when grafted onto the interspecific hybrid 'PS1313,' whereas an opposite trend was recorded for the TA (Colla et al., 2013).

Aroma profile

Besides taste, the effect of grafting on aroma was also quantitated in cucumber. In a recent study, Guler et al. (2013)demonstrated that grafting affected the aroma profile of both the peel and flesh of cucumber cv. Cengelköy in response to the use of a bottle gourd rootstock. Thus, grafting caused a substantial increase in the alcohol content [(*Z*)-6-Nonenol, (*E,Z*)-2,6-nonadienol, 1-nonanol, and (*Z,Z*)-3,6-nonadienol], a decrease in the aldehyde content [(*E,Z*)-2,6-nonadienal] with no significant influence on ketones, terpenes and hydrocarbons in both cucumber peel and flesh tissues. The authors concluded that the bottle gourd '33-41' could be considered a promising rootstock for improving major volatile components identified in cucumber.

Solanaceae

Tomato (Solanum lycopersicum L.)

Nowadays tomato production under protected environment is resorting to the use of grafted plants. Reservations against their use relate to their higher price being considered non-affordable by growers, the use of speciality cultivars or particular problems associated with disease outbreaks, such as the novel tobamoviruses (Luria et al., 2017). However, the number of commercial rootstocks offered by breeding companies has burst and their widespread use is becoming highly visible. Not only rootstocks from the species *Solanum lycopersicum* L. are available but also interspecific hybrids, such as *S. lycopersicum* × *Solanum habrochaites* S. Knapp & D.M. Spooner, and rootstocks from other species, such as *Solanum torvum* L. or *Solanum melongena* L. Therefore, growers often follow breeders' recommendation and use same-company scions and rootstocks. Even in the field, the application of grafted plants has started, e.g., as a means to protect the scion from invasive weeds, such as broomrape. Compared to the Cucurbitaceae representatives, the effect of rootstocks on fruit quality traits seems less intense and reports on reduced quality are mainly related to reduced sweetness. The effects of grafting on most tomato quality characteristics have been variable, strongly influenced by the rootstock–scion combination. Moreover, the effect of the grafting combination interacts with other factors, such as climate, cultural practice, duration and intensity of stress, water and nutrient disposability and not to least with the sampling strategy (Riga, 2015; see chapter 'Methodological approaches').

Morphometric characteristics

Grafting tomato often results in significant increase in fruit weight and consequently in fruit diameter and size compared with non- or self-grafted plants (Passam et al., 2005; Moncada et al., 2013; Riga, 2015). This was reported for many different rootstock–scion combinations resulting in total yield increase. However, yield gain may be also attributed to an increase in the number of fruits rather than an increase in mean fruit weight (Savvas et al., 2011). The effect of grafting on fruit weight and size depends on grafting combinations (Khah et al., 2006; Leonardi and Giuffrida, 2006; Schwarz et al., 2013). Larger fruit size seems to be attained when vigorous rootstocks are used, such as 'Maxifort' (Krumbein and Schwarz, 2013; Schwarz et al., 2013), 'Beaufort' (Romano et al., 2000; Pogonyi et al., 2005; Turhan et al., 2011), 'Heman' (*S. habrochaites*), 'Joint,' 'P1614,' and 'RS1427' (Romano et al., 2000), or 'Star Fighter' (Theodoropoulou et al., 2007). This phenomenon is particularly recognized when scions have smaller fruit sizes, e.g., cherry tomato with less than 40 g (Schwarz et al., 2013). In some cases, grafting may reduce fruit size when less vigorous rootstocks are used, such as 'Brigeor' (Schwarz et al., 2013), 'Energy,' 'Firefly,' 'Linea9243,' 'Nico' (Romano et al., 2000). Based on the same reasons, fruit size of two different scion cultivars was significantly reduced when a salt tolerant goji berry (*Lycium chinense* Mill.) served as rootstock (Huang et al., 2015).

Fruit shape has seldom been assessed in grafted tomato despite indications of its differentiation once the fruit size is affected (Schwarz et al., 2013). Increase in shape index, measured as the ratio of fruit diameter to maximal height, was reported as corresponding to increase in fruit size (Turhan et al., 2011). However, rootstock 'Beaufort' raised the fruit shape index, compared to non-grafted tomato, irrespective of the scion cultivar ('Yeni Talya,' 'Swanson,' 'Beril'), while rootstock 'Arnold' only increased it in combination with the scion 'Yeni Talya.' This indicates a similar dependence of fruit shape to rootstock vigor as already mentioned for the fruit size.

Fruit color was in certain cases affected by grafting (Di Gioia et al., 2010; Brajović et al., 2012) but in others not (Krumbein and Schwarz, 2013; Schwarz et al., 2013). As in the case of watermelon (see chapter 'Watermelon'), changes pertained particularly to color component a^* (redness) which is associated with lycopene content. Thus, also for tomato, an effect of grafting on color seems to be significant if a rootstock influences the fruit lycopene content (Miskovic et al., 2016). However, color as well as texture assessment, are often presented with the difficulty of obtaining sufficient and uniform fruit samples in terms of development and harvest maturity to constitute a representative sample. Failure to control sampling procedures effectively may lead to misleading or inconsistent results (see also chapter 'Methodological approaches').

Physiological defects

Although physiological disorders related to grafted plants have not received much attention in the literature, they are not uncommon. Blossom end-rot (BER), the most typical tomato disorder (Ho and White, 2005), was invariably reduced in tomato grafted on rootstocks 'Brigeor,' 'Maxifort,' and LA1777 (*S. habrochaites*), and under different environmental conditions shaped by factors such as salinity, potassium nutrition, sub-optimal temperature, and light conditions (Fan et al., 2011; Krumbein and Schwarz, 2013; Schwarz et al., 2013; Ntatsi et al., 2014). Moreover, reduction in BER incidence with grafting was pronounced under stress conditions in comparison to both non- and self-grafted scions. Decrease in BER was mainly related to the rootstock genotype; e.g., the BER incidence in 'Classy,' a medium round type tomato of ~70 g, was more diminished when grafted on the rootstock 'Brigeor' compared to 'Maxifort' or self-grafted plants. The BER reduction was also influenced by rootstock–scion interaction; e.g., it was decreased to a greater extent when cherry tomato 'Piccolino' was used as a scion. Under certain conditions, the use of a rootstock may raise the BER incidence. This was the case in an experiment where two cultivation systems were compared during summer: hydroponics vs. soil (Takasu et al., 1996); the improved nutrient and water uptake facilitated by grafting did not cope sufficiently with the very fast fruit growth under high radiation conditions. Also, BER increased in trials involving rootstock 'Edkawi,' or eggplant rootstocks (e.g., 'EG203,' 'VFR Takii') (Oda et al., 1996; Poudel and Lee, 2009; Fan et al., 2011). Here possible reasons are justified by the characteristics of the rootstocks selected. Rootstock 'Edkawi,' although known as salinity-tolerant, as well as eggplant rootstocks, lower the uptake and transport of Ca-ions into the fruits compared with self-grafted tomato. Results indicate that the incidence of BER is reduced by grafting when Ca uptake and transport into the fruits is improved (Fan et al., 2011; Savvas et al., 2017). Increased fruit Ca concentration may lead to strengthening of cell walls and cellular integrity and improvement of fruit firmness (Dorais, 2007; Schwarz et al., 2013).

Textural characteristics

Attributes of texture are seldom considered in grafted tomato. Cultivar Jack grown under Mediterranean conditions as a scion grafted onto nine rootstocks typified rootstock effects: e.g., 'Alligator' tended to reduce, 'Maxifort' did not affect and 'King Kong' enhanced firmness (Riga, 2015). Other reports corroborate these findings although loss of firmness seems as the predominant effect. Thus, fruits of the cultivars 'Classy' and 'ASVEG10' obtained from plants grafted onto 'Brigeor' or 'Maxifort' and grown under potassium deficiency but also fruits from plants grafted on eggplant rootstock were less firm and scored higher maximum deformation than fruits from self-grafted tomato (Poudel and Lee, 2009; Schwarz et al., 2013). The reasons are not clear but K^+/Ca^{2+} interaction was not implicated in the differences in fruit firmness. Independently, it could be clearly demonstrated that fruit Ca content was increased by grafting (Khah et al., 2006; Fan et al., 2011; Savvas et al., 2017). However, as in the case of Khah et al. (2006) it did not affect fruit firmness. While Riga (2015) did not find differences in fruit firmness between non- and self-grafted 'Jack' tomato, Rahmatian et al. (2014)found significantly lower firmness in fruits from self-grafted compared to non-grafted 'Synda' tomato. However, the use of a rootstock (cv. King Kong) independently of simple or double grafting did not affect firmness.

Sweetness and acidity

Results concerning the variation in taste of grafted tomato fruit, comprising sugars (glucose, fructose), SSC as a non-specific sweetness parameter, and TA are also very contradictory and seem to be affected by the same parameters as mentioned above. In several experiments, the use of a rootstock did not change fruit taste attributes (Matsuzoe et al., 1996; Khah et al., 2006; Savvas et al., 2011; Barrett et al., 2012). However, decrease and increase in the main components of taste were also observed, as explained below, and based on these findings grafting appears not to constitute a reliable tool for improving tomato fruit taste. This conclusion was confirmed by the results of one of the rare consumer sensory tests performed during a 2-year cultivation of the heirloom tomato 'Brandywine' as non-, self-grafted and grafted onto 'Survivor' and 'Multifort.' While in the first year the rootstock 'Survivor' scored significantly lower than the non-grafted 'Brandywine' in appearance, acceptability, and flavor, no differences were observed between these treatments in the second year (Barrett et al., 2012).

The main sugars in mature tomato are glucose and fructose in equal shares and the total sugar concentration ranges from about 20 to 100 $g \cdot kg^{-1}$ fresh mass, depending on cultivar and growing conditions. Improvement of fruit sweetness related to grafting is rather seldomly reported. Such cases described were with tomato grafted onto 'Fanny,' 'King Kong,' 'LA1777' (*S. habrochaites*), or onto scarlet eggplant rootstocks (e.g., 'EG 203'), whereby the enhanced SSC content was associated with the effect of water deficiency which lowered plant growth and yield and decreased fruit water content (Oda et al., 1996; Fernández-García et al., 2004a,b; Poudel and Lee, 2009; Ntatsi et al., 2014; Rahmatian et al., 2014). The same association occurred when grafted plants grew under saline or drought conditions or when using a drought tolerant cultivar as a rootstock (Flores et al., 2010; Sánchez-Rodríguez et al., 2012a) or when grafting onto a medicinal plant (*L. chinense*; Huang et al., 2015). However, in many grafting combinations, rootstocks reportedly decreased SSC and sugar concentration in the scion fruits (Pogonyi et al., 2005; Qaryouti et al., 2007; Turhan et al., 2011; Barrett et al., 2012; Nicoletto et al., 2013a,b; Schwarz et al., 2013; Gajc-Wolska et al., 2015; Kumar et al., 2015; Riga, 2015). Nevertheless, the decline caused by grafting is very low compared to the potential increase procured by employing a selected scion that might at least double the fruit sugar concentration. The decline in sugars incurred with grafting is reported to account for approximately not more than 16% (Riga, 2015), which does not exceed the range of maximum decline proposed for consumer acceptability (Kader, 1999; Maynard et al., 2002). The reasons for a lower carbohydrate content in grafted tomato may stem indirectly through rootstock effect on scion vigor, timing of flowering, fruit load, yield and, ultimately, fruit maturation, as fruit sugar concentration is highly dependent on fruit maturity at harvest (Rouphael et al., 2010; Soteriou and Kyriacou, 2015). In this respect, grafting may be considered a high-input production method, with a prevalent tendency for increasing crop load and potentially suppressing fruit sugar content (Davis et al., 2008b; Soteriou and Kyriacou, 2015). Moreover, vigorous rootstocks may act as additional sinks for assimilates and thus, reduce assimilate flow to the fruits (Xu et al., 2006; Martínez-Ballesta et al., 2010). Alternatively, water uptake-efficient rootstocks may increase fruit water content even if sufficient assimilates are available, thus, leading to a reduced fruit sugar concentration (Turhan et al., 2011; Krumbein and Schwarz, 2013). Fruits of grafted scions are often larger than fruits of the same non- or self-grafted scion, and though the fruit sugar/acid ratio might remain unaffected, a decline in soluble carbohydrates may be incurred as a dilution effect (Tieman et al., 2017).

While sugars may decrease in grafted tomato, acid content expressed as TA is on the contrary enhanced (Turhan et al., 2011; Sánchez-Rodríguez et al., 2012a; Nicoletto et al., 2013a,b; Schwarz et al., 2013; Krumbein and Schwarz, 2013; Huang et al., 2015; Kumar et al., 2015; Riga et al., 2016). Total organic acids in tomato fruit are usually in the range of 0.2 to 1.7 $g \cdot kg^{-1}$ fresh mass, with citric and malic acid being the main components of sourness. Grafting accounts for an increase in TA up to 15% reported under a range of different environmental conditions which indicates a direct rootstock effect. Comparing fruits from 'Classy' and 'Piccolino' self-grafted or grafted onto 'Maxifort' and 'Brigeor' resulted under different experimental conditions almost always in the highest TA produced by 'Maxifort' followed by 'Brigeor' and self-grafted. Independent of the presence or absence of water stress, tomato fruits from the drought sensitive cultivar Josefina grafted onto the drought tolerant cv. Zarina had always higher TA contents, particularly citric acid, compared with non-, self-grafted or the reciprocal grafting combination (Sánchez-Rodríguez et al., 2012a).

The mechanisms involved in grafting-elicited increase of fruit TA have not been thoroughly investigated; however, organic acids constitute a direct substrate for respiratory demands and their increased *de novo* synthesis in developing fruits might be a plausible mechanism for coping with the sugar deficit incurred on the heavy crop load supported by vigorous rootstocks. Moreover, the capability of a vigorous root system to enhance the uptake of nutrients, such as K, could be another reason (Ruiz and Romero, 1999; Leonardi and Giuffrida, 2006; Albacete et al., 2009). Potassium is positively related to the acid concentration in tomato fruits, and plays a role in maintaining electroneutrality of acids in the fruit. However, K transport depends not only on the rootstock but also on growing conditions, such as the current K concentration in the root zone, and on climatic factors (Albacete et al., 2009). Interestingly, differences in K concentration were found not to be significant between fruits of self-grafted and grafted plants, but increase in fruit TA was significant (Schwarz et al., 2013). To complicate matters further, an exceptional report demonstrated that when cv. Lemance was grafted onto 'Beaufort,' fruit organic acid concentration was lower compared with fruits from non-grafted plants (Pogonyi et al., 2005). Therefore, the enhanced TA in fruits of grafted tomato warrants further investigation (Leonardi and Giuffrida, 2006).

Aroma profile

Odor-active volatiles contribute to tomato flavor (Krumbein and Auerswald, 1998) but their influence on sensory properties awaits further treatise. Six major volatiles which contribute to tomato flavor were evaluated by a consumer panel in Florida (USA): 2-butylacetate, *cis*-3-hexen-1-ol, 3-methyl-1-butenol, 2-methylbutanal, 1-octen-3-one, *trans,trans*-2,4-decadienal (Tieman et al., 2012; Zhang B. et al., 2016). Krumbein and Schwarz (2013) found that for two different scion cultivars, 'Piccolino' – a cherry, and 'Classy' – a round type, grafting on rootstocks 'Brigeor' and 'Maxifort' induced a general enhancement of three aroma volatiles: methyl salicylate, guaiacol and eugenol, with oily, sweet and spicy odors, respectively; but the concentrations of three other aroma volatiles with almondy odor (benzaldehyde), violet-like odor (β-ionone) and tomato-like flavor (geranylacetone) were decreased by grafting. The variation of the carotenoid content in tomato (see below) affects the carotenoid-derived volatiles responsible for tomato flavor, such as the violet-like odor (β-ionone) and the tomato-like flavor (geranylacetone) (Krumbein and Schwarz, 2013). However, the actual sensory contribution of these volatiles to changes in tomato flavor was not assessed, which remains critical particularly under the light of the findings of Tieman et al. (2012).

Functional compounds

Carotenoid content of tomato fruit, mainly lycopene and β-carotene, can be influenced by grafting, but it is subject to significant rootstock–scion interaction which indicates that graft combination plays an important role. Moreover, as Riga et al. (2016) demonstrated, the comparison of the grafting combination to either the non-grafted or self-grafted scion is very important. According to several authors, lycopene concentration in tomato fruits tends to decrease with grafting (Helyes et al., 2009; Brajović et al., 2012; Nicoletto et al., 2013b); e.g., most out of 15 rootstocks investigated, including 'Maxifort,' 'Beaufort,' and 'King Kong,' decreased the fruit lycopene concentration of tomato scion 'Jeremy' and 'Jack' (Miskovic et al., 2009; Riga et al., 2016). Similar results have been reported for tomato scion 'Cecilia' grafted onto 'Beaufort' and 'Heman' (Mohammed et al., 2009), for scion 'Macarena' grafted onto 'Maxifort' (Gajc-Wolska et al., 2010), as well as for 'Classy' grafted onto 'Brigeor' (Krumbein and Schwarz, 2013; Schwarz et al., 2013). Though in the latter experiments total carotenoids were diminished due to lycopene decrease, under specific conditions of grafting the lycopene/carotenoid content may increase. When eggplant rootstock 'Madonna' was used, tomato lycopene concentration increased (Miskovic et al., 2016). In another experiment, nutritional stress caused by low potassium in the nutrient solution was applied to 'Classy' grafted onto 'Maxifort' resulting in enhanced β-carotene (Schwarz et al., 2013), as well as in another grafting combination ('Amati' grafted onto 'Robusta' or 'Body') under non-stressed conditions (Brajović et al., 2012). The mechanisms for these opposite responses remain unclear. In other studies, authors did not find any grafting effect on carotenoids (Khah et al., 2006; Vinkovic-Vrcek et al., 2011) of 'Big Red' tomato grafted onto 'Heman' and 'Primavera' (*S. lycopersicum*) under open-field and greenhouse conditions and of 'Tamaris' grafted onto 'Heman,' 'Efiato,' and 'Maxifort' always comparing with fruits from non-grafted cultivars.

Tomato fruit contains significant amounts of ascorbic acid, and several studies showed that fruit content strongly reduced by grafting both in greenhouse and field studies (Fernández-García et al., 2004a,b; Arvanitoyannis et al., 2005; Di Gioia et al., 2010; Vinkovic-Vrcek et al., 2011; Djidonou et al., 2016; Riga et al., 2016). Fruit vitamin C content was reduced in soil cultivation of different tomato scions grafted onto 'Heman,' 'Spirit,' 'Arnold,' 'Beaufort' (Qaryouti et al., 2007; Turhan et al., 2011) and in hydroponics using 'Maxifort,' 'Interpro,' or 'King Kong' rootstocks (Riga et al., 2016). The lower ascorbic acid content could be explained by the higher plant/shoot biomass in grafted plants compared with non-grafted ones or by the fact that grafted plants were initially subjected to stress following the grafting operation. Ascorbic acid is known to control cell differentiation (Arrigoni, 1994) and to promote callus division and growth (Tabata et al., 2001). The decreased total vitamin C content of the fruits from grafted plants could therefore be a resultant of redistribution or accumulation of vitamin C in other parts of grafted plants (Wadano et al., 1999). Alternatively, changes in ascorbic acid content can be influenced by the choice of rootstock, as shown for tomato grafted onto 'King Kong,' 'Beaufort,' or 'Maxifort' rootstocks, which exhibited higher ascorbic acid content compared to the same plants self-grafted or grafted onto 'Arnold' and 'Brigeor' rootstock, respectively (Turhan et al., 2011; Schwarz et al., 2013; Rahmatian et al., 2014). A similar increase in Vitamin C was analyzed when tomato were grafted on *L. chinense* (Huang et al., 2015).

Abundant flavonoids in tomato fruits are the hydroxycinnamic acids and their derivatives (Gómez-Romero et al., 2010; Sánchez-Rodríguez et al., 2012b; Riga et al., 2016), as well as naringenin, chalcone and rutin (quercetin-3-*O*-rutinoside) (Slimestad et al., 2008; Sánchez-Rodríguez et al., 2012b), which are natural antioxidants. The choice of cultivar (Steward et al., 2000) as well as abiotic and agronomic factors are major contributing factors to the total content of phenolics in tomato (Tomas-Barberan and Espin, 2001). Under water stress the combination with a drought tolerant rootstock (cv. Zarina) resulted in the highest value in total flavonoids, hydroxycinnamic acids and rutin compared with not or self-grafted 'Zarina.' Nicoletto et al. (2013b) found also a higher phenolic acid content for another grafting combination with 'Profitto' grafted onto 'Beaufort' compared with non-grafted plants, but this was not observed with rootstock 'Big Power.' However, Vinkovic-Vrcek et al. (2011) reported that grafting significantly reduced the total phenolic content of tomato cv. Tamaris grafted onto 'Heman,' 'Efiato,' and 'Maxifort,' while no significant differences were found among these rootstocks. Comparing nine different mainly commercial rootstocks, Riga et al. (2016) confirmed that the reduction or increase in flavonoids clearly depends on the selection of the rootstock when the same scion cultivar was used. Thus, relative to tomato from non-grafted 'Jack,' soluble and total phenolics were reduced when grafted onto 'King Kong' but increased when grafted onto 'Brigeor.' The trigger for the rootstock to affect flavonoid concentration remains unclear. Although as indicated by the drought experiment, rootstocks better adapted to stress conditions responsible for higher flavonoid production may improve total flavonoids in the whole plant (Sánchez-Rodríguez et al., 2012b).

Among other functional compounds, serotonin concentration in fruits was found lower after grafting 'Jack' onto different commercial rootstocks independent of the cultivars selected (Riga et al., 2016).

Eggplant (Solanum melongena L.)

Eggplant and its relatives constitute an important source of rootstocks for the production of not only eggplant itself but also of tomato. By far the most common rootstock for eggplant is *S. torvum* (Lee et al., 2010). However, numerous other rootstock species and interspecific hybrids have also been tested as rootstocks for eggplant, including *S. incanum*, *S. incanum* × *S. melongena*, *S. melongena* × *S. aethiopicum*, *S. macrocarpon*, *S. sisymbriifolium*, *S. torvum* × *S. sanitwongsei*, *S. integrifolium* syn., *S. aethiopicum* gr. *Aculeatum* × *S. melongena*, S. *lycopersicum*, *S. lycopersicum* × *S. lycopersicum*, *S. habrochaites*, *S. lycopersicum* × *S. habrochaites* and *S. melongena* (Lee et al., 2010; Gisbert et al., 2011a,b; Khah, 2011; Moncada et al., 2013; Marsic et al., 2014; Sabatino et al., 2016). Current reports on the changes conferred by grafting on eggplant fruit quality provide conflicting information. This could be attributed in part to the environment in which experiments were ran (greenhouse vs. open-field), possible rootstock–scion interaction underscoring graft combinations, and differences stemming from failure to standardize fruit harvest maturity (Rouphael et al., 2010; Kyriacou et al., 2016).

Morphometric characteristics

Based on recent studies, the effect of grafting on eggplant mean fruit weight tends to be non-significant, compared to non- and self-grafted plants. For instance, when cultivar Black Beauty was cultivated non-grafted, self-grafted or grafted onto *S. torvum*, *S. incanum* × *S. melongena* and *S. melongena* × *S. aethiopicum* similar mean fruit weights were observed (Gisbert et al., 2011b). Similar findings were also recorded when *S. melongena* landraces 'Bianca,' 'Sciacca,' 'Marsala,' and 'Sicilia' were grafted onto *S. torvum* under open field conditions (Sabatino et al., 2016). Khah (2011) also confirmed these results when eggplant cv. Rima was cultivated non-grafted, self-grafted, or grafted onto two hybrid tomato rootstocks, 'Heman' and 'Primavera' under both greenhouse and open-field conditions. Exceptional increase (29%) in fruit weight was reported for eggplant 'Black Bell' when grafted onto *S. torvum* and grown in a soilless system (Cassaniti et al., 2011).

Eggplant fruit shape is highly heritable and subject to strong genetic control (Gisbert et al., 2011b). Several studies revealed that the effect of grafting on shape index has been circumstantial and mostly non-significant or minimal (4%) when the following rootstocks were used: *S. incanum*, *S. incanum* × *S. melongena* and *S. torvum* (Cassaniti et al., 2011; Gisbert et al., 2011a,b). Information on fruit physical properties of grafted eggplants, such as peel color, is conflicting but generally considered as having a negative effect (Moncada et al., 2013). For instance, the calyx of 'Brigah' fruits from non-grafted plants exhibited higher values of lightness (L*) and more vivid color saturation (chroma) in comparison to those from plants grafted onto *S. torvum*; however, in other similar works such differences between fruits of grafted and non-grafted plants were not observed (Cassaniti et al., 2011; Gisbert et al., 2011b). The most likely source of this disparity could be the difficulty of standardizing sampling practices based on optimal harvest maturity for eggplant.

Textural characteristics

Negative effects on eggplant fruit textural properties amounting to loss of firmness were reported when the *S. melongena*cultivars Black Bell and Tsakoniki were grafted onto *S. torvum* and *S. sisymbriifolium* rootstocks, respectively (Arvanitoyannis et al., 2005; Cassaniti et al., 2011). The greater fruit external and pulp internal firmness of non-grafted plants observed by Arvanitoyannis et al. (2005) could be attributed to the fact that the pest and disease pressures were more pronounced in this treatment. Therefore, it is likely that restriction of water uptake efficiency in non-grafted plants resulted fruits with lower water content and tougher texture.

Sweetness and acidity

Information on taste compounds of eggplant fruits in relation to grafting remains conflicting and conclusive trends may be difficult to deduce currently, however, the reporting of positive effects is the one mostly absent. For example, according to Lee et al. (2010) *S. torvum* rootstock had no effect on eggplant fruit sugar content. Moreover, only non-significant differences in the SSC, in TA, and in juice pH were recorded among fruits from non-grafted, self-grafted and plants grafted onto *S. habrochaites* and *S. lycopersicum* rootstocks (Khah, 2011). In line with the previous work, Arvanitoyannis et al. (2005) observed that grafted plants yielded less sweet fruits with lower ratings of sensory acceptability than non-grafted plants. The reduced fruit sugar concentration in the fruits of grafted plants may be attributed to several mechanisms, including (i) the reduction of assimilate flow to the reproductive organs since vigorous rootstocks may act as additional sinks for assimilates, and (ii) the increased water uptake by rootstocks which could reduce fruit dry matter content and consequently sugar content (Martínez-Ballesta et al., 2010; Rouphael et al., 2010).

Functional compounds

Eggplant is among the most important vegetables in terms of oxygen radicals scavenging capacity, which is a quality trait associated with its high content of phenolic antioxidants (Cao et al., 1996). Gisbert et al. (2011b) observed a higher total phenolic content only in fruits of eggplant 'Cristal' grafted onto *S. macrocarpon* rootstock. Furthermore, Sabatino et al. (2016) showed that grafting eggplant onto *S. torvum* increased total polyphenol fruit content in three out of four Sicilian landraces grown under open-field conditions, whereas an opposite trend was observed by Moncada et al. (2013), wherein the total phenolic content was greater in the non-grafted plants. Moreover, changes in fruit phenolic contents and other important flavonoids, notably anthocyanins, can be highly influenced by the rootstock–scion combination which is often subject to significant interaction (Marsic et al., 2014). However, the latter study also highlighted the importance of environmental parameters such as solar radiation in the same respect, as fruits from the same landrace/rootstock combination behaved differently in two growing seasons, with the first season being characterized by lower solar radiation compared to second. The higher vigor of grafted plants may have a negative effect on the concentration of anthocyanins, therefore grafted plants should be properly pruned under low solar conditions to improve light interception since the accumulation of anthocyanins in eggplant fruit epidermis is strongly dependent on light exposure (Awad et al., 2001).

Pepper (Capsicum annuum L.)

Pepper is currently the least grafted among the solanaceous crops, especially compared to tomato and eggplant, presumably because the commercial rootstocks currently available provide modest benefits (Lee et al., 2010). Accordingly, an urgent need exists for developing new rootstocks that can augment efforts to meet growing demands for fresh sweet pepper. The most popular rootstocks currently used are intraspecific hybrids or cultivars of *C. annuum*, however, accessions of the cultivated *Capsicum* species, including *C. baccatum* L., *C. chacoense* Hunz., C. *chinense* Jacq., and *C. frutescens* L. and their interspecific hybrids *C. annuum* × *C. chinense*, have also been tested as rootstocks for pepper (Lee et al., 2010). As might be expected, the main reason for grafting pepper has been the resistance to soilborne pathogens and nematodes but also to abiotic stresses (Schwarz et al., 2010; Penella et al., 2016), and very limited work has yet been conducted to address the implications of grafting for pepper fruit quality.

Morphometric characteristics

Several reports in the scientific literature indicated strong rootstock specificity in the responses of pepper to grafting (Rouphael et al., 2010). For instance, Doñas-Uclés et al. (2014) demonstrated an increase in fruit weight when cv. Palermo was grafted onto the *C. annuum* rootstock 'Tesor,' whereas the use of rootstocks 'Oscos' and 'AR40' incurred a minimal increase in fruit weight. Similarly, Leal-Fernández et al. (2013) showed that the mean fruit weight was higher when sweet pepper 'Triple star' was grafted onto chili pepper rootstock 'AR96029,' in comparison to non-grafted plants. In the case of F1 hybrids 'Edo' and 'Lux,' fruit weight was not influenced by grafting onto *C. annuum* rootstocks of the cultivars Snooker, Tresor, RX360, DRO8801, and 97.9001 (Colla et al., 2008). By contrast Gisbert et al. (2010) identified a general trend for reduction of fruit weight in greenhouse trials of grafted pepper (cvs. Almuden and Coyote), based, however, only on two hybrid rootstocks, 'Charlot' and 'Foc.' Decrease in fruit weight against non-grafted control is usually an indicator of rootstock–scion incompatibility.

Moreover, absence of defects in particular blossom end rot (BER) is another important quality consideration for peppers. This physiological disorder of the pepper fruit could be ascribed to a local shortage of Ca and is manifested as a leathery brown patch at the blossom-end of the fruit. However, to date there is no information in the international literature on whether the incidence of BER is reduced by grafting. The incidence of BER in grafted pepper plants could be influenced positively by rootstocks able to improve uptake and translocation of Ca to the fruits, thus strengthening cell walls and cellular integrity, or could be exacerbated by vigorous rootstocks of high nitrogen-uptake efficiency that may encourage fast growth whose demands in calcium might be difficult to meet. Therefore, research in this field is currently a prime necessity.

Sweetness and acidity

The SSC and TA of pepper fruit is in general not highly compromised by grafting on most commercial *C. annuum*rootstocks (Colla et al., 2008; López-Marín et al., 2013). In the former, two studies neither SSC nor TA were affected when pepper plants, cultivated under greenhouse conditions, were grafted onto the following *Capsicum* rootstocks: 'Snooker,' 'Tresor,' 'RX360,' 'DRO8801,' '97.90001,' 'Atlante,' 'Creonte,' and 'Terrano' (Colla et al.,

2008; López-Marín et al., 2013). Contrarily, positive effects were observed in the TA and SSC of 'Herminio' grafted onto 'Atalante' under both full and deficit irrigation conditions (López-Marín et al., 2017). The contradictory results pertaining to these taste compounds could relate to differential environments and cultural practices, as well as to possible rootstock–scion interaction.

Functional compounds

Pepper fruit carotenoid content, in particular lycopene and β-carotene which is a precursor of vitamin A, can be affected by grafting and is strongly dependent on the choice of rootstock. For instance, red cultivar Fascinato and yellow cultivar Jeanette when grafted onto the rootstock 'Terrano' incurred increase in fruit antioxidant capacity and β-carotene content, but not in lycopene content (Chávez-Mendoza et al., 2013). Polyphenols, which constitute a large family of secondary metabolites that act as major antioxidants in the neutralization of free radicals, are abundant in pepper (Colla et al., 2013). Two studies conducted by Spanish researchers showed that grafting effects on the levels of total phenolics in pepper were non-significant (Chávez-Mendoza et al., 2013; López-Marín et al., 2013; Sánchez-Torres et al., 2016).

Pepper fruit contains significant important amounts of ascorbic acid, however, currently available studies have presented conflicting results concerning the variation in vitamin C content in response to grafting (Gisbert et al., 2010; Chávez-Mendoza et al., 2013; López-Marín et al., 2013; Sánchez-Torres et al., 2016). For example, Chávez-Mendoza et al. (2013)observed a significant enhancement of ascorbic acid in grafted pepper plants; but the same effect was not confirmed by López-Marín et al. (2013). The former authors concluded that variation in vitamin C depends on both scion–rootstock combinations and growing conditions, such as plant shading. Nevertheless, other authors reported that grafting had no effect on pepper content in ascorbic acid, such as Gisbert et al. (2010) and Sánchez-Torres et al. (2016) who found no differences when two commercial pepper hybrid cultivars (Almuden and Coyote) were grafted onto two rootstocks ('Foc' and Charlot'). In light of the above studies, it might be inferred that high genotypic dependence of this quality trait in pepper scions likely confounds more limited rootstock effects.

METHODOLOGICAL APPROACHES AND POSTULATES IN ASSESSING GRAFTING EFFECTS

Homeografting vs. Heterografting

In attempting to discern the effects of various rootstocks on the fruit quality of annual crops, an implicit postulate is whether the observed responses stem not entirely from the rootstock but partly from the grafting process itself. In a number of studies, focused mainly on melon, this postulate has been addressed by using homeografts, i.e., self-grafted controls, apart from non-grafted controls. In a study assessing melon transplant growth for 10 days in a hydroponic system, Aloni et al. (2011) found that self-grafting reduced salinity-induced oxidative stress and improved the growth of homeografts compared to both non-grafted control and heterografts on interspecific hybrid TZ148; spanning, however, only a brief vegetative period this could be considered a transient post-transplanting effect. At 30 days after planting, Edelstein et al. (2011) found no difference in shoot and root dry weights between self-grafted and non-grafted treatments of either Galia melon or pumpkin. Additionally, Galia homeografts yielded no differences against non-grafted control in instrumental measurements of quality (SSC and firmness) but only sporadic differences in sensory evaluation, whereas in the case of honeydew melon self-grafting and non-grafting showed no differences in any respect of quality (Guan et al., 2014, 2015); moreover, differences in yield parameters of either scion type were not identified, as was also the case with a wide range of homeografts and non-grafted melon cultivars tested by Schultheis et al. (2015). Notwithstanding the above findings, it cannot be precluded that the grafting process in itself affects plant physiological responses to the growth environment; for instance, improved growth of homeografted melon under salinity stress (Orsini et al., 2013), and improved water relations and xylem water transport efficiency (Agele and Cohen, 2009). Nevertheless, the potential effects of homegrafting seem to pertain chiefly to the early vegetative stages of grafted transplants, as there is no convincing evidence of a lasting effect expressed at the reproductive stage on the quality characteristics of the fruit, which are unequivocally rootstock-mediated. Further to the numerous reports on rootstock mediation of fruit quality discussed in the context of the current review, the effect of heterografting was recently highlighted by high throughput sequencing which revealed that 787 and 3485 genes, associated with primary and secondary metabolism, hormone signaling, transcription factor regulation, transport, and responses to stimuli, were differentially expressed in watermelon when grafted onto bottle gourd and squash rootstocks, respectively, as opposed to self-grafted watermelon (Liu et al., 2016).

Confounding Harvest Maturity with Rootstock Effects on Quality

Quality is configured in the course of post-anthesis ontogeny and ripening, hence harvesting at optimum maturity is particularly critical for non-climacteric annual fruits (e.g., watermelon, honeydew melon, cucumber, eggplant, and bell pepper), the quality of which is configured while on the plant and steadily deteriorates postharvest at a temperature-dependent rate; whereas the quality of climacteric fruits (e.g., muskmelon and tomato), provided they are harvested physiologically mature, will improve postharvest with the onset of the climacteric and ethylene-induced changes in physicochemical composition (Kader, 1999, 2008). Harvest maturity is a major parameter of quality configuration in annual fruit crops which owed to be standardized before sound conclusions can be drawn on the effects of grafting thereon. Most studies reporting rootstock-mediated effects on fruit quality have relied on an implicit assumption of synchronous ripening behavior in grafted and non-grafted plants, and either did not explicitly monitor harvest maturity or have implicitly relied on crop-specific empirical maturity indices, such as skin color development, formation of abscission layer, or axillary tendril wilting and ground spot formation, which may provide only limited standardization of maturity (Reid, 2002); however, satisfactory standardization must rely principally on the age of the fruit monitored in days post-anthesis (Kyriacou et al., 1996, 2016). The simultaneous harvest of grafted and non-grafted plants is inherently problematic as it overlooks the potential effect of grafting on fruit ripening behavior and may yield misleading results regarding rootstock effects on quality (Davis et al., 2008a). This may partly explain contradictory reports on rootstock-mediated effects on quality and widespread rootstock–scion interaction. The significant effect of vigorous commercial rootstocks, especially of interspecific hybrids, on the yield characteristics of grafted plants indicates that grafting may mediate source–sink relations in the course of ripening. Recent work has demonstrated that grafting watermelon on vigorous rootstocks can increase crop load and retard ripening events responsible for physicochemical changes in fruit composition (Soteriou et al., 2014; Kyriacou et al., 2016). In this case, the apparent effect of grafting on key quality traits, such as the concentration of non-structural carbohydrates and the SSC, was found insignificant and differences between grafted and non-grafted treatments were sourced to the interaction of grafting with maturity due to asynchronous ripening. The synthesis of key pigments responsible for fruit color development, such as lycopene, is also highly dependent on the stage of maturity. Monitoring pigment levels and colourimetric values during the course of fruit ripening has revealed significant grafting × maturity interaction which, in the absence of standardized sampling, might be taken as mere grafting effect (Soteriou et al., 2014). Further complications in interpreting grafting effects might be compiled by recurrent harvests from the same plants and from non-discriminate data analysis on fruits sampled from different orders of fruit clusters.

BIOLOGICAL MECHANISMS AFFECTING QUALITY IN GRAFTED ANNUAL FRUIT CROPS

The interactions between rootstock and scion are highly complex, but increasing investigations in this field have recently shed considerable light on the biological mechanisms involved (Goldschmidt, 2014; Wang et al., 2016). It is widely accepted that metabolic substances could be transferred from one grafting partner to the other, including signaling molecules that may cause large biological effects. Hormonal signaling is implicated in graft union formation, rootstock–scion communication, growth, yield, and potentially flowering and fruit quality (Aloni et al., 2010). Specific studies have documented that grafting also enables long-distance movement of RNA through the phloem (Lucas et al., 2001), the functional importance of which, however, needs to be determined individually. For example, long-distance movement of mutant mRNA from the rootstock to the wild-type tomato scions caused obvious change in leaf morphology, suggesting that translocated RNAs were functional (Kim et al., 2001). Many other phloem-mobile mRNAs have been identified (Harada, 2010), and recent work with grafted grapevines suggests that genomic-scale mRNA exchange across graft junctions is widespread in grafted fruit and vegetable species (Yang et al., 2015). From the different mRNA patterns, it might be concluded that the profile of mobile mRNAs has specific genotype- and environment-dependent characteristics able to modulate plant performance (Yang et al., 2015). But what determines that an mRNA is selected for long-distance movement? Current knowledge is increasing regarding RNA motifs that trigger mobility, the extent of mRNA transport, and the potential for post-transport translation of mRNAs into functional proteins. Long-distance transport of gibberellic acid insensitive-RNA via the phloem altered leaf morphology and raised the question whether RNA delivery may be regulated by sequence motifs conserved between plant families (Haywood et al., 2005). Further studies indicated that coding sequences, 3′ untranslated regions, and also the structure of the RNA might be factors to target for RNA long-distance movement (Huang and Yu, 2009). A recent study exploring the motifs triggering mobility of mRNA demonstrated that tRNA-derived sequences with specific structures are sufficient to mediate mRNA transport and seem necessary for the mobility of a large number of endogenous transcripts that can move through graft junctions (Zhang W.N. et al., 2016). However, it must be considered that the great number of mobile mRNAs identified by combining interspecific grafting with high throughput RNA sequencing, indicate that a postulated tissue-specific gene expression profile might not be predictive for the actual plant body part in which a transcript exerts its function (Thieme et al., 2015).

Furthermore, it also has to be taken into account that grafting itself induces differential gene expression. For example, transcriptomic analysis of grapevine scions demonstrated extensive transcriptional re-programming after heterografting onto two different genotypes (Cookson and Ollat, 2013). While the choice of rootstock genotype had little effect on gene expression in the shoot apex, it was concluded that homeografting and heterografting was the major factor regulating gene expression (Cookson and Ollat, 2013). Besides, heterografting with non-self rootstocks induced genes involved in stress responses at the graft interface when compared with homeografted controls (Cookson et al., 2014). Genome-wide investigation using high-throughput sequencing and comparative analysis of grafting-responsive mRNA in watermelon grafted onto bottle gourd and squash rootstocks identified genes associated with primary and secondary metabolism, hormone signaling, transcription factors, transporters, and response to stimuli, which provide an excellent resource to further elucidate the molecular mechanisms underlying grafting-induced physiological processes (Liu et al., 2016). In addition to protein-encoding mRNAs, various non-coding small RNAs have been shown to move long distances via phloem sap in grafts. Some specifically accumulate in response to nutrient deprivation (Buhtz et al., 2010) with potential signaling role in long distance regulation of gene expression (Pant et al., 2008). Furthermore, it was reported that transgene derived small RNAs from endogenous inverted repeat loci are mobile through the graft union with direct epigenetic modification in recipient cells (Molnar et al., 2010). However, it also has to be considered that grafting itself induces differential expression of microRNAs, as aptly demonstrated by high-throughput sequencing in watermelon grafted onto different rootstocks (Liu et al., 2013). This leads to the suggestion that microRNAs playing an important role in diverse biological and metabolic processes might regulate plant development and adaptation to stress by grafting-induced alterations (Liu et al., 2013).

Despite the mobility of RNA, the transport of various macromolecules through the phloem has received increasing interest following the discovery that FLOWERING LOCUS T protein moves from leaves to the shoot apical meristem where it induces flowering (Corbesier et al., 2007). Paultre et al. (2016) further addressed movement of proteins through the phloem and showed that many proteins in companion cells can get swept away by the translocation stream without resembling a specific protein signal (Paultre et al., 2016). These data reveal that proteins are lost constitutively to the translocation stream, making the identification of unique systemic phloem signals a difficult challenge for the future. However, movement of proteins across graft unions is not restricted to the phloem path as it was demonstrated in transgrafting pathogen resistant, genetically engineered rootstocks with wild type scions. Rootstocks expressing transgenic polygalacturonase inhibiting protein (PGIP) as components of the defense against invasion with pathogens, onto which non-expressing scions were grafted, do not export the respective encoding nucleic acid rather than the PGIP protein itself via the xylem system (Aguero et al., 2005). Furthermore, the PGIP protein in the wild-type scion tissue grafted onto PGIP-expressing genetically engineered rootstocks reduced pathogen damage in scion tissues (Haroldsen et al., 2012). Thus, defense factors in roots can be made available to scions via grafting, improving the vigor, quality, and pathogen resistance of the food-producing scion and its crop (Guan et al., 2012).

It has long been questioned whether grafting might stimulate heritable changes in the scion. Studies have documented that grafting enables the exchanges of DNA molecules between the grafting partners, thus providing a molecular basis for grafting-induced genetic variation (Stegemann and Bock, 2009). By grafting sexually incompatible species, it was further shown that complete chloroplast genomes can travel across the graft junction from one species into another (Stegemann et al., 2012). Additionally, it has been demonstrated that upon grafting entire nuclear genomes can be transferred between plant cells (Fuentes et al., 2014). Although these alterations are localized to the contact zone between scion and rootstock it indicates that the changes may become heritable via lateral shoot formation from the graft site. Hence, it demonstrates that large DNA pieces or entire plastid genomes can travel into the scion as a prerequisite of graft hybridization (Liu et al., 2010). Heritable changes in the scion might also be the result of epigenetic effects associated with grafting. Wu and co-workers demonstrated that in solanaceous plants heterografting causes extensive alteration of DNA methylation patterns in a locus-specific manner, especially in the scions (Wu et al., 2013). They further detected that altered methylation patterns could be inherited to sexual progenies with some sites showing additional alterations or revisions. Such putatively heritable changes in the DNA methylation pattern of solanaceous scion genomes were extended to the Cucurbitaceae. Using methylation-sensitive amplified polymorphism markers, global DNA methylation changes in scions of cucumber, melon and watermelon heterografted onto pumpkin rootstocks were observed (Avramidou et al., 2015). The differential epigenetic marking in different rootstock–scion combinations will enable the development of epi-molecular markers for generation and selection of superior quality grafted vegetables in the future (Avramidou et al., 2015).

CONCLUDING REMARKS AND THE CHALLENGES AHEAD

Regarded primarily as a phytoprotective measure and as a means to alleviate abiotic plant stress, the grafting of annual fruit crops carries significant, crop-specific implications for fruit quality and nutritive value. The positive effects of vigorous interspecific rootstocks on scion performance are often reflected on fruit size, particularly in crops such as watermelon, cucumber, and tomato, whereas fruit shape constitutes a trait predominantly governed by the scion genotype. Similarly, grafting effect on exocarp and mesocarp thickness is limited and inferior to that of the scion genotype, moreover it interacts with fruit maturity. Variation in the epidermal and pulp colouration of annual fruits, determined by changes in pigment concentrations, can be influenced by grafting directly and indirectly through its interaction with fruit ripening behavior; such an interaction is common for watermelon while colouration effects on tomato, melon and pepper appear strongly rootstock-specific.

Fruit texture can be highly affected by grafting as manifested most consistently in the case of watermelon grafted on interspecific cucurbit rootstocks which generally increase pulp firmness; whereas loss of firmness in melon can reflect latent rootstock–scion incompatibility. Arguably the most important sensorial attribute is fruit sweetness, elicited by soluble carbohydrates whose concentration is liable to the effects of grafting. Rootstock-mediated changes in sweetness may also encompass changes in melon starch content and in the relative proportions of hexoses to sucrose. Decrease in sugars is not an infrequent response to grafting, but the increments of reported decrease are in general not highly critical for overall quality and marketability. Nevertheless, additional work is warranted across fruit crops to elucidate widespread rootstock–scion interactions regarding sugar content. While advances have been made with regards to grafting effects on fruit aroma profile and the levels of secondary bioactive phytochemicals, these areas remain largely uncharted, underscored by conflicting reports and warranting further research before grafting may constitute a reliable tool for improving fruit sensorial and nutritional quality.

Disparate results on critical quality attributes such as sugar content and aroma profile often reflect a wider effect of grafting on flowering behavior and post-anthesis ripening events partly mediated by changes in crop load.

Further complications can be compounded by sampling practices such as recurrent harvests from the same plants and non-standardization of harvest maturity. From a physiological standpoint, the grafting process in itself may modulate plant responses to the growth environment, but these effects of homeografting appear concerted mainly in the early vegetative stages following graft union formation; unlike heterografting whose effects may pervade the reproductive stage configuring fruit quality characteristics. Hormonal signaling, however, is implicated in graft union formation, rootstock–scion communication, growth, yield, and potentially flowering and fruit quality. Moreover, the long-distance phloem transport of genomic-scale mRNA across graft unions is widespread in grafted fruit and vegetable species. Yet, additional knowledge is required on RNA motifs that trigger mobility, the extent of mRNA transport, and the potential for its post-transport translation into functional and tissue-specific proteins. The identification of systemic phloem signals, including noncoding microRNAs and proteins with diverse roles in post-grafting biological and metabolic processes, will prove valuable in understanding grafting effects on fruit quality. Ultimately, the identification of inheritable locus-specific alterations in scion DNA methylation patterns may enable the development of epi-molecular markers for generation and selection of superior quality grafted vegetables in the future.

FUNDING

The work of DS was supported by the German Federal Ministry of Food, Agriculture and Consumer Protection, the Brandenburg State Ministry of Infrastructure and Agriculture and the Thuringia State Ministry for Agriculture, Nature and Environment Protection.

REFERENCES

1. Agele, S., and Cohen, S. (2009). Effect of genotype and graft type on the hydraulic characteristics and water relations of grafted melon. *J. Plant Interact.* 4, 59–66. doi: 10.1080/17429140802311671
2. Aguero, C. B., Uratsu, S. L., Greve, C., Powell, A. L. T., Labavitch, J. M., Meredith, C. P., et al. (2005). Evaluation of tolerance to Pierce's disease and *Botrytis* in transgenic plants of *Vitis vinifera* L. expressing the pear PGIP gene. *Mol. Plant Pathol.* 6, 43–51. doi: 10.1111/j.1364-3703.2004.00262.x
3. Akashi, K., Miyake, C., and Yokota, A. (2001). Citrulline, a novel compatible solute in drought-tolerant wild watermelon leaves, is an efficient hydroxyl radical scavenger. *FEBS Lett.* 508, 438–442. doi: 10.1016/s0014-5793(01)03123-4
4. Alan, O., Ozdemir, N., and Gunen, Y. (2007). Effect of grafting on watermelon plant growth, yield and quality. *J. Agron.* 6, 362–365. doi: 10.3923/ja.2007.362.365

5. Albacete, A., Martínez-Andújar, C., Ghanem, M. E., Acosta, M., Sánchez-Bravo, J., Asins, M. J., et al. (2009). Rootstock-mediated changes in xylem ionic and hormonal status are correlated with delayed leaf senescence, and increased leaf area and crop productivity in salinized tomato. *Plant Cell Environ.* 32, 928–938. doi: 10.1111/j.1365-3040.2009.01973.x
6. Alexopoulos, A. A., Kondylis, A., and Passam, H. C. (2007). Fruit yield and quality of watermelon in relation to grafting. *J. Food Agric. Environ.* 5, 178–179.
7. Allwood, J. W., Cheung, W., Xu, Y., Mumm, R., De Vos, R. C. H., Deborde, C., et al. (2014). Metabolomics in melon: a new opportunity for aroma analysis. *Phytochemistry* 99, 61–72. doi: 10.1016/j.phytochem.2013.12.010
8. Aloni, B., Cohen, R., Karni, L., Aktas, H., and Edelstein, M. (2010). Hormonal signaling in rootstock-scion interactions. *Sci. Hortic.* 127, 119–126. doi: 10.1016/j.scienta.2010.09.003
9. Aloni, B., Karni, L., Deventurero, G., Cohen, R., Katzir, N., Edelstein, M., et al. (2011). The use of plant grafting and plant growth regulators for enhancing abiotic stress tolerance in vegetable transplants. *Acta Hortic.* 898, 255–264. doi: 10.17660/actahortic.2011.898.31
10. Arrigoni, O. (1994). Ascorbate system in plant development. *J. Bioenerg. Biomemb.* 26, 407–419. doi: 10.1007/bf00762782
11. Arvanitoyannis, I. S., Khah, E. M., Christakou, E. C., and Bletsos, F. A. (2005). Effect of grafting and modified atmosphere packaging on eggplant quality parameters during storage. *Intern. J. Food Sci. Techn.* 40, 311–322. doi: 10.1111/j.1365-2621.2004.00919.x
12. Avramidou, E., Kapazoglou, A., Aravanopoulos, F. A., Xanthopoulou, A., Ganopoulos, I., Tsaballa, A., et al. (2015). Global DNA methylation changes in Cucurbitaceae inter-species grafting. *Crop Breed. Appl. Biotechnol.* 15, 112–116. doi: 10.1590/1984-70332015v15n2n20
13. Awad, M. A., de Jager, A., van der Plas, L. H. W., and van der Krol, A. R. (2001). Flavonoid and chlorogenic acid changes in skin 'Elstar' and 'Jonagold' apples during development and ripening. *Sci. Hortic.* 90, 69–83. doi: 10.1016/s0304-4238(00)00255-7
14. Bai, Y., and Lindhout, P. (2007). Domestication and breeding of tomatoes: what have we gained and what can we gain in the future? *Ann. Bot.* 100, 1085–1094. doi: 10.1093/aob/mcm150
15. Barrett, C. E., Zhao, X., Sims, C. A., Brecht, J. K., Dreyer, E. Q., and Gao, Z. F. (2012). Fruit Composition and sensory attributes of organic Heirloom tomatoes as affected by grafting. *HortTechnology* 22, 804–809.
16. Beaulieu, J. C., and Lea, J. M. (2006). Characterization and semiquantitative analysis of volatiles in seedless watermelon varieties using solid-phase microextraction. *J. Agric. Food Chem.*54, 7789–7793. doi: 10.1021/jf060663l
17. Beaulieu, J. C., and Lea, J. M. (2007). Quality changes in cantaloupe during growth, maturation, and in stored fresh-cut cubes prepared from fruit harvested at various maturities. *J. Am. Soc. Hortic. Sci.* 132, 720–728.
18. Borgognone, D., Colla, G., Rouphael, Y., Cardarelli, M., Rea, E., and Schwarz, D. (2013). Effect of nitrogen form and nutrient solution pH on growth and mineral composition of self-grafted and grafted tomatoes. *Sci. Hortic.* 149, 61–69. doi: 10.1016/j.scienta.2012.02.012
19. Brajović, B., Kastelec, D., Šircelj, H., and Maršić, N. K. (2012). The effect of scion/rootstock combination and ripening stage on the composition of carotenoids and some carpometric characteristics of tomato fruit. *Eur. J. Hortic. Sci.* 77, 261–271. doi: 10.1093/jxb/erf059
20. Bramley, P. M. (2002). Regulation of carotenoids formation during tomato fruit ripening and development. *J. Exper. Bot.* 53, 2107–2113. doi: 10.1093/jxb/erf059
21. Brown, A. C., and Summers, W. L. (1985). Carbohydrate accumulation and color development in watermelon. *J. Am. Soc. Hortic. Sci.* 110, 683–687. doi: 10.2174/1874256400903010015

22. Bruton, B. D., Fish, W. W., Roberts, W., and Popham, T. W. (2009). The influence of rootstock selection on fruit quality attributes of watermelon. *Open Food Sci. J.* 3, 15–34. doi: 10.2174/1874256400903010015
23. Buhtz, A., Pieritz, J., Springer, F., and Kehr, J. (2010). Phloem small RNAs, nutrient stress responses, and systemic mobility. *BMC Plant Biol.* 10:64. doi: 10.1186/1471-2229-10-64
24. Çandır, E., Yetişir, H., Karaca, F., and Üstün, D. (2013). Phytochemical characteristics of grafted watermelon on different bottle gourds (*Lagenaria siceraria*) collected from the Mediterranean region of Turkey. *Turkish J. Agric. For.* 37, 443–456. doi: 10.3906/tar-1207-21
25. Cao, G. H., Sofic, E., and Prior, R. L. (1996). Antioxidant capacity of tea and common vegetables. *J. Agric. Food Chem.* 44, 3426–3431. doi: 10.1021/jf9602535
26. Cassaniti, C., Giuffrida, F., Scuderi, D., and Leonardi, C. (2011). Effect of rootstock and nutrient solution concentration on eggplant grown in a soilless system. *J. Food Agric. Environ.* 9, 252–256.
27. Causse, M., Saliba Colombani, V., Lecomte, L., Duffé, P., Rousselle, P., and Buret, M. (2002). QTL analysis of fruit quality in fresh market tomato: a few chromosome regions control the variation of sensory and instrumental traits. *J. Exp. Bot.* 53, 2089–2098. doi: 10.1093/jxb/erf058
28. Chávez-Mendoza, C., Sánchez, E., Carvajal-Millán, E., Munõz-Márquez, E., and Guevara-Aguillar, A. (2013). Characterization of the nutraceutical quality and antioxidant activity in bell pepper in response to grafting. *Molecules* 18, 15689–15703. doi: 10.3390/molecules181215689
29. Chisholm, D. N., and Picha, D. H. (1986). Effect of storage temperature on sugar and organic acid contents of watermelon. *HortScience* 21, 1031–1033.
30. Chuan-qiang, X., Yu-xue, H., and Lin, Y. (2011). Effects of grafting on four characteristic ester aromas contents and related enzyme activities in muskmelon fruit. *North. Hortic.* 16, 41–44.
31. Colla, G., Rouphael, Y., Cardarelli, M., and Rea, E. (2006a). Effect of salinity on yield, fruit quality, leaf gas exchange, and mineral composition of grafted watermelon plants. *HortScience*41, 622–627.
32. Colla, G., Rouphael, Y., Cardarelli, M., Massa, D., Salerno, A., and Rea, E. (2006b). Yield, fruit quality and mineral composition of grafted melon plants grown under saline conditions. *J. Hortic. Sci. Biotechnol.* 81, 146–152. doi: 10.1080/14620316.2006.11512041
33. Colla, G., Rouphael, Y., Cardarelli, M., Temperini, O., Rea, E., Salerno, A., et al. (2008). Influence of grafting on yield and fruit quality of pepper (*Capsicum annuum* L.) grown under greenhouse conditions. *Acta Hortic.* 782, 359–363. doi: 10.17660/actahortic.2008.782.45
34. Colla, G., Rouphael, Y., Jawad, R., Kumar, P., Rea, E., and Cardarelli, M. (2013). The effectiveness of grafting to improve NaCl and CaCl2 tolerance in cucumber. *Sci. Hortic.* 164, 380–391. doi: 10.1016/j.scienta.2013.09.023
35. Colla, G., Rouphael, Y., Leonardi, C., and Bie, Z. (2010a). Role of grafting in vegetable crops grown under saline conditions. *Sci. Hortic.* 127, 147–155. doi: 10.1016/j.scienta.2010.08.004
36. Colla, G., Suãrez, C. M. C., Cardarelli, M., and Rouphael, Y. (2010b). Improving nitrogen use efficiency in melon by grafting. *HortScience* 45, 559–565.
37. Colla, G., Rouphael, Y., Mirabelli, C., and Cardarelli, M. (2011). Nitrogen-use efficiency traits of mini-watermelon in response to grafting and nitrogen-fertilization doses. *J. Plant Nutr. Soil Sci.* 174, 933–941. doi: 10.1002/jpln.201000325
38. Colla, G., Rouphael, Y., Rea, E., and Cardarelli, M. (2012). Grafting cucumber plants enhance tolerance to sodium chloride and sulfate salinization. *Sci. Hortic.* 135, 177–185. doi: 10.1016/j.scienta.2011.11.023
39. Commission Implementing Regulation (EU) No 543/2011 (2011). Commission Implementing Regulation (EU) No 543/2011 of 7 June 2011 laying down detailed rules for the application of Council Regulation (EC) No 1234/2007 in respect of the

fruit and vegetables and processed fruit and vegetables sectors. *Off. J. Eur. Union* L 157, 1–163.

40. Condurso, C., Verzera, A., Dima, G., Tripodi, G., Crinò, P., Paratore, A., et al. (2012). Effects of different rootstocks on aroma volatile compounds and carotenoid content of melon fruits. *Sci. Hortic.* 148, 9–16. doi: 10.1016/j.scienta.2012.09.015
41. Cookson, S. J., Moreno, M. J. C., Hevin, C., Mendome, L. Z. N., Delrot, S., Magnin, N., et al. (2014). Heterografting with nonself rootstocks induces genes involved in stress responses at the graft interface when compared with autografted controls. *J. Exp. Bot.* 65, 2473–2481. doi: 10.1093/jxb/eru145
42. Cookson, S. J., and Ollat, N. (2013). Grafting with rootstocks induces extensive transcriptional re-programming in the shoot apical meristem of grapevine. *BMC Plant Biol.* 13:147. doi: 10.1186/1471-2229-13-147
43. Corbesier, L., Vincent, C., Jang, S. H., Fornara, F., Fan, Q. Z., Searle, I., et al. (2007). FT protein movement contributes to long-distance signaling in floral induction of *Arabidopsis*. *Science* 316, 1030–1033. doi: 10.1126/science.1141752
44. Corey, K. A., and Schlimme, D. V. (1988). Relationship of rind gloss and grounds pot colour to flesh quality of watermelon fruits during maturation. *Sci. Hortic.* 34, 211–218. doi: 10.1016/0304-4238(88)90094-5
45. Crinò, P., Lo Bianco, C., Rouphael, Y., Colla, G., Saccardo, F., and Paratore, A. (2007). Evaluation of rootstock resistance to *Fusarium* wilt and gummy stem blight and effect on yield and quality of a grafted 'Inodorus' melon. *HortScience* 42, 521–525.
46. Curis, E., Nicolis, I., Moinard, C., Osowska, S., Zerrouk, N., Bénazeth, S., et al. (2005). Almost all about citrulline in mammals. *Amino Acids* 29, 177–205. doi: 10.1007/s00726-005-0235-4
47. Cushman, K. E., and Huan, J. (2008). Performance of four triploid watermelon cultivars grafted onto five rootstock genotypes: yield and fruit quality under commercial growing conditions. *Acta Hortic.* 782, 335–337. doi: 10.17660/actahortic.2008.782.42
48. Davis, A. R., and Perkins-Veazie, P. (2005). Rootstock effects on plant vigor and watermelon fruit quality. *Cucurbit Gen. Coop. Rep.* 2, 39–42.
49. Davis, A. R., Perkins-Veazie, P., Hassell, R., Levi, A., King, S. R., and Zhang, X. (2008a). Grafting effects on vegetable quality. *HortScience* 43, 1670–1672.
50. Davis, A. R., Perkins-Veazie, P., Sakata, Y., Lopez-Galarza, S., Maroto, J. V., Lee, S. G., et al. (2008b). Cucurbit grafting. *Crit. Rev. Plant Sci.* 27, 50–74. doi: 10.1080/07352680802053940
51. Davis, A. R., Webber, C. L., Perkins-Veazie, P., Ruso, V., Lopez Galarza, S., and Sakata, Y. (2008c). "A review of production systems on watermelon quality," in *Proceedings of the IXth EUCARPIA Meeting on Genetics and Breeding of Cucurbitaceae*, ed. M. Pitrat (Avignon: INRA).
52. Davis, A. R., Webber, C. L., and Fish, W. W. (2011). L-Citrulline levels in watermelon cultigens tested in two environments. *HortScience* 46, 1572–1575.
53. Di Gioia, F., Serio, F., Buttaro, D., Ayala, O., and Santamaria, P. (2010). Vegetative growth, yield, and fruit quality of 'Cuore di Bue', an heirloom tomato, as influenced by rootstock. *J. Hortic. Sci. Biotechnol.* 85, 477–482. doi: 10.1080/14620316.2010.11512701
54. Dima, G., Tripodi, G., Condurso, C., and Verzera, A. (2014). Volatile constituents of mini-watermelon fruits. *J. Essent. Oil Res.* 26, 323–327. doi: 10.1080/10412905.2014.933449
55. Djidonou, D., Simonne, A. H., Koch, K. E., Brecht, J. K., and Zhao, X. (2016). Nutritional quality of field-grown tomato fruit as affected by grafting with interspecific hybrid rootstocks. *HortScience* 51, 1618–1624. doi: 10.21273/hortsci11275-16
56. Doñas-Uclés, F., Jiménez-Luna, M. M., Gongora-Corral, J. A., Pérez-Madrid, D., Verde-Fernandez, D., and Camacho-Ferre, F. (2014). Influence of three rootstocks on yield and commercial quality of 'Italian sweet' pepper. *Ciênc. Agrotecnol.* 38, 538–545. doi: 10.1590/s1413-70542014000600002

57. Dorais, M. (2007). Effect of cultural management on tomato fruit health qualities. *Acta Hortic.* 744, 279–293. doi: 10.17660/actahortic.2007.744.29
58. Dos-Santos, N., Bueso, M. C., and Fernández-Trujillo, J. P. (2013). Aroma volatiles as biomarkers of textural differences at harvest in non-climacteric near-isogenic lines of melon. *Food Res. Intl.* 54, 1801–1812. doi: 10.1016/j.foodres.2013.09.031
59. Edelstein, M., Plaut, Z., and Ben-Hur, M. (2011). Sodium and chloride exclusion and retention by non-grafted and grafted melon and *Cucurbita* plants. *J. Exper. Bot.* 62, 177–184. doi: 10.1093/jxb/erq255
60. Elmstrom, G. W., and Davis, P. L. (1981). Sugars in developing and mature fruits of several watermelon cultivars. *J. Amer. Soc. Hort. Sci.* 106, 330–333.
61. Fan, M., Bie, Z., Krumbein, A., and Schwarz, D. (2011). Salinity stress in tomatoes can be alleviated by grafting and potassium depending on the rootstock and K-concentration employed. *Sci. Hortic.* 130, 615–623. doi: 10.1016/j.scienta.2011.08.018
62. FAO (2012). *Food and Agricultural Organization.* Available at: www.fao.org [accessed December 12, 2016].
63. Fernández-García, N., Martínez, V., and Carvajal, M. (2004a). Effect of salinity on growth, mineral composition, and water relations of grafted tomato plants. *J. Plant Nutr. Soil Sci.* 167, 616–622. doi: 10.1002/jpln.200420416
64. Fernández-García, N., Martínez, V., Cerda, A., and Carvajal, M. (2004b). Fruit quality of grafted tomato plants grown under saline conditions. *J. Hortic. Sci. Biotechnol.* 79, 995–1001. doi: 10.1080/14620316.2004.11511880
65. Fish, W. W., and Davis, A. R. (2003). The effects of frozen storage conditions on lycopene stability in watermelon tissue. *J. Agric. Food Chem.* 51, 3582–3585. doi: 10.1021/jf030022f
66. Fita, A., Pico, B., Roig, C., and Nuez, F. (2007). Performance of *Cucumis melo* ssp. agrestis as a rootstock for melon. *J. Hortic. Sci. Biotechnol.* 82, 184–190. doi: 10.1080/14620316.2007.11512218
67. Flores, F. B., Sanchez-Bel, P., Estan, M. T., Martinez-Rodriguez, M. M., Moyano, E., Morales, B., et al. (2010). The effectiveness of grafting to improve tomato fruit quality. *Sci. Hortic.* 125, 211–217. doi: 10.1016/j.scienta.2010.03.026
68. Fredes, A., Roselló, S., Beltrán, J., Cebolla-Cornejo, J., Pérez-de-Castro, A., Gisbert, C., et al. (2017). Fruit quality assessment of watermelons grafted onto citron melon rootstock. *J. Sci. Food Agric.* 97, 1646–1655. doi: 10.1002/jsfa.7915
69. Fuentes, I., Stegemann, S., Golczyk, H., Karcher, D., and Bock, R. (2014). Horizontal genome transfer as an asexual path to the formation of new species. *Nature* 511, 232–235. doi: 10.1038/nature13291
70. Gajc-Wolska, J., Kowalczyk, K., Marcinkowska, M., Radzanowska, J., and Bujalski, D. (2015). Influence of growth conditions and grafting on the yield, chemical composition and sensory quality of tomato fruit in greenhouse cultivation. *J. Elementol.* 20, 73–81. doi: 10.5601/jelem.2014.19.4.565
71. Gajc-Wolska, J., Łyszkowska, M., and Zielony, T. (2010). The influence of grafting and biostimulators on the yield and fruit quality of greenhouse tomato cv. *(Lycopersicon esculentum Mill.) grown in the field. Veg. Crops Res. Bull.* 72, 63–70. doi: 10.2478/v10032.010-0006-y
72. Gisbert, C., Prohens, J., and Nuez, F. (2011a). Performance of eggplant grafted onto cultivated and wild materials and hybrids of eggplant and tomato. *Intern. J. Plant Prod.* 5, 367–380.
73. Gisbert, C., Prohens, J., Raigón, M. D., Stommel, J., and Nuez, F. (2011b). Eggplant relatives as sources of variation for developing new rootstocks: effects of grafting on yield and fruit apparent quality and composition. *Sci. Hortic* 128, 14–22. doi: 10.1016/j.scienta.2010.12.007
74. Gisbert, C., Sánchez-Torres, P., Dolores Raigón, M., and Nuez, F. (2010). *Phytophthora capsici* resistance evaluation in pepper hybrids: agronomic performance and fruit quality of pepper grafted plants. *J. Food Agric. Environ.* 8, 116–121.

75. Goldschmidt, E. E. (2014). Plant grafting: new mechanisms, evolutionary implications. *Front. Plant Sci.* 5:725. doi: 10.3389/fpls.2014.00727
76. Gómez-Romero, M., Segura-Carretero, A., and Fernández-Gutiérrez, A. (2010). Metabolite profiling and quantification of phenolic compounds in methanol extracts of tomato fruit. *Phytochemistry* 71, 1848–1864. doi: 10.1016/j.phytochem.2010.08.002
77. Goreta Ban, S., Zanic, K., Dumiciv, G., Raspudic, E., Vuletin Selak, G., and Ban, D. (2014). Growth and yield of grafted cucumbers in soil infested with root-knot nematodes. *Chilean J. Agric. Res.* 74, 29–34. doi: 10.4067/s0718-5839201 4000100005
78. Guan, W., Zhao, X., Dickson, D. W., Mendes, M. L., and Thies, J. (2014). Root-knot nematode resistance, yield, and fruit quality of specialty melons grafted onto *Cucumis metulifer. HortScience* 49, 1046–1051.
79. Guan, W. J., Zhao, X., Hassell, R., and Thies, J. (2012). Defense mechanisms involved in disease resistance of grafted vegetables. *HortScience* 47, 164–170.
80. Guan, W., Zhao, X., Hubler, J. H., Huber, J. D., and Sims, A. C. (2015). Instrumental and sensory analyses of quality attributes of grafted specialty melons. *J. Sci. Food. Agric.* 95, 2989–2995. doi: 10.1002/jsfa.7050
81. Guler, Z., Karaca, F., and Yetisir, H. (2013). Volatile compounds in the peel and slesh of cucumber (*Cucumis sativus* L.) grafted onto bottle gourd (*Lagenaria siceraria*) rootstocks. *J. Hortic. Sci. Biotechnol.* 88, 123–128. doi: 10.1080/14620316.2013.11512945
82. Harada, T. (2010). Grafting and RNA transport via phloem tissue in horticultural plants. *Sci. Hortic.* 125, 545–550. doi: 10.1016/j.scienta.2010.05.013
83. Harker, F. R., Gunson, F. A., and Jaeger, S. R. (2003). The case for fruit quality: an interpretive review of consumer attitudes, and preferences for apples. *Postharvest Biol. Technol.* 28, 333–347. doi: 10.1016/s0925-5214(02)00215-6
84. Haroldsen, V. M., Szczerba, M. W., Aktas, H., Lopez-Baltazar, J., Odias, M. J., Chi-Ham, C. L., et al. (2012). Mobility of transgenic nucleic acids and proteins within grafted rootstocks for agricultural improvement. *Front. Plant Sci.* 3:39. doi: 10.3389/fpls.2012.00039
85. Haywood, V., Yu, T. S., Huang, N. C., and Lucas, W. J. (2005). Phloem long-distance trafficking of Gibberellic acid-insensitive RNA regulates leaf development. *Plant J.* 42, 49–68. doi: 10.1111/j.1365-313x.2005.02351.x
86. Helyes, L., Lugasi, A., Pogonyi, A., and Pek, Z. (2009). Effect of variety and grafting on lycopene content of tomato (*Lycopersicon lycopersicum* L. Karsten) fruit. *Acta Aliment.* 38, 27–34. doi: 10.17660/actahortic.2006.712.62
87. Ho, L., and White, P. J. (2005). A cellular hypothesis for the induction of blossom-end rot in tomato fruit. *Ann. Bot.* 95, 571–581. doi: 10.1093/aob/mci065
88. Huang, N. C., and Yu, T. S. (2009). The sequences of Arabidopsis GA-INSENSITIVE RNA constitute the motifs that are necessary and sufficient for RNA long-distance trafficking. *Plant J.*59, 921–929. doi: 10.1111/j.1365-313x.2009.03918.x
89. Huang, W., Liao, S., Lv, H., Khaldun, A. B. M., and Wang, Y. (2015). Characterization of the growth and fruit quality of tomato grafted on a woody medicinal plant, *Lycium chinense. Sci. Hortic.* 197, 447–453. doi: 10.1016/j.scienta.2015.10.005
90. Huang, Y., Tang, R., Cao, Q., and Bie, Z. (2009). Improving the fruit yield and quality of cucumber by grafting onto the salt tolerant rootstock under NaCl stress. *Sci. Hortic.* 122, 26–31. doi: 10.1016/j.scienta.2009.04.004
91. Huitrón, M. V., Diaz, M., Dianez, F., and Camacho, F. (2007). The effect of various rootstocks on triploid watermelon yield and qualify. *J. Food Agric. Environ.* 5, 344–348.
92. Huitrón, M. V., Ricárdez, M., Dianez, F., and Camacho, F. (2009). Influence of grafted watermelon plant density on yield and quality in soil infested with melon necrotic spot virus. *HortScience* 44, 1838–1841.

93. Huitrón, M. V., Rodriguez, N., Diaz, M., and Camacho, F. (2008). Effect of different rootstocks on the production and quality of watermelon cv. Reina de Corazones. *Acta Hortic.* 797, 437–442. doi: 10.17660/actahortic.2008.797.63

94. Hwang, J. M., Lee, W. H., and Hwang, G. S. (1992). Yield performance test of cucumber (*Cucumis sativus* L.) grafted with the rootstocks, *Sicyos angulatus* L. grown in the farm plastic house. *Korean J. Hortic. Sci. Technol.* 10, 34–35.

95. Kader, A. A. (1999). Fruit maturity, ripening and quality relationships. *Acta Hortic.* 485, 203–208. doi: 10.17660/actahortic.1999.485.27

96. Kader, A. A. (2008). Flavor quality of fruits and vegetables - Perspective. *J. Sci. Food Agric.* 88, 1863–1868. doi: 10.1002/jsfa.3293

97. Khah, E. M. (2011). Effect of grafting on growth, performance and yield of aubergine (*Solanum melongena* L.) in greenhouse and open-field. *Intern. J. Plant Prod.* 5, 359–366.

98. Khah, E. M., Kakava, E., Mavromatis, A., Chachalis, D., and Goulas, C. (2006). Effect of grafting on growth and yield of tomato (*Lycopersicon esculentum* Mill.) in greenhouse and open-field. *J. Appl. Hortic.* 8, 3–7.

99. Kim, M., Canio, W., Kessler, S., and Sinha, N. (2001). Developmental changes due to long-distance movement of a homeobox fusion transcript in tomato. *Science* 293, 287–289. doi: 10.1126/science.1059805

100. Klee, H. J., and Tieman, D. M. (2013). Genetic challenges of flavor improvement in tomato. *Trends Gen.* 29, 257–262. doi: 10.1016/j.tig.2012.12.003

101. Kourkoutas, D., Elmore, J. S., and Mottram, D. S. (2006). Comparison of the volatile compositions and flavor properties of cantaloupe, Galia and honeydew muskmelons. *Food Chem.* 97, 95–102. doi: 10.1016/j.foodchem.2005.03.026

102. Krumbein, A., and Auerswald, H. (1998). Characterization of aroma volatiles in tomatoes by sensory analyses. *Food* 42, 395–399. doi: 10.1002/(sici)1521-3803(199812)42:06<395::aid-food395<3.3.co;2-z

103. Krumbein, A., and Schwarz, D. (2013). Grafting: A possibility to enhance health-promoting and flavour compounds in tomato fruits of shaded plants? *Sci. Hortic.* 149, 97–107. doi: 10.1016/j.scienta.2012.09.003

104. Kumar, P., Lucini, L., Rouphael, Y., Cardarelli, M., Kalunke, R. M., and Colla, G. (2015). Insight into the role of grafting and arbuscular mycorrhiza on cadmium stress tolerance in tomato. *Front. Plant Sci.* 6:477. doi: 10.3389/fpls.2015.00477

105. Kyriacou, M. C., Polking, G. F., Hannapel, D. J., and Gladon, R. J. (1996). Activity of 5-aminolevulinic acid dehydratase declines during tomato fruit development and ripening. *J. Am. Soc. Hortic. Sci.* 121, 91–95.

106. Kyriacou, M. C., and Soteriou, G. A. (2012). Postharvest change in compositional, visual and textural quality of grafted watermelon cultivars. *Acta Hotic.* 934, 985–992. doi: 10.17660/actahortic.2012.934.131

107. Kyriacou, M. C., and Soteriou, G. A. (2015). Quality and postharvest behavior of watermelon fruit in response to grafting on inter-specific cucurbit rootstocks. *J. Food Qual.* 38, 21–29. doi: 10.1111/jfq.12124

108. Kyriacou, M. C., Soteriou, G. A., Rouphael, Y., Siomos, A. S., and Gerasopoulos, D. (2016). Configuration of watermelon fruit quality in response to rootstock-mediated harvest maturity and postharvest storage. *J. Sci. Food Agric.* 96, 2400–2409. doi: 10.1002/jsfa.7356

109. Lanchun, N., Jifang, C., Xueying, Z., and Bao, D. (2010). Changes in carbohydrates and carbohydrate metabolizing enzymes during watermelon fruit set. *Acta Hortic. ISHS* 871, 313–318. doi: 10.17660/actahortic.2010.871.42

110. Laur, L. M., and Tian, L. (2011). Provitamin A and vitamin C contents in selected California-grown cantaloupe and honeydew melons and imported melons. *J. Food Comp. Anal.* 24, 194–201. doi: 10.1016/j.jfca.2010.07.009

111. Leal-Fernández, C., Godoy-hernández, H., Núñezcolín, C. A., Anaya-López, J. A., Villalobos-Reyes, S., and Castellanos, J. Z. (2013). Morphological response and fruit yield of sweet pepper (*Capsicum annuum* L.) grafted onto different commercial rootstocks. *Biol. Agric. Hortic.* 29, 1–11. doi: 10.1080/01448765.2012.746063

112. Lee, J. M. (1994). Cultivation of grafted vegetables, I. Current status, grafting methods, and benefits. *HortScience* 29, 235–239.
113. Lee, J. M., Bang, H. J., and Ham, H. S. (1999). Quality of cucumber fruit as affected by rootstock. *Acta Hortic.* 483, 117–123. doi: 10.17660/actahortic.1999.483.12
114. Lee, J. M., Kubota, C., Tsao, S. J., Biel, Z., Hoyos Echevaria, P., Morra, L., et al. (2010). Current status of vegetable grafting: diffusion, grafting techniques, automation. *Sci. Hortic.* 127, 93–105. doi: 10.1016/j.scienta.2010.08.003
115. Leonardi, C., and Giuffrida, F. (2006). Variation of plant growth and macronutrient uptake in graftedtomatoes and eggplants on three different rootstocks. *Eur. J. Hortic. Sci.* 71, 97–101.
116. Leskovar, D. I., Bang, H., Crosby, K. M., Maness, N., Franco, J. A., and Perkins-Veazie, P. (2004). Lycopene, carbohydrates, ascorbic acid and yield components of diploid and triploid watermelon cultivars are affected by deficit irrigation. *J. Hortic. Sci. Biotechnol.* 79, 75–81. doi: 10.1080/14620316.2004.11511739
117. Lewinsohn, E., Sitrit, Y., Bar, E., Azulay, Y., Meir, A., Zamir, D., et al. (2005). Carotenoid pigmentation affects the volatile composition of tomato and watermelon fruits, as revealed by comparative genetic analyses. *J. Agric. Food Chem.* 53, 3142–3148. doi: 10.1021/jf047927t
118. Liu, B., Ren, J., Zhang, Y., An, J., Chen, M., Chen, H., et al. (2015). A new grafted rootstock against root-knot nematode for cucumber, melon, and watermelon. *Agron. Sustain. Dev.* 35, 251–259. doi: 10.1007/s13593-014-0234-5
119. Liu, H. Y., Zhu, Z. J., Diao, M., and Guo, Z. P. (2006). Characteristic of the sugar metabolism in leaves and fruits of grafted watermelon during fruit development. *Plant Phys. Commun.* 42, 835–840.
120. Liu, N., Yang, J. H., Fu, X. X., Zhang, L., Tang, K., Guy, K. M., et al. (2016). Genome-wide identification and comparative analysis of grafting-responsive mRNA in watermelon grafted onto bottle gourd and squash rootstocks by high-throughput sequencing. *Mol. Genet. Genomics* 291, 621–633. doi: 10.1007/s00438-015-1132-5
121. Liu, N., Yang, J. H., Guo, S. G., Xu, Y., and Zhang, M. F. (2013). Genome-wide identification and comparative analysis of conserved and novel MicroRNAs in grafted watermelon by high-throughput sequencing. *PLoS ONE* 8:e57359. doi: 10.1371/journal.pone.0057359
122. Liu, Y. S., Wang, Q. L., and Li, B. Y. (2010). New insights into plant graft hybridization. *Heredity* 104, 1–2. doi: 10.1038/hdy.2009.115
123. López-Marín, J., Gálvez, A., del Amor, F. M., Albacete, A., Fernandez, J. A., Egea-Gilabert, C., et al. (2017). Selecting vegetative/generative/dwarfing rootstocks for improving fruit yield and quality in water stressed sweet peppers. *Sci. Hortic.* 214, 9–17. doi: 10.1016/j.scienta.2016.11.012
124. López-Marín, J., González, A., Pérez-Alfocea, F., Egea-Gilabert, C., and Fernandez, J. A. (2013). Grafting is an efficient alternative to shading screens to alleviate thermal stress in greenhouse-grown sweet pepper. *Sci. Hotic.* 149, 39–46. doi: 10.1016/j.scienta.2012.02.034
125. López–Galarza, S., San Batista, A., Perez, D. M., Miquel, A., Baixauli, C., Pascual, B., et al. (2004). Effects of grafting and cytokinin – induced fruit setting on colour and sugar – content traits in glasshouse – grown triploid watermelon. *J. Hortic. Sci. Biotechnol* 79, 971–976. doi: 10.1080/14620316.2004.11511875
126. Louws, F. J., Rivard, C. L., and Kubota, C. (2010). Grafting fruiting vegetables to manage soilborne pathogens, foliar pathogens, arthropods and weeds. *Sci. Hortic.* 127, 127–146. doi: 10.1016/j.scienta.2010.09.023
127. Lucas, W. J., Yoo, B. C., and Kragler, F. (2001). RNA as a long-distance information macromolecule in plants. *Nat. Rev. Mol. Cell Biol.* 2, 849–857. doi: 10.1038/35099096
128. Luria, N., Smith, E., Reingold, V., Bekelman, I., Lapidot, M., Levin, I., et al. (2017). A new Israeli *Tobamovirus* isolate infects tomato plants harboring Tm-22 resistance genes. *PLoS ONE*12:e0170429. doi: 10.1371/journal.pone.0170429

129. Magwaza, L. S., and Opara, U. L. (2015). Analytical methods for determination of sugars and sweetness of horticultural products – A review. *Sci. Hortic.* 184, 179–192. doi: 10.1016/j.scienta.2015.01.001
130. Marsic, N. K., Mikulik-Petkovsek, M., and Stampar, F. (2014). Grafting influences phenolic profile and carpometric traits of fruits of greenhouse-grown eggplant (*Solanum melongena* L.). *J. Agric. Food Chem.* 62, 10504–10514. doi: 10.1021/jf503338m
131. Martínez-Ballesta, M. C., Alcaraz-Lopez, C., Muries, B., and Carvajal, M. (2010). Physiological aspects of rootstock–scion interactions. *Sci. Hortic.* 127, 112–118. doi: 10.1016/j.scienta.2010.08.002
132. Matsuzoe, N., Hiromi, A., Hanada, K., Ali, M., Okubo, H., and Fujideda, K. (1996). Fruit quality of tomato plants grafted on *Solanum* rootstocks. *J. Jpn. Soc. Hortic. Sci.* 65, 73–80. doi: 10.2503/jjshs.65.73
133. Maynard, D. N., Dunlap, A. M., and Sidoti, B. J. (2002). Sweetness in diploid and triploid watermelon fruit. *Cucurbit Genet. Coop. Rep.* 25, 32–35.
134. Miguel, A., Maroto, J. V., San Bautista, A., Baixauli, C., Cebolla, V., Pascual, B., et al. (2004). The grafting of triploid watermelon is an advantageous alternative to oil fumigation. *Sci. Hortic.* 103, 9–17. doi: 10.1016/j.scienta.2004.04.007
135. Miskovic, A., Ilic, O., Bacanovic, J., Vujasinovic, V., and Kukiæ, B. (2016). Effect of eggplant rootstock on yield and quality parameters of grafted tomato. *Acta Sci. Polon. Hortic. Cul.* 15, 149–159.
136. Miskovic, A., Ilin, Z., and Markovic, V. (2009). Effect of different rootstock type on quality and yield of tomato fruits. *Acta Hortic.* 807, 619–624. doi: 10.17660/actahortic.2009.807.92
137. Mohammed, S. T. M., Humidan, M., Boras, M., and Abdalla, O. A. (2009). Effect of grafting tomatoon different rootstocks on growth and productivity under glasshouse conditions. *Asian J. Agric. Res.* 3, 47–54. doi: 10.3923/ajar.2009.47.54
138. Molnar, A., Melnyk, C. W., Bassett, A., Hardcastle, T. J., Dunn, R., and Baulcombe, D. C. (2010). Small silencing RNAs in plants are mobile and direct epigenetic modification in recipient cells. *Science* 328, 872–875. doi: 10.1126/science.1187959
139. Moncada, A., Miceli, A., Vetrano, F., Mineo, V., Planeta, D., and D'Anna, F. (2013). Effect of grafting on yield and quality of eggplant (*Solanum melongena* L.). *Sci. Hortic.* 149, 108–114. doi: 10.1016/j.scienta.2012.06.015
140. Morishita, M. (2001). Origin and characteristics of the local cucumber variety 'Kema' in the Osaka district. *Bull. Osaka Prefectural Agric. For. Res. Center* 37, 27–34.
141. Nicoletto, C., Tosini, F., and Sambo, P. (2013a). Effect of grafting and ripening conditions on some qualitative traits of 'Cuore di bue' tomato fruits. *J. Sci. Food Agric.* 93, 1397–1403. doi: 10.1002/jsfa.5906
142. Nicoletto, C., Tosini, F., and Sambo, P. (2013b). Effect of grafting on biochemical and nutritional traits of 'Cuore di bue' tomatoes harvested at different ripening stages. *Acta Agric. Scand. B* 63, 114–122. doi: 10.1080/09064710.2012.729606
143. Nissinen, R., Paimela, L., Julkunen, H., Tienari, P. J., Leirisalo-Repo, M., Palosuo, T., et al. (2003). Peptidylarginine deiminase, the arginine to citrulline converting enzyme, is frequently recognized by sera of patients with rheumatoid arthritis, systemic lupus erythematosus and primary Sjögren syndrome. *Scand. J. Rheumatol.* 32, 337–342. doi: 10.1080/03009740410004990
144. Ntatsi, G., Savvas, D., Huntenburg, K., Druege, U., and Schwarz, D. (2014). A study on ABA involvement in the response of tomato to suboptimal root temperature using reciprocal grafts with notabilis, a null mutant in the ABA-biosynthesis gene LeNCED1. *Environ. Exp. Bot.* 97, 11–21. doi: 10.1016/j.envexpbot.2013.09.011
145. Obando-Ulloa, M., Moreno, E., Garcia-Mas, J., Nicolai, B., Lammertync, J., Monforte, A. J., et al. (2008). Climacteric or non-climacteric behavior in melon fruit: 1. Aroma volatiles. *Postharvest Biol. Technol.* 49, 27–37. doi: 10.1016/ j.postharvbio. 2007.11.004

146. Oda, M., Nagata, M., Tsuji, K., and Sasaki, H. S. O. (1996). Effects of scarlet eggplant rootstock on growth, yield, and sugar content of grafted tomato fruits. *J. Jpn. Soc. Hortic. Sci.* 65, 531–536. doi: 10.2503/jjshs.65.531
147. Orsini, F., Sanoubar, R., Oztekin, G. B., Kappel, N., Tepecik, M., Quacquarelli, C., et al. (2013). Improved stomatal regulation and ion partitioning boostssalt tolerance in grafted melon. *Funct. Plant Biol.* 40, 628–636. doi: 10.1071/FP12350
148. Özdemir, A., Çandir, E., Yetişir, H., Aras, V., Arslan,Ö., Baltaer, Ö., et al. (2016). Effects of rootstocks on storage and shelf life of grafted watermelons. *J. Appl. Bot. Food Qual.* 9, 191–201. doi: 10.5073/JABFQ.2016.089.024
149. Pant, B. D., Buhtz, A., Kehr, J., and Scheible, W. R. (2008). MicroRNA399 is a long-distance signal for the regulation of plant phosphate homeostasis. *Plant J.* 53, 731–738. doi: 10.1111/j.1365-313x.2007.03363.x
150. Park, D. K., Son, S. H., Kim, S., Lee, W. M., Lee, H. J., Choi, H. S., et al. (2013). Selection of melon genotypes with resistance to *Fusarium* wilt and *Monosporascus* root rot for rootstocks. *Plant Breed. Biotechnol.* 1, 277–282. doi: 10.9787/pbb.2013.1.3.277
151. Passam, H. C., Stylianoy, M., and Kotsiras, A. (2005). Performance of eggplant grafted on tomato and eggplant rootstocks. *Eur. J. Hortic. Sci.* 70, 130–134.
152. Paultre, D. S. G., Gustin, M. P., Molnar, A., and Oparka, K. J. (2016). Lost in transit: long-distance trafficking and phloem unloading of protein signals in Arabidopsis homografts. *Plant Cell* 28, 2016–2025. doi: 10.1105/tpc.16.00249
153. Pech, J. C., Bouzayen, M., and Latché, A. (2008). Climacteric fruit ripening: ethylene-dependent and independent regulation of ripening pathways in melon fruit. *Plant Sci.* 175, 114–120. doi: 10.1016/j.plantsci.2008.01.003
154. Penella, C., Landi, M., Giudi, L., Nebauer, S. G., Pelligrini, E., San Bautista, A., et al. (2016). Salt tolerant rootstock increases yield of pepper under salinity through maintenance of photosynthetic performance and sinks strength. *J. Plant Phsyiol.* 193, 1–11. doi: 10.1016/j.jplph.2016.02.007
155. Perkins-Veazie, P., and Collins, J. K. (2006). Carotenoid changes of intact watermelons after storage. *J. Agric. Food Chem.* 54, 5868–5874. doi: 10.1021/jf0532664
156. Perkins-Veazie, P., Collins, J. K., Pair, S. D., and Roberts, W. (2001). Lycopene content differs among red-fleshed watermelon cultivars. *J. Sci. Food Agric.* 81, 983–987. doi: 10.1002/jsfa.880
157. Perkins-Veazie, P., Zhang, X., Lu, G., and Huan, J. (2007). Grafting increases lycopene in seedless watermelon. *HortScience* 42, 959.
158. Petropoulos, S. A., Olympios, C., Ropokis, A., Vlachou, G., Ntatsi, G., Paraskevopoulos, A., et al. (2014). Fruit volatiles, quality, and yield of watermelon as affected by grafting. *J. Agric. Sci. Technol.* 16, 873–885.
159. Pogonyi,Á., PéK, Z., Helyes, L., and Lugasi, A. (2005). Effect of grafting on the tomato's yield, quality and main fruit components in spring forcing. *Acta Aliment.* 34, 453–462. doi: 10.1556/aalim.34.2005.4.12
160. Poudel, S. R., and Lee, W. S. (2009). Response of eggplant (*Solanum melongena* L.) as rootstock for tomato (*Solanum lycopersicum* L.). *Hortic. NCHU* 34, 39–52.
161. Proietti, S., Rouphael, Y., Colla, G., Cardarelli, M., De Agazio, M., Zacchini, M., et al. (2008). Fruit quality of mini-watermelon as affected by grafting and irrigation regimes. *J. Sci. Food Agric.* 88, 1107–1114. doi: 10.1002/jsfa.3207
162. Qaryouti, M. M., Qawasmi, W., Hamdan, H., and Edwan, M. (2007). Tomato fruit yield and quality as affected by grafting and growing system. *Acta Hortic.* 741, 199–206. doi: 10.17660/actahortic.2007.741.22
163. Rahmatian, A., Delshad, M., and Salehi, R. (2014). Effect of grafting on growth, yield and fruot quality of single and double stemmed tomato plants grown hydroponically. *Hortic. Environ. Biotechnol.* 55, 115–119. doi: 10.1007/s13580-014-0167-6
164. Reid, M. S. (2002). "Maturation and maturity indices," in *Postharvest Technology of Horticultural Crops*, 3rd Edn, ed. A. A. Kader (Oakland, CA: University of California).

165. Riga, P. (2015). Effect of rootstock on growth, fruit production and quality of tomato plants grown under low temperature and light conditions. *Hortic. Environ. Biotechnol.* 56, 626–638. doi: 10.1007/s13580-015-0042-0
166. Riga, P., Benedicto, L., Garcia-Flores, L., Villano, D., Medina, S., and Gil-Izquierdo, A. (2016). Rootstock effect on serotonin and nutritional quality of tomatoes produced under low temperature and light conditions. *J. Food Compost. Anal.* 46, 59–59. doi: 10.1016/j.jfca.2015.11.003
167. Rimando, A., and Perkins-Veazie, P. (2005). Determination of citrulline in watermelon rind. *J. Chromatogr. A* 1078, 196–200. doi: 10.1016/j.chroma .2005.05.009
168. Romano, D., Paratore, A., and Vindigni, G. (2000). "Caratteristiche dei frutti di pomodoro in rapporto a diversi portinnesti," in *Proceedings of the Workshop on 'Applicazione di Tecnologie Innovative per il Miglioramento dell' Orticoltura Meridionale'*, Rome: Consiglio Nazionale delle Ricerche, 113–114.
169. Romero, L., Belakbir, A., Ragala, L., and Ruiz, M. (1997). Response of plant yield and leaf pigments to saline conditions: effectiveness of different rootstocks in melon plants (*Cucumis melo*L.). *Soil Sci. Plant Nutr.* 43, 855–862. doi: 10.1080/00380768.1997.10414652
170. Ronen, G., Carmel, G. L., Zamir, D., and Hirschberg, J. (2000). An alternative pathway to beta-carotene formation in plant chromoplasts discovered by map-based cloning of Beta and old-gold color mutations in tomato. *Proc. Nat. Acad. Sci. U.S.A.* 97, 11102–11107. doi: 10.1073/pnas.190177497
171. Rouphael, Y., Cardarelli, M., and Colla, G. (2008). Yield, mineral composition, water relations, and water use efficiency of grafted mini-watermelon plants under deficit irrigation. *HortScience* 43, 730–736.
172. Rouphael, Y., Cardarelli, M., Rea, E., and Colla, G. (2012). Improving melon and cucumber photosynthetic activity, mineral composition, and growth performance under salinity stress by grafting onto *Cucurbita* hybrid rootstocks. *Photosynthetica* 50, 180–188. doi: 10.1007/s11099-012-0002-1
173. Rouphael, Y., Rea, E., Cardarelli, M., Bitterlich, M., Schwarz, D., and Colla, G. (2016). Can adverse effects of acidity and aluminum toxicity be alleviated by appropriate rootstock selection in cucumber? *Front. Plant Sci.* 7:1283. doi: 10.3389/fpls.2016.01283
174. Rouphael, Y., Schwarz, D., Krumbein, A., and Colla, G. (2010). Impact of grafting on product quality of fruit vegetables. *Sci. Hortic.* 127, 172–179. doi: 10.1016/j.scienta.2010.09.001
175. Ruiz, J. M., and Romero, L. (1999). Nitrogen efficiency and metabolism in grafted melon plants. *Sci. Hortic.* 81, 113–123. doi: 10.1016/s0304-4238(98)00200-3
176. Sabatino, L., Iapichino, G., Maggio, A., D'anna, E., Bruno, M., and D'Anna, F. (2016). Grafting affects yield and phenolic profile of *Solanum melongena* L. landraces. *J. Integr. Agric.* 15, 1017–1024. doi: 10.1016/s2095-3119(15)61323-5
177. Saftner, R., Luo, Y., McEvoy, J., Abbott, J. A., and Vinyard, B. (2007). Quality characteristics of fresh-cut watermelon slices from non-treated and 1-methylcyclopropene- and/or ethylene-treated whole fruit. *Postharvest Biol. Technol.* 44, 71–79. doi: 10.1016/j.postharvbio.2006.11.002
178. Sakata, Y., Ohara, T., and Sugiyama, M. (2007). The history and present state of grafting of cucurbitaceous vegetables in Japan. *Acta Hortic.* 731, 159–170. doi: 10.17660/actahortic.2007.731.22
179. Sánchez-Rodríguez, E., Leyva, R., Constan-Aguilar, C., Romero, L., and Ruiz, J. M. (2012a). Grafting under water stress in tomato cherry: improvingthe fruit yield and quality. *Ann. Appl. Biol.* 161, 302–312. doi: 10.1111/j.1744-7348.2012.00574.x
180. Sánchez-Rodríguez, E., Ruiz, J. M., Ferreres, F., and Moreno, D. A. (2012b). Phenolic profiles of cherry tomatoes as influenced by hydric stress and rootstock technique. *Food Chem.* 134, 775–782. doi: 10.1016/j.foodchem.2012.02.180

181. Sánchez-Torres, P., Raigón, M. D., Gammoudi, N., and Gisbert, C. (2016). Effects of grafting combinations on the nutritional composition of pepper fruit. *Fruits* 71, 249–256. doi: 10.1051/fruits/2016014
182. Satoh, S. (1996). Inhibition of flowering of cucumber grafted on rooted squash rootstocks. *Physiol. Plantarum.* 97, 440–444. doi: 10.1034/j.1399-3054.1996.970304.x
183. Savvas, D., Colla, G., Rouphael, Y., and Schwarz, D. (2010). Amelioration of heavy metal and nutrient stress in fruit vegetables by grafting. *Sci. Hortic.* 127, 156–161. doi: 10.1016/j.scienta.2010.09.011
184. Savvas, D., Öztekin, G. B., Tepecik, M., Ropokis, A., Tüzel, Y., Ntatsi, G., et al. (2017). Impact of grafting and rootstock on nutrient-to-water uptake ratios during the first month after planting of hydroponically grown tomato. *J. Hortic. Sci. Biotechnol.* 92, 294–302. doi: 10.1080/14620316.2016.1265903
185. Savvas, D., Savva, A., Ntatsi, G., Ropokis, A., Karapanos, I., Krumbein, A., et al. (2011). Effects of three commercial rootstocks on mineral nutrition, fruit yield, and quality of salinized tomato. *J. Plant Nutr Soil Sci.* 174, 154–162. doi: 10.1002/jpln.201000099
186. Schofield, A., Vasantha Rupasinghe, H. P., and Gopinadhan, P. (2008). "Isoprenoid Biosynthesis in fruits and vegetables," in *Postharvest Biology and Technology of Fruits, Vegetables and Flowers*, eds G. Paliyath, D. P. Murr, A. K. Handa, and S. Lurie (Ames: Wiley-Blackwell Publishing).
187. Schreiner, M., Korn, M., Stenger, M., Holzgreve, L., and Altmann, M. (2013). Current understanding and use of quality characteristics of horticulture products. *Sci. Hortic.* 163, 63–69. doi: 10.1016/j.scienta.2013.09.027
188. Schultheis, J., Thompson, W., and Hassel, R. (2015). Specialty melon yield and quality response to grafting in trials conducted in the Southeastern United States. *Acta Hortic.* 1086, 269–278. doi: 10.17660/actahortic.2015.1086.34
189. Schwarz, D., Öztekin, G. B., Tüzel, Y., Brückner, B., and Krumbein, A. (2013). Rootstocks canenhance tomato growth and quality characteristics at low potassium supply. *Sci. Hortic.* 149, 70–79. doi: 10.1016/j.scienta.2012.06.013
190. Schwarz, D., Rouphael, Y., Colla, G., and Venema, J. H. (2010). Grafting as a tool to improve tolerance of vegetables to abiotic stresses: thermal stress, water stress and organic pollutants. *Sci. Hortic.* 127, 162–171. doi: 10.1016/j.scienta.2010.09.016
191. Slimestad, R., Fossen, T., and Verhuel, M. J. (2008). The flavonoids of tomatoes. *J. Agric. Food Chem.* 56, 2436–2441. doi: 10.1021/jf073434n
192. Soteriou, G. A., and Kyriacou, M. C. (2014). Rootstock mediated effects on watermelon field performance and fruit quality characteristics. *Intern. J. Vegetable Sci.* 21, 344–362. doi: 10.1080/19315260.2014.881454
193. Soteriou, G. A., and Kyriacou, M. C. (2015). Rootstock-mediated effects on watermelon field performance and fruit quality characteristics. *Int. J. Vegetable Sci.* 21, 344–362. doi: 10.1080/19315260.2014.881454
194. Soteriou, G. A., Kyriacou, M. C., Siomos, A. S., and Gerasopoulos, D. (2014). Evolution of watermelon fruit physicochemical and phytochemical composition during ripening as affected by grafting. *Food Chem.* 165, 282–289. doi: 10.1016/j.foodchem.2014.04.120
195. Soteriou, G. A., Papayiannis, L. C., and Kyriacou, M. C. (2016). Indexing melon physiological decline to fruit quality and vine morphometric parameters. *Sci. Hortic.* 203, 207–215. doi: 10.1016/j.scienta.2016.03.032
196. Stegemann, S., and Bock, R. (2009). Exchange of genetic material between cells in plant tissue grafts. *Science* 324, 649–651. doi: 10.1126/science.1170397
197. Stegemann, S., Keuthe, M., Greiner, S., and Bock, R. (2012). Horizontal transfer of chloroplast genomes between plant species. *Proc. Nat. Acad. Sci. U.S.A* 109, 2434–2438. doi: 10.1073/pnas.1114076109
198. Stepansky, A., Kovalski, I., Schaffer, A. A., and Peri-Treves, R. (1999). Variation in sugar levels and invertase activity in mature fruit representing a broad spectrum

of *Cucumis melo*genotypes. *Genet. Res. Crop Evol.* 46, 53–62. doi: 10.1023 /A:1008636732481

199. Steward, A. J., Bozonnet, S., Mullen, W., Jenkins, G. I., Lean, M. E. J., and Crozier, A. (2000). Occurrence of flavonols in tomatoes and tomato-based products. *J. Agric. Food Chem.* 48, 2663–2669. doi: 10.1021/jf000070p
200. Strang, J., Bale, K., Snyder, J., Carpenter, D., and Smigell, C. (2007). *Specialty Melon Variety Evaluations (Kentucky).* Lexington, KY: University of Kentucky.
201. Tabata, K., Oba, K., Suzuki, K., and Esaka, M. (2001). Generation and properties of ascorbic acid-deficient transgenic tobacco cells expressing antisense RNA for L-galactono-1,4-lactonedehydrogenase. *Plant J.* 27, 139–148. doi: 10.1046/j.1365-313x.2001.01074.x
202. Takasu, E., Wu, D. L., and Imai, H. (1996). Effects of grafting on blossom end rot incidence insummer tomato production. *TVIS Newsletter* 1, 22–23.
203. Tarazona-Díaz, M. P., Viegas, J., Moldao-Martinsc, M., and Aguayoa, E. (2011). Bioactive compounds from flesh and by-product of fresh-cut watermelon cultivars. *J. Sci. Food Agric.* 91, 805–812. doi: 10.1002/jsfa.4250
204. Theodoropoulou, A., Giotis, C., Hunt, J., Gilroy, J., Toufexi, E., Liopa-Tsakalidis, A., et al. (2007). "Effect of variety choice and use of resistant rootstock on crop yield and quality parameters of tomato plants grown in organic, low input and conventional production systems/growth media," in *Proceedings of the 3rd QLIF Congress: Improving Sustainability in Organic and Low Input Food Production System*, eds U. Niggli, C. Leifert, T. Alföldi, L. Lück, and H. Willer (Stuttgart: University of Hohenheim), 177–180.
205. Thieme, C. J., Rojas-Triana, M., Stecyk, E., Schudoma, C., Zhang, W. N., Yang, L., et al. (2015). Endogenous Arabidopsis messenger RNAs transported to distant tissues. *Nat. Plants*1:15025. doi: 10.1038/nplants.2015.25
206. Tieman, D., Bliss, P., McIntyre, L. M., Blandon-Ubeda, A., Bies, D., Odabasi, A. Z., et al. (2012). The Chemical interactionsunderlying tomato flavor preferences. *Curr. Biol.* 22, 1035–1039. doi: 10.1016/j.cub.2012.04.016
207. Tieman, D., Zhu, G., Resende, M. F. Jr, Lin, T., Nguyen, C., Bies, D., et al. (2017). A chemical genetic roadmap to improved tomato flavor. *Science* 355, 391–394. doi: 10.1126/science.aal1556
208. Tomas-Barberan, F. A., and Espin, J. C. (2001). Phenolic compounds and related enzymes as determinants of quality in fruits and vegetables. *J. Sci. Food Agric.* 81, 853–876. doi: 10.1002/jsfa.885
209. Traka-Mavrona, E., Koutsika-Sotiriou, M., and Pritsa, T. (2000). Response of squash (*Cucurbita* spp.) as rootstock for melon (*Cucumis melo* L.). *Sci. Hortic.* 83, 353–362. doi: 10.1016/s0304-4238(99)00088-6
210. Trionfetti-Nisini, P., Colla, G., Granati, E., Temperini, O., Crinò, P., and Saccardo, F. (2002). Rootstock resistance to *Fusarium* wilt and effect on fruit yield and quality of two muskmelon cultivars. *Sci. Hortic.* 93, 281–288. doi: 10.1016/s0304-4238(01)00335-1
211. Turhan, A., Ozmen, N., Serbeci, M. S., and Seniz, V. (2011). Effects of grafting on different rootstocks on tomato fruit yield and quality. *Hortic. Sci.* 38, 142–149. doi: 10.1007/s00122-015-2462-8
212. Verzera, A., Dima, G., Tripodi, G., Condurso, C., Crinò, P., and Romano, D. (2014). Aroma and sensory quality of honeydew melon fruits (*Cucumis melo* L. subsp. *Melo* var. inodorus H. Jacq.) in relation to different rootstocks. *Sci. Hortic.* 169, 118–124. doi: 10.1016/j.scienta.2014.02.008
213. Verzera, A., Dima, G., Tripodi, G., Ziino, M., Lanza, C. M., and Mazzaglia, A. (2011). Fast quantitative determination of aroma volatile constituents in melon fruits by headspace solid-phase microextraction and gas chromatography- mass spectrometry. *Food Anal. Meth.* 4, 141–149. doi: 10.1007/s12161-010-9159
214. Vinkovic-Vrcek, I., Samobor, V., Bojic, M., Medic-Saric, M., Vukobratovic, M., Erhatic, R., et al. (2011). The effect of grafting on the antioxidant properties of tomato

(*Solanum lycopersicum* L.). *Spanish J. Agric. Res.* 9, 844–851. doi: 10.5424/sjar/20110903-414-10

215. Wadano, A., Azeta, M., Itotani, S., Kanda, A., Iwaki, T., Taira, T., et al. (1999). Change of ascorbic acid level after grafting of tomato seedlings. *Z. Naturforsch.* 54, 830–833. doi: 10.1515/znc-1999-9-1032
216. Wang, J., Jiang, L., and Wu, R. (2016). Plant grafting: how genetic exchange promotes vascular reconnection. *New Phytol.* 214, 56–65. doi: 10.1111/nph.14383
217. Wu, R., Wang, X. R., Lin, Y., Ma, Y. Q., Liu, G., Yu, X. M., et al. (2013). Inter-species grafting caused extensive and heritable alterations of DNA methylation in Solanaceae plants. *PLoS ONE* 8:e61995. doi: 10.1371/journal.pone.0061995
218. Xu, C. Q., Li, T. L., and Qui, H. Y. (2006). Effects of grafting on development, carbohydrate content, and sucrose metabolizing enzymes activities of muskmelon fruit. *Acta Hortic.* 33, 773–778.
219. Yajima, I., Sakakibara, H., Ide, J., Yanai, T., and Hayashi, K. (1985). Volatile flavor components of watermelon (*Citrullus vulgaris*). *Agric. Biol. Chem.* 49, 3145–3150. doi: 10.1271/bbb1961.49.3145
220. Yamaguchi, M., Hughes, D. L., Yabumoto, K., and Jennings, W. G. (1977). Quality of cantaloupe muskmelons: variability and attributes. *Sci. Hortic.* 6, 59–70. doi: 10.1016/0304-4238(77)90079-6
221. Yamasaki, A., Yamashita, M., and Furuya, S. (1994). Mineral concentrations and cytokinin activity in the xylem exudate of grafted watermelons as affected by rootstocks and crop load. *J. Jpn. Soc. Hortic. Sci.* 62, 817–826. doi: 10.2503/jjshs.62.817
222. Yang, Y. Z., Mao, L. Y., Jittayasothorn, Y., Kang, Y. M., Jiao, C., Fei, Z. J., et al. (2015). Messenger RNA exchange between scions and rootstocks in grafted grapevines. *BMC Plant Biol.*15:251. doi: 10.1186/s12870-015-0626-y
223. Yativ, M., Hanary, I., and Wolf, S. (2010). Sucrose accumulation in watermelon fruits: genetic variation and biochemical analysis. *J. Plant Physiol.* 167, 589–596. doi: 10.1016/j.jplph.2009.11.009
224. Yetisir, H., and Sari, N. (2003). Effect of different rootstock on plant growth, yield and quality of watermelon. *Austr. J. Exp. Agric.* 43, 1269–1274. doi: 10.1071/ea02095
225. Yetisir, H., Sari, N., and Yncel, S. (2003). Rootstock resistance to Fusarium wilt and effect on watermelon fruit yield and quality. *Phytoparasitica* 31, 163–169. doi: 10.1007/bf02980786
226. Zhang, B., Tieman, D., Jiao, C., Xu, Y., Chen, K., Fei, Z., et al. (2016). Chilling-induced tomato flavor loss is associated with altered volatile synthesis and transient changes in DNA methylation. *Proc. Nat. Acad. Sci. U.S.A.* 113, 12580–12585. doi: 10.1073/pnas.1613910113
227. Zhang, W. N., Thieme, C. J., Kollwig, G., Apelt, F., Yang, L., Winter, N., et al. (2016). tRNA-related sequences trigger systemic mRNA transport in plants. *Plant Cell* 28, 1237–1249. doi: 10.1105/tpc.15.01056
228. Zhao, X., Guo, Y., Huber, D. J., and Lee, J. (2011). Grafting effects on postharvest ripening and quality of 1-methylcyclopropene-treated muskmelon fruit. *Sci. Hortic.* 130, 581–587. doi: 10.1016/j.scienta.2011.08.010

CHAPTER 5

Soil Degradation and Prevention in Greenhouse Production

Y Liang[1,2*], X Lin[2], S Yamada[3], M Inoue[4], K Inosako[3]

[1]Institute of Soil & Water Conservation, Northwest A&F University, Yangling,Shaanxi, China

ABSTRACT

Soil degradation has been a very serious problem for sustainable production, especially by a re-cropping of greenhouse-cultivated cucumber (*Cucumis statirus* L.). The aim of this research was to expound the actuality for soil degradation, at the same time, put forward some suggestion for preventing from soil degradation and maintain sustainable production in greenhouse basic on the two experiments conducted in a solar greenhouse during 2001-2008 suburb area of Yan'an, Shaanxi province in North China. The result shown that cucumber fruit productivity increased as the increasing of re-cropping years, but decreased after 5years continuously cropping. As increasing of re-cropping years, the population of fungus and bacteria increased, which was assumingly main factor of soil degradation. There was significant difference in cropping models on soil bio-characteristics and system productivity. The productivity were the highest in cropping model between cucumber and greengrocery, cucumber and cowpea *(Vigna sinensis L)*, the second higher were in cropping model between cucumber and maize (*Zea mays*) for green manure, cucumber and kidney bean *(Phaseolus vulgaris)*. That was the best way to reduce soil bacteria and epiphyte amount to follow lasting three or four months during summer season after cucumber harvest, the better method was planting cowpea or other leguminous crops. Basic on the experiment, the optimums approaches to preventive soil degradation were put forward.

Solar greenhouse vegetable cultivation has been proven to be a good farming practice in various areas of different countries [1, 2, 3], and has been developed rapidly during recent years because of the comparatively higher economic benefit [4, 5, 6, 7].

Soil degradation is the universal problem in greenhouse soil. It affects the sustainable use of greenhouse soil and also endangers vegetable security.

ACTUALITY OF SOIL DEGRADATION IN GREENHOUSE

Soil salinity

Greenhouse vegetable cultivation often requires a greater degree of management and larger input of nutrients and irrigation. However, over-fertilization and over-irrigation for vegetable cultivation resulted in nutrient accumulation in soil and led to soil salinity [8, 9]. Salinity reduces phosphate uptake and accumulation in crops grown in soils primarily by reducing phosphate availability. Salinity can directly affect nutrient uptake, such as Na^+ reducing K^+uptake or by Cl^- reducing NO_3^- uptake. Salinity can also cause a combination of complex interactions that affect plant metabolism, susceptibility to injury or internal nutrient requirement. Salinity has been described as the 'AIDS of the soil' and its influence is spreading throughout society where crop production has been seriously affected and caused economic hardship.

Continuous cropping

As the development of specialized vegetable production, continuous cropping is inevitable in greenhouses. A lot of vegetable such as tomato (*Lycopersicum esculentum*), pepper (*Capsicum frutescens*), eggplants (*Solanum melongena*), and cucumber have frequently suffered continuous cropping obstacle [10]. At the same time, poorly differentiated plant mass getting into the soil under continuous cropping negatively affects soil organism communities [11]. This negative action leads to the change in metabolism of microbes from the primary to the secondary [12]. Continuous cropping obstacle has became a main problem for sustainable production of vegetable, for example, serious plant diseases and fall of yield, which intimidated facility vegetables production sustainable development, especially in North China [5, 13, 14, 15, 16, 17].

Soil acidification

A number of agricultural practices have expanded the areas of acidification soils. The main causal factor is the growth of plants that use large amounts of basic ions (e.g. legumes); particularly when fertilizers that leaves acidic residues (such as Superphosphate) are used.

Nutrition accumulation

At present, fertilizer application rates in intensive agricultural systems have increased dramatically in recent years, especially in greenhouse vegetable production systems [18]. These high fertilizer inputs and the extremely low crop recoveries of fertilizer nutrients lead to marked deterioration in soil and groundwater quality and the systems are clearly unsustainable [19]. Soil chemical and biological properties in greenhouse vegetable production may change dramatically after several years of continually high inputs of fertilizers and irrigation water and planting of shallow rooting vegetable crops compared with open field cereal production systems. Soil OM, alkalihydrolyzable N, and available P and K have been observed to accumulate significantly in vegetable greenhouses [5]. High concentrations of P and imbalances of N, P and K often occur in the intensively managed soils. High fertilizer application rates may lead to increasing salinity in the surface soils of the plastic film greenhouses in which vegetables are grown [20].

Compaction

Some researcher found that bulk density was higher in greenhouse soils than uncovered soil. In recent years, there has been increasing interests in intensive vegetable cultivation, this caused the soil compacted. Compaction of soils causes a reduction in soil pore space. This reduces the rate at which water can infiltrate and drain through the soil. It also reduces the available space for oxygen in the plant root zones. For this reason, some of the major consequences of compaction are poor drainage, poor aeration, and hard pan surfaces which cause runoff. Repeated cultivation of some soils leads to a breakdown of soil structure and this also increases the likelihood of compaction.

Chemical residue

Although not as large a problem as some of the other types of soil degradation, the presence of chemical residues can be quite a problem on a local scale. These residues derive almost entirely from long term accumulation after repeated use of pesticides, etc., or of use of pesticides or other chemicals with long residual effects. Some problems that result from chemical residues include toxic effects on crop species and contamination of workers, livestock and adjacent streams.

The studies on the soil degradation of greenhouse soil and its affection mechanism on vegetables are highly important for us to maintain high land capacity. For a sustainable management of soil in the greenhouse, it is urgent to improve the fertilizer and irrigation management for vegetable production, especially for the winter season under greenhouse conditions. Excessive nitrogen application can be effectively avoided by site-specific fertilizer application based on soil and plant analysis.

Prevention from soil degradation in greenhouse

Materials and methods

The two experiment were conducted in a solar greenhouse during 2001-2008 suburb area of Yan'an, Shaanxi province in North China, a typical loess hill-gullied region of the Loess Plateau Longitude E109°19'23", Latitude N36°51'30", with an elevation of 1068-1309 m, mean temperature of 8.8°C, the annual precipitation of 500 mm. In the experiment 1, in 2001, the plots were filled with continuously cropped typical sandy soils of winter-spring season cucumber for 0year (hereafter referred to as yr), 1yr, 4yr, or 8yrs, respectively with 4 replications. The plots area was 15.6 m^2. A grafted cucumber with the squash root was transplanted to the continuous cropped soil. The cucumber fruit productivity and species of soil microbes were analyzed. In the experiment 2, the 7 kinds cropping models between winter-spring season cucumber and summer other different crops on the productivity and soil bio-characteristics were studied. The cropping models were: cucumber and greengrocery, cucumber and cowpea, cucumber and maize for green manure, cucumber and kidney bean, cucumber and tomato (*Lycopersicum esculentum*), cucumber and dark bean (*Phaseolus*) for green manure, cucumber and fallow. The plots area was 39.6 m^2 with three replications.

The soil moisture was maintained 75-90% as relative soil water capacity using drip irrigation system. The cucumber fruits were harvested sequentially from the early of December till the next year early of July. Microbe was determined by dilution-plate assay method, PDA broth medium used for soil fungus measured, beef extract peptone broth medium used for soil bacteria, starch-ammonium and martin broth medium used for soil actinamices [21]. To improve the precision of the results, every sample was measured three times, and the average value was calculated.

The experimental data was analyzed using SPSS software, tested with F-test means. Results were compared based on Duncan method (L.S.R).

RESULTS

Cucumber fruit productivity in different re-cropping year

Table 1 showed that there was significant difference in cucumber productivity among re-cropping years, the treatment of 1yr, 2yr, 3yr and 4yr showed significant higher yield, the treatment after 5yr showed lower yield, and this difference had reached the statistics significant level ($p = 0.01$). It means that the productivity increased as re-cropping years increasing before re-cropping 4yr, but the yield decreased obviously after 5 yrs re-cropping, indicting that degradation of soil quality was occurred.

Table 1 The cucumber fruit yield in different re-cropping year (kg/m^2)

Treat.	2002	2003	2004	2005	2006	2007	2008
0yr	12.76 B*	15.31 A	13.83 A	12.41 A	12.39 A	10.03 A	9.36 A
1yr	15.32 A	15.74 A	13.69 A	12.05 A	11.36 A	9.12 B	8.62 AB
4yr	13.95 B	11.23 B	9.65 B	8.32 B	8.16 B	8.28 C	8.23 B
8yr	11.86 C	10.23 B	9.31 B	8.36 B	8.21 B	8.13 C	7.86 B

* Notes: the letters in the same column show significant difference at 0.01 levels.

Soil microbe in different re-cropping year

In the soil, fungus and bacteria increased greatly, but actinamices did not show a big change as re-cropping years increasing (Table 2), which led to a rise of occurrence rate of pathogenic microbe and soil-borne, and lead to a decrease of fruit productivity. Additionally, the change in microbe group affected nutrient conversion in the soil, which possibly became an obstacle for nutrient absorption by crops.

Table 2 Soil fungus, bacteria and actinamices content in different re-cropping year

Treat.	Bacteria $10^6.g^{-1}ds$	Significant	Fungus $10^2.g^{-1}ds$	Significant	Actinamices $10^4.g^{-1}ds$	Significant
0yr	1.57	D*	3.8	D	20.75	A
1yr	1.48	D	1.2	E	11.13	B
3yr	1.03	E	5.4	C	10.49	B
5yr	2.95	C	4.2	D	8.19	C
7yr	3.32	B	6.3	B	3.53	D
9yr	8.90	A	7.2	A	3.43	D

* Notes: the letters in the same column show significant difference at 0.01 levels.

There are some means of soil quality rehabilitation. Reasonable rotation is the best way, for example, use of rotation system of cucumber with maize or dark-bean as green manure, rotation system of cucumber with cowpea. These crops can absorb different nutrients from soil, and abate the occurrence of diseases infected by soil through changing stubble [22]. Maintenance of adequate balance of fertilizers, addition of organic manure, and soil sterilization are secondary recommended for soil quality rehabilitation.

Cucumber fruit productivity in different cropping models

The significant difference in cucumber fruit productivity among cropping models was showed in Table 3, during 2002-2008, the cropping models of cucumber with greengrocery, cucumber with cowpea showed significant higher yield, the second higher yield appeared in the cropping models between cucumber and maize for green manure, cucumber and tomato, cucumber and kidney bean, the cropping models of cucumber with dark-bean for green manure, cucumber with fallow showed lower yield, the difference had reached the significant level ($p = 0.01$).

Table 3 The cucumber fruit yield in different cropping models (kg/m^2)

Treat.	2002-2003	2003-2004	2004-2005	2005-2006	2006-2007	2007-2008
Kidney bean	16.24 B*	16.17 AB	13.34 BC	17.30 B	16.70 B	16.40 C
Tomato	15.20 B	15.06 B	14.30 B	17.80 B	15.30 BC	15.10 CD
Fallow	13.40 C	11.91 C	11.63 C	13.90 C	12.50 C	14.80 D
Maize for green manure	16.23 B	13.27 BC	13.84 B	17.60 B	17.80 B	18.50 B
Dark-bean for green manure	14.02 C	10.92 C	12.62 C	15.30 C	14.10 C	16.60 C
Cowpea	17.91 A	17.26 A	14.97 B	18.60 B	20.80 AB	18.60 B
Greengrocery	20.05 A	19.24 A	20.44 A	21.50 A	22.30 A	20.80 A

* Notes: the letters in the same column show significant difference at 0.01 levels.

Soil microbe in different cropping models

In the soil, the number of fungus and bacteria changed greatly when planting different crops during summer season after winter-spring season cucumber harvested (Table 4). The number of bacteria in the soil was the highest after planting tomato during summer season, the second higher number were in the soil after planting greengrocery, maize for green manure and dark-bean for green manure during summer season, the lowest number of bacteria were in the soil after planting kidney bean, cowpea and fallow. The number of fungus in the soil was higher after planting maize for green manure, dark-bean for green manure and kidney bean, the second higher were in the soil after planting tomato and greengrocery, the lowest number were in the soil after planting cowpea and fallow during summer season.

Table 4 Soil fungus and bacteria in different cropping models (18 Oct. 2002)

Treat	Bacteria $10^6.g^{-1}ds$	5% Significant	Fungus $10^2.g^{-1}ds$	1% Significant
Maize for green manure	56.94	b	88.97	A
Tomato	101.36	a	46.97	B
Cowpea	46.43	c	35.71	C
Kidney bean	43.59	c	95.89	A
Greengrocery	62.79	b	40.70	B
Dark-bean for green manure	48.84	bc	79.70	A
Fallow	41.21	c	25.45	C

These results indicated that it was effective method to use different cropping models for decreasing number of microbe in soil, to fallow three or four month or to plant leguminous rotation system during summer season.

CONCLUSIONS

The productivity increased as re-cropping years increasing, but the yield decreased obviously after 5 yrs re-cropping, this indicted that degradation of soil quality was occurred. The increase of fungus and bacteria number became main factors of soil degradation as re-cropping years increasing. The cropping models of cucumber and greengrocery, cucumber and cowpea showed significant higher cucumber yield, the second higher yield appeared in the cropping models between cucumber and maize for green manure, cucumber and tomato, cucumber and kidney bean. Therefore, that was effective method to use different cropping rotation models for decreasing number of microbe in soil, to prevent from soil degradation in greenhouse to fallow three or four month or to plant leguminous during summer season after winter-spring season cucumber harvested.

DECLARATIONS

The publication costs for this article were funded by Scientific & Technical Development Inc.

This article has been published as part of SpringerPlus Volume 2 Supplement 1, 2013: Proceedings of the 2010 International Conference on Combating Land Degradation in Agricultural Areas (ICCLD'10). The full contents of the supplement are available online at http://www. springerplus.com/ supplements/2/S1.

ACKNOWLEDGEMENTS

The research was supported by the Knowledge Innovation Program of CAS (KZCX2-YW-443) (2009-2011), JSPS Core University Program "Researches on Combating Desertification and Developmental Utilization in Inland China" (2001-2010), and west-action program of the Chinese Academy of Sciences (KZCX2-XB2-05-01).

DECLARATIONS

The publication costs for this article were funded by Scientific & Technical Development Inc.

This article has been published as part of *SpringerPlus* Volume 2 Supplement 1, 2013: Proceedings of the 2010 International Conference on Combating Land Degradation in Agricultural Areas (ICCLD'10). The full contents of the supplement are available online at http://www. springerplus. com/supplements/2/S1.

REFERENCES

1. Li W, Luo HY, Liu JF: **The material and energy fluxes in greenhouse ecosystem and the evaluation of its benefits.** *Ecology Agric Res (in Chinese)* 1996, **3:** 53-55.

2. Nakano A, Yamauchi A, Uehara Y: **Effects of application of low-sulfate slow-release fertilizer (LSR) on shoot and root growth and fruit yield of tomato.** *Jpn Agric Res Qua* 2003, **l37:** 121-127.
3. Darwish T, Atallah T, Moujabber ME, Khatib N: **Salinity evolution and crop response to secondary soil salinity in two agro-climatic zones in Lebanon.** *Agric Water Manage* 2005, **78:** 152-164. 10.1016/j.agwat.2005.04.020
4. Liao YC, Wang LX: **Facility Agriculture and the Building of Chinese Agricultural Modernization.** *Research of Agricultural Modernization (Chinese)* 1999,**20**(1):5-8.
5. Guo LS, Yang WC: **The understanding and fulfillment of Facility Agriculture.** *Sciences Research Management (Chinese)*2000,**21**(3):56-59.
6. Feng GH: **Discussion on Facility Agriculture.** *Facility Agriculture (Chinese)* 2000, **2:** 5-6.
7. Liang YL, Chen ZJ, Wang ZM: **The status and function of Facility Agriculture on Ecological Environment constructing.***Journal of Soil and Water Conservation* 2002,**16**(5):32-35.
8. Stigter TY, Ooijen SPJ, Post VEA, Applello CAJ, Dill AMMC: **A hydro geological and hydro chemical explanation under irrigated land in a Mediterranean environment.** *Algarve, Portugal J Hydrol (Amst)* 1998, **208:** 262-279. 10.1016/S0022-1694(98)00168-1
9. Chen Q, Zhang XS, Zhang HY, Christie P, Li XL, Horlacher D, *et al.*: **Evaluation of current fertilizer practice andsoil fertilizer in vegetable production in the Beijin gregion.** *Nutr Cycl Agroecosyst* 2004, **69:** 51-58.
10. Takahashi K: **Injury by continuous cropping in vegetables: various problems in the cultivation using grafted plants.***Yasaishikenjo Kenkyu Shiryo* 1984, **18:** 87-89. (In Japanese)
11. Lech Szajdak: **Impact of crop rotation and phonological periods on rhodanese activity and free sulfuric amino acids concentrations in soils under continuous rye cropping and crop rotation.** *Environment International* 1996,**22**(5):563-569. 10.1016/0160-4120(96)00053-0
12. Bu'Lock JD: *Mycotoxins as secondary metabolism. In: Steyn, P.S.,ed. Biosynthesis of mycotoxins.* New York, NY: Academic Press; 1980:70-85.
13. Liu D, Wu FZ, *et al.*: **Effects of the re-cropping on the roots activity and the photosynthesis rate of the plastic greenhouse cucumbers.** *Journal of Northeast Agricultural University* 1998,**29**(3):219-223.
14. Liang YL, Chen ZJ, Xu FL, *et al.*: **The effect of re-cropping year on physiological characteristics of cucumber in sunlight greenhouse.** *Acta botanical boreali-occidentalia sinica* 2003,**23**(8):1467-1470.
15. Liang YL, Chen ZJ, Xu FL, Zhang CE, Du SN, Yan YG: **Soil Re-cropping Obstacles in Facility Agriculture on Loess Plateau.** *Journal of Soil and Water Conservation* 2004,**18**(4):134-136.
16. Yu JQ, Du YS: **The Re-cropping Obstacles Problems among The Sustainable Development of Vegetable Facility Gardening.** *Journal of Shenyang Agriculture University (Chinese)* 2000,**31**(1):124-126.
17. Wu FZ, Zhao FY, Liu YY: **Analysis of the problems in continues cropping system of protective vegetable and the controlling ways.** In *Fertilizing for Sustainable Production of High Quality Vegetables.* Edited by: Li, X.L., Zhang, F.S., Mi, G.H. China Agricultural University Publisher, Beijing; 2000:157-163.
18. Ju XT, Wan YJ, Kou CL, Zhang FS: **Nitrate accumulation in agricultural soils and groundwater of the North China Plain.**In *Plant Nutrition for Food Security, Human Health and Environmental Protection.* Edited by: Li. Tsinghua University Press, Beijing; 2005:1056-1057.
19. Zhu JH, Li XL, Christie P, Li JL: **Environmental implications of low nitrogen use efficiency in excessively fertilized hot pepper (Capsicum frutescent L.) Cropping systems. Agriculture.** *Ecosystems and Environment* 2005, **111:** 70-80. 10.1016/j.agee.2005.04.025

20. Wu FZ, *et al.*: **On the reasons of re-cropping obstacles in vegetable facility gardening.** *Journal of Northeast Agriculture* 2000, **13:** 63-65.

21. Zhou DQ: **Microbiology experiment Booklet.** *Shanghai, Science and technology press of Shanghai* 1983.

22. Liang YL, You HX, Chen ZJ, Gao JY, Du SN, Zhou MJ, Chen JR, Xiong YM: **System productivity and soil bio-characteristics of different cropping model in Facility Agriculture.** *Journal of Soil and Water Conservation* 2006, **20**(4):163-165.

CHAPTER 6

Tomato Production with Cover Crops in Greenhouse

Hajime Araki

[1]*Field Science Center for Northern Biosphere, Japan*

ABSTRACT

One of the ways to reduce chemical fertilizer application is the use of cover crops, which improve soil properties and supply nutrition to subsequent crops. Hairy vetch (*Vicia villosa* R.; HV) is one of the processing legume cover crops. A similar yield of fresh marketable tomato (*Solanum lycopersicum* L.) was obtained in the soil with HV mulch and incorporation even if the reduction of chemical N fertilizer input compared with the conventional production with 240 kg-N/ha fertilizer in the greenhouse from 2006 to 2012. Using 15N-labeling method, HV residue incorporated into soil was decomposed rapidly for about 1 month and N released from HV residue was absorbed into the tomato plant. Nitrogen was absorbed by tomato through out production period. The rate of N uptake derived from HV to total N uptake in tomato plants ($\%N_{dfhv}$) in the small amount N fertilizer was higher than that with high amount of N fertilizer application. It ranged from 24.8% in 240 kg-N/ha to 37.1% in no N fertilizer. The nitrogen use efficiency (NUE) from HV-derived N by tomato plant reached about 50% during the tomato production with HV incorporation. Other 50% of HV-derived N remained in the soil and 4% of were absorbed by tomato in the next year's production. HV has the possibility of alternative material for basal N fertilizer to ensure the tomato growth of early period after transplanting, and continuous supply of N is necessary to late stage of tomato. The combined system of incorporation of HV cultivated at the seeding density of 20–50 kg/ha before tomato planting and the slow released N fertilizer was established for the reduction N fertilizer application and obtaining conventional tomato yield in plastic house.

Keywords: cover cropping, hairy vetch, nitrogen, nitrogen dynamics, tomato

1. INTRODUCTION

1.1. Management of Fertilizer Application in Tomato Production in Plastic House

More than 70% of fruit vegetables including tomato were repeatedly produced with the application of much amount of chemical fertilizer in plastic house in Japan, and salt accumulation and injury by continuous cropping became serious problems. Excessive N input causes extra N accumulation in the soil [1], as a result of environmental pollution, such as nitrate leaching and greenhouse gas emission occur [2]. For the establishment of sustainable greenhouse production, reducing fertilizer inputs and proper applications are required.

Because tomatoes are produced with N application by basal and side-dressing fertilizer in practice, the dynamics of N derived from basal, side-dressing fertilizer, and soil in the plant are complicated, and real-time nutritional diagnosis was developed [3, 4].

Side dressings of fertilizer combined with diagnosis of nutrient conditions are popularly carried out in production areas [4, 5]. Leaf petioles below the first fruit cluster or upper fruit cluster with 2–4 cm size tomatoes are used as material for the nitrate analysis. Nitrate concentration of ca 3000–4000 ppm was reported to be an appropriate concentration for vegetative growth in tomato.

Controlled release fertilizer, the so-called slow-release fertilizer, is applied at the beginning of tomato production to reduce application work. These fertilizer management methods contribute to proper nutrient application in tomato production. However, for sustainable tomato production in greenhouses or plastic houses, soil properties have to be improved and be healthy in addition to proper N fertilization.

1.2. Advantage of Cover Cropping

Planting a cover crop is one of the biological tools used for sustainable crop production, which has the effects of N supply [6], increasing organic soil carbon [7], improving soil physical properties [8], and so on. Among the various kinds of cover crops, hairy vetch (*Vicia villosa* R.; HV), legume crop, is one of the important species. HV residues can supply plenty of N due to its efficiency in biological nitrogen fixation [9, 10]. HV also has other advantages, such as adaptability to low temperatures, resistance to pests, delaying senescence, covering ground surface effectively, and fitness for vegetable production, particularly in rotation with tomatoes [11]. Araki et al. [7] also reported that utilizing HV could reduce chemical N fertilizer inputs to half of the recommended amount without yield reduction in tomato production. Such results showed that HV had potential as an alternative fertilizer to chemical N fertilizers.

The N release pattern from inorganic or organic fertilizer has to be synchronized with crop demands for proper application [12]. Yaffa et al. [13] have shown the synchronization of tomato N uptake and N release from cover crops; the tomato N uptake increased after the increase of soil inorganic N, followed by a decline. On the other hand, Kumar et al. [11] reported that because N release from white clover residues was faster under rotary hoeing treatment, it was not synchronized with the N demand of wheat during its early growth period and resulted in minimum N-benefits. We have to investigate the N supply pattern when HV was introduced in tomato cropping in plastic house.

Non-legume cover crops are higher in carbon than legume cover crops. Because of their high carbon content, grasses break down more slowly than legumes, resulting in longer-lasting residue. As grasses mature, the carbon-to-nitrogen ratio (C:N) increases. This has two tangible results: The higher carbon residue is harder for soil microbes to break down, so the process takes longer, and the nutrients contained in the cover crop residue usually are less available to the next crop.

2. TOMATO PRODUCTION WITH COVER CROPS, 2006–2012

2.1. Field Preparation and Chemical Fertilizer Application

Tomatoes were grown in the rows with or without cover crop residue mulch shown in **Table 1**, in high plastic tunnels, adding nitrogen fertilizer 120 kg and 240 kg/ha, 2006 and 2007.

Hairy vetch (*Vicia villosa* R.) and wild oat (*Avena sterigosa* L.) were used for cover crop. They were drill planted at 10–15 cm distance, each alone or together on April 5. Seeding density was 50 kg/ha in hairy vetch, 100 kg/ha in wild oat, and 35 and 50 kg/ha in hairy vetch and wild oat, respectively. Cover crops were grown until May 26 and mowed by cutter 2 cm above the ground surface for making residue mulch of cover crops. The plots with cover crops were left untilled to maintain the residue mulch, and bare plot without cover crops were slightly tilled by hand (**Table 2**, **Figure 1**).

TABLE 1. Field operation in the tomato production with cover crop, 2007.

Date	Operation
March 22	Cover plastic film
April 4	Surface tillage in each row
April 5	Sowing of cover crop
May 26	Mowing of cover crop
	Application of fertilizers
	Making an organic mulch
May 29	Planting of tomato seedings
	Measurement of nitrate in leaf petiole and growth index
July 7–	Tomato harvest every 2–3 days
October 15	
November 16	Measurement of Soil N and C

Source: Araki et al., Hort. Environ.Biotechnol. 50:324-328. 2009.

TABLE 2. Hairy vetch treatment and fertilizer application in tomato plots from 2006 to 2012.

Year Examined	Mark	Cover crop		Fertilizer		
		Spices	Treatment	N[z] (kg/ha)	P_2O_5 (kg/ha)	K_2O (kg/ha)
2006–2007	N240	Nothing	Bare	240	200	200
	N120	Nothing	Bare	120	200	200
	HV	HV	Mulch	120	200	200
	Oats	Oats	Mulch	120	200	200
	Mix	HV + oats	Mulch	120	200	200
2008	N240	Nothing	Bare	240	200	200
	N80	Nothing	Bare	80	200	200
	HV	HV	Mulch	80	200	200
	Oats	Oats	Mulch	80	200	200
	Mix	HV + oats	Mulch	80	200	200
2009–2012	N240	Nothing	Bare	240	200	200
	HV	HV	Incorporation	80	200	200
	HV	HV	Mulch	80	200	200
	Oats	Oats	Mulch	80	200	200
	Mix	HV + oats	Mulch	80	200	200

[z]In both N rates, 20 and 80% of total N fertilizer were applied by fast-release (ammonium sulfate) and slow-release fertilizer (LPS100, N40%), respectively.
Source: Araki et al., Hort. Environ. Biotechnol. 50:324-328. 2009.

FIGURE 1. Cover crops grown in plastic high tunnel (1), hairy vetch reside mulch plot (2) and bare plot (3).

Chemical fertilizers for tomato were applied on the ground surface on May 27, 2006. Two rates of nitrogen fertilizer of 120 and 240 kg/ha were applied in the bare rows, and 120 kg N/ha fertilizer was done in rows with cover crops. In both N rates, 20 and 80% of total N fertilizer were applied by fast-release (ammonium sulfate) and slow-effect fertilizer (LP40, Chisso ASAHI Co. Ltd.), respectively (**Table 2**). In every plot, 200 kg of P_2O_5 and K_2O was added per ha.

Each plot was 0.8 m wide and 3 m long, 0.6 m between tomato lines and 0.5 m between plants. Planting density was 22,220 plants/ha. Treatments were arranged in a randomized block design with three replications.

Tomato production with cover crops was continued till 2012. N fertilizer application in cover crop plots reduced to 80 kg/ha from 2008, and the plot of HV incorporation was set up from 2009.

2.2. Biomass Production of Cover Crops

Wild oat grew to heading stage; however, seeds were not developed at the mowing time. Flowering occurred in some HV plants. At the mowing, late in May, contents of N and C were 4.3 and 41.3% in HV (C/N: 10.1), and 1.4 and 37.1% in oats (C/N: 32.3) in average from 2007 to 2012 (**Table 3**).

TABLE 3. Biomass production of cover crops, hairy vetch and oats, cultivated in the plastic house for 2 months, April and May. 2007–2012.

Cover	Concentration		N,C	FW			DW			N Contents (kg/ha)
Crop	N (%)	C (%)	C/N ratio	HV (kg/ha)	Oats (kg/ha)	Total (kg/ha)	HV (kg/ha)	Oats (kg/ha)	Total (kg/ha)	
HV	4.3	41.3	10.1	33,000	–	33,000	4,709		4,709	203
Oats	1.4	37.1	32.3	–	28,270	28,270	–	5,301	5,301	87
Mix				11,570	26,270	37,840	1,595	4,950	6,545	148

Data was the average of values obtained from 2007 to 2012.

Aboveground biomass (dry weight) was 4709 kg/ha in HV and 5301 kg/ha in wild oat late in May. However, in mix-culture, it was 1595 kg/10a in HV and 4950 kg/ha in wild oat.

Sainju et al. [14] reported that bi-culture of legume and non-legume cover crops had greater biomass yield, and also N and C contents in cover crops than monoculture of each species in the southeast area of USA. On the other hand, aboveground and underground biomass yield, and N and C contents of cover crops were varied by year and experimental location.

Experimental location of this examination, Sapporo, Japan, was snow cover region. More than 1 m high snow cover is observed in winter. HV is a typical winter annual legume crops originally, but few plants survive after long snow cover [15]. Wild oat also cannot overwinter under snow cover more than 3 month in Sapporo. HV and wild oat were usually sown in early spring in Sapporo. Greater biomass in bi-culture of wild oat and HV was not observed than monoculture of wild oat because of short growing period of cover crops.

2.3. Tomato Growth and Yield, 2007

2.3.1. Nitrate in petiole sap

There was a difference in nitrate concentration in petiole sap in leaf below first fruit cluster among treatments in July and August. Nitrate in HV plot showed 3650 and 3550 ppm in July and August (**Figure 2**). Those in Bare + N240 kg plot were 3280 and 2966 ppm, respectively. However, nitrate concentrations in other treatments were smaller than those in HV and Bare + N240 kg plots. That in oat plots was smallest. Though nitrate value decreased from July to August in bare plot (N120 kg), it increased in oat and mix (HV + Oat) plots. Such observation accorded with the result that N content became high in the leaves of tomatoes produced with HV mulch [16].

2.3.2. Growth And Yield Of Tomato

Similar total marketable yield was shown in HV and Bare + N240 kg plots, 78.2 t/ha in the former and 79.5 t/ha in the latter (**Figure 3**). Yield in first and second fruit cluster in Bare + N240 kg plot was larger than those in HV plot. Yields in Bare + N120 kg and mix plots were 68.8 t and 69.6 t/ha, smaller than those in HV and Bare + N240 kg plots. Oat plot showed smallest yield, 61.5 t/ha.

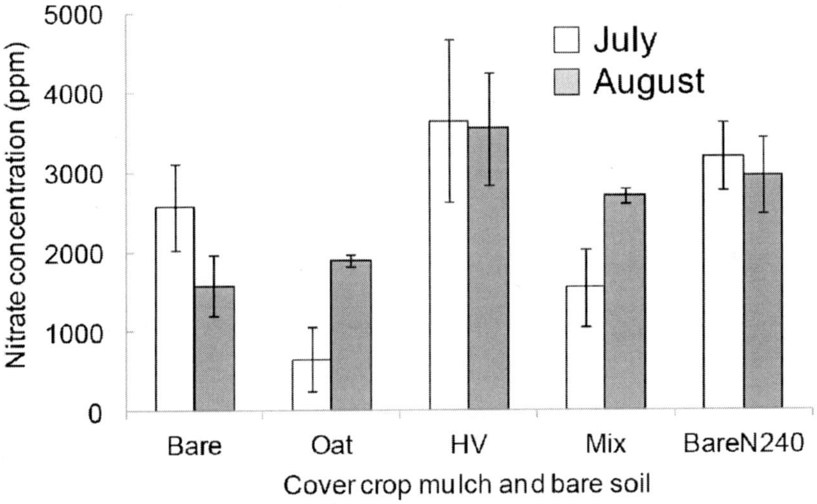

FIGURE 2. Effect of cover crop residue mulch on nitrate concentration of petiole sap on July 9 and August 10, 2007. 240 kg N fertilizer per ha was added into BareN240 plot and 120 kg N fertilizer per ha was done into other plots. Vertical bars represents SE (n = 3).

To consider fruit yield, nitrogen absorption/application rate, and nitrate nitrogen concentration, the proper range of petiole sap nitrate concentration was 4000–7000 ppm [4]. This has been recognized as recommendable value of nitrate concentration for current yield in Hokkaido prefecture. In this examination, nitrate concentrations in petiole sap in HV and Bare + N240 kg plots were close to the recommendable value in July and August. These plots have possibility to obtain the current yield. Those in other plots were obviously lower than recommended value, and it is necessary to apply the N fertilizer for the current yield.

In bare plots, especially adding with N120 kg, nitrate concentration in petiole tended to decrease from July to August; however, it increased in oat and mix plots in spite of same rate of N fertilizer. It is caused by decomposing the organic matter of cover crop residue applied in 2006 and 2007.

Nitrate concentration in petiole sap was concerned with growth index (GI), indicator of vegetative growth. GI values in Bare + N240 kg and HV plots were higher than other cover crop plots. HV, typical legume crops, decomposes faster than non-legume crops because of low C/N ratio [17], and released nitrogen is absorbed into tomato plant. However, C/N ratio of wild oat is high, as much as 24 in our previous observation. Nitrogen absorbance into tomato plants was restricted due to immobilization in soil.

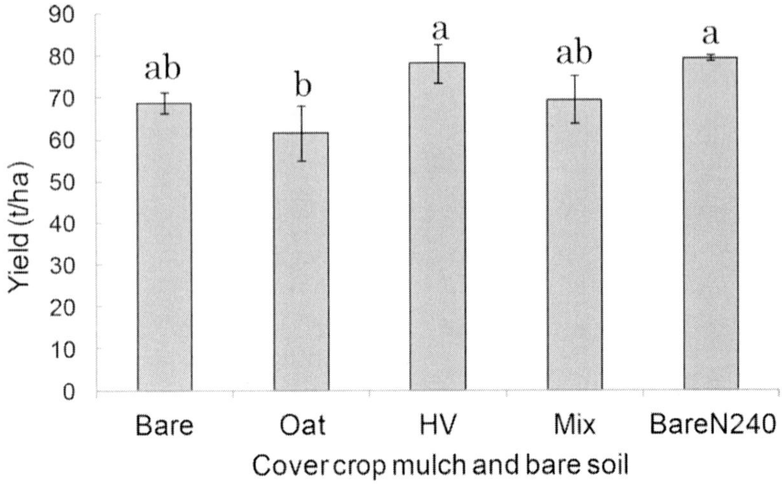

FIGURE 3. Effect of cover crop residue mulch on total fresh marketable yield of tomato 2007. 24 kg N fertilizer per 10a was added into BareN24 plot and 12 kg N fertilizer per 10a was done into other plots. Vertical bars represent SE (n = 3). Means followed by different letters are significantly different at 5%, Tukey's test.

Same trend in marketable tomato yield was observed as GI, with larger yield in Bare + N240 kg and HV plots, medium yield in Bare + N120 kg and mix plots and smallest in oat plots. The examined location is a cool summer region of which air temperature usually decreases from September. It is necessary to obtain good vegetative growth until August for getting much yield.

Even if N fertilizer decreases to half of conventional (240 kg/ha), current tomato yield was obtained in the HV plots. Such tomatoes are recognized as organically grown agro-product in combined with the reduction of fungicide and pesticide application.

2.4. Long-Term Evaluation of Tomato Yield, 2007–2012

Such tomato production system continued till 2012 (**Table 4**). The vegetative growth (GI) and marketable yield showed same trend as 2007. That is, highest GI and yield were shown in Bare-N240 plot. However, there was no significant difference between Bare-N240 and HV (N80 kg)-incorporation. GI and marketable yield of HV mulch plot were a little smaller than those of Bare-N240 and HV (N80 kg)-incorporation. From the long-term examination, 2008–2010, it was clarified HV supports the vegetative growth and tomato yield even if N fertilizer reduces to half or one third of conventional application in greenhouse.

TABLE 4. Effects of cover crops and N fertilizer on the growth and the yield of tomato: 2007–2012[z].

Mark (cover crops)	Treatment of cover crops	N fertilizer[y] (kg/ha)	Growth index[x]		Marketable yield (t/10a)	
N240	Bare	240	40514	a[w]	4.7	a
HV	Incorporation	80	38424	a	4.3	a
HV	Mulch	80	33200	b	4.1	a
Oats	Mulch	80	25365	c	2.6	b
Mix	Mulch	80	29 336	bc	3.0	b
Tukey's test[w]			*		*	

[z]Data was average of the values obtained from 2007 to 2012.
[y]In both N rates, 20 and 80% of total N fertilizer were applied by fast-release (ammonium sulfate) and slow-release fertilizer (LPS100), respectively.
[x]Growth index (GI) was measured 6 weeks after transplanting, just before the first harvest.
[w]Means followed by different letters and asterisk are significantly different among the treatments at 5%, Tukey's test. NS: not significant.

2.5. Soil N and C Content after Tomato Production

Increase of inorganic N increased near soil surface, 0–5 cm depth, in the rows with cover crops in the case of N 120 kg/ha application, 2007. Considering the high nitrate in petiole, high GI and large yield in plots with HV, it is thought that released N from cover crops was absorbed into tomato plants in near soil surface. There was no significant difference in soil total N among the plots though the examination period (**Table 5**).

As to soil carbon, the change of soil C content was observed in of 0–5 cm depth soil because of leaving cover crop residue on the soil surface. There was tendency to increase in mix plot (oats and HV); however, significant difference was recognized only 2 years, 2007 and 2008.

Winter cover crops have the potential to increase soil organic C in agricultural soils [18]. Komatsuzaki and Mu [19] evaluated the effects of tillage in continuous field rice cropping with rye and hairy vetch cover crops in Kanto region, non-snow cover region in Japan. Soil organic carbon in the top soil, 0–2.5 cm depth, increased compared with winter fallow 2 years later adopting cover cropping; however, other soil layer did not show any change in their observation.

TABLE 5. Content of total nitrogen and total carbon in the soil of the plot cultivated cover crops for 7 years (2006–2012), at the depth 5 cm and 15 cm.

Mark (cover crops)	Treatment of cover crops	0.5 cm				15 cm			
		N (%)		C (%)		N (%)		C (%)	
		2007	2012	2007	2012	2007	2012	2007	2012
N240	Bare	0.28	0.22	3.89	3.57	0.24	0.22	3.54	3.50
HV	Incorporation	0.24	0.26	3.45	3.51	0.23	0.22	3.35	3.50
HV	Mulch	0.30	0.24	4.24	3.87	0.24	0.23	3.70	3.32
Oats	Mulch	0.27	0.23	4.08	3.69	0.25	0.23	3.57	3.55
Mix	Mulch	0.30	0.27	4.61	4.07	0.26	0.25	3.77	3.78

Data in 2007 and 2012 were presented.

Organic matter level in the soils under mixed cover crops improved as much as 8.8% after 3 years of cover crop use [20]. Interestingly, soil under the legume-only cover actually dropped slightly in organic matter content after 3 years, probably because the lower C/N ratio of the incorporated organic matter caused more rapid microbial breakdown. Soil C varies from year-to-year as a result of weather-affected changes in crop residue inputs or decomposition of residues and organic matter [21].

As to one of the reasons of little change of soil C in our examination, ash soil is distributed around plot area and soil C content is originally high, more than 3%. If the examination was performed in the soil with low content of soil C, soil C will increase using of cover crop, especially no legume cover crops. Content of soil C affects the biological properties of soil. Some methods such as active carbon and SIR (substrate-induced respiratory) are applied for the estimation of diversity of microorganisms.

3. N DYNAMICS IN TOMATO PRODUCTION WITH HV

3.1. Stable Isotope Technique

These results mentioned previous section showed that HV could be alternative fertilizer for crop production instead of chemical N fertilizer and become a useful tool to solve the high input problem in N management. It is important to clarify the tomato uptake of N mineralized from HV residue for the establishment of effective N management in the cropping system with cover crops.

The method using ^{15}N-labelled plant materials has been effective for direct estimation of N uptake from the cover crops. Earlier studies have found that the N uptake by the subsequent crop was 6–25% of N applied by ^{15}N-labelled cover crops, HV, ryegrass, etc [10, 22–25]. The efficiency differed depending on subsequent crop species, cover crop species, or cultivation circumstances.

In tomato production with cover crops, Thönnissen et al. [26] reported that when legumes were incorporated or mulched into the soil, tomato absorbed 8.9 or 9.6% of soybean-derived N and 10.0 or 15.0% of indigofera-derived N, respectively. In other report, the higher recovery rate of cover crop-derived N was also reported; 56% of HV-derived N was recovered by rice [27]. ^{15}N-labeling method was used in the evaluation of N recovery from cover crops in these reports.

The application effect of legume cover crop, hairy vetch (*Vicia villosa* R.; HV), on N dynamics in fresh market tomato (*Solanum lycopersicum* L.), "House Momotaro," was investigated using ^{15}N-labelling method [28]. Tomato seedlings were transplanted in the 1/2000a Wagner pot at 0, 80, 240 kg ha^{-1} of N application (N0HV, N80HV and N240HV) on June 9. Before transplanting, the labeled HV (0.86 atom% excess) and chemical fertilizer were incorporated into soil.

3.2. Absorption and Distribution N Derived From HV

HV-derived N uptake was recognized mainly in first 4 WAT. Especially, in N240HV, the uptake of HV-derived N ceased at 4 WAT. The uptake amounts of HV-derived N at 10 WAT were 587, 657, and 729 mg $plant^{-1}$ in N240HV, N80HV, and N0HV, respectively, and there were significant differences among three treatments (**Figure 4**). The ratio of N uptake derived from HV to total N uptake in tomato plants ($\%N_{dfhv}$) was the highest at 2 WAT, and $\%N_{dfhv}$ in N80HV (52.1%) and N0HV (51.5%) were significantly higher than in N240HV (43.6%) (**Figure 5**). After 2 WAT, $\%N_{dfhv}$ was decreased gradually in all N rates as tomatoes grew, and it was decreased to 24.8, 34.4, and 37.1% in N240HV, N80HV, and N0HV, respectively, until 12 WAT (**Figure 6**). The nitrogen use efficiency (NUE) by tomato plant from HV-derived N was the highest at 10 WAT, and N0HV (55.3%) was significantly higher than N240HV (44.5%) and N80HV (49.8%) (**Table 5**).

The partition rate of HV-derived N into fruits was 63.9 and 39.7% of HV-derived N was partitioned into low fruit clusters, first and second. The partition rate of N derived from soil and fertilizer into fruits was 57.9%, significantly lower than that of HV-derived N. From these results, it was clarified that (1) HV-derived N was used effectively in small rate of chemical N fertilizer and (2) the N supply effect from HV was expressed in early period of tomato growth (**Table 6**).

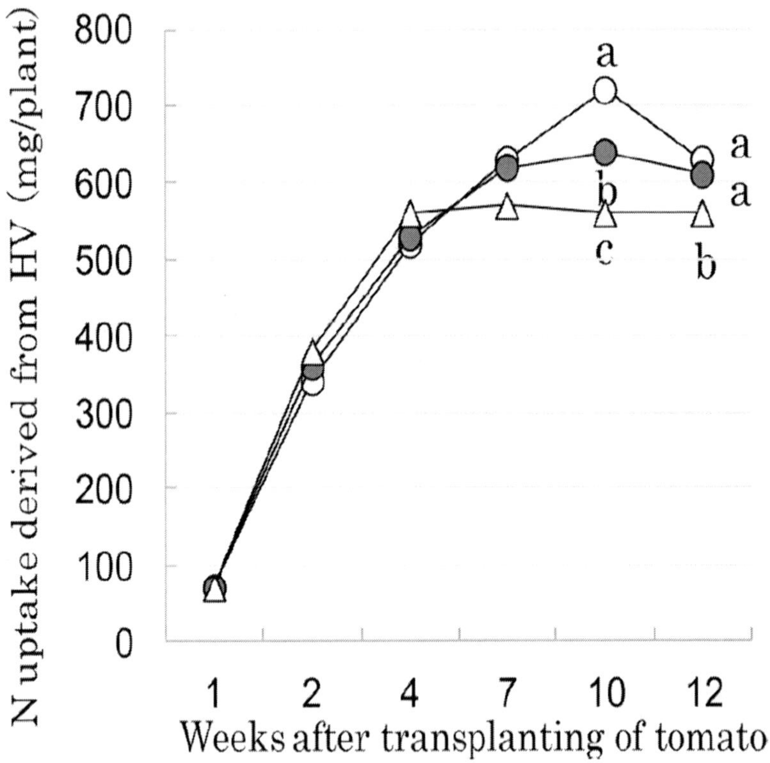

FIGURE 4. Effect of N fertilizer and HV treatment on the increase of the amount of tomato plant N uptake derived from HV. The plots of N0HV (○), N80HV (•), and N240HV (Δ) were fertilized 0, 80, and 240 kg N ha^{-1}, respectively, and applied HV (1319 mgN pot^{-1}). Means followed by different letters are significantly different at 5%, Tukey's test.

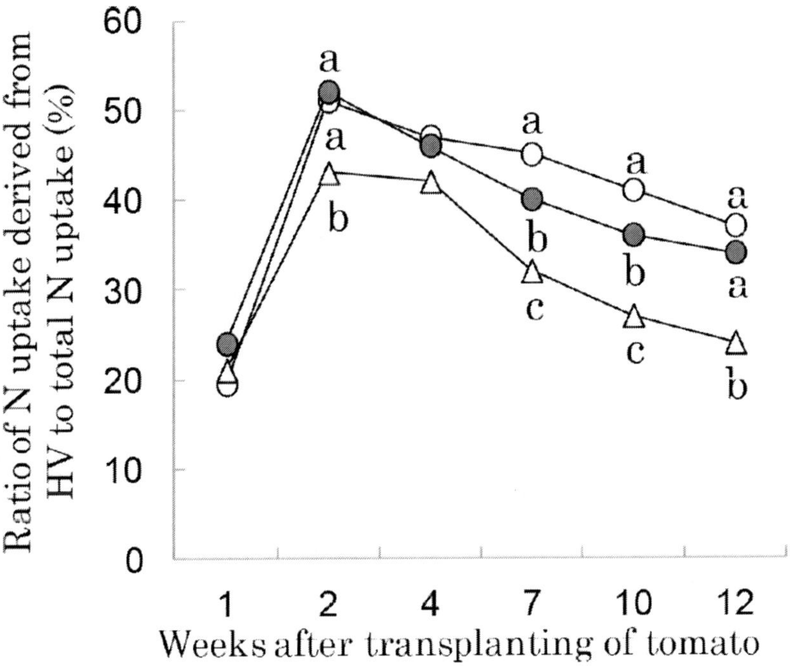

FIGURE 5. Effect of N fertilizer and HV treatment on the change of the ratio of plant N uptake derived from HV to total N uptake. The plots of N0HV (○), N80HV (●), and N240HV (Δ) were fertilized 0, 80 and 240 kg N ha^{-1}, respectively, and applied HV (1319 mgN pot^{-1}). Means followed by different letters are significantly different at 5%, Tukey's test.

TABLE 6. Effect of N fertilizer and HV treatment on nitrogen use efficiency (NUE) derived from HV.

Treatment[z]	Nitrogen use efficiency (%)							
	Weeks after transplant (WAT)							
	1	2	4	7	10		12	
N240HV	5.1	29.4	43.4	45.2	44.5	c[y]	44.4	b
N80HV	4.8	27.8	41.2	48.7	49.8	b	47.5	ab
N0HV	4.4	25.9	40.3	49.9	55.6	a	49.4	a
Tukey's test[y]	ns[y]	ns	ns	ns	*		*	

[z]The plots of N240HV, N80HV and N0HV were fertilized 240, 80 and 0 kg N ha^{-1}, respectively, and applied HV (1,319mgN pot^{-1}).
[y]Means followed by different letters and asterisk are significantly different among the treatments at 5%, Tukey's test.
NS: not significant.
Source: Sugihara et al., J. Japan. Soc. Hort. Sci. 82: 30-38. 2013. Available online at www.jstage.jst.go.jp/browse/jjshs1

3.3. Combination of N Fertilizer, HV, Fast-Release N and Slow-Release N

In order to improve the use efficiency of both hairy vetch (*Vicia villosa* R.; HV) and chemical N fertilizer, N release and uptake patterns from HV, fast-release N fertilizer (FF), and slow-release N fertilizer (SF) in fresh market tomato (*Solanum lycopersicum* L.) production were investigated using the ^{15}N-labeling method. There was no difference in tomato fruit yield between FF + SF and SF-only. HV-derived N was up taken by the tomatoes mainly until 4 weeks after transplant (WAT). The uptake amount of HV-derived N (N_{dfhv}) was same in the pot with FF + SF and SF-only. The rate of N uptake derived from HV to total N uptake in tomato plants (%N_{dfhv}) was 43% (480/1116 mg) in SF-only, higher than that in FF + SF(35%, 430/1204 mg) at 4 WAT; however, such difference disappeared after 4 WAT (**Table 7**). N uptake by tomato plants was continued 12 WAT. From these results, HV has possibility of alternative material for FF to ensure the tomato growth of early period after transplanting and continuous supply of N is necessary to late stage of tomato.

TABLE 7. Effects of N fertilizer and HV treatments on N uptake by tomato plants.

Treatments[z]	N uptake (mg/plant)													
	2 WAT[y]			4 WAT			8 WAT			12 WAT				
	Total[x]	N_{dfsf}	N_{dfhv}	Total	N_{dfsf}	N_{dfhv}	Total	N_{dfsf}	N_{dfhv}	Total		N_{dfsf}		N_{dfhv}
N240HV (FF + SF)	767	508	259	1204	774	430	2010	1518	492	2497	a[w]	1983	a	513
N240HV (SF-only)	745	419	326	1116	636	480	1930	1435	495	2497	a	1959	a	538
N240 (FF + SF)	–	–	–	–	–	–	–	–	–	1748	b	1748	b	–
N240(SF-only)	–	–	–	–	–	–	–	–	–	1786	b	1786	b	–
t-test	NS[v]	*	NS	NS	*	NS	NS	NS	NS	–		–		NS
Tukey's test	–	–	–	–	–	–	–	–	–	*		*		–

[z]Plots of N240HV were fertilized with 240 kg $N \cdot ha^{-1}$, and HV was applied (1007 mg N/pot^{-1}). Plots of N240 were fertilized with 240 kg $N \cdot ha^{-1}$ without HV. (FF + SF) contained 20% fast-release fertilizer + 80% slow-release fertilizer, and (SF-only) contained 100% slow-release fertilizer.
[y]WAT: weeks after transplant.
[x]Total = N_{dfsf} + N_{dfhv}. N_{dfsf}: N derived from soil and fertilizer, N_{dfhv}: N derived from HV.
[w]Means followed by different letters and asterisk are significantly different among the treatments at 5%, *t*-test and Tukey's test. NS: Not significant.
Source: Sugihara et al., J. Japan. Soc. Hort. Sci. 83: 222–228. 2014. doi: 10.2503/jjshs1.CH-061

3.4. Absorption of HV-N Remained in Soil in The Following Year

After the tomato cultivation in 2011, the soil was stored in a greenhouse without any water and fertilizer. Tomatoes were cultivated again in the Wagner pots in which contained used soil of 2011 and were added same rate of N fertilizer (0, 80, and 240 kg ha^{-1} of N) and unlabeled HV (935 mgN/pot) in 2012. Total N uptake of tomato plant was higher in N240HV (2377 mg/plant), followed by N80HV (1760 mg/plant), N0HV (1498 mg/plant). On the other hand, the uptake of N derived from HV applied in 2011 (HV_{2011}, 1319 mgN/pot) was not different among the treatments (57.7 mg/plant on average), so nitrogen use efficiency derived from HV_{2011} in 2012 was 4.4% on average (**Figure 6**). This value was much lower than that in 2011 (47.1% on average), but HV_{2011}-N also remained in the soil yet after the tomato cultivation in 2012 (500 mgN/pot). These results showed that although the N supplying effect of HV was small in the following year, HV could be available for not only short-term N source, but also long-term N source, and HV-derived N applied in the previous year was absorbed by tomato plant during relatively early growth period in the following year.

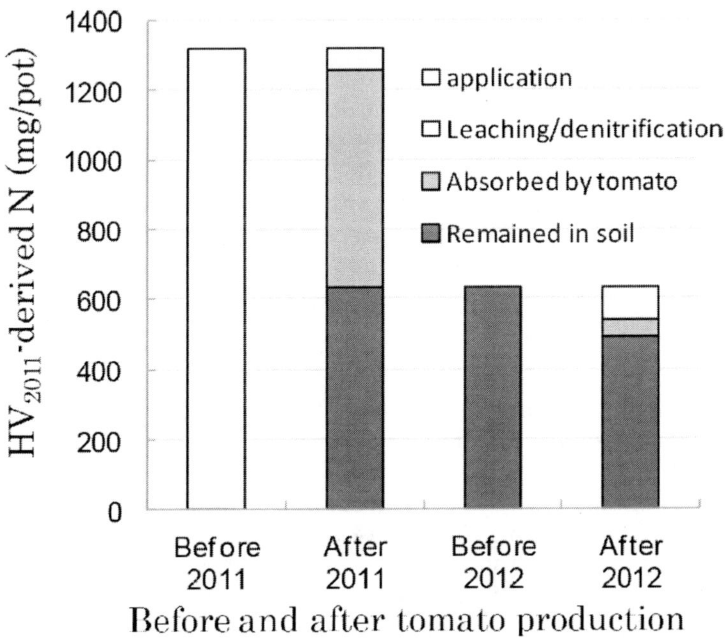

FIGURE 6. Dynamics of HV_{2011}-N's input and output. The values show an average of three treatments.

HV-derived N is probably available as alternate of fast-release fertilizer, but in order to cultivate tomato healthy, it is supposed that the application of additional fertilizers for late period growth is needed. Although reducing chemical fertilizer application improves the HV-derived N use efficiency, the excess reducing fertilizer could lead to nitrogen deficiency during late period. Therefore, it is important to balance reducing chemical fertilizers and ensuring normal tomato fruit yield in order to establish the low input sustainable cropping system using HV as a cover crop in tomato cultivation.

From the experiment 2012, the N derived from HV applied in the previous year contributed slightly to the tomato growth, especially in the early growth stage. Although the contribution rate of N derived from HV applied in the previous year was much lower than that from HV applied in the current year, it was suggested that HV could be used for not only short-term N source but also long-term N source as same as other organic materials. The result of this study will be able to use as one of the knowledge to establish the HV-tomato rotation cropping system.

4. PROOF STUDY OF TOMATO PRODUCTION WITH HV

The positive responses of tomatoes grown under hairy vetch residues have economic and environmental importance. From the results mentioned above, HV has the possibility of alternative material for basal N fertilizer to ensure the tomato growth of early period after transplanting and continuous supply of N is necessary to late stage of tomato. A proof study to utilize HV in a real house is necessary because the examinations on N dynamics were carried out in Wagner pot.

4.1. Fertilizer Design

The application of ammonium sulfate (AS; 100-N kg/ha), HV (two seeding density; 20 kg and 50 kg/ha) and nothing, was for basal fertilizer management (**Table 8**). AS was applied into soil, May 31, 2015, the previous day of tomato planting. HV grew for over 2 months; it was mowed in late May. At the mowing time, flowering was observed in some plants. Aboveground biomass (dry weight) of HV at densities 50 and 20 kg/ha were 7.2 and 5.9 t/ha, respectively. This biomass accounts for the incorporation of organic nitrogen of 252 and 309 kg/ha of equivalency into soil in HV20 and HV50, respectively. The C:N ratio of HV was low, 9.6, suggesting rapid decomposition and mineralization of organic residues after incorporation into soil.

LPS100 (Long player N fertilizer, slow release type, Chisso ASAHI Co. Ltd) was added in all plots for side dressing. Similarly, in every plot, 200 kg/ha of P_2O_5 as fused magnesium phosphate and 200 kg/ha of K_2O as potassium sulfate were added. Transplanting occurred on June 1, 2015.

TABLE 8. Design of nitrogen application using HV and chemical fertilizer and marketable yield of fresh tomato.

Fertilizer or hairy vetch	Basal N		Top N LPS100	Yield of marketable	
	HV seeding (kg/ha)	AS-N application (kg/ha)	application (kg/ha)	tomato (t/ha)	
Control	0	0	150	97	c
AS100	0	100	150	114	b
HV20	20	0	150	129	a
HV50	50	0	150	130	a

Yield means followed by different letters are significantly different among the treatments at 5%, Tukey's test.

4.2. Soil Inorganic Nitrogen and Tomato Yield

AS100 plots exhibited the highest soil inorganic nitrogen (9 mg/100 g) in the second week after transplanting (2 WAT); however, it showed a decreasing trend (**Table 9**). The soil inorganic nitrogen in HV20 and HV50 plots got to increase from 4 WAT until 6 WAT.

TABLE 9. Change of inorganic nitrogen in the after tomato planting.

Plots	Inorganic nitrogen (mg/100g)		
	WAT		
	2	6	12
Control	2.6	3.1	1.0
AS100	9.0	3.1	1.1
HV2	2.8	5.0	1.6
HV5	5.4	4.0	2.1

Similar and higher marketable yields were found in HV plots, 130 t/ha in HV50 and 129 t/ha in HV20. These yields were higher than those in AS100 and control plots,114 and 97 t/ha, respectively (**Table 8**).

Because of high N accumulation, HV increased soil inorganic nitrogen, tomato yield, and N uptake [29]. The HV residues increased soil inorganic N in the early stages of tomato cultivation. HV released more nitrogen in the first 6 weeks while ammonium sulfate provided more nitrogen in the first 4 weeks after transplanting. The results were in agreement with previous studies [28, 30–32]. The increased level of inorganic N with the HV at 15–42 days after residue incorporation may indicate that hairy vetch is suitable to be used as N source for the early stages of tomato cultivation.

From this study, 40% of conventional N fertilizers were reduced by incorporating hairy vetch residues in soil, even though good vegetative growth and high marketable yield were obtained.

5. ACKNOWLEDGEMENTS

The author thanks to technical staff of Experiment Farm, Field Science Center for Northern Biosphere, Hokkaido University, for their managing the cover crop and tomato production. The author also thanks to Mr. Y. Sugihara and Mr. R. A. Muchanga, Master course students, Graduated school of Environmental Science, Hokkaido University, for carrying out the examination.

REFERENCES

1. Hao Z. P., Christie P., Zheng F., Li J. L., Chen Q., Wang J. G. and Li X. L. 2009. Excessive nitrogen inputs in intensive greenhouse cultivation may influence soil microbial biomass and community composition. Commun. Soil Sci. Plant Anal. 40:2323–2337.
2. McNeal B. L., Stanley C. D., Graham W. D., Gilreath P. R., Downey D. and Creighton J. F. 1995. Nutrient loss trends for vegetable and citrus fields in west central Florida, I: Nitrate. J. Environ. Qual. 24:95–100.
3. Yamada R., Kato T., Ido Y., Seki M. and Hayakawa I. 1995. Rational manuring management of greenhouse tomatoes based on real-time nutritional diagnosis of plant and soil 1. Diagnostic standard according to nitrate concentration of petiole juice. Res. Bull. Aichi Agric. Res. Cent. 27:205–211.
4. Sakaguchi M., Hikasa Y. and Nakazumi H. 2004. Diagnostic technique for nitrogen nutrition of summer–autumn harvest culture green-house tomato, Japan. J. Soil Sci. Plant Nutr. 75:29–35.
5. Tanaka T. 2003. Nutritional diagnosis indicator in drip-fertigation on tomato plants in greenhouse culture. Res. Bull. Aichi Agric. Res. Cent. 35:73–78.
6. Ruffo M. L. and Bollero G. A. 2003. Modeling rye and hairy vetch residue decomposition as a function of degree-days and decomposition-days. Agron. J. 95:900–907.
7. Araki H., Hane S., Hoshino Y. and Hirata T. 2009. Cover crop use in tomato production in plastic high tunnel. Hortic. Environ. Biotechnol. 50:1–5.
8. Blevins R. L. and Frye W. W. 1993. Conservation tillage: An ecological approach to soil management. Adv. Agron. 51:33–78
9. Acosta J. A. A., Amado T. J. C., Neergaard A., Vinther M., Silva L. S. and Nicoloso R. S. 2011. Effect of 15N-labeled hairy vetch and nitrogen fertilization on maize nutrition and yield under no-tillage. R. Bras. Ci. Solo. 35:1337–1345.
10. Seo J. H., Meisinger J. J. and Lee H. J. 2006. Recovery of nitrogen-15-labeled hairy vetch and fertilizer applied to corn. Agron. J. 98:245–254.

11. Kumar V., Abdul-Baki A. A., Anderson J. D. and Mattoo A. K. 2005. Cover crop residues enhance growth, improve yield, and delay leaf senescence in greenhouse-grown tomatoes. HortScience. 40:1307–1311.
12. Kramer A. W., Doaneb T. A., Horwath W. R. and van Kessel C. 2002. Combining fertilizer and organic inputs to synchronize N supply in alternative cropping systems in California. Agric. Ecosyst. Environ. 91:233–243.
13. Yaffa S., Sainju U. M., Singh B. P. and Reddy K. C. 2000. Fresh market tomato yield and soil nitrogen as affected by tillage, cover cropping, and nitrogen fertilization. HortScience. 35:1258–1262.
14. Sainju U. M., Whitehead W. F. and Singh B. P. 2005. Biculture legume–cereal cover crops for enhanced biomass yield and carbon and nitrogen. Agron. J. 97:1403–1412.
15. Araki H., Hatano Y., Horimoto S., Fujii Y and Ito M. 2007. Biomass production and weed control in some winter cover crops in Hokuriku district. Japan. J. Farm. Work. Res. 42:111–122.
16. Abdul-Baki A. A., Teasdale J. R. and Korcak R. F. 1997. Nitrogen requirements of fresh-market tomatoes on hairy vetch and black polyethylene mulch. HortScience. 32(2):217–221.
17. Ranells N. N. and Wagger M. G. 1996. Nitrogen release from grass and legume cove r crop monocultures and bicultures. Agron. J. 88:777–782.
18. Jarecki M. K. and Lal R. 2003. Crop management for soil carbon sequestration. Crit. Rev. Plant. Sci. 22:471–502.
19. Komatsuzaki M. and Mu Y. 2005. Effects of tillage system and cover cropping on carbon and nitrogen dynamics. Proceeding and abstract of ecological analysis and control of greenhouse gas emission from agriculture in Asia. September 2005, Ibaraki, Japan, pp. 62–67.
20. Gliessman S. R. 1987. Species interactions and community ecology in low external-input agriculture. Am. J. Altern. Agric. 11:160–165.
21. Campbell C. A., Zenter R. P., Selles F., Biederbeck V. O., McConkey B. G., Blomert B. and Jefferson P. G. 2000. Quantifying short-term effects of crop rotations on soil organic carbon in south western Saskatchewan. Can. J. Soil. Sci. 80:193–202.
22. Bergström L. and Kirchmann H. 2004. Leaching and crop uptake of nitrogen from nitrogen-15-labeled green manures and ammonium nitrate. J. Environ. Qual. 33:1786–1792.
23. Harris G. H., Hesterman O. B., Paul E. A., Peter S. E. and Janke R. R. 1994. Fate of legume and fertilizer nitrogen-15 in a long-term cropping systems experiment. Agron. J. 86:910–915.
24. Jackson L. E. 2000. Fate and losses of nitrogen from a nitrogen-15-laveled cover crop in an intensively managed vegetable system. Soil. Sci. Soc. Am. J. 64:1404–1412.
25. Kumar K, Goh. K. M, Scott W. R. and Frampton C. M. 2001. Effects of 15N-labelled crop residues and management practices on subsequent winter wheat yields, nitrogen benefits and recovery under field conditions. J. Agric. Sci. 136:35–53.
26. Thönnissen C., Midmore D. J., Ladha J. K., Holmer R. J. and Schmidhalter U. 2000. Tomato crop response to short-duration legume green manures in tropical vegetable system. Agron. J. 92:245–253.
27. Asagi N. and Ueno H. 2009. Nitrogen dynamics in paddy soil applied with various 15N-labelled green manures. Plant. Soil. 322:251–262.
28. Sugihara Y., Ueno H., Hirata T. and Araki H. 2013. Uptake and distribution of nitrogen derived from hairy vetch used as a cover crop by tomato plant. J. Jpn. Soc. Hortic. Sci. 82:30–38.
29. Sainju U. M., Singh B. P. and Yaffa S. 1999. Tomato yield and soil quality as influenced by tillage, cover cropping, and nitrogen fertilization. In Hook J. E. (ed.). Proceedings of the 22nd Annual Southern Conservation Tillage Conference for Sustainable Agriculture. Tifton, GA.
30. Stute J. K. and Posner J. L. 1995. Synchrony between legume nitrogen release and corn demand in the upper Midwest. Agron. J. 87:1063–1069.

31. Kuo S., Sainju U. M. and Jullum E. J. 1997. Winter cover cropping influence on nitrogen in soil. Soil Sci. Soc. Am. J. 61:1392–1399.

32. Sainju U. M., Singh B. P. and Whitehead W. F. 2000. Cover crops and nitrogen fertilization effect on soil carbon and nitrogen and tomato yield. Can. J. Soil Sci. 80:523–532.

CHAPTER 7

Fluorescence Imaging in the Red and Far-Red Region during Growth of Sunflower Plantlets. Diagnosis of the Early Infection by the Parasite *Orobanchecumana*

Carmen M. Ortiz-Bustos[1], María L. Pérez-Bueno[2], Matilde Barón[2] and Leire Molinero-Ruiz[1*]

[1]*Department of Crop Protection, Institute for Sustainable Agriculture – Spanish National Research Council, Cordoba, Spain*

[2]*Estación Experimental delZaidín, Spanish National Research Council, Granada, Spain*

ABSTRACT

Broomrape, caused by the root holoparasite *Orobanchecumana*, is the main biotic constraint to sunflower oil production worldwide. By the time broomrape emerges, most of the metabolic imbalance has been produced by *O. cumana* to sunflower plants. UV-induced multicolor fluorescence imaging (MCFI) provides information on the fluorescence emitted by chlorophyll (Chl) *a* of plants in the spectral bands with peaks near 680 nm (red, F680) and 740 nm (far-red, F740). In this work MCFI was extensively applied to sunflowers, either healthy or parasitized plants, for the first time. The distribution of red and far-red fluorescence was analyzed in healthy sunflower grown in pots under greenhouse conditions. Fluorescence patterns were analyzed across the leaf surface and throughout the plant by comparing the first four leaf pairs (LPs) between the second and fifth week of growth. Similar fluorescence patterns, with a delay of 3 or 4 days between them, were obtained for LPs of healthy sunflower, showing that red and far-red fluorescence varied with the developmental stage of the leaf. The use of F680 and F740 as indicators of sunflower infection by *O. cumana* during underground development stages of the parasite was also evaluated under similar experimental conditions. Early increases in F680 and F740 as well as decreases in F680/F740 were detected upon infection by *O. cumana*. Significant differences between inoculated and control plants depended on the LP that was considered at any time. Measurements of Chl contents and final total Chl content supported the results of MCFI, but they were less sensitive in differentiating healthy from inoculated plants. Sunflower infection was confirmed by the presence of broomrape nodules in the roots at the end of the experiment. The potential of MCFI in the red and far-red region for an early detection of *O. cumana*infectionin sunflower was revealed. This technique might have a particular interest for early phenotyping in sunflower breeding programs. To our knowledge, this is the first work where the effect of a parasitic plant in its host is analyzed by means of fluorescence imaging in the red and far-red spectral regions.

Keywords: broomrape, chlorophyll content, *Helianthus annuus* L., leaf pair, pre-symptomatic detection

INTRODUCTION

Orobanchecumana Wallr. is a holoparasitic plant that parasitizes sunflower (*Helianthus annuus* L.) and limits crop yield in all the countries of Southern Europe and areas around the Black Sea where sunflowers are grown, as well as in North Africa, Israel, and China (Molinero-Ruiz et al., 2015). On average, sunflower seed losses caused by *O. cumana* (broomrape) can be quantified as being above 50% when susceptible hybrids are grown, and they frequently reach 100% in heavily infested fields (Domínguez, 1996; Jestin et al., 2014). Crop yield reductions due to infection by *O. cumana* depend on several factors, such as aggressiveness of the parasite, sunflower genotype, earliness of broomrape emergence, soil depth and, particularly, the soil infestation level or seed-bank in the soil (Molinero-Ruiz et al., 2009; Jestin et al., 2014). Germination of *O. cumana*seeds in the soil is stimulated by chemical compounds exuded in the rhizosphere by surrounding sunflower roots (Joel et al., 2011; Yoneyama et al., 2011; Raupp and Spring, 2013; Ueno et al., 2014). The seed produces a radicle that grows toward the host root, reaches it, and adheres to the host plant. *Orobanche* spp. obtain water and inorganic compounds from the host through a xylem to xylem contact, and, unlike the vast majority of parasitic plants, they also utilize the carbon fixed by the host plants, which is transferred through phloem continuity between host and parasite (Molinero-Ruiz et al., 2015). Penetration of the intrusive cells into the xylem triggers their division, leading to the formation of a tubercle, a callus-shaped bulge of parasite cells outside the root (Molinero-Ruiz et al., 2015). In *O. cumana*-sunflower interaction, the utilization of host photoassimilates by the parasite results in depletion of resources which are necessary for the growth of sunflower and for the optimal development of the seeds.

The complete disruption of the host vascular system of sunflower forced by the parasitism of *O. cumana*, begins during the initial weeks of the crop. In fact, and as early as 3 weeks after inoculation, a decrease in secondary metabolites is observed in the host upon infection by the parasite (Pérez-Bueno et al., 2014). After underground development of tubercles parasite stems emerge aboveground at around flowering-time of sunflower plants (Melero-Vara and Alonso, 1988). By that time most of the metabolic imbalance has been produced irreversibly by the parasite to sunflower. From that moment onward, the impact of *O. cumana* on the crop yield is evident through absent or small sized capitula, their low number and small seeds, and even plant death (Molinero-Ruiz et al., 2015). Because the parasite establishes itself in the host plant in underground stages, non-destructive methods for diagnosing that establishment are greatly needed. Additionally, diagnosis tools used as early as possible after plant sowing and parasite infection will have a wide range of applications in the search for efficient control approaches, such as genetic resistance or treatments with potential biocontrol agents.

Multicolor fluorescence imaging (MCFI) is a technique based on induction of fluorescence emission through UV excitation. It provides leaf images of the autofluorescence emitted by plants in four spectral bands peaking at around 440 nm (blue, F440), 520 nm (green, F520), 690 nm (red, F690), and 740 nm (far-red, F740; Buschmann and Lichtenthaler, 1998; Buschmann et al., 2000). The first two fluorescence signals (F440 and F520) are related to secondary metabolites (Harris and Hartley, 1976; Morales et al., 1996) while the other two (F680 and F740) are emitted by Chl *a* (Buschmann et al., 2000). Also of particular interest are the fluorescence ratios; the red/far-red (F680/F740) ratio was found to be a good indicator of stress-induced decreases in Chl content (Hák et al., 1990; Buschmann, 2007). Therefore, MCFI is considered to be a powerful tool for the research on plant physiology related to development (Lang et al., 1994; Cerovic et al., 1999; Subhash et al., 1999). Moreover, this imaging technique is especially valuable for detecting early symptoms in leaves as those caused by *Tobacco mosaic virus* in tobacco plants (Chaerle et al., 2007) and *Pseudomonas syringae* in bean plants (Pérez-Bueno et al., 2015).

On the other hand, the use of Chl meters for quantifying leaf Chl levels has become common in recent years. Unlike traditional leaf Chl content estimation techniques consisting mainly of pigments extraction by organic solvents followed by *in vitro* measurements with a spectrophotometer, Chl content meters allow to asses this leaf parameter in a fast and non-destructive way. Furthermore, previous studies using Chl meters have shown that they provide good estimates of leaf Chl content (Fanizza et al., 1991; Brito et al., 2011). Other useful applications of Chl meters are the estimation of foliar N concentration (Chang and Robison, 2003; Liu et al., 2012) and crop disease detection (Calderón et al., 2014).

The objectives of this study were to: (a) analyze the distribution of red and far-red fluorescence emission during early growth stages of healthy sunflower plants, (b) evaluate the use of fluorescence imaging in the red and far-red region as an indicator of sunflower infection by *O. cumana*during underground development stages of the parasite, and (c) compare fluorescence imaging in the red and far-red region with measurements of total leaf Chl content as indicator of early infection by *O. cumana.*

MATERIALS AND METHODS

Distribution of Red and Far-Red Fluorescence Emission in Healthy Sunflower Plantlets

Healthy Plant Material

Four sunflower seeds of the inbred line NR5 were surface-disinfected by soaking them in 20% household bleach (50 g of active chlorine per liter) for 5 min, then seeds were rinsed in deionised water for 5 min and kept in a dark incubator at 25°C at saturation humidity until radicles were 2–5 mm long. After germination, seeds were individually transplanted into small pots with 250 g of a soil mixture (sand:silt:peat moss 2:1:1, *V*). Plants were grown in a glasshouse at 12–22°C without additional lighting for 5 weeks and watered as needed. When plants were 2-weeks-old and once a week for the next 2 weeks of growth, they were fertilized with 15 ml/pot of a nutrient solution with N:P:K (7:5:6).

Multicolor Fluorescence Imaging

Multicolour fluorescence imaging was carried out using an Open FluorCamFC 800-O and analyzed using the Fluorcam7 software (Photon Systems Instruments, Brno, Czech Republic) according to Pérez-Bueno et al.

(2015). Fluorescence images in red (F680) and far-red (F740) regions of the spectrum (which are emitted by chlorophyll *a*) were acquired sequentially after excitation with UV light (360 nm). Images corresponding to the F680/F740 ratio were calculated by the mentioned software. Fully developed leaf pairs (LPs) were measured from 2 weeks after transplanting. Thereafter, they were measured twice a week for 3 weeks until the end of the experiment. Measurements were always taken at the same time of day on attached and unshaded leaves.

The F680 and F740 emission by the leaf of healthy sunflower plants was studied by defining three different leaf areas (base, apex, and a region in the middle of the leaf). In addition, progress of chlorophyll fluorescence emission on the first four LPs of the plants was analyzed.

Statistical Analyses

The F680 and F740 emission in leaf areas of healthy sunflower was analyzed in a completely randomized design. Eight replicates were considered and one blade of an LP was regarded as a replicate. Significant differences in the intensity of F680 and F740 between leaf areas was calculated using analysis of variance (ANOVA). If significant effects were obtained, Fisher's protected LSD tests ($P = 0.05$) were used as the comparison procedure. In the case of the time course of the F680 and F740 in LP, means and standard errors of 8–12 replicates were considered. Both experiments were repeated twice and ANOVA analysis was conducted to determine significant differences between the whole set of data. As no significant differences were found, data were pooled.

Red and Far-Red Fluorescence Emission, and Leaf Chl Content in Sunflower Plantlets upon Inoculation with *O. cumana*

Orobanchecumana Inoculation and Sunflower Plant Growth

Five sunflower seeds (replications) of the inbred line NR5, which is the differential for race F (Molinero-Ruiz et al., 2015), were grown in the greenhouse for 5 weeks under the same conditions as those previously described and used as controls. In the case of inoculated plants, five NR5 seedlings were individually transplanted into a soil mixture uniformly infested with 10 mg of *O. cumana* seeds of LPA13 population previously characterized as race F (García-Carneros et al., 2014). From then on, the experiment was carried out under the same conditions described in the section on healthy sunflower plantlets.

At the end of the experiment, sunflower plants were uprooted and their roots carefully cleaned to remove soil, washed and air-dried. Nodules of *O. cumana* in each plant were counted using a stereo microscope and weighed with a precision balance.

Measurements of the Total Leaf Chlorophyll Content

The total content on Chl was determined by an Chl content meter (CL-01 Hansatech Instruments, Ltd., Kings Lynn, UK), which provides a rapid and direct way to determine the relative content of Chl in leaves. Measurements were taken twice a week, from 2 and until 5 wai, in control and in inoculated plants. In each plant, one measurement (replicate) was made in the middle of all fully developed leaves, avoiding the main vein. For each time point, all LPs were individually analyzed. Between 12 and 20 replicates (measurements) were considered for each LP.

Spectrophotometric Chlorophyll Quantification

At the end of the experiment (5 wai), and before plants were uprooted, Chl *a* andChl *b* content were determined spectrophotometrically for the second, third and fourth LP of four control and four inoculated plants. Two 1 cm diameter leaf disks were collected from each LP, weighted and immediately placed in vials and frozen in liquid nitrogen. Each sample was ground, the total pigments extracted with 4 ml of a solution of 80% acetone (v/v). The absorbance was measured at 470, 647 and 663 nm with a Shimadzu UV-1800 spectrophotometer (Shimadzu Corporation, Tokyo, Japan). Total Chl content and the Chl *a*/*b* ratio were calculated according to Lichtenthaler and Buschmann (2001).

Statistical Analyses

Sunflower plants were inoculated in two experiments which were set up as a factorial on a completely randomized design. Since ANOVA analysis did not detect any significant difference of data between them, the data were pooled. For each time point and LP, data of MCFI and Chl content included a maximum of 20 measurements. Sometimes, MCFI measurements were not taken on completely horizontal leaves resulting in a reduction to 14 replications for that particular time point. Similarly, if Chl content was determined on leaves that were too small to be clamped or had a very irregular surface, measurements were omitted and the replication number was reduced to 12. In any case, and for each time point, the same number of replications of control and inoculated plants was used in the statistical analyses. Significant differences in fluorescence parameters, broomrape incidence, total Chl content and Chl *a*/*b* ratio between inoculated and control plants were analyzed using the same statistical procedures as explained for the healthy sunflowers.

RESULTS

MCFI Distribution Pattern on Healthy Sunflower Plants

Figure 1 shows the average values of F680, F740 and their ratio for the first four LPs of healthy sunflower plants from the second to the fifth week of growth. Values of F680 and F740 of the first LP were significantly higher in the third week as compared to those recorded 1 week earlier. However, the F680/F740 ratio of the same LP showed a slight decrease in the third week. Concerning the second LP, no significant changes were detected in any of the fluorescence parameters analyzed. The third LP showed a significant increase in the F680/F740 ratio during the measurement period. Conversely, the other two variables decreased slightly throughout the same period. The three fluorescence parameters of the fourth LP remained constant during the measuring week (Figure 1).

The pattern of fluorescence across the leaves was found to be very homogeneous and therefore not dependent on measurement location on the leaf (data not shown). Consequently, a similar size area was selected in the middle of the leaf for the subsequent image analysis of healthy vs. *O. cumana* infected plants.

Alterations in MCFI Distribution Pattern on Sunflower Plants upon Infection with *O. cumana*

Figures 2 and 3 show the progress of F680 and F740, respectively in the first four LPs of sunflower plants inoculated with *O. cumana* compared with that of control plants.

The most informative parameters were those of the first LP from the second to the third week of growth and the second LP from the fourth week on. Between 2 and 3 wai, F680 and F740 of the first LP was significantly higher ($P < 0.05$) in the inoculated sunflowers as compared to the controls. In the same period of time, F680 and F740 signals in the second LP of inoculated plants were significantly higher than those of the controls ($P < 0.05$), as well as at 4 wai ($P < 0.05$). F680 and F740 signals of the third and fourth LP of inoculated plants were significantly higher than those of the controls at 4 wai in the third LP ($P < 0.05$) and half a week later in the fourth LP ($P < 0.05$).

Representative images of the two first LPs from control and inoculated sunflower are shown in Figure 4.

Figure 5 displays the values of the F680/F740 ratio in inoculated and control sunflower plants in the first four LPs as well as significant differences between treatments. A significant effect ($P < 0.05$) of the inoculation by *O. cumana* was observed in F680/F740 of the second LP, being lower in inoculated plants at 2.5 and 3 wai ($P < 0.05$), and higher during the rest of the measurements ($P < 0.05$). Significant differences ($P < 0.05$) of F680/F740 between treatments were obtained in the fourth LP (averages 0.73 and 0.71 for inoculated and control plants, respectively). Although no significant differences ($P > 0.05$) were obtained between inoculated and control plants when the F680/F740 ratio of the third LP was analyzed, similar trends were observed.

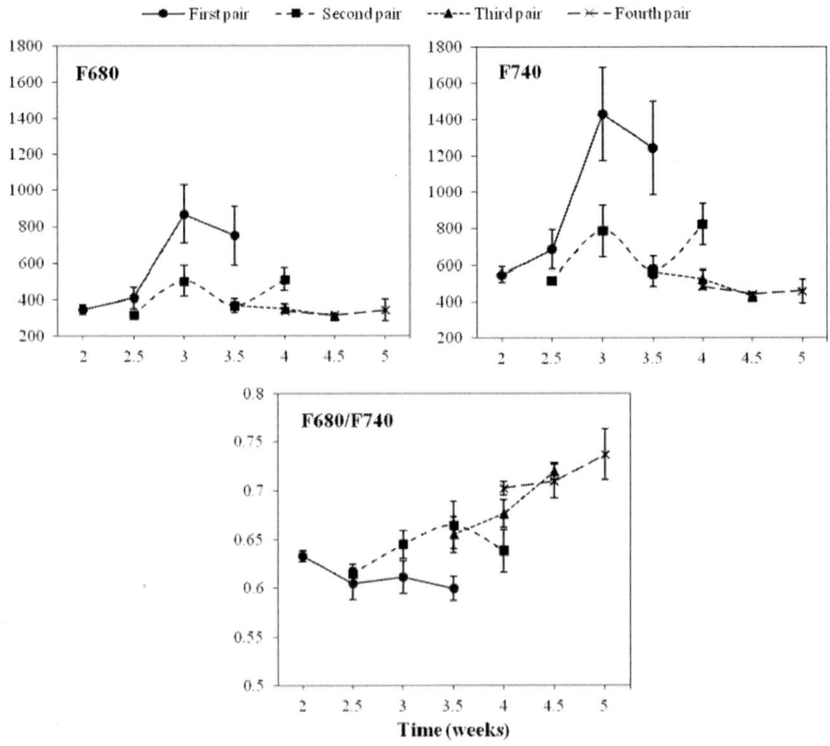

FIGURE 1. Progress of chlorophyll fluorescence emission on the first, second, third and fourth pair of true leaves of a healthy sunflower plant. Error bars represent the standard error of the mean of 8–12 replications. Time is expressed as weeks after transplanting.

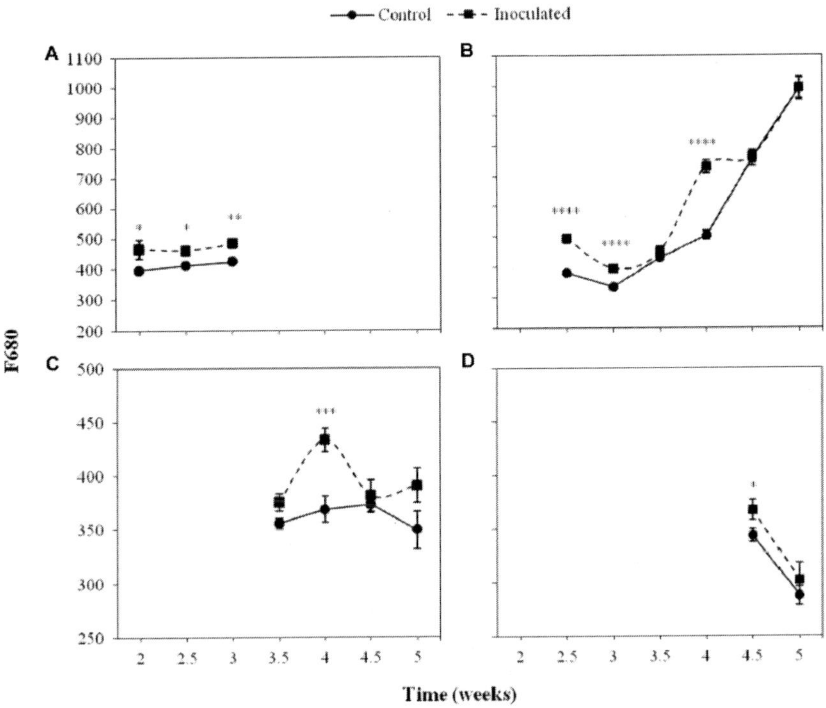

FIGURE 2. Mean measurements of F680 of the first (A), second (B), third (C), and fourth (D) leaf pair (LP) of sunflower plants inoculated with Orobanchecumana and control plants. Vertical bars represent the standard error of the mean of 14–20 replications. Analyses of variance were conducted and asterisks indicate significant differences (∗P < 0.05; ∗∗P < 0.01; ∗∗∗P < 0.001; ∗∗∗∗P < 0.0001).

Chlorophyll Content in Sunflower Plants upon Infection with *O. cumana*

The time course of the total Chl content of the first four LPs (A to D) is presented in Figure 6. Differences between the Chl content of inoculated and control plants depended on the LP analyzed, being significantly low ($P < 0.05$) in the second LP of inoculated plants at nearly all time points (7.70 arbitrary units as compared to 8.94 in the controls). The measurements conducted on the fourth LP resulted also in significantly ($P < 0.05$) lower values of inoculated plants as compared to the controls (5.38 and 6.98 arbitrary units respectively). In the case of the first and the third LP, no significant ($P > 0.05$) differences in total Chl content were detected between inoculated and control plants.

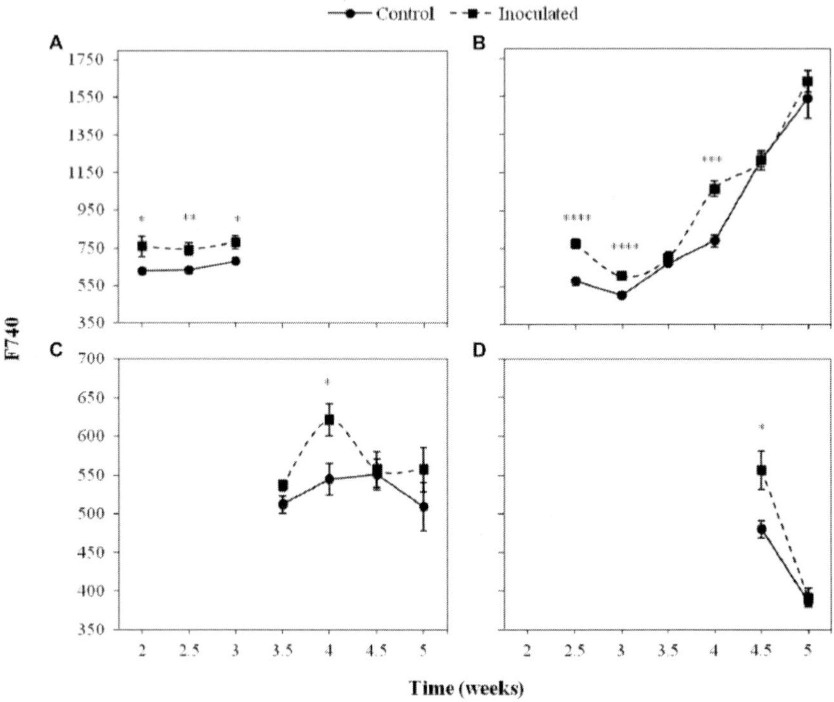

FIGURE 3. Mean measurements of F740 of the first (A), second (B), third (C), and fourth (D) LP of sunflower plants inoculated with O. cumana and control plants. Vertical bars represent the standard error of the mean of 14–20 replications. Analyses of variance were conducted and asterisks indicate significant differences (*P < 0.05; **P < 0.01; ***P < 0.001; ****P < 0.0001).

Spectrophotometric measurements of Chl content are presented in Supplementary Figure S1, which shows the differences in the total Chl content of each LP between control and inoculated plants. A significant ($P < 0.05$) increase in the total Chl content of the second LP (Supplementary Figure S1) was observed in inoculated plants as compared to the controls (1659.90 and 1460.19 μg/g fresh weight, respectively). On the contrary, measurements of the fourth LP resulted in significantly ($P < 0.05$) lower total Chl content in inoculated plants than in control plants (1050.90 and 1639.31 μg/g fresh weight, respectively). No significant ($P > 0.05$) differences of total Chl content were detected in the third LP. Neither differences in the Chl a/b ratio of LPs were observed between inoculated and control plants (data not shown).

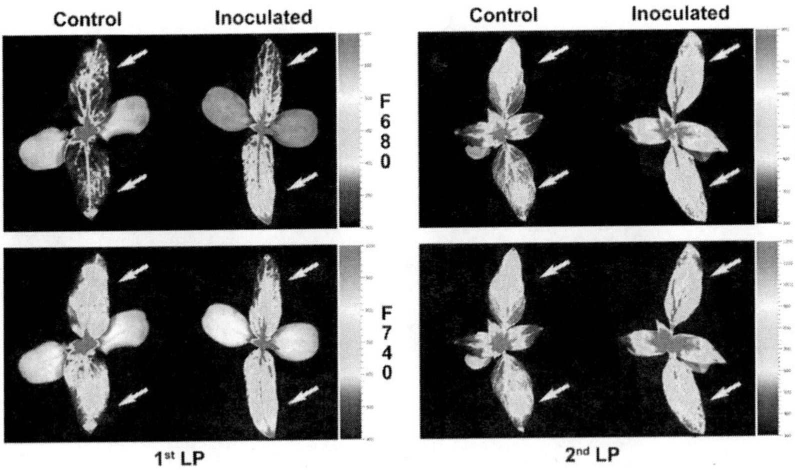

FIGURE 4. Red fluorescence (F680) and far- red fluorescence (F740) images of the two first LP of sunflower plants inoculated with O. cumana and control plants at 2 and 2.5 growth weeks respectively. First LPs are marked with an arrow.

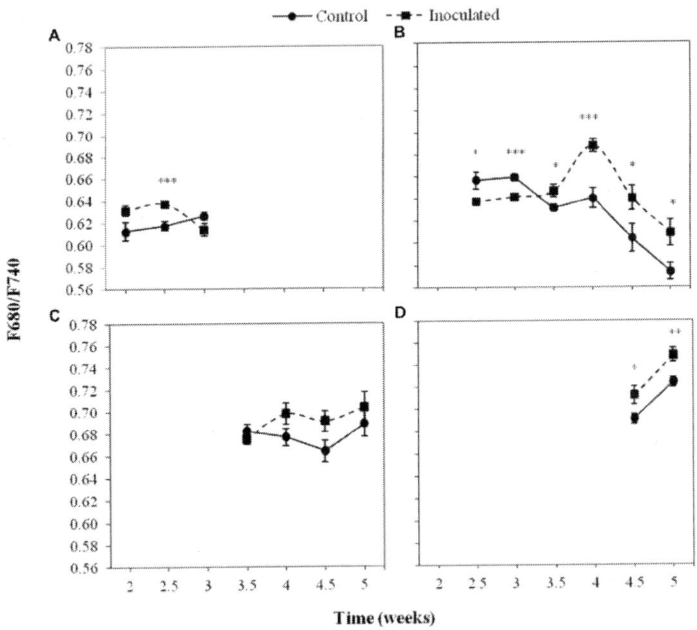

FIGURE 5. Mean measurements of the F680/F740 ratio of the first (A), second (B), third (C), and fourth (D) LP of sunflower plants inoculated with O. cumana and control plants. Vertical bars represent the standard error of the mean of 14–20 replications. Analyses of variance were conducted and asterisks indicate significant differences (*$P < 0.05$; **$P < 0.01$; ***$P < 0.001$).

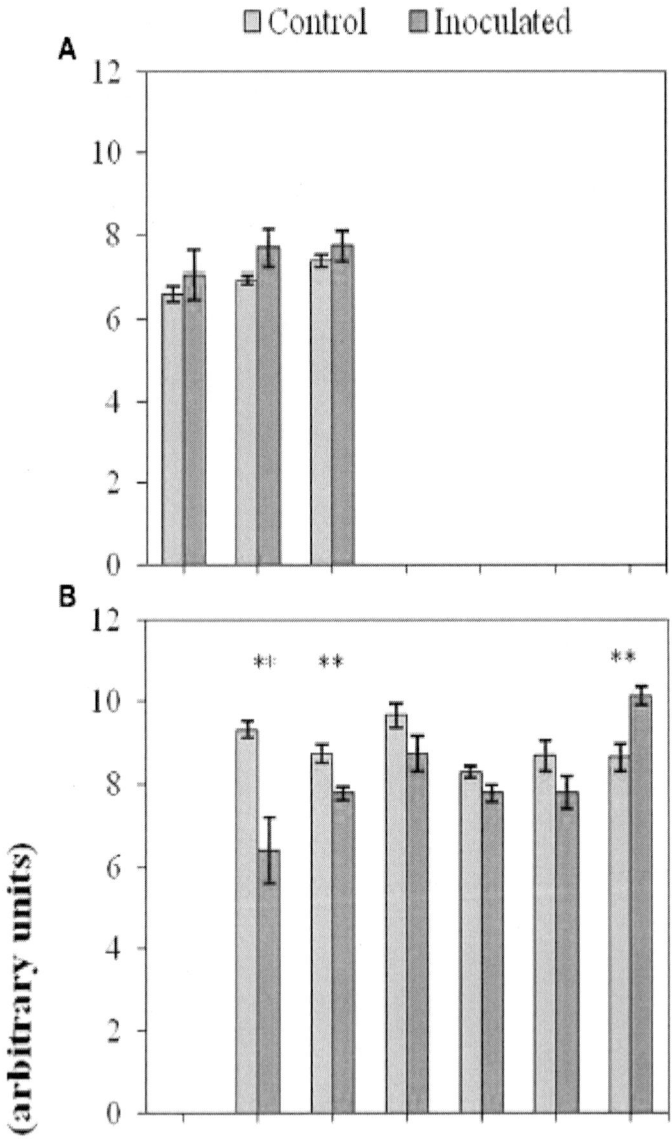
Control
Inoculated
A
12
10
8
6
4
2
0
B
12
10
8
6
4
2
0
**
**
**
(arbitrary units)

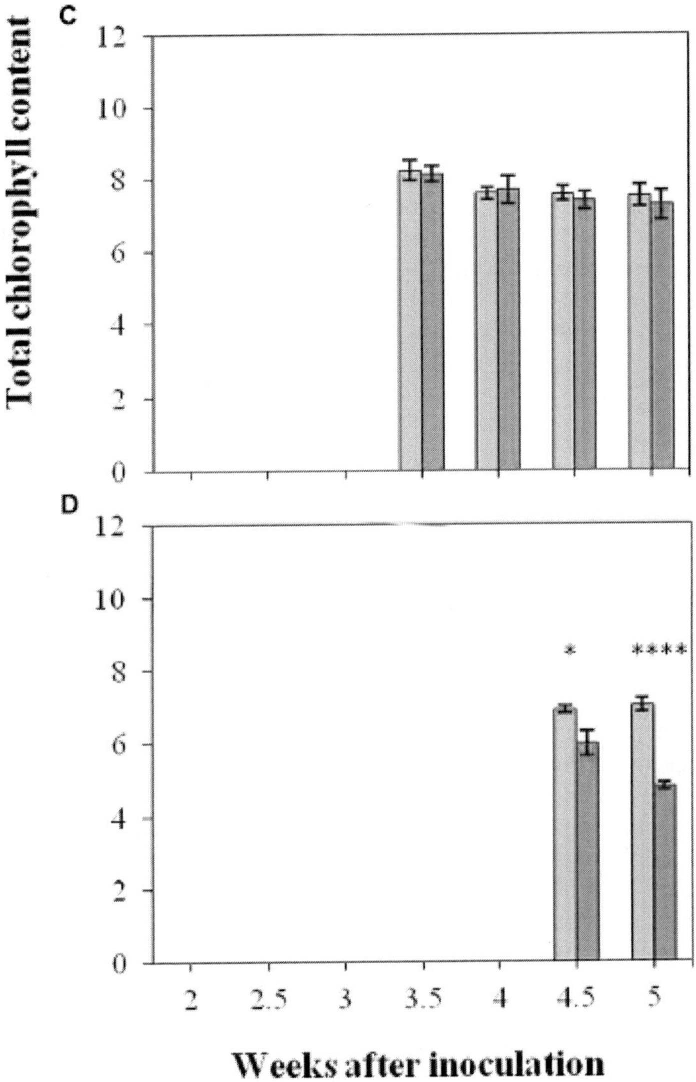

FIGURE 6. Total chlorophyll content of the first (A), second (B), third (C), and fourth (D) LP of control plants and plants inoculated with O. cumana. Vertical bars represent the mean of 12–20 independent measurements with their standard error. Analyses of variance were conducted at P < 0.05 and asterisks indicate significant differences (∗P < 0.05; ∗∗P < 0.01; ∗∗∗∗P < 0.0001).

Determination of Final Degree of Sunflower Infection by *O. cumana*

When sunflower plants were inoculated with *O. cumana*, both inoculated and control plants developed four LPs at the end of the experiment (5 wai). No symptoms of parasite infection were observed in the inoculated plants throughout the experiment. Nevertheless, the presence of the parasitic plant in inoculated sunflowers was confirmed by the existence of tubercles in the roots at the end of the experiment. While no tubercles were observed in the controls, means of 24.4 tubercles/plant and 0.46 g tubercles/plant were obtained in inoculated sunflowers (data not shown).

DISCUSSION

During our study, the emission of red and far-red fluorescence in healthy sunflower depended on the LP analyzed. F680 and F740 nm showed the same trend in all the LPs analyzed, with higher F740 values compared to those of F680, probably due to the partial reabsorption of the red fluorescence emitted by leaf Chl *a* (Gitelson et al., 1998). Since F680/F740 was inversely correlated with the Chl content of leaves (Hák et al., 1990; Buschmann and Lichtenthaler, 1998), low values found in the first LP during early stages of the growth of sunflower, compared with the other three LPs under study, were indicative of higher Chl contents. A similar trend of low F680/F740 was observed in the case of the second LP, although with some delay in the time. In our study, the role of an LP as leader in Chl content was sequentially acquired by the one immediately above, and this happened approximately every 1–2 weeks. A different physiological activity between cotyledons and the first LP of sunflower, in terms of photosynthesis, has been reported by Jucknischke and Kutschera (1998). Also, differences in Chl content and photosynthesis have been reported as being dependent on leaf position in sunflower (Connor and Sadras, 1992; Subrahmanyam and Rathore, 2004). The correlation between photosynthetic activity and Chl content in individual LPs upon infection with *O. cumana* should be investigated in the future.

Increases in both F680 and F740 upon infection of *O. cumana* were observed in all LPs analyzed compared to that of the healthy controls in agreement with previous results by our group (Pérez-Bueno et al., 2014). The evolution of F680/F740 throughout the time was similar to that of F680 and F740 in the same LP. Values of F680/F740 increased for the second to fourth LPs upon infection by *O. cumana* and at different times in each LP, which could be indicative of a decrease in the Chl content upon infection assuming a similarity in the results reported by Gitelson et al.

(1999), where the ratio F700/F730 was presented as an indicator of Chl content. Similarly, to those observed in F680 and F740 of LP from healthy sunflowers, the decreases were detected in different LPs sequentially in the time. In inoculated plants, higher values of F680/F740 in the first LP until the third week after inoculation indicated lower Chl contents. From the third week on, a decrease in Chl content as compared to that of the controls occurred in the second LP of the plants. The same trend was observed in the third and fourth LPs when measurements of inoculated and control plants were compared later on in the time. The alteration in F680 and F740 due to *O. cumana* infection shows the usefulness of the MCFI as an early indicator of parasite infection, as was suggested in a previous work (Pérez-Bueno et al., 2014). Furthermore, our results showed that the best indicators of parasite infection were the measurements of F680, F740, and F680/F740 in the first and second LPs from the second to the third week of plant growth. These findings confirm previous results by our research group about early detection of parasite infection by means of fluorescence emission of the first LP of sunflower. Even 3 wai proposed by Pérez-Bueno et al. (2014) as the earliest time for discriminating infected from uninfected plants, was reduced in this work to 2 wai. Therefore, the non-destructive pre-symptomatic disease diagnosis might be feasible as early as 15 days after sunflower inoculation.

Low Chl contents in LP from infected plants at certain times were also shown by outputs of the portable Chl-meter, which showed significantly low Chl contents in the second LP 4.5 weeks after inoculation and in the fourth LP from that moment on. Our results also reveal the greater sensitivity and accuracy of MCFI in diagnosing sunflower infection by *O. cumana* as compared to the direct measurements of Chl in intact leaves, because trends observed when using the hand-held Chl-meter were statistically significant according to red and far-red Chl fluorescence emissions. The positive correlation between outputs of portable Chl-meters and Chl fluorescence measurements has been reported by other authors (Netto et al., 2002, 2005; Nyi et al., 2012; Hu and Fang, 2013). In our work, F680 and F740 in parasitized plants were in agreement with direct measurements of Chl, particularly when the second and fourth LPs were considered. In addition, the spectrophotometric measurements of Chl content at 5 wai provided further evidence that, upon parasite infection, the Chl content in sunflower leaves decreased. The positive correlation between outputs of portable Chl-meters and photosynthetic pigments extracted from leaves has been reported by many authors (Fanizza et al., 1991; Netto et al., 2005; Brito et al., 2011). Finally, the greater sensitivity of MCFI compared to measurements of Chl content might be related to the fact that fluorescence intensity is not only dependent on the Chl content, but also on other factors such as leaf architecture or chloroplast organization

(Cerovic et al., 1999). The effect of parasite infection on them might be explored in the future.

Early studies detected a decrease in Chl content in leaves of *Trifoliumrepens* upon infection by *O. minor*, although no effect on the host photosynthesis was observed (Dale and Press, 1998). Some authors, such as Hibberd et al. (1999) and Watling and Press (2001), suggested that the high competitiveness of holoparasites as a source forces an increase in host photosynthesis. On the contrary, other authors reported low rates of photosynthetic activity of the host upon infection by the holoparasite, such as in the case of tomato plants infected by the broomrape *O. ramosa* (Mauromicale et al., 2008) or *Mikaniamicrantha* infected by *Cuscutacampestris* (Shen et al., 2007). It is widely known that Chl fluorescence is an excellent approach to photosynthetic efficiency, which can directly or indirectly reflect the impact of biotic and abiotic stresses on plants. Hence, a number of studies have used chlorophyll fluorescence imaging to examine the impact of plant pathogens, i.e., virus, bacteria, and fungi, on host photosynthesis, as excellently reviewed by Rolfe and Scholes (2010). In future research, imaging measurements of Chl fluorescence kinetics in sunflower upon infection by *O. cumana* might explore on the plant photosynthesis alteration as a consequence of the holoparasite infection.

Generally, MCFI has been applied to the study of local symptoms in leaves caused by infections of airborne pathogens (Chaerle et al., 2007; Konanz et al., 2014; Pérez-Bueno et al., 2015). However, symptoms in leaves can as well be a consequence of an infection that occurred far beyond those organs, such as roots in the case of soilborne pathogens (Granum et al., 2015). In the latter, the responses observed in leaves clearly relate to plant defense mechanisms and the alteration of the host physiology due to the infection, since any direct damage in them can be discarded. Up to now, plant infections by soilborne pathogens have only been addressed by means of MCFI in Granum et al. (2015). Rousseau et al. (2015) have recently proved the potential of Chl fluorescence to detect the effect of *Phelipancheramosa* on *Arabidopsis thaliana*. However, to the best of our knowledge, this is the first time that the impact of a root parasite on its plant host is examined by using fluorescence imaging in the red and far-red region. This inexpensive, non-destructive technic has proved to be an effective diagnostic tool in the early detection of sunflower plants parasitized by *O. cumana*. Moreover, the parameters F680 and F740 detect the parasite infection as early as 2 weeks after inoculation, and this might have a particular interest for early phenotyping in sunflower breeding programs.

ACKNOWLEDGMENTS

Financial support for this research was provided by "Consejería de Economía, Innovación y Ciencia" of the Andalusian Government (P12-AGR1281 to LM-R and P12-AGR370 to MB) and the European Social Fund, and RECUPERA 2020 (grant number 20134R060 to MB) and FEDER Funds.

REFERENCES

1. Brito, G. G., Sofiatti, V., Brandão, Z. N., Silva, V. B., Silva, F. M., and Silva, D. A. (2011). Non-destructive analysis of photosynthetic pigments in cotton plants. *ActaScie.Agro.* 33, 671–678. doi: 10.4025/actasciagron.v33i4.10926
2. Buschmann, C. (2007). Variability and application of the chlorophyll fluorescence emission ratio red/far-red of leaves. *Photosynth. Res.* 92, 261–271. doi: 10.1007/ s11120-007-9187-8
3. Buschmann, C., Langsdorf, G., and Lichtenthaler, H. K. (2000). Imaging of the blue, green, and red fluorescence emission of plants: an overview. *Photosynthetica* 38, 483–491. doi: 10.1023/A:1012440903014
4. Buschmann, C., and Lichtenthaler, H. K. (1998).Principles and characteristics of multicolour fluorescence imaging of plants.*J. Plant Physiol.* 152, 297–314. doi: 10.1016/S0176-1617(98)80144-2
5. Calderón, R., Lucena, C., Trapero-Casas, J. L., Zarco-Tejada, P. J., and Navas-Cortés, J. A. (2014). Soil temperature determines the reaction of olive cultivars to *Verticilliumdahliae*Pathotypes. *PLoS ONE* 9: e110664. doi: 10.1371/ journal. pone.0110664
6. Cerovic, Z. G., Samson, G., Morales, F., Tremblay, N., and Moya, I. (1999). Ultraviolet-induced fluorescence for plant monitoring: present state and prospects. *Agronomie* 19, 543–578. doi: 10.1051/agro:19990701
7. Chaerle, L., Lenk, S., Hagenbeek, D., Buschmann, C., and Van Der Straeten, D. (2007). Multicolor fluorescence imaging for early detection of the hypersensitive reaction to tobacco mosaic virus.*J. Plant Physiol.* 164, 253–262. doi: 10.1016/j. jplph.2006.01.011
8. Chang, S. X., and Robison, D. J. (2003).Nondestructive and rapid estimation of hardwood foliar nitrogen status using the SPAD-502 chlorophyll meter. *For. Ecol. Manage.* 181, 331–338. doi: 10.1016/S0378-1127(03)00004-5
9. Connor, D. J., and Sadras, V. O. (1992).Physiology of yield expression in sunflower. *Field Crop Res.* 30, 333–389. doi: 10.1016/0378-4290(92)90006-U
10. Dale, H., and Press, M. C. (1998). Elevated atmospheric CO2 influences the interaction between the parasitic angiosperm *Orobanche minor* and its host *Trifoliumrepens. New Phytol.* 65–73.doi: 10.1046/j.1469-8137.1998.00247.x
11. Domínguez, J. (1996). "Estimating effects on yield and other agronomic parameters in sunflower hybrids infested with the new races of sunflower broomrape," in: *Symposium I: Disease tolerance in Sunflower*, ed. A. Pouzet (Int. Sunflower Assoc: Paris), 118–123.
12. Fanizza, G., Ricciardi, L., and Bagnulo, C. (1991). Leaf greenness measurements to evaluate water stressed genotypes in *Vitisvinifera. Euphytica* 55, 27–31. doi: 10.1007/BF00022556
13. García-Carneros, A. B., Dedic, B., Miladinovic, D., and Molinero-Ruiz, L. (2014). "Pathogenic comparison of highly virulent *O. cumana* affecting sunflower in Moldova, the South of Russian Federation, Serbia and Spain," in *Proceedings of the*

Third International Symposium of Orobanche spp. of Sunflower, ed. International Sunflower Association (Paris: International Sunflower Association), 173–177.

14. Gitelson, A. A., Buschmann, C., and Lichtenthaler, H. K. (1998). Leaf chlorophyll fluorescence corrected for re-absorption by means of absorption and reflectance measurements. *J. Plant Physiol.* 152, 283–296. doi: 10.1016/S0176-1617(98)80143-0
15. Gitelson, A. A., Buschmann, C., and Lichtenthaler, H. K. (1999).The chlorophyll fluorescence ratio F 735/F 700 as an accurate measure of the chlorophyll content in plants. *Remote Sens. Environ.* 69, 296–302. doi: 10.1016/S0034-4257(99)00023-1
16. Granum, E., Pérez-Bueno, M. L., Calderón, C. E., Ramos, C., de Vicente, A., Cazorla, F. M., et al. (2015). Metabolic responses of avocado plants to stress induced by *Rosellinianecatrixanalysed by fluorescence and thermal imaging. Eur. J. Plant Pathol.* 142, 625–632. doi: 10.1007/s10658-015-0640-9
17. Hák, R., Lichtenthaler, H. K., and Rinderle, U. (1990). Decrease of the chlorophyll fluorescence ratio F690/F730 during greening and development of leaves. *Radiat. Environ. Biophys.* 29, 329–336. doi: 10.1007/BF01210413
18. Harris, P. J., and Hartley, R. D. (1976).Detection of bound ferulic acid in the cell walls of the Gramineae by ultraviolet fluorescence microscopy. *Nature* 259, 508–510. doi: 10.1038/259508a0
19. Hibberd, J. M., Quick, W. P., Press, M. C., Scholes, J. D., and Jeschke, W. D. (1999). Solute fluxes from tobacco to the parasitic angiosperm *Orobanchecernua* and the influence of infection on host carbon and nitrogen relations. *Plant Cell Environ.* 22, 937–947. doi: 10.1046/j.1365-3040.1999.00462.x
20. Hu, H., and Fang, S. (2013). "Relation of leaf image, chlorophyll fluorescence, reflectance and SPAD in rice and barley," in *Proceedings of the Second International Conference on Agro-Geoinformatics (Agro-Geoinformatics)* (Rome: IEEE), 28–31.
21. Jestin, C., Lecomte, V., and Duroueix, F. (2014). "Current situation of sunflower broomrape in France," in *Current Situation of Sunflower Broomrape Around the World, Proceedings of the Third International Symposium on Broomrape* (Orobanche spp.) *in Sunflower*, ed. International Sunflower Association (Paris: International Sunflower Association), 28–31.
22. Joel, D. M., Chaudhuri, S. K., Plakhine, D., Ziadna, H., and Steffens, J. C. (2011).Dehydrocostus lactone is exuded from sunflower roots and stimulates germination of the root parasite *Orobanchecumana*. *Phytochemistry* 72, 624–634. doi: 10.1016/j.phytochem.2011.01.037
23. Jucknischke, A., and Kutschera, U. (1998).The role of the cotyledons and primary leaves during seedling establishment in sunflower. *J. Plant Physiol.* 153, 700–705. doi: 10.1016/S0176-1617(98)80223-X
24. Konanz, S., Kocsányi, L., and Buschmann, C. (2014).Advanced multi-color fluorescence imaging system for detection of biotic and abiotic stresses in leaves. *Agriculture* 4, 79–95. doi: 10.3390/agriculture4020079
25. Lang, M., Lichtenthaler, H. K., Sowinska, M., Summ, P., and Heisel, F. (1994).Blue, green and red fluorescence signatures and images of tobacco leaves. *Bot. Acta* 107, 230–236. doi: 10.1111/j.1438-8677.1994.tb00790.x
26. Lichtenthaler, H. K., and Buschmann, C. (2001). Chlorophylls and carotenoids: measurement and characterization by UV-VIS spectroscopy. *Curr.Protoc.Food Anal.Chem.* (Suppl. 1), F4.3.1–F4.3.8.doi: 10.1002/0471142913.faf0403s01
27. Liu, Z. A., Yang, J. P., and Yang, Z. C. (2012).Using a chlorophyll meter to estimate tea leaf chlorophyll and nitrogen contents. *J. Soil Sci. Plant Nutr.* 12, 339–348. doi: 10.4067/S0718-95162012000200013
28. Mauromicale, G., Monaco, A. L., and Longo, A. M. (2008).Effect of branched broomrape (*Orobancheramosa*) infection on the growth and photosynthesis of tomato. *Weed Sci.* 56, 574–581. doi: 10.1614/WS-07-147.1
29. Melero-Vara, J. M., and Alonso, L. C. (1988). "Las Enfermedadesdel Girasol," in *Enfermedades y Daños de Herbicidas en el Cultivo del Girasol*, ed. Koipesol (Madrid: ArtesGráficas EMA), 33–39.

30. Molinero-Ruiz, L., Delavault, P., Pérez-Vich, B., Pacureanu-Joita, M., Bulos, M., Altieri, E., et al. (2015). History of the race structure of *Orobanchecumana* and the breeding of sunflower for resistance to this parasitic weed: a review. *Span. J. Agric. Res.* 13: e10R01. doi: 10.5424/sjar/2015134-8080
31. Molinero-Ruiz, M. L., García-Ruiz, R., Melero-Vara, J. M., and Domínguez, J. (2009). *Orobanchecumana* race F: performance of resistant sunflower hybrids and aggressiveness of populations of the parasitic weed. *Weed Res.* 49, 469–478. doi: 10.1111/j.1365-3180.2009.00708.x
32. Morales, F., Cerovic, Z. G., and Moya, I. (1996). Time resolved blue-green fluorescence of sugar beet (*Beta vulgaris* L.) leaves: spectroscopic evidence for the presence of ferulic acid as the main fluorophore of the epidermis. *BBA Bioenergetics* 1273, 251–262. doi: 10.1016/0005-2728(95)00153-0
33. Netto, A. T., Campostrini, E., de Oliveira, J. G., and Bressan-Smith, R. E. (2005). Photosynthetic pigments, nitrogen, chlorophyll a fluorescence and SPAD-502 readings in coffee leaves. *Sci. Hortic.* 104, 199–209. doi: 10.1016/j. scienta.2004.08.013
34. Netto, A. T., Campostrini, E., de Oliveira, J. G., and Yamanishi, O. K. (2002).Portable chlorophyll meter for the quantification of photosynthetic pigments, nitrogen and the possible use for assessment of the photochemical process in *Carica papaya* L. *Braz. J. Plant Physiol.* 14, 203–210.doi: 10.1590/S1677-042020020003 00004
35. Nyi, N., Sridokchan, W., Chai-arree, W., and Srinives, P. (2012).Nondestructive measurement of photosynthetic pigments and nitrogen status in *Jatropha* (*Jatrophacurcas* L.) by chlorophyll meter. *Philipp. Agric. Sci.* 95, 83–89.
36. Pérez-Bueno, M. L., Barón, M., Garcia-Carneros, A. B., and Molinero-Ruiz, L. (2014). Diagnosis of the infection of sunflower by *Orobanchecumana* using multicolour fluorescence imaging. *Helia* 37, 161–174. doi: 10.(1515)/helia-2014-0015
37. Pérez-Bueno, M. L., Pineda, M., Díaz-Casado, E., and Barón, M. (2015).Spatial and temporal dynamics of primary and secondary metabolism in *Phaseolus vulgaris* challenged by *Pseudomonas syringae*. *Physiol. Plant.* 153, 161–174. doi: 10.1111/ppl.12237
38. Raupp, F. M., and Spring, O. (2013). New sesquiterpene lactones from sunflower root exudate as germination stimulants for *Orobanchecumana*. *J. Agric. Food Chem.* 61, 10481–10487. doi: 10.1021/jf402392e
39. Rolfe, S. A., and Scholes, J. D. (2010).Chlorophyll fluorescence imaging of plant-pathogen interactions. *Protoplasma* 247, 163–175. doi: 10.1007/s00709-010-0203-z
40. Rousseau, C., Hunault, G., Gaillard, S., Bourbeillon, J., Montiel, G., Simier, P., et al. (2015).Phenoplant: a web resource for the exploration of large chlorophyll fluorescence image datasets. *Plant Methods* 11, 24.doi: 10.1186/s13007-015-0068-4
41. Shen, H., Hong, L., Ye, W., Cao, H., and Wang, Z. (2007). The influence of the holoparasitic plant *Cuscutacampestris* on the growth and photosynthesis of its host *Mikaniamicrantha*. *J. Exp. Bot.* 58, 2929–2937.doi: 10.1093/jxb/erm168
42. Subhash, N., Wenzel, O., and Lichtenthaler, H. K. (1999).Changes in blue-green and chlorophyll fluorescence emission and fluorescence ratios during senescence of tobacco plants. *Remote Sens. Environ.* 69, 215–223. doi: 10.1016/S0034-4257(99)00029-2
43. Subrahmanyam, D., and Rathore, V. S. (2004).14CO2 assimilation and 14C-photosynthate translocation in different leaves of sunflower. *Photosynthetica* 42, 313–316. doi: 10.1023/B:PHOT.0000040607.67963.38
44. Ueno, K., Furumoto, T., Umeda, S., Mizutani, M., Takikawa, H., Batchvarova, R., et al. (2014).Heliolactone, a non-sesquiterpene lactone germination stimulant for root parasitic weeds from sunflower. *Phytochemistry* 108, 122–128. doi: 10.1021/jf2024193
45. Watling, J. R., and Press, M. C. (2001).Impacts of infection by parasitic angiosperms on host photosynthesis. *Plant Biol.* 3, 244–250. doi: 10.1055/s-2001-15195

46. Yoneyama, K., Xie, X. N., Kisugi, T., Nomura, T., Sekimoto, H., Yokota, T., et al. (2011).Characterization of strigolactones exuded by Asteraceae plants. *Plant Growth Regul.* 65, 495–504. doi: 10.1007/s10725-011-9620-z

CHAPTER 8

Trends of Soybean Yields under Climate Change Scenarios

Frank Eulenstein[1,2,*], Marcos Lana[1], Sandro Schlindwein[3], Askhad Sheudzhen[2], Marion Tauschke[1], Axel Behrend[1], Edgardo Guevara[4] and Santiago Meira[4]

[1] *Leibniz-Zentrum für Agrarlandschaftsforschung (ZALF) Müncheberg, Eberswalder Straße 84,Müncheber15374,Germany*

[2] *Department Agro-Chemistry, Kuban State Agrarian University, Krasnodar 350044, Russia*

[3] *Departamento de Engenharia Rural, Universidade Federal de Santa Catarina, Florianópolis 88034-000, Brazil*

[4] *Department of Crop Production INTA—Instituto Nacional de Tecnologia Agropecuaria, Pergamino2700,Argentina*

ABSTRACT

Soybean is a very important crop, cultivated mainly as stock feed for animal production, but also for other uses such as biodiesel. Brazil is the second largest producer of soybeans, and the main exporter. About 10% of total Brazilian production is aimed for biodiesel production. The aim of this work was to assess the impact of climate change scenarios on soybean yield and evaluate two simple adaptation strategies: cultivar and planting date. Tests were done for soil profiles from two important producing regions: Chapecó-Red Oxisol, and Passo Fundo-Rodic Hapludox. Two commercial soybean cultivars (CD202 and CD204) and seven regional circulation models (RCM) were used. All simulations were done with DSSAT. After model calibration, eleven planting dates were run for two periods (2011–2040 and 2071–2100) using the RCM's. There were no differences between cultivars. For Chapecó, the majority of RCM's projected yield reductions, with few RCM's projecting increments, and for only few planting dates (November). The response pattern for both time periods were identical, although the end-of-century period presented a further yield reduction. The main reason was reduced water holding capacity in soil, high temperatures, and changes in rainfall distribution along the cropping season. For Passo Fundo, 2011–2040 yields were distinct, depending on the RCM. Simulated yields tended to follow the actual yield pattern along the different planting dates, besides discrepancies. For 2071–2100, all but one RCM indicated yields equal or lower to actual levels. Regarding planting dates, no significant changes were identified, although reductions were observed for the early planting dates (August–September). The scenarios suggest that soybean yields will be reduced, jeopardizing the viability of this crop and biodiesel production in the studied regions.

Keywords: climate change; crop model; efficiency use

1. INTRODUCTION

Climate change is a worldwide concern, and regardless of all the scientific improvements to comprehend and forecast changes, the determination of future climate is still a very hard task. The high level of complexity and the nature of climatic interactions is a challenge to forecasting, although there are scenarios that point to possible directions of change.

The impact of climate change on agricultural production is actually the core issue of several investigations. Rising seasonal temperatures are expected to increase more than the annual averages, with reduced precipitation expected to accompany higher temperatures in some regions. Additionally, heat waves are expected to increase in frequency, intensity, and duration [1]. End-of-century growing season temperatures in the tropics and subtropics may exceed even the most extreme seasonal temperatures measured to date [2]. Not considering all the inherent variability of crop production factors, all climate changes described above can lead to modifications of crop yields, posing a threat to agricultural systems that will affect the whole production and consumption chain, impacting especially agroecosystems and populations with low availability of or access to financial and natural resources. The global food and financial crises of 2007 and 2008, which have pushed an additional 115 million into hunger, highlight the severity of the hunger and poverty crisis that has challenged the world for decades [3]. Price volatility remains a concern, with weather-related yield variability the main threat as long as stocks remain low [4]. This risky situation will be worsened by the present effects of drought on soybean yields of USA [5], which will impact the whole world's food supply.

Increasing the prediction capacity of climate change impacts for stakeholders has become a major challenge in Southern Brazil, where economic wealth strongly depends on agriculture [6,7]. In this region, the agricultural landscape has faced major changes during the last 30 years due to new technologies for crops, to a strong increase in cereal and oil crop world demands, and also to favorable climate conditions with increases of about 20%–30% in annual precipitation over large parts of the region [8].

Crop models can be a useful tool to assess the influence of climatic and other environmental or management factors on crop development and yield [9]. The Decision Support System for Agrotechnology Transfer—DSSAT v. 4.5 contains the CROPGRO-Soybean model [10], and is used to (a) determine best planting dates [11], (b) for fertilization timing [12], (c) in precision agriculture [13], and (d) also for detecting/investigating potential impacts of climate change on agriculture [14]. In the embedded model the development and growth of the crop is simulated on a daily basis from the planting until physiological maturity. The model calculations are based on environmental and physiological processes that control the phenology and dry matter accumulation in the different organs of the plant. The DSSAT also has other embedded models that can simulate the flow of nutrients and water balance in the soil. The minimum data set necessary to run DSSAT [15] consists of daily weather data of maximum and minimum temperature, rainfall and solar radiation, soil chemical and physical parameters for each layer, genetic coefficients for each cultivar with information about development and biomass accumulation, and management information, such as soil preparation, planting dates, plant density, fertilization amounts and timing or other agricultural practices. Experimental data like soil available water, plant phenology, biomass partitioning and other morphological components like leaf area index are necessary to calibrate the genetic coefficients and check the accuracy of the model [16].

2. EXPERIMENTAL SECTION

In order to run simulations for soybeans, data from field experiments and literature were used. For simulation in the Brazilian sites, data from literature was obtained from Dallacort et al. [10], who performed experiments in Parana State evaluating four soybean cultivars. The cultivars were characterized, calibrated and validated for the CROPGRO-Soybean. The four cultivars, namely CD 202, CD 204, CD 206 and CD 210, were tested for both Brazilian sites using census data and generic agronomic management. The two cultivars with lowest RMSE for yield were selected to run further analyses.

After calibrating and validating the genetic parameters and the model itself, scenarios provided by CLARIS LPB Project WP5 (2011–2040 and 2071–2100 periods) were downloaded and formatted for the DSSAT standard using Weatherman Software [17]. From the CLARIS-LPB Project Data Archive Center, seven weather series of RCM's (and matching the same location of the study sites weather stations) were downloaded, converted and adjusted to be used as weather input for DSSAT using Weatherman software. The RCM's are RCA1, RCA2 and RCA3, from the Rossby Centre Regional Climate model [18]; PROMES, from Universidad de Castilla-La Mancha [19]; LMDZ version 4 Configuration South America with IPSLA1B and EC5OM-R3 boundaries, from Laboratoire de Meteorologie Dynamique [20]; and ETA, from Instituto Nacional de Pesquisas Espaciais [21].

The crop model was run with each one of the seven RCM's for the target periods (2011–2040 and 2071–2100). Simulations of the impact of RCM's scenarios on soybean cultivars (CD202 and CD204) planted on eleven different dates, in two locations (Chapecó and Passo Fundo), and two time periods (2011–2040 and 2071–2100, summing 29 cropping seasons each). Black lines represent median yields simulated with RCM's while black bars represent the standard error of each planting date for the 29 seasons; the grey lines represent actual yields with respective planting dates and standard error.

3. RESULTS AND DISCUSSION

The analyses (Figure 1) showed the impact of seven RCM's on the yield of the soybean cultivars CD202 and CD204 in two locations and two time periods (2011–2040 and 2071–2100). It is important to mention that both soybean cultivars, besides having differences in genetic coefficients, presented very similar results.

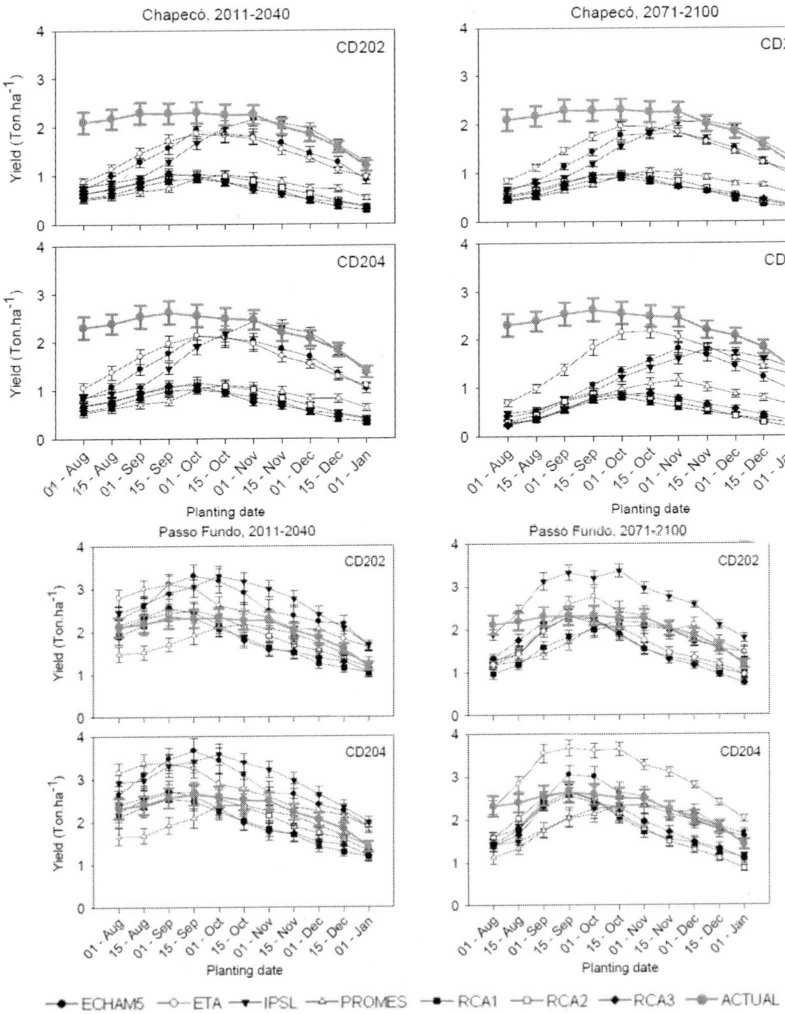

Figure 1. Simulations of the impact of RCM's scenarios on soybean cultivars (CD202 and CD204) planted on eleven different dates, in two locations (Chapecó and Passo Fundo), and two time periods (2011–2040 and 2071–2100): black lines represent yields simulated with RCM's and black bars represent the standard error of each planting date; the grey lines represent actual yields with respective planting dates and standard error.

For the Chapecó 2011–2040 period, the majority of RCM's projected very low yields when compared with actual yields. Only ETA, IPSL and ECHAM5 presented a trend of increase in yields, and after the 1 October planting date. Even so, only IPSL could mimic the actual yields for the late planting dates. This assessment is also applicable for the 2071–2100 period, but with a further reduction of projections of all RCMs. An integrated analysis indicates with high level of agreement that early planting dates—prior to 1 October—will generate lower yields; planting after 1 October shows that three out of seven RCM's (namely, ETA, ECHAM5 and IPSL) have a tendency to follow the actual yields, while the others remain with very low yields, jeopardizing the viability of this crop in the region.

The results presented for Passo Fundo showed significant difference from the ones of Chapecó, with RCM yields following the trend of actual yield. It also presents a situation where RCMs project even significant increments in yield for the 2011–2040 period. This can be observed especially in the early planting dates, where all but one RCM are equal or significantly higher than the actual yield. For the end-of-century period a generalized reduction of yield was calculated, with exception of IPSL, which showed significant increases. Though a trend of yield reduction, all RCMs presented at least one planting date that did not differ significantly from the actual best yields.

4. CONCLUSIONS

Both genotypes tested (CD202 and CD204) did not present remarkable differences among them when in the same region. Unfortunately, no other suitable soybean data sets are available to calibrate and validate the crop model in the study region, undermining the assessment of the role of cultivar as adaptation strategy. The impact of climate scenarios on soybean yield was directly influenced by location: in Chapecó region yields tend to decrease, while for Passo Fundo region yields could even increase.

ACKNOWLEDGMENTS

This work in the CLARIS LPB Project was supported the Environmental Risks Unit, Directorate Environment, DG-Research, European Commission (http://www.claris-eu.org).

REFERENCES

1. Tebaldi, C.; Hayhoe, K.; Arblaster, J.M.; Meehl, G.A. Going to the extremes. *Clim. Chang.* 2006, *79*, 185
2. Battisti, D.S.; Naylor, R.L. Historical warnings of future food insecurity with unprecedented seasonal heat. *Science*2009, *323*, 240–244
3. Viatte, G.; De Graaf, J.; Demeke, M.; Takahatake, T.; Rey de Arce, M. *Responding to the Food Crisis: Synthesis of Medium-Term Measures Proposed in Inter-Agency Assessments*; FAO: Rome, Italy, 2009
4. *OECD-FAO Agricultural Outlook 2012*; OECD: Paris, France, 2012
5. University of Nebraska-Lincoln Crop Watch 2012. Available online: http://cropwatch.unl.edu/2012archive (accessed on 18 May 2015)
6. FAO Global Information System of Water and Agriculture. Available online: http://www.fao.org (accessed on 13 May 2015)
7. Eulenstein, F.; Lana, M.; Schlindwein, S.L.; Sheudzhen, A.K.; Tauschke, M.; Guevara, E.; Meira, S. Impact of climate scenarios on soybean yields in southern Brazil. In Proceedings of the Moving Farm Systems to Improved Nutrient Attenuation: 28th Annual Fertilizer and Lime Research Centre Workshop, Palmerston North, New Zealand, 10–12 February 2015; p. 89
8. Magrin, G.O.; Travasso, M.I.; Rodríguez, G.R. Changes in climate and crop production during the 20th century in Argentina. *Clim. Chang.* 2005, *72*, 229–249
9. Reidsma, P.; Ewert, F.; Lansink, A.O.; Leemans, R. Adaptation to climate change and climate variability in European agriculture: The importance of farm level responses. *Eur. J. Agron.* 2010, *32*, 91–102
10. Dallacort, R.; Freitas, P.S.L.; Faria, R.T.; Goncalves, A.C.A.; Rezende, R.; Guimaraes, R.M.L. Simulation of bean crop growth, evapotranspiration and yield in Paraná State by the CROPGRO-Drybean model. *Acta Sci. Agric.* 2011, *33*, 429–436
11. Banterng, P.; Hoogenboom, G.; Patanothai, A.; Singh, P.; Wani, S.P.; Pathak, P.; Tongpoonpol, S.; Atichart, S.; Srihaban, P.; Buranaviriyakul, S.; et al. Application of the Cropping System Model (CSM)—CROPGRO—Soybean for determining optimum management strategies for soybean in tropical environments. *J. Agron. Crop Sci.* 2010, *196*, 231–242
12. D'orgeval, T.; Boulanger, J.P.; Capalbo, M.J.; Guevara, E.; Penalba, O.; Meira, S. Yield estimation and sowing date optimization based on seasonal climate information in the three CLARIS sites. *Clim. Chang.* 2010, *98*, 565–580
13. Asadi, M.E.; Clemente, R.S. Simulation of maize yield and N uptake under tropical conditions with the CERES-Maize model. *Trop. Agric.* 2001, *78*, 211–217
14. Thorp, K.R.; DeJonge, K.C.; Kaleita, A.L.; Batchelor, W.D.; Paz, J.O. Methodology for the use of DSSAT models for precision agriculture decision support. *Comput. Electron. Agric.* 2008, *64*, 276–285
15. Fischer, G.; Shah, M.; Tubiello, F.N.; Van Velhuizen, H. Socio-economic and climate change impacts on agriculture: An integrated assessment, 1990–2080. *Philos. Trans. R. Soc. B Biol. Sci.* 2005, *360*, 2067–2083
16. Jones, J.W.; Hoogenboom, G.; Porter, C.H.; Boote, K.J.; Batchelor, W.D.; Hunt, L.A.; Wilkens, P.W.; Singh, U.; Gijsman, A.J.; Ritchie, J.T. The DSSAT cropping system model. *Eur. J. Agron.* 2003, *18*, 235–265
17. Wilkens, P.W. DSSAT v4 weather data editing program (Weatherman). In *Data Management and Analysis Tools—Decision Support System for Agrotechnology Transfer Version 4.0: DSSAT v4: Data Management and Analysis Tools*; University of Hawaii: Honolulu, HI, USA, 2004; Volume 2, pp. 92–151
18. Samuelsson, P.; Jones, C.G.; Willen, U.; Ullrstig, A.; Gollvik, S.; Hansson, U.; Jansoon, C.; Kjellström, E.; Nikulin, G.; Wyser, K. The rossby centre regional climate model RCA3: Model description and performance. *Tellus A* 2011, *63*, 4–23

19. Domínguez, M.; Gaertner, M.; de Rosnay, P.; Losada, T. A regional climate model simulation over West Africa: Parameterization tests and analysis of land-surface fields. *Clim. Dyn.* 2010, *35*, 249–265
20. Hourdin, F.; Musat, I.; Bony, S.; Braconnot, P.; Codron, F.; Dufresne, J.L.; Fairhead, L.; Filiberti, M.A.; Friedlingstein, P.; Grandpeix, J.Y. The LMDZ4 general circulation model: Climate performance and sensitivity to parametrized physics with emphasis on tropical convection. *Clim. Dyn.* 2006, *27*, 787–813
21. Marengo, J.; Chou, S.; Kay, G.; Alves, L.; Pesquero, J.; Soares, W.; Santos, D.; Lyra, A.; Sueiro, G.; Betts, R.; et al. Development of regional future climate change scenarios in South America using the Eta CPTEC/HadCM3 climate change projections: Climatology and regional analyses for the Amazon, São Francisco and the Paraná River basins. *Clim. Dyn.* 2012, *38*, 1829–1848

CHAPTER 9

Cucumber (Cucumis sativus L.) Nitric Oxide Synthase Associated Gene1 (CsNOA1) Plays a Role in Chilling Stress

Xingwang Liu [1,2†], BinLiu [1,2†], ShudanXue[1 2], YanlinqCai[1,2], WenzhuQi[1,2], ChenJian[1,2], Shuo Xu[1,2], TingWang[1,2] and HuazhongRen[1,2*]

[1]*CollegeofHorticultureandDevelopmentalRegulationforProtectedVegetableCrops,ChinaAgriculturalUniversity,Beijing, China,*

[2]*BeijingKeyLaboratoryofGrowthandDevelopmentalRegulationforProtectedVegetableCrops,ChinaAgricultural University,Beijing,China*

ABSTRACT

Nitric oxide (NO) is a gaseous signaling molecule in plants, transducing information as a result of exposure to low temperatures. However, the underlying molecular mechanism linking NO with chilling stress is not well understood. Here, we functionally characterized the cucumber (*Cucumis sativus L.*) nitric oxide synthase-associated gene, *NITRIC OXIDE ASSOCIATED 1 (CsNOA1).* Expression analysis of *CsNOA1,* using quantitative real-time PCR, in situ hybridization, and a promoter::β-glucuronidase (GUS) reporter assay, revealed that it is expressed mainly in the root and shoot apical meristem (SAM), and that expression is up-regulated by low temperatures. A CsNOA1-GFP fusion protein was found to be localized in the mitochondria, and ectopic expression of *CsNOA1*in the *A.thaliana noa1* mutant partially rescued the normal phenotype. When overexpressing *CsNOA1* in the *Atnoa1* mutant under normal condition, no obvious phenotypic differences was observed between its wild type and transgenic plants. However, the leaves from mutant plant grown under chilling conditions showed hydrophanous spots and wilting. Physiology tolerance markers, chlorophyll fluorescence parameter (Fv/Fm), and electrolyte leakage, were observed to dramatically change, compared mutant to overexpressing lines. Transgenic cucumber plants revealed that the gene is required by seedlings to tolerate chilling stress: constitutive over-expression of *CsNOA1* led to a greater accumulation of soluble sugars, starch, and an up-regulation of Cold-regulatory C-repeat binding factor3 (CBF3) expression as well as a lower chilling damage index (CI). Conversely, suppression of *CsNOA1* expression resulted in the opposite phenotype and a reduced NO content compared to wild type plants. Those results suggest that *CsNOA1* regulates cucumber seedlings chilling tolerance. Additionally, under normal condition, we took several classic inhibitors to perform, and detect endogenous NO levels in wild type cucumber seedling. The results suggest that generation of endogenous NO in cucumber leaves occurs largely independently in the (CsNOA1) and nitrate reductase (NR) pathway.

Keywords: cucumber,nitricoxide, *CsNOA1,* nitratereductase, transgenicplants, chillingstress

Abbreviations: CBF3, C-repeatbindingfactor3; GFP, greenfluorescentprotein; GUS, b glucuronidase; NO,nitricoxide; NOA1, nitricoxideassociated1; NR,nitratereductase; OE,overexpressing; RNAi, RNAinterference; WT,wildtype.

INTRODUCTION

Plants are frequently exposed to adverse environmental conditions that can limit growth and development, among which low temperature is a key factor. Chilling stress can results in poor seed germination, stunted seedling growth, delayed crop heading, and increased pollen sterility (Xiong and Zhu, 2001; Beck et al., 2004; Minami et al., 2005). To withstand chilling damage, plants have developed multiple survival strategies. For example, upon exposure to cold stress, a set of signals are immediately triggered, including the calcium and reactive oxygen species burst, which activates the MAPK signal cascade, and ultimately initiates the downstream cold-responsive transcriptional cascade (Cheng et al., 2015). In addition, plants can set up physiological and biochemical adaptations to cope with the stress challenge (Ruelland et al., 2009; Theocharis et al., 2012). These adaptations include changes in membrane composition, the induction of anti-oxidative systems, and the synthesis of protective molecules. In addition, it has been reported that transcriptional regulation of the carbohydrate metabolic pathway *Arabidopsis thaliana* is essential for the accumulation of specific carbohydrates that represent an important factor in improved tolerance to chilling stress (Cook et al., 2004; Maruyama et al., 2009, 2014). Previous studies have also provided evidence for cold exposure triggering a large remodeling of plant metabolism, which, at least partly, depended on modifications in gene expression (Ruelland et al., 2009; Theocharis et al., 2012). Consequently there is interest in identifying and characterizing the signaling network underlying the low temperature stress and there is growing evidence that nitric oxide (NO) is an important signal for transducing information related to low temperature exposure.

A number of studies have shown that NO can alleviate low-temperature stress (Zhao et al., 2009; Liu et al., 2011; Yang et al., 2011a; Tan et al., 2013a). Additionally, Zhao et al. (2009) reported that cold acclimation in *A. thaliana* was associated with an increase in endogenous NO production, and Guillas et al. (2011) observed an immediate increase in NO synthesis in response to cold stress, which regulated the expression of cold-responsive genes, as well as with novel downstream elements identified as phosphosphingolipid metabolic species. In mammals, NO is synthesized via an oxidative mechanism involving NO synthase (NOS) enzymes, which oxidize arginine to generate citrulline and NO (Mayer and Hemmens, 1997; Wendehenne et al., 2001). In the algal species, *Ostrococcus tauri*, a NOS-like enzyme has been reported to synthesize NO (Foresi et al., 2010). Although NOS activity has been detected in plants, a novel plant NOS has not yet been identified. The *Arabidopsis thaliana* NO synthase1 (AtNOS1) was initially documented as a putative plant NOS, but was later renamed *AtNOA1* (*Arabidopsis thaliana NO-ASSOCIATED PROTEIN 1*), as it was suggested as a circular permuted GTPase of YIqF family (Moreau et al., 2008). The defective arginine-dependent NO synthesis activity in recombinant *AtNOS1* protein and the contradictory NO accumulation responses in NOA1-silenced mutants have confirmed that AtNOS1 is not an authentic NOS (Kwan et al., 2014). The indirect regulatory effect of NOA1 in NO production was demonstrated with impaired NO accumulation in NOA1-silenced *A. thaliana* (Guo et al., 2003). While others have suggested that the enzyme, nitrate reductase (NR), is more important to product NO in plants (Rockel et al., 2002; Planchet et al., 2005; Stöhr and Stremlau, 2006; Wilson et al., 2008). It has been reported that the *A. thaliana* double NR mutant, nia1nia2, fails to accumulate NO, or to mediate certain NO responses (Desikan et al., 2002). Moreover, NR inhibitors, such as sodium azide and tungstate have been shown to inhibit NO production in plants (Bright et al., 2006; Sang et al., 2008). Less well explored is the possibility that reductive pathways for NO production include the mitochondrial electron transport system and the enzyme, xanthine dehydrogenase/oxidase (Gupta and Kaiser, 2010). In addition, under aerobic conditions, exogenous hydroxylamine application to NR-deficient tobacco cell cultures resulted in the release of NO (Rümer et al., 2009).

We previously used exogenous application of an NO donor to elucidate the role of NO under chilling stress in cucumber (Liu et al., 2011), a warm-season species that requires protection from cold temperatures (Kuk and Shin, 2007). However, the role of NO in chilling stress tolerance and the source(s) of endogenous NO in cucumber seedlings have still not been resolved and these questions are addressed in this current study.

MATERIALS AND METHODS

Plant Material, Growth Conditions, and Treatments

Cucumber (*Cucumis sativus L.*) line ZN407 (collected from Ren lab in China Agricultural University) seedlings were grown in a growth chamber under a 16/8 h and 25/18°C day/night photoperiod. *A. thaliana* wild type (WT) Col-0 and *Atnoa1 mutant* (Columbia ecotype background) plants were kindly provided by the Zhang laboratory (Zhao et al., 2009). WT, *Atnoa1* and transgenic CsNOA1 seedlings were grown at 22°C under a long-day (16 h of light/8 h of dark) photoperiod at 100 μmol m^{-2} s^{-1} with 50–70% relative humidity in turf substrate. For chilling stress, 7-week-old seedlings were subjected to a 10 day chilling regime at 4°C.

Cucumber seedlings at the three leaf stage were used for the experimental treatments (Liu et al., 2011). Sodium nitroprusside (SNP, Sigma, USA) was used as an NO donor and the potassium salt of 2-(4-carboxyphenyl)-4, 4, 5, 5-tetramethylimidazoline-1-oxyl-3-oxide (cPTIO, Sigma, USA) was used as an NO scavenger. Sodium azide (NaN_3) and tungstate were used as NR inhibitors and N^G-Nitro-L-arginine Methyl Ester (L-NAME, Selleck, USA) and Nω-Nitro-L-arginine (L-NNA, Selleck, USA) as NOA1 inhibitors. Specific doses of NO donor, scavenger and inhibitors were sprayed onto the leaves of cucumber seedlings before exposure to 72 h of cold treatment at 4°C under continuous light (520 μmolm^{-2} s^{-1}). Each treatment was repeated at least three times using 10 seedlings per treatment. After treatment, the third leaves were collected and stored at –80°C until further use.

Endogenous Distribution of Cellular NO

The cellular distribution of endogenous NO in cucumber leaf sections (1.5 × 2.0 cm) was detected using confocal laser scanning microscopy (Nikon A1R-si) in combination with the fluorescent dye, 4-amino-5-methylamino-2,′7′-difluorofluoresceindiacetate (DAF-FM DA, Calbiochem) according to Zhao et al. (2009) with slight modifications. Briefly, after washing the excised leaves with buffer solution (20 mM HEPES-NaOH, pH 7.4), they were incubated in a buffer solution containing 25 μM DAF-FM DA for 1 h at 25°C. The incubated leaves were visualized using laser confocal scanning microscope after having been washed thoroughly with buffer solution to remove excess fluorophore. Excitation was at 488 nm and emission was at 515 nm.

Quantification of Endogenous NO

The concentration of endogenous NO was determined according to Zhao et al. (2009) with modifications. Fresh cucumber leaves (500 mg) were incubated with 100 U of Catalase (EC1.11.1.6, CAT) and 100 U of Superoxide Dismutase (EC1.15.1.1, SOD) for 5 min to remove endogenous reactive oxygen species (ROS) before addition of 5 ml oxyhaemoglobin (5 mM). After a 3 min incubation, the absorbance of the solution at 577 nm and 591 nm were used to calculate the amounts of oxyhaemoglobin and methaemoglobin, respectively, and the endogenous NO content was calculated based on the conversion ratio between oxyhaemoglobin and methaemoglobin.

NR Activity Assay

NR activity was assayed based on a modified protocol from Rockel et al. (2002). One gram of cucumber leaf tissue was ground in liquid N_2 and re-suspended in an extraction buffer containing 100 mM HEPES-KOH pH 7.5, 1 mM EDTA, 15% glycerol (v/v), 10 mM dithiothreitol, 0.1% Triton X-100, 1 mM phenylmethylsulfonyl fluoride, 20 μM FAD, 1 μM leupeptin, 5 μM Na_2MoO_4. The samples were centrifuged at 12,000 g for 15 min at 4°C. Five-hundred μl of each sample was then added to 250 μl of 1% (w/v) sulfanilamide (3 N HCl) and 250 μl 0.02% (w/v) N-(1-naphthyl) ethylene diamine and samples left for 20 min at room temperature. The samples were centrifuged at 16,000 g for 10 min, and the nitrite formed was assessed by colorimetric determination at OD546 nm using the Tecan Infinite M1000 plate reader. Absorbance was normalized to the protein concentration of the sample, as determined by the bicinchoninic acid (BCA) protein assay (ThermoFisher Scientific), using bovine serum albumin as a standard. To test whether NR could produce NO *in vitro*, previously published methods were used (Rockel et al., 2002).

Determination of CsNOA1 Activity

NOA1 activity determination was performed according to Zhao et al. (2007b). Approximately 1 g of leaf tissue, together with 50 mg of polyvinylpolypyrrolidone, were ground in liquid N_2 and then re-suspended in extraction buffer (50 mM Tris-HCl, pH 7.4, 1 mM EDTA, 320 mM Sucrose,

1 mM dithiothreitol, 1 μM leupeptin, 1 μM pepstatin, and 1 mM phenylmethylsulfonyl fluoride). After centrifugation at 10,000 g for 30 min at 4°C, the supernatant was used for NOA1 activity determination using the citrulline assay and the NOA1 assay kit (Cayman Chemical). The reaction mixture (50 μl) contained 25 mM Tris-HCl, pH7.4, 3 μM tetrahydrobiopterin, 1 μM FAD, 1 μM FMN, 1 mM NADPH, 0.6 mM $CaCl_2$, 0.1 μM Calmodulin, 0.3 μM (1 μl Ci) [^{3}H] Arg (Amersham Biosciences), and 10 mL enzyme extract. After incubation for 30 min at 37°C, the reaction was stopped with 400 μl stop buffer (50 mM HEPES, pH 5.5, 5 mM EDTA). A 100 μl resin slur was added to the reaction mixture and the resin was removed by centrifugation at 10,000 g. Flow through (400 μl) was added to 5 mL of scintillation liquid and radioactivity was counted (LS 6000, Beckman). The protein contents in the supernatant were determined according to the Bradford method (1976) with bovine serum albumin as a standard.

Transcriptional Analysis by Real Time RT-PCR and Semi-Quantitative PCR

Total RNA was extracted from cucumber third true leaves using the Promega SV Total RNA Isolation System (http://promega.bioon.com.cn/), and cDNA was synthesized using MultiScribe™ reverse transcriptase (Applied Biosystems). Quantitative real-time RT-PCR was performed using SYBR®*Premix Ex Taq*™ from TaKaRa (China) and an Applied Bio-systems 7500 real-time PCR system. To determine the relative fold differences in template abundance for each sample, the $-2^{-\Delta\Delta Ct}$ method was used. For semi-quantitative PCR, the conditions were as follows: 10 min at 94°C, then 25 cycles of 30 s at 94°C, 30 s at 50°C and 30s at 72°C, followed by 7 min at 72°C. The RT-PCR products were separated using a 2.0% (m/v) agarose gel. The gene specific primers are presented in Supplementary Table 3.

In situ Hybridization Assay

The shoot apices of 10-day-old seedlings and roots from greenhouse grown plants were fixed, embedded, sectioned, and hybridized as described by Zhang et al. (2013). Digoxigenin-labeled sense and antisense RNA probes were generated using SP6 and T7 RNA polymerases (Roche), respectively, and PCR amplification. The primer pairs are listed in Supplementary Table 3.

Transformation of A. thaliana and Cucumber

To generate the *CsNOA1* over-expressing cucumber plants, the full length *CsNOA1* cDNA was cloned and inserted into the pCAMBIA1305.1

vector between *Xba*I (5′end) and *Sma*I (3′end) restriction sites (Zhang et al., 2014). To generate the *CsNOA1*-RNAi transgenic lines, two *CsNOA1* fragments were amplified using specific primers containing *Asc*I (5′end) and *Swa*I (3′end) endonuclease sites, and *Spe*I (5′end) and *BamH*I (3end) endonuclease sites. Two fragments were inversely inserted into the pFGC1008 vector (Zhang et al., 2014). For subcellular localization of *CsNOA1*, the coding region of *CsNOA1* without the stop codon was cloned and fused upstream of the EGFP (enhanced green fluorescent protein) sequence between the *Sai*I and *Kpn*I sites of the pEZS-NL vector to generate 35S:*CsNOA1*:GFP. To make the *CsNOA1*:GUS (β-glucuronidase) construct, a 1.8 kb sequence upstream from the ATG start site of the *CsNOA1* coding sequence was cloned and inserted into the PBI121 vector between the *Xba*I and *Bam H*I sites. *A. thaliana* transformation was performed as previously described (Clough and Bent, 1998), using the *Agrobacterium tumefaciens* strain GV3101. For cucumber transformation, all the resulting constructs and corresponding empty vectors were introduced into *A. tumefaciens* strain LBA4404 by electroporation and cucumber line NZ407 was transformed using the cotyledon transformation method (Zhang et al., 2014). Primers used are listed in Supplementary Table 3.

Subcellular Localization

Stable expression of the GFP control, CsNOA1-GFP, and Mito-tracker Red fusion proteins in T1 transgenic plants was examined and imaged using a LSM510 META confocal microscope (Zeiss). Measurements of the GFP and blue fluorescence of the chloroplasts were acquired with a 488-nm laser excitation and a band pass emission filter of 490–530 nm (GFP channel) and 650–750 nm (blue channel). For co-localization experiments with Mito-tracker Red and CsNOA1-GFP, multi-tracks were configured with the GFP channel setting as above and an additional red channel with an excitation of 543 nm and a band pass emission filter of 560–650 nm (red channel). Images of the GFP control and CsNOA1-GFP/Mito-tracker Red were collected using the same settings.

GUS Histochemical Assay

GUS staining of the WT/Pro$_{CsNOA1}$:GUS transgenic line (T2) was performed as previously described with slight modifications (Jefferson et al., 1987). Briefly, fresh tissue samples were fixed in ice cold 90% (v/v) acetone for 1 h and then vacuum filtrated for 30 min. Samples were vacuum infiltrated with staining buffer (50 mM PO_4 buffer with 0.2% [v/v] tritonX-100, 100 mM $K_3Fe(CN)_6$, and 100 mM$K_4Fe(CN)_6$ for10 min on ice followed by vacuum infiltration in staining buffer containing 2 mM 5-bromo-4chloro-3-

indolyl-ß-D-glucuronic acid for 15 min on ice. Samples were then incubated at 37°C for 30 min to 24 h followed by a wash with 75% (v/v) ethanol. The stained tissues were viewed under a stereo microscope and photographed.

Transmission Electron Microscopy

Young leaves from transgenic plants and wild type were fixed in 2.5% (w/v) glutaraldehyde and rinsed thoroughly with a 0.1 M phosphate buffer (pH 6.8). Samples were post-fixed with 1% osmic acid, washed in 0.1 M phosphate buffer (pH 6.8), dehydrated through an acetone series (using the sequence 30, 50, 70, 80, 90, and 100%), and then embedded in Spurr's resin. Thin cross-sections were made with a UC6I microtome (Leica) and examined with a JEM-1230 scanning transmission electron microscope (Liu et al., 2016).

Electrolyte Leakage Assay

Electrolyte leakage was assayed according to Liu et al. (2011). Briefly, tubes containing six to eight *A. thalianaleaves*, detached from 20-days old plants and chilled at 4°C for 10 days, and cucumber leaves detached from 25-days old plants and chilled at 4°C for 72 h, were placed in a low-temperature bath (Grant) set at 0°C. The bath temperature was lowered at a rate of 2°C h^{-1}. Tubes were then remove at the defined temperatures and thawed overnight at 4°C in the dark, before incubation with 5 mL of deionized water at 25°C for 2 h with gentle shaking (150 rpm). Electrical conductivity in the bathing solution was first determined (C1), before samples were heated to 100°C for 30 min and the second electrical conductivity (C2) of the bathing solution determined. Relative ion leakage was expressed as a percentage of the total conductivity after heating to 100°C. Relative ion leakage % = C1/C2 × 100.

Determination of Starch, Soluble Sugars, and Proline and the Iodine Staining Assay

Starch and total soluble sugar content in cucumber leaves were measured as previously described (Liu et al., 2011). Proline accumulation in cucumber leaves was determined as method previously described (Zhao et al., 2009) using L-Pro as standard. Briefly, leaves from the lines to be tested were harvested, weighed and extracted in 3% sulfosalicylic acid. An aliquot of each extract (2 mL) was incubated with 2 mL of ninhydrin reagent (2.5% [w/v] ninhydrin, 60% [v/v] glacial acetic acid, 40% 6 M phosphoric acid) and 2 mL of glacial acetic acid at 100°C for 45 min, and the reaction was terminated in an ice bath. Toluene (5 mL) was added, followed by vortexing and incubation at 23°C for 24 h. The absorbance was measured at 520 nm. For the iodine staining assay, cucumber leaves were harvested at the end of the light phase (24 h without darkness) from plants grown for 5 weeks in a chamber as described above. Different doses of NO donor, scavenger, and inhibitors were sprayed onto the leaves of cucumber seedling before exposing them in to an I_2/IK solution for 24 h and then destaining for 15 min in water (Zhang et al., 2008).

Statistical Analysis

All data were obtained from at least three independent experiments with three replicates each. Data were analyzed using one-way analysis of variance (ANOVA) followed by Duncan's multiple range test ($P < 0.05$).

Fv/Fm Assay and Chilling Damage Index

The chilling damage index was measured following the methods described by Liu et al. (2011). The Fv/Fm was measured using a portable fluorometer (FMS2; Hansatech, Kings's Lynn, UK). Cucumber and *Arabidopsis thaliana* leaves were dark-acclimated for 30 min before Fv/Fm measurement. The maximal efficiency of PSII photochemistry in the dark-acclimated leaves was calculated according to the formula: Fv/Fm = (Fm-Fo)/Fm.

RESULTS

Identification of the Cucumber CsNOA1 Gene

Basic Local Alignment Search Tool (BLAST) analysis of the Cucumber Genome Database (Huang et al.,2009) revealed a single *NOA1-like* gene, named *CsNOA1* (*Csa5M168870*). *CsNOA1* has 12 exons and 11 introns, which is different to the *A.thaliana* homolog, *AtNOA1,* which contains 13 exons, and 12 introns. The full-length *CsNOA1* cDNA is predicted to encode a protein of 556 amino acids (Supplementary Figure S1A). An alignment of the CsNOA1 amino acid sequence with apparent NOA1 homologs from other plant species (Supplementary Figure S1B) showed that CsNOA1, AtNOA1, and OsNOA1 (from rice, *Oryza sativa*) share three domains that are typical of the GTPase family: the zinc-binding domain (ZBD), the circularly permuted G-domain (CPG), and the C-terminal domain (CTD) (Moreau et al., 2008; Sudhamsu et al., 2008; Anand et al., 2009) (Supplementary Figure S1B). Over the full length sequence, CsNOA1 shows 66 and 64% identity to

AtNOA1 and OsNOA1, respectively, indicating that the NOA1 sequence is well conserved between monocots and dicots. To better understand the evolutionary relationship between CsNOA1 and other NOA1 homologs, a phylogenetic analysis was performed using the neighbor-joining (NJ) method (Saitou and Nei, 1987; Figure 1). The phylogenetic tree divides the NOA1 homologs into two clades: the dicotyledon (green line) and monocotyledon (red line) group, with CsNOA1 in the dicotyledon clade as expected.

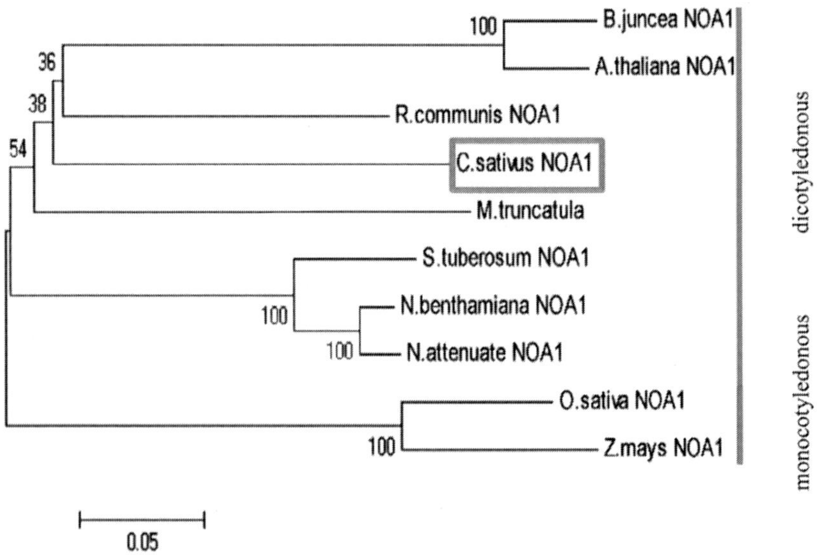

FIGURE 1. PHYLOGENETIC ANALYSIS OF NOA1 HOMOLOGS IN VARIOUS SPECIES. This phylogenetic tree was constructed using the Neighbor—Joining (NJ) method through MEGA 5.0 software. Ten species were used for this analysis and formed two main groups: dicotyledonous group and monocotyledon group. A NOA1 homolog from cucumber is indicated in the red box. Gene ID for each of the NOA1 protein used for this analysis is listed in the "accession numbers."

Low Temperature Increases the Expression of CsNOA1

To obtain insights into the biological function of *CsNAO1*, its expression was examined in different organs with highest levels in the shoot apex (Figures 2A,B). Since exogenous NO has been shown to alleviate chilling stress injury in cucumber seedlings (Liu et al., 2011), we hypothesized that *CsNOA1* may be regulated by low temperatures. *CsNOA1* mRNA was detected by *in situ* hybridization analysis throughout the inflorescence meristem and floral meristem, as shown in Figure 2C, with higher expression in the im, fm and root apex under chilling conditions. We also analyzed *CsNOA1* expression using a *CsNOA1* promoter::GUS transcriptional fusion and histochemical staining. GUS labeling was strongest in the root and stem (Figure 2D) and the expression in these organs was further enhanced when plants were grown at low temperatures, indicating an induction of CsNOA1 expression under these conditions.

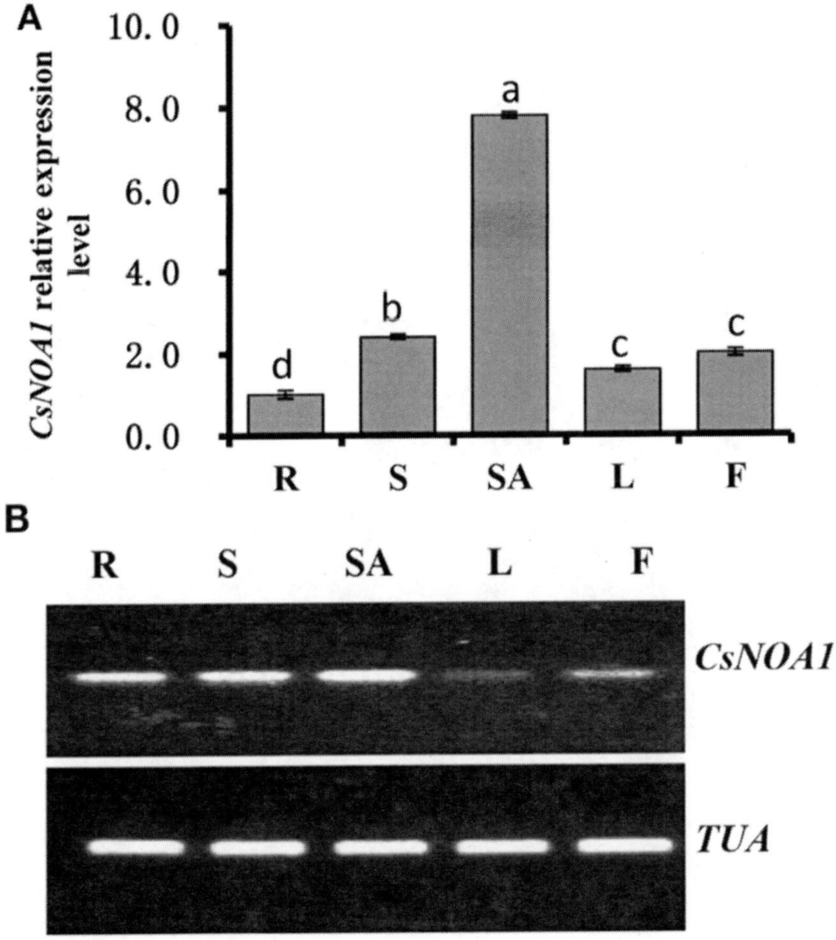

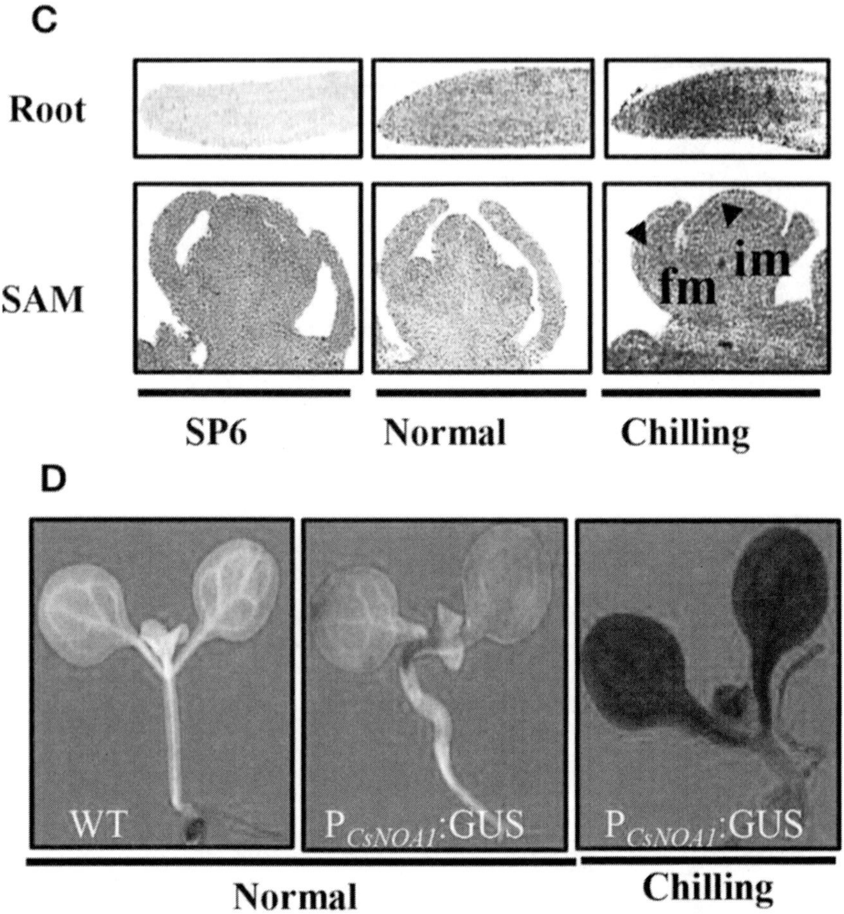

FIGURE 2. CSNOA1 EXPRESSION PROFILE ANALYSES. (A) Relative expression of *CsNOA1* in cucumber tissues. R, root; S, stem; SA, shoot apical; L, leaf; F, flower. **(B)** Semi-quantitative RT-PCR analyses of CsNOA1 expression in cucumber, the cucumberα-*TUBULIN (TUA)* was used as an internal control, and three biological replicates were performed for these experiments. **(C)** *In situ* hybridization of *CsNOA1* in plants grown under normal and chilling conditions. Hybridization with the *CsNOA1* sense probe gave no signal in the SP6 control. im, inflorescence meristem; fm, floral meristem. **(D)** Transgenic plants (10 days-old seedlings) harboring Pro_{CsNOA1}:GUS were stained for GUS activity in seedlings under normal and chilling condition. The mean values of three independent samples and standard errors are presented; the same letter above the column indicates no significant difference at $P < 0.05$.

Subcellular Localization of CsNOA1

To determine the subcellular localization of CsNOA1, green fluorescent protein (GFP) fused to the full-length *CsNOA1* sequence was stably expressed in cucumber seedlings under the control of the 35S promoter. As shown in Figures 3A–D, GFP signal was observed in small subcellular vesicles in the guard cells, but did not overlap with the blue fluorescence of the chloroplasts, indicating that CsNOA1 may be localized in the mitochondria or Golgi apparatus. We therefore used a mitochondria-specific stain, Mito-tracker Red (Matre et al., 2009) to evaluate the *CsNOA1-GFP* transgenic lines (Figures 3E–H), and this analysis showed co-localization with the CsNOA1-GFP fusion protein (Figures 3E–H), suggesting that CsNOA1 is localized in the mitochondria. This is congruent with the previously reported subcellular localization of AtNOA1 and TaNOA1 from *A.thaliana* and wheat, respectively (Guo and Crawford, 2005; Hao et al., 2010).

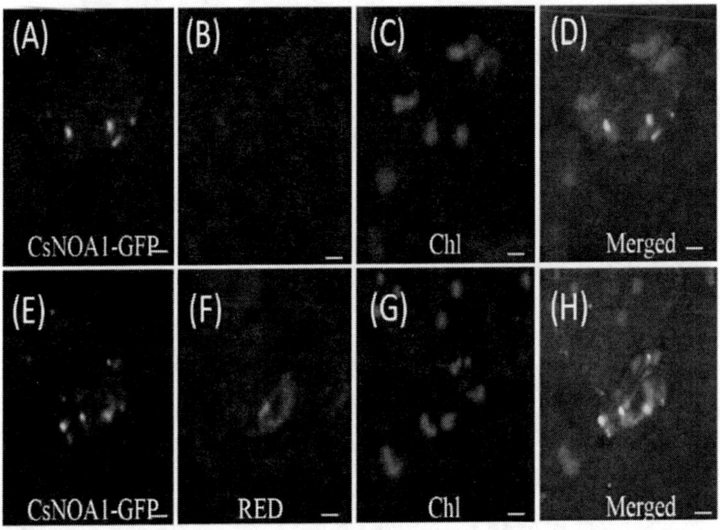

FIGURE 3. SUBCELLULAR LOCALIZATION OF CSNOA1 PROTEIN CUCUMBER LEAF STOMATA. (A) CsNOA1-GFP fusion protein was stably expressed in cucumber leaf stomata and observed by confocal laser scanning microscopy (as a control). **(B)** No Mito-tracker Red shown in **(A)** could be imaged using Red channel setting of LSM510 META. **(C)** Leaf stomata fluorescence (Chl) could be imaged using Blue channel setting ofLSM510 META. **(D)** Merged image of **(A–C)**. **(E)** CsNOA1-GFP fusion protein was stably expressed in cucumber leaf stomata and observed by confocal laser scanning microscopy. **(F)** Mito-tracker Red shown in **(E)** could be imaged using Red channel setting of LSM510 META. **(G)** Leaf stomata fluorescence could be imaged in **(E)** using Blue channel setting ofLSM510 META. **(H)** Merged image of **(E–G)**. Scale bar = 50 μm.

Ectopic Expression of CsNOA1 in *A.thaliana* Results in Increased Chilling Stress Tolerance

To investigate the biological role of *CsNOA1*, we ectopically expressed the full-length *CsNOA1* cDNA under the control of a 35S promoter in an *Arabidopsis thaliana nitric oxide associated* 1 (*Atnoa1*) mutant. A total of 19 independent transgenic lines were obtained, all of which showed partial rescue of the phenotypes of the Atnoa1mutant similar phenotypes. As shown in Figure 4, normal plant size and green coloration of the leaves in transgenic plants were restored as compared with those in the *noa1* mutant (Figures 4A–C). We chose six transgenic lines to analyze expression of *CsNOA1* and determined that it was expressed at different levels in these lines (Figure 4D).

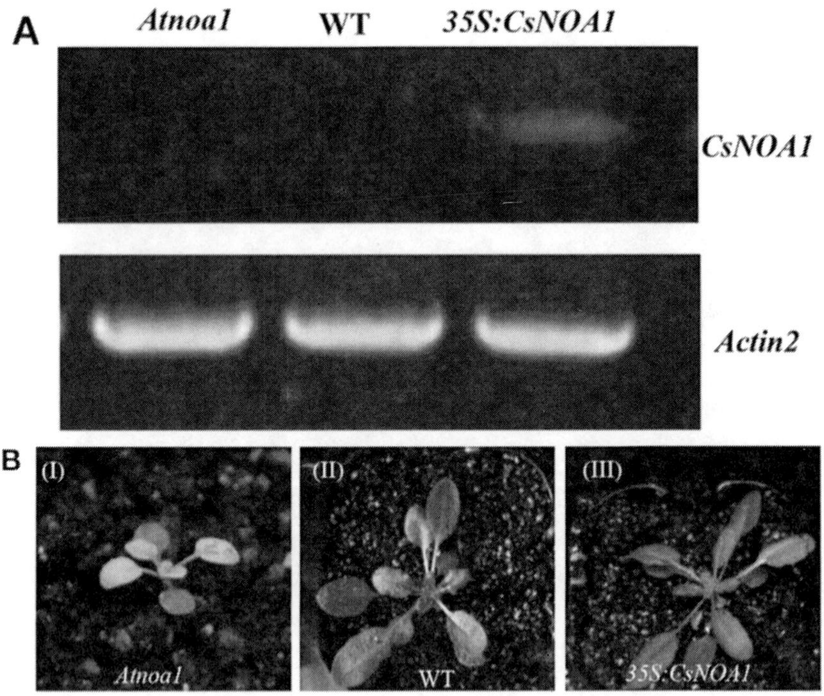

C

D

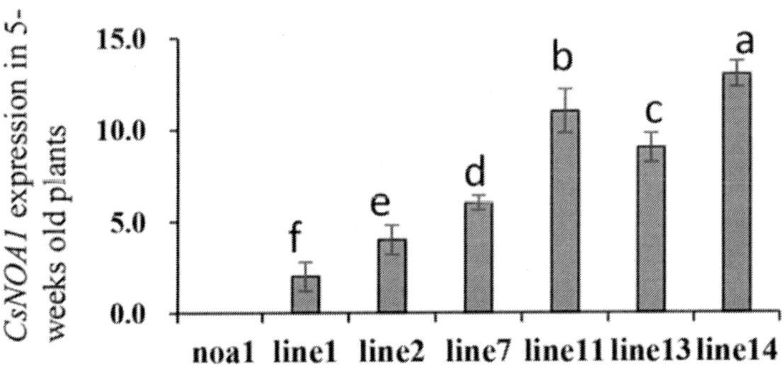

FIGURE 4. PHENOTYPIC ANALYSIS OF ARABIDOPSIS THALIANA ECTOPICALLY EXPRESSING CSNOA1. (A) Semi-quantitative PCR analyses of *CsNOA1* in transgenic plants in the *Atnoa1* background. The *Arabidopsis thaliana ACTIN2* gene was used as an internal control. **(B)** Phenotype of transgenic *Arabidopsis thaliana* of the (I) *Atnoa1* mutant phenotype under normal conditions (20 days); (II) Wild type grown under the same conditions; (III) Ectopically expressing *CsNOA1* in the Atnoa1 background. **(C)** Phenotype of transgenic *Arabidopsis thaliana* of the (I) *Atnoa1* mutant phenotype under normal conditions (7-weeks); (II)Wild type grown under the same conditions; (III) Ectopically expressing *CsNOA1* in the Atnoa1 background. **(D)** Real time RT-PCR analyses of *CsNOA1* in six selected transgenic lines in the *noa1* background. The Arabidopsis *actin2* was used as an internal control, three biological replicates were performed for these experiments. Error bars indicate the standard errors. Different letters above the column indicate significant difference at $P< 0.05$.

Seven-week-old T2 lines L7 and L11, as well as WT and *Atnoa1* mutant plants, were exposed to chilling conditions (4°C) for 10 days. No obvious phenotypic differences were observed between the WT and transgenic plants; however, the leaves from mutant plants grown under chilling conditions showed hydrophanous spots and wilting (Figure 5A). A quantitative assay further showed that endogenous NO levels in leaves of the transgenic *A. thaliana* lines were higher than in those of the WT and the mutant plants (Figure 5C). *CsNOA1* expression in the L11 line was more than 3.5 fold higher than that of the L7 line (Figure 5B), and the transcript levels of the cold tolerance marker gene, CBF3, were higher in the L11 line than in the L7 line (Figure 5D). Moreover, in both the L11 and L7 lines, a higher Fv/Fm value was observed in plants under chilling stress compared to WT plants and the *Atnoa1*mutant, even though this was still lower than in the control (Figure 5E). Electrolyte leakage under normal and chilling stress conditions was also tested, and was observed to increase significantly, with the *Atnoa1*mutant lines showing exacerbated damage under chilling stress (Figure 5F). While the chilled WT plants showed an almost 2 fold increase in leakage, this value was lower in the over-expressing (OE) plants and was more than 3-fold higher in the mutant lines. Taken together, these results indicated that over-expression of *CsNOA1* in *A. thaliana* enhanced chilling stress tolerance.

CsNOA1 is Involved in Chilling Stress Tolerance in Cucumber

To verify that *CsNOA1* indeed confers chilling stress tolerance, several overexpressing and knockdown cucumber lines were generated. These were evaluated by real-time quantitative PCR, and *CsNOA1* transcript levels in the knockdown lines (RNAi1, RNAi6, and RNAi9) were shown to be reduced by approximately 70% compared to WT, while *CsNOA1* transcript levels in the OE lines were more than 15-fold higher than in WT plants (Figure 6A). Seedlings of both OE and RNAi lines were phenotypically similar to WT plants under normal conditions; however, after chilling at 4°C for 72 h, the WT plants displayed symptoms such as mildly wilted leaves (Figure 6B), and this phenotype was greater in the RNAi lines. In contrast, no such stress symptoms were observed in the OE lines following the identical chilling treatment (Figure 6B). A chilling damage index (CI) was also used to evaluate the stress tolerance of the transgenic lines. As shown in Table 1, the CI of the OE lines was substantially lower than that of the WT plants, while values were highest in the RNAi lines. The chilling tolerance of the OE lines was therefore greater than that of the RNAi lines, which is consistent with *CsNOA1* playing a role in low temperature stress tolerance.

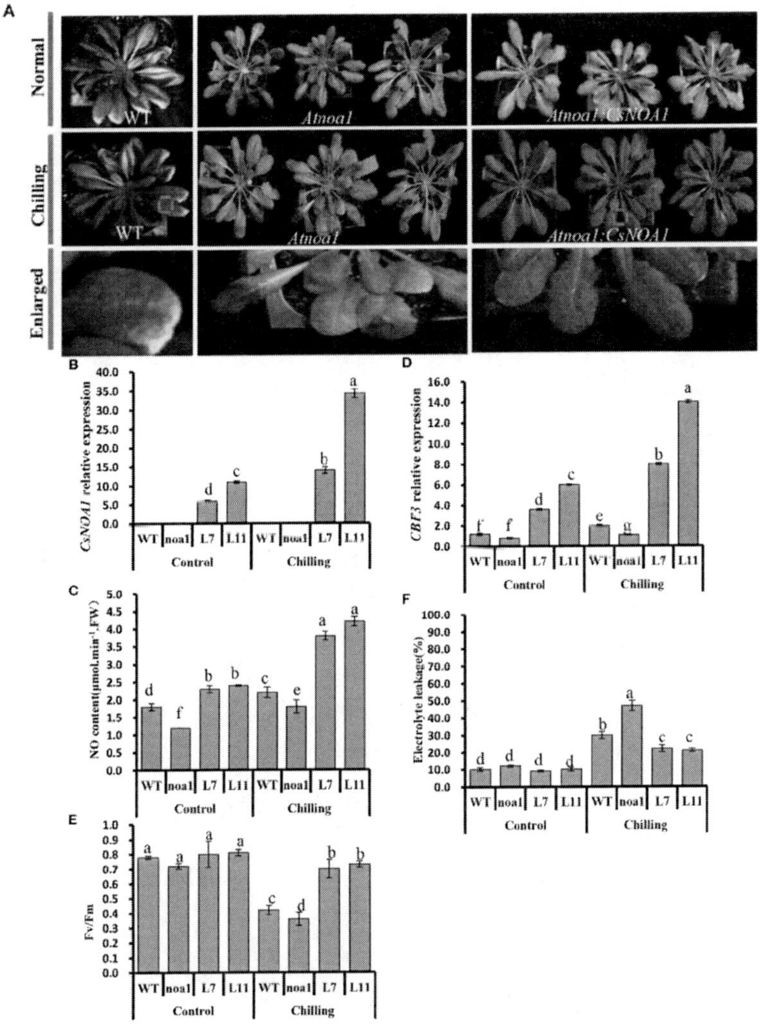

FIGURE 5. OVER-EXPRESSION OF CSNOA1 IN ARABIDOPSIS THALIANA CONFERS TOLERANCE OF CHILLING STRESS. **(A)** Phenotypes of transgenic Arabidopsis thaliana lines subjected to chilling stress. **(B)** *CsNOA1*expression levels in wild type (WT), mutant and transgenic lines under normal and chilling conditions. **(C)**Endogenous NO (nitric oxide) content under normal and chilling conditions. **(D)** *CBF3* expression under different temperature conditions. **(E)** Effect of chilling treatment on Fv/Fm in WT, mutant and transgenic *Arabidopsis thaliana* lines. **(F)** Effect of chilling treatment on electrolyte leakage in *A.thaliana*. L7 and L11 are two independent *CsNOA1* overexpression lines. Normal growth condition, 25°C, chilling treatment, 4°Cfor 10-days. Three independent experiments were performed and error bars indicate SD (n = 3). Different letters above the column indicate significant difference at (P < 0.05).

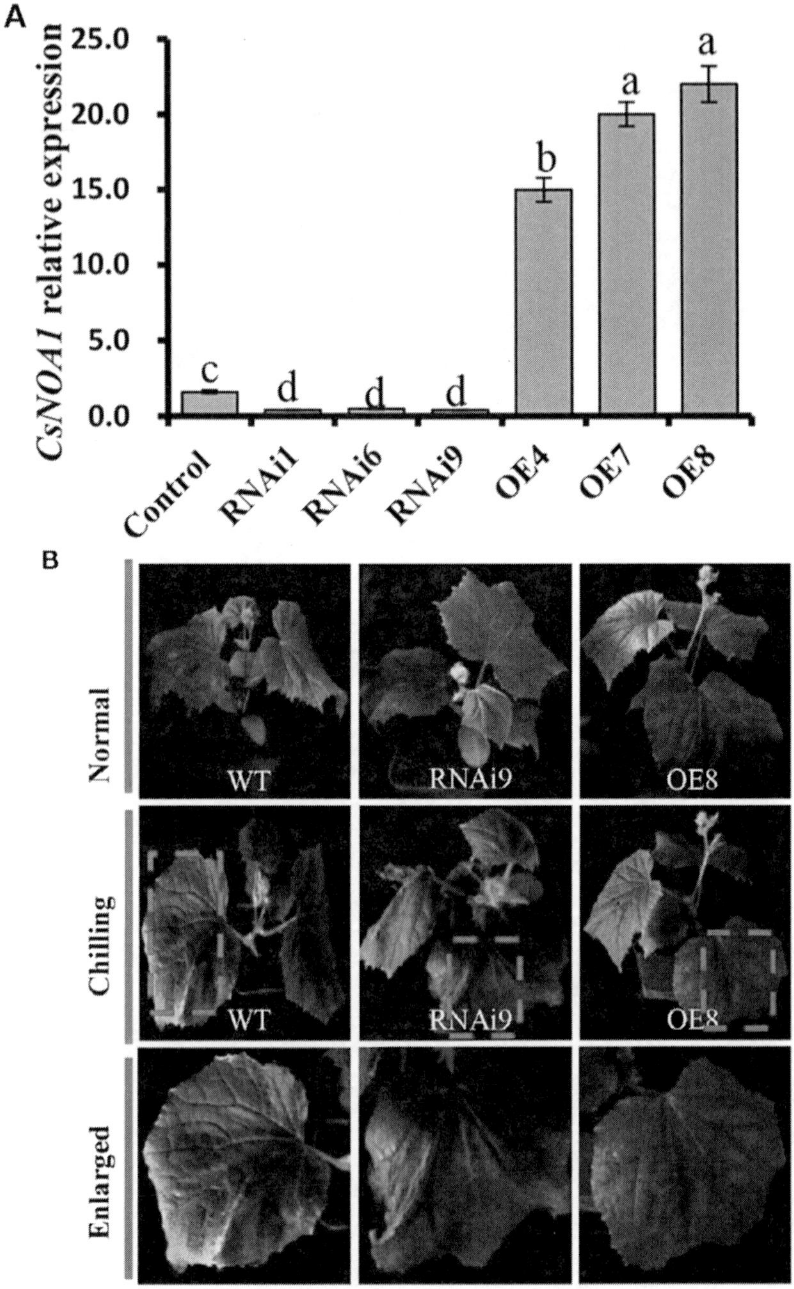
A
CsNOA1 relative expression
25.0
20.0
15.0
10.0
5.0
0.0
c
d
d
d
b
a
a
Control
RNAi1
RNAi6
RNAi9
OE4
OE7
OE8
B
Normal
Chilling
Enlarged
WT
RNAi9
OE8
WT
RNAi9
OE8

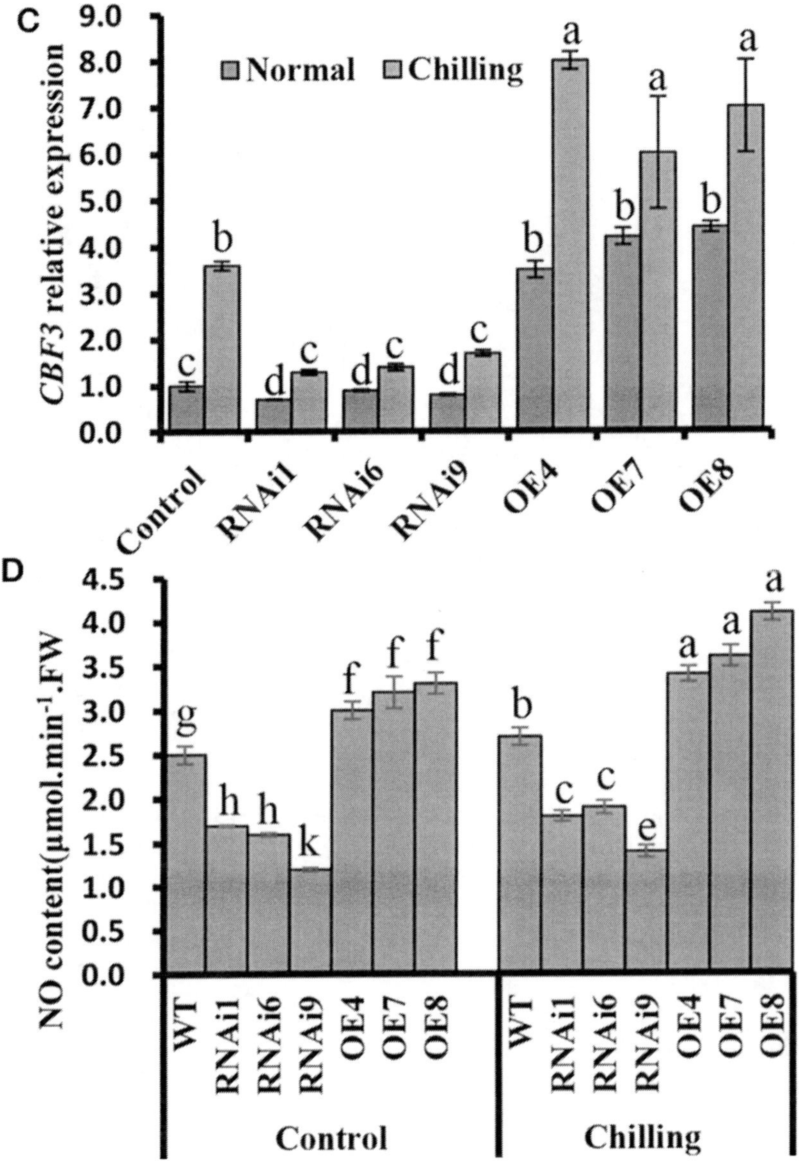
C
CBF3 relative expression
Normal
Chilling
9.0
8.0
7.0
6.0
5.0
4.0
3.0
2.0
1.0
0.0
a
a
a
b
b
b
b
c
c
c
c
d
d
d
Control
RNAi1
RNAi6
RNAi9
OE4
OE7
OE8
D
NO content(μmol.min-1.FW
4.5
4.0
3.5
3.0
2.5
2.0
1.5
1.0
0.5
0.0
g
h
h
k
f
f
f
b
c
c
e
a
a
a
WT
RNAi1
RNAi6
RNAi9
OE4
OE7
OE8
WT
RNAi1
RNAi6
RNAi9
OE4
OE7
OE8
Control
Chilling

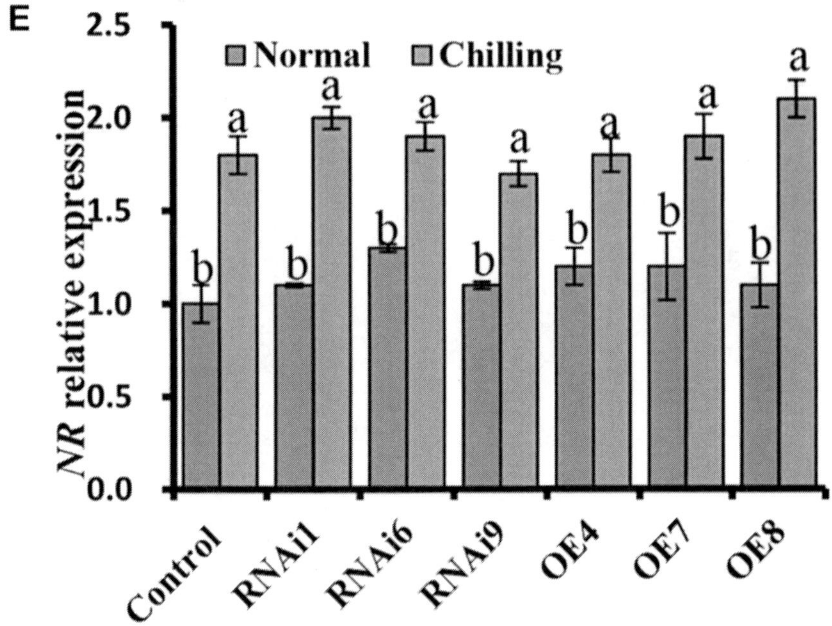

FIGURE 6. CHILLING TOLERANCE OF CSNOA1-OE AND CSNOA1-RNAI CUCUMBER LINES. (A) *CsNOA1*expression analysis in wild-type (WT) plants, transgenic overexpressing (OE) and RNAi lines. **(B)**Phenotypes of WT, OE, and RNAi plants at the same growth stage under normal and chilling stress conditions. **(C)** *CBF3* expression profiles under different temperature conditions in WT, OE, and RNAi lines. **(D)** The effect of chilling stress on NO content in WT, OE, and RNAi lines. **(E)** NR expression profile in WT and transgenic cucumber lines under different growth conditions. Normal growth condition, 25°C, chilling treatment, 4°C for 72 h. Three independent experiments were performed and error bars indicate SD (n = 3). Different letters above the column indicate significant difference at ($P < 0.05$).

We further determined that the expression of CBF3, a gene that has been shown to contribute to enhancing chilling stress tolerance (Miura et al., 2007; Liu et al., 2010), was significantly up-regulated in the OE lines (OE4, OE7, and OE8), while its expression was greatly reduced in the RNAi lines (Figure 6C).

TABLE 1. Chilling damage index (ci) in wild type and transgenic lines affected by chilling stress.

Line	Normal condition	Chilling condition
Control	0	0.43 ± 0.01c
RNAi1	0	0.54 ± 0.04a
RNAi6	0	0.57 ± 0.02a
RNAi9	0	0.59 ± 0.02a
OE4	0	0.21 ± 0.04b
OE7	0	0.24 ± 0.06b
OE8	0	0.22 ± 0.04b

Different letters in the third column indicate significant difference at $P < 0.05$.

Next, a quantitative assay was performed to detect endogenous NO in leaves. We observed that the NO content of OE line leaves was higher than that of in WT and RNAi plants grown under normal condition (Figure 6D). Under chilling stress condition, the NO levels in all test plants was higher than that of in WT, OE, and RNAi plants grown under normal condition. For example, the NO content of OE8 was 1.5 times higher than that of the WT under chilling conditions, suggesting that *CsNOA1* contributed to the production of endogenous NO (Figure 6D). When the expression level of NR was analyzed by real time RT-PCR, no obvious change was observed between the transgenic cucumber lines and WT; however, NR transcript levels were generally higher under chilling conditions than under normal conditions (Figure 6E). We infer from these results that *CsNOA1*produced NO independent of NR expression.

Chloroplast Ultrastructure was Altered in Transgenic Cucumber Leaves

WT cucumber leaves had smaller shrunken chloroplasts when subjected to chilling stress, unlike non-stressed leaves (Figures 7A,B). The chloroplasts in the OE8 line appeared rounded, with large starch granules, in contrast to those of the WT under the same conditions (Figures 7A–D), while small starch granules were observed in the chloroplasts of the RNAi3 leaves (Figures 7E,F). Starch content was measured and found to increase from 6.5 to 7.3% of fresh weight in WT leaves after 4 days of chilling at 4°C (Figure 7H). Incubation under the same conditions led to an increase of starch content in the OE8 line from 7.5 to 9.2% of fresh weight, while a dramatically decrease was observed in the RNAi9 line (Figure 7H). Given that the NO level was lower in the RNAi lines than in the WT plants, both under non-chilling stress (control) and chilling stress conditions (Figure 6D), we hypothesized that the accumulation of starch may be regulated by endogenous NO. To confirm this, we investigated the relationship between endogenous NO levels and starch content in the leaves of WT plants grown under chilling conditions. This study was performed by manipulating endogenous NO levels using an NO scavenger and inhibitor. Figures 7G,I show that low concentration (<1000 μM) exogenous nitric oxide donor SNP stimulated starch accumulation under normal conditions, while the increased starch content induced by chilling was markedly inhibited by treatment with cPTIO and L-NAME. A higher starch accumulation in leaves with treatment of SNP and its inhibitor, was evident by a darker color when stained with potassium iodide (Figure 7J). These results indicate that endogenous NO may function as a trigger to promote starch accumulation as a consequence of chilling stress. We used real time RT-PCR to analyze the expression levels of the starch synthesis related genes *SS1* and *SS3*, as well as that of *SA1*, which is related to starch degradation. In Figures 8A,C, the expression of the *SS1* and *SS3* genes in OE lines were higher than in WT and the RNAi lines under normal conditions; however, chilling resulted in an obvious increase in SS1 and SS3 expression in OE lines (Figures 8B,D). The expression of the starch degradation related gene, *SA1,* was also higher in WT than in the RNAi lines under normal conditions; however, chilling resulted in a decrease in SA1 expression in WT plants (Figures 8E,F), while expression increased in the RNAi lines under chilling stress. These results indicate that an up-regulation of *SS1* and *SS3* and a down-regulation of *SA1* expression may confer greater starch accumulation in the OE lines and WT under chilling stress.

CsNOA1-Overexpressing Plants Accumulate High Levels of Soluble Sugars, but not of Proline

Under normal conditions, the levels of soluble sugars in the transgenic lines showed no significant difference compared to WT (Figure 9A), while an increase was observed upon exposure to chilling stress in both WT and transgenic plants. However, the increase in soluble sugars in the *CsNOA1*-overexpressing plants was substantially higher than in WT plants. No significant difference in proline content was observed between WT and transgenic lines both under normal and chilling conditions (Figure 9B). The expression of genes associated with proline biosynthesis (*CsP5CS1*) and proline transport (*CsProT1*) was investigated and, as shown in Supplementary Figures S2A,B, after exposure to chilling stress, there was no significant difference in their transcript levels among the OE lines, the

WT and the RNAi lines. We therefore concluded that *CsNOA1* affects soluble sugar accumulation but not proline levels.

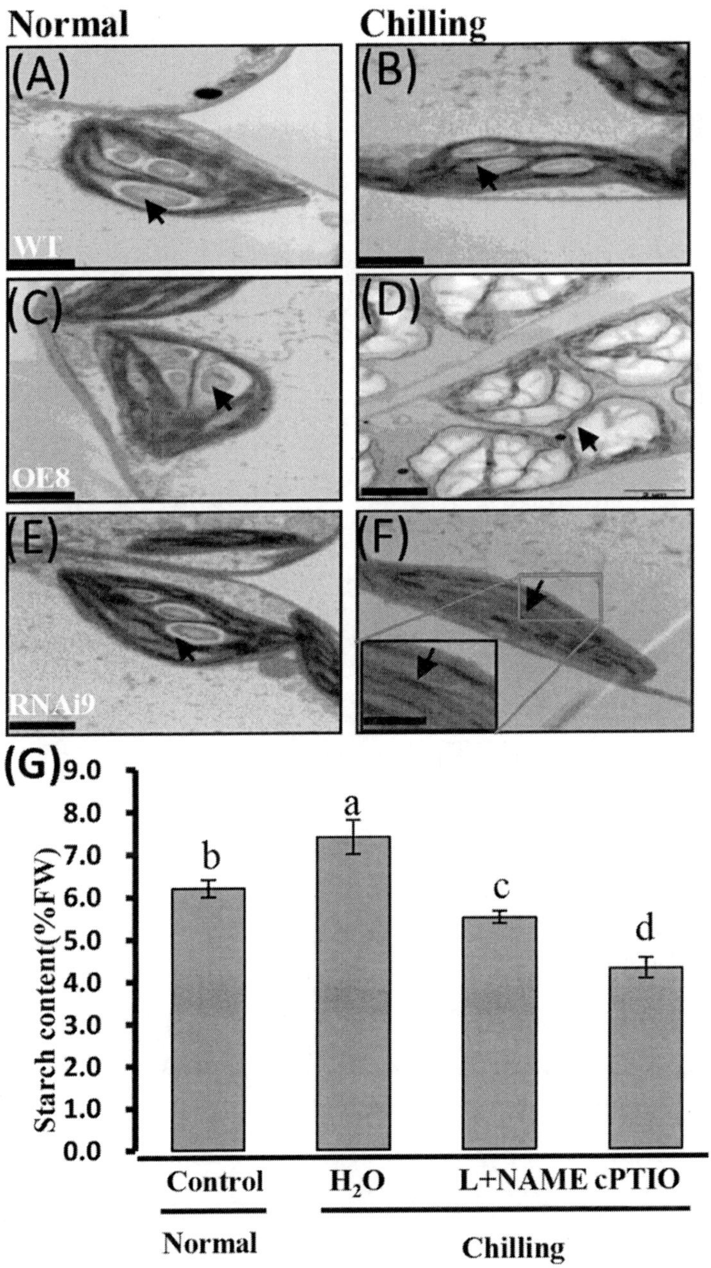

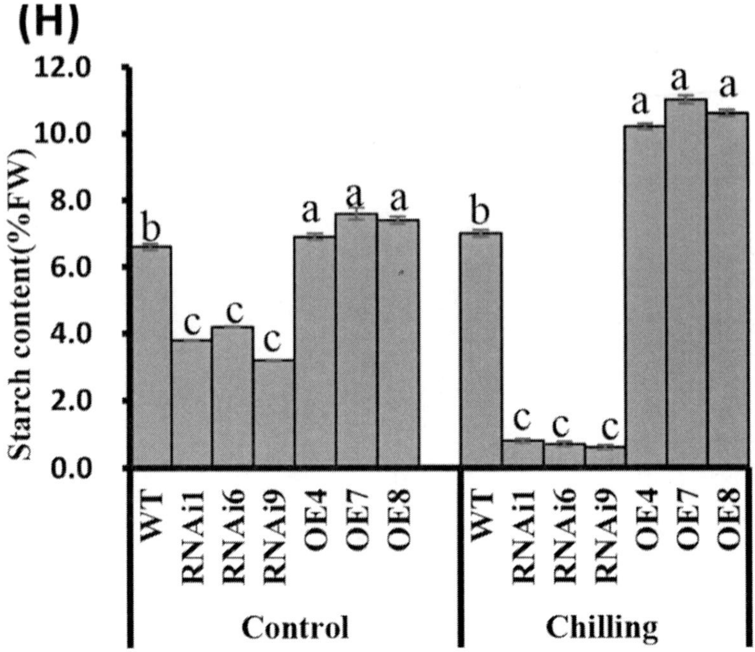
(H)
Starch content(%FW)
12.0
10.0
8.0
6.0
4.0
2.0
0.0
b
c
c
c
a
a
a
b
c
c
c
a
a
a
WT
RNAi1
RNAi6
RNAi9
OE4
OE7
OE8
WT
RNAi1
RNAi6
RNAi9
OE4
OE7
OE8
Control
Chilling

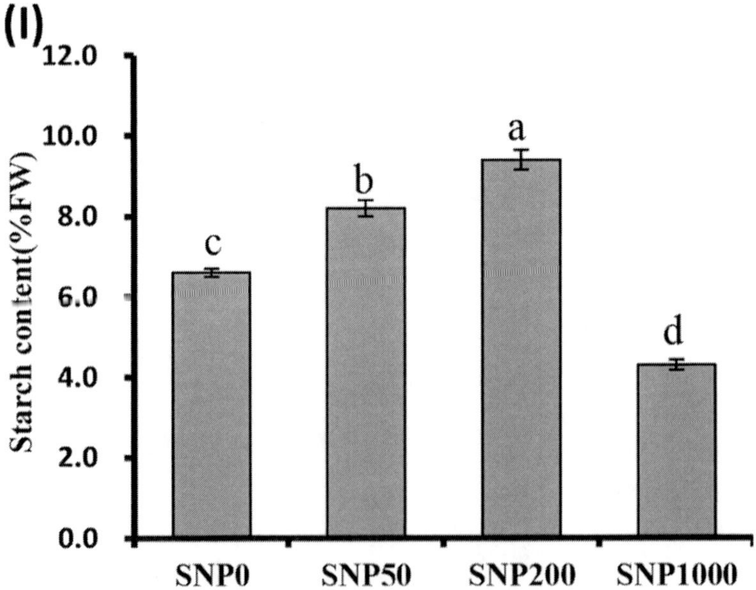
(I)
Starch content(%FW)
12.0
10.0
8.0
6.0
4.0
2.0
0.0
c
b
a
d
SNP0
SNP50
SNP200
SNP1000

(J)

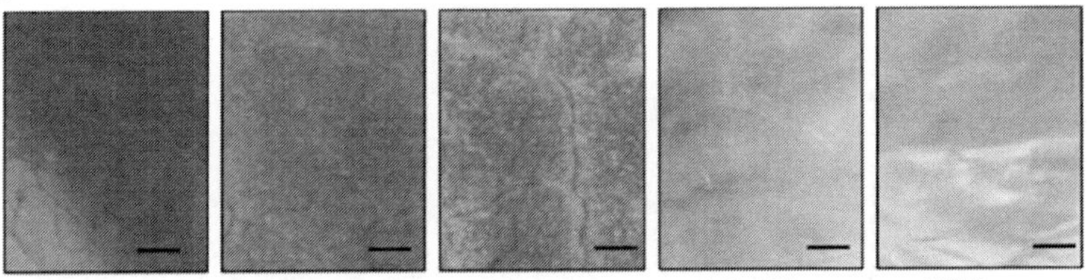

FIGURE 7. ANALYSES OF CHILLING INDUCED ULTRASTRUCTURE OF CHLOROPLASTS AND STARCH CONTENT IN WILD TYPE (WT) AND CSNOA1 TRANSGENIC CUCUMBER LINES. (A-F) Transmission electron microscopy (TEM) images of the ultrastructural changes of chloroplasts during normal and chilling stress conditions. The black arrows indicates starch granules. Scale bar = 2 µm. **(G)** Starch content in WT cucumber leaves under chilling stress at 4°C for 2 days with or without treatment with 200 µM L-NAME, 400 µM cPTIO. **(H)** Starch content changes in WT and transgenic cucumber lines under normal and chilling conditions. **(I)** Changes of starch content in WT leaves treated with varying concentrations (0, 50, 200, and1000 µM) of SNP for 2 days at 25°C. **(J)** Cucumber leaves stained for the presence of starch with potassium iodide under different SNP concentrations. Scale bar = 2 mm. Experiments were repeated three times. Mean values of three independent samples are shown and the error bars indicate the standard errors. Different letters above the column indicate significant differences at $P < 0.05$.

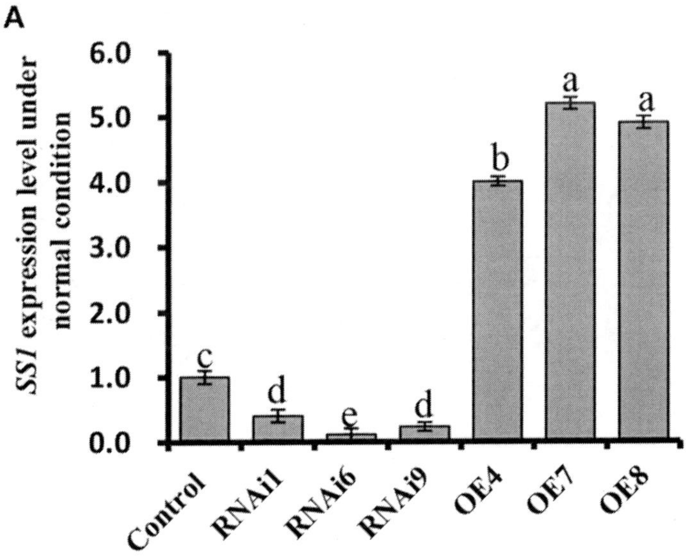

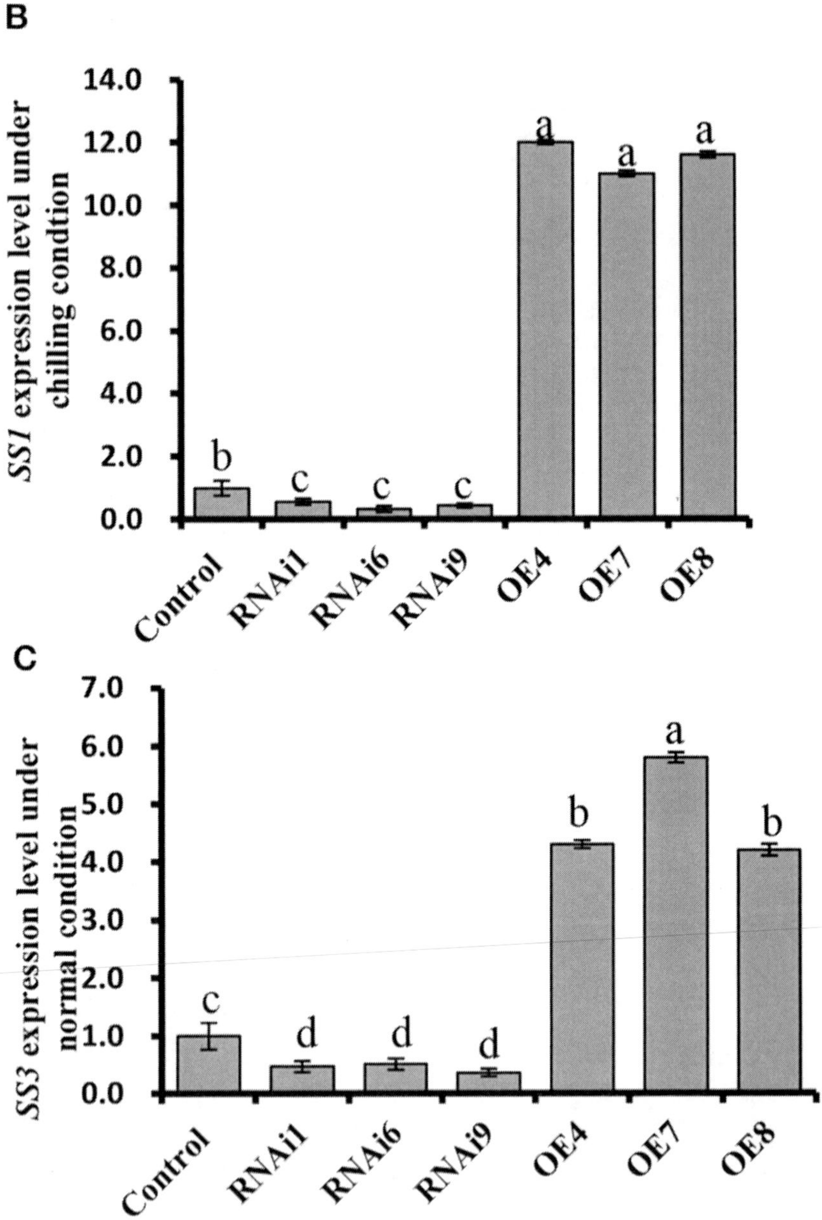
B
SS1 expression level under chilling condtion
14.0
12.0
10.0
8.0
6.0
4.0
2.0
0.0
a
a
a
b
c
c
c
Control
RNAi1
RNAi6
RNAi9
OE4
OE7
OE8
C
SS3 expression level under normal condition
7.0
6.0
5.0
4.0
3.0
2.0
1.0
0.0
a
b
b
c
d
d
d
Control
RNAi1
RNAi6
RNAi9
OE4
OE7
OE8

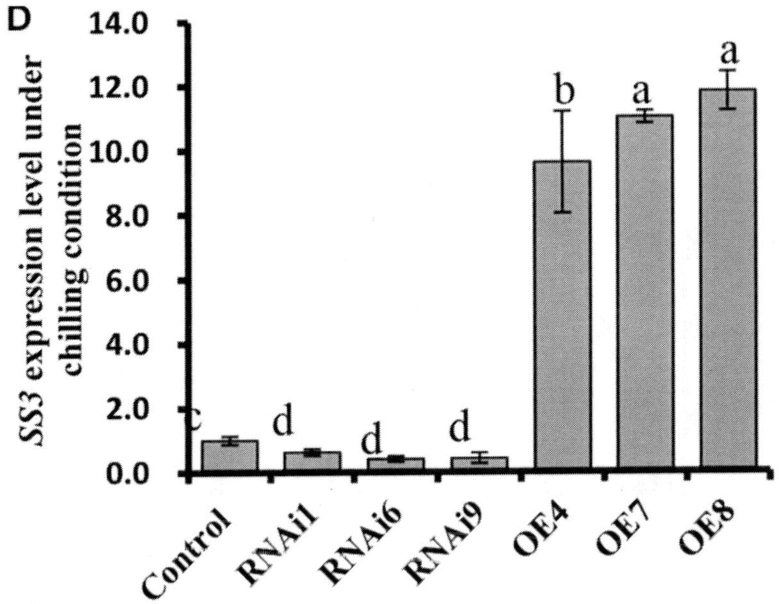
D
14.0
12.0
10.0
8.0
6.0
4.0
2.0
0.0
SS3 expression level under chilling condition
c
d
d
d
b
a
a
Control
RNAi1
RNAi6
RNAi9
OE4
OE7
OE8

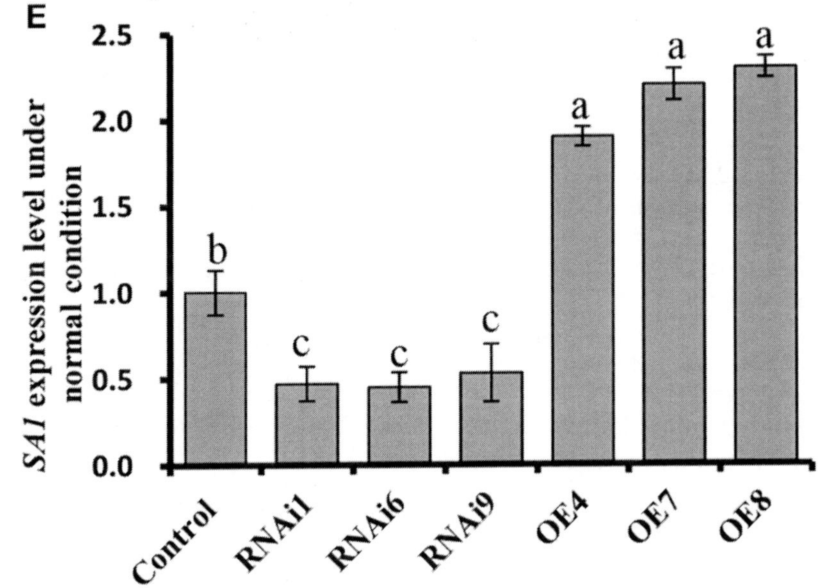
E
2.5
2.0
1.5
1.0
0.5
0.0
SA1 expression level under normal condition
b
c
c
c
a
a
a
Control
RNAi1
RNAi6
RNAi9
OE4
OE7
OE8

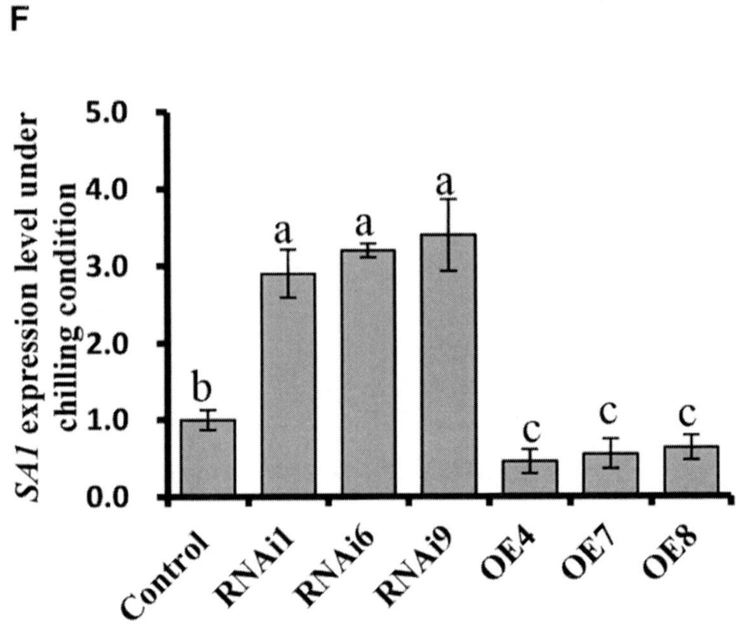

FIGURE 8. EXPRESSION PROFILES OF THE STARCH SYNTHESIS RELATED GENES, SS1, AND SS3, AND THE STARCH DEGRADATION RELATED GENE, SA1, IN CUCUMBER LEAVES. (A,C,E) Expression patterns of *SS1*, *SS3*, and *SA1* determined by real time RT-PCR in wild type (WT) and *CsNOA1* transgenic lines grown under normal condition. **(B,D,F)** Expression patterns of *SS1*, *SS3*, and *SA1* determined by real time RT-PCR in WT and *CsNOA1* transgenic lines grown under chilled condition. The cucumber α-*TUBULIN (TUA)* gene was used as an internal control. The experiments were repeated three times and the mean values of three independent samples and standard errors are presented. Different letters above the column indicate significant differences at $P < 0.05$.

Two Independent Pathways NOA1- and NR- may Generate Endogenous NO in Cucumber Leaves

In order to determine whether endogenous NO is present in cucumber seedlings, we used the fluorochrome DAF-2DA to detect NO at the three leaf stage. Figure 10 shows that under confocal laser scanning microscopy (CLSM) an intense green fluorescence was present in the leaf stomata in fluorochrome treated samples, but no fluorescence was observed when the fluorescent probe was omitted. Likewise, NO-derived fluorescence was detected in cucumber leaf sections pre-incubated with the NOS1/NOA1 activity inhibitor, L-NAME, and we observed that the green fluorescent spots were not brighter than those in the control samples. As with the L-

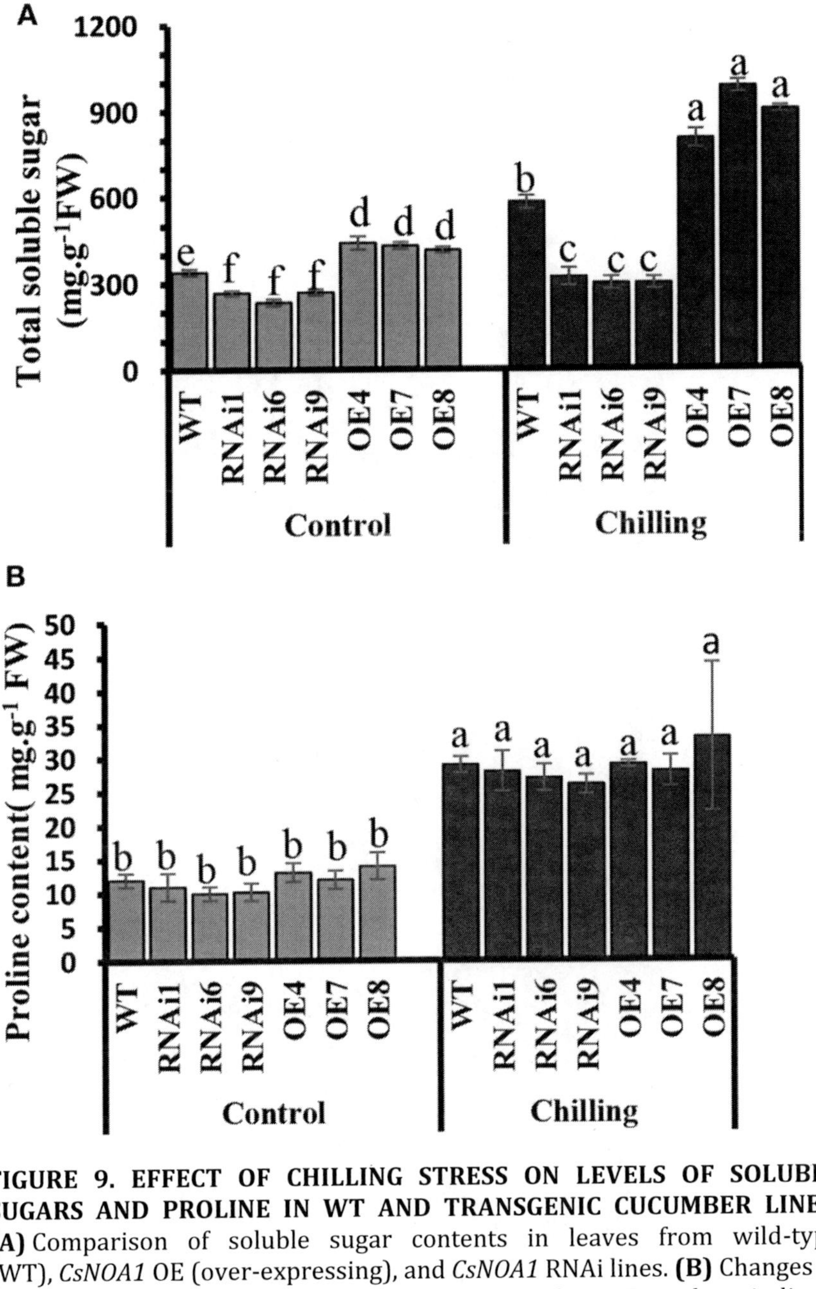

FIGURE 9. EFFECT OF CHILLING STRESS ON LEVELS OF SOLUBLE SUGARS AND PROLINE IN WT AND TRANSGENIC CUCUMBER LINES. **(A)** Comparison of soluble sugar contents in leaves from wild-type (WT), *CsNOA1* OE (over-expressing), and *CsNOA1* RNAi lines. **(B)** Changes in proline content in leaves from WT, OE, and RNAi lines. Error bars indicate the standard errors and different letters above the column indicate significant differences at $P < 0.05$.

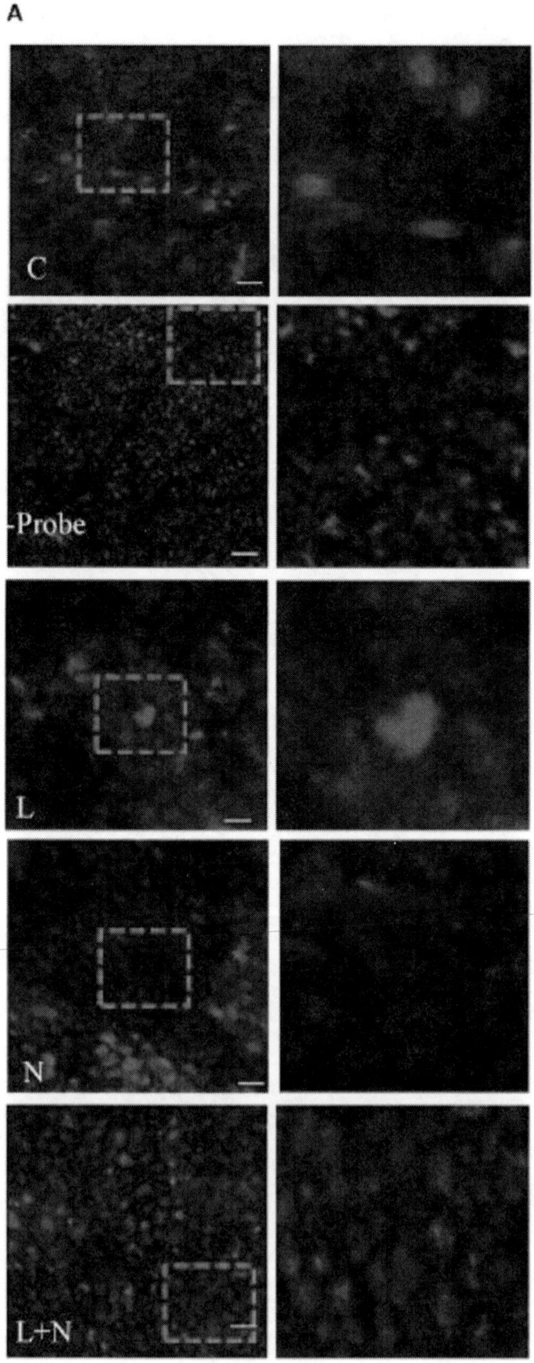
A
C
-Probe
L
N
L+N

B

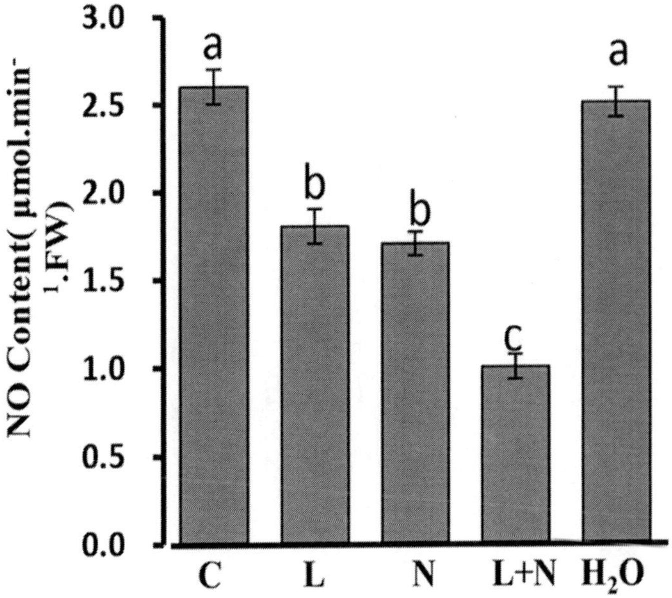
NO Content(µmol.min-1.FW)
3.0
2.5
2.0
1.5
1.0
0.5
0.0
a
b
b
c
a
C
L
N
L+N
H2O

C

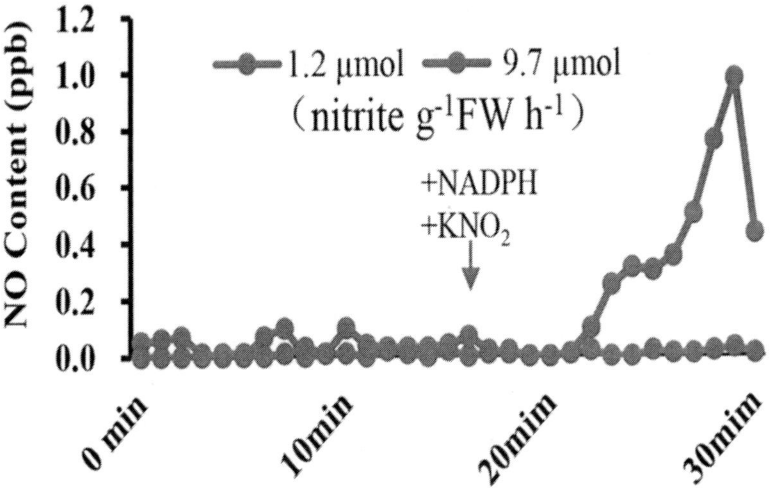
NO Content (ppb)
1.2
1.0
0.8
0.6
0.4
0.2
0.0
1.2 µmol
9.7 µmol
(nitrite g-1FW h-1)
+NADPH
+KNO2
0 min
10min
20mim
30mim

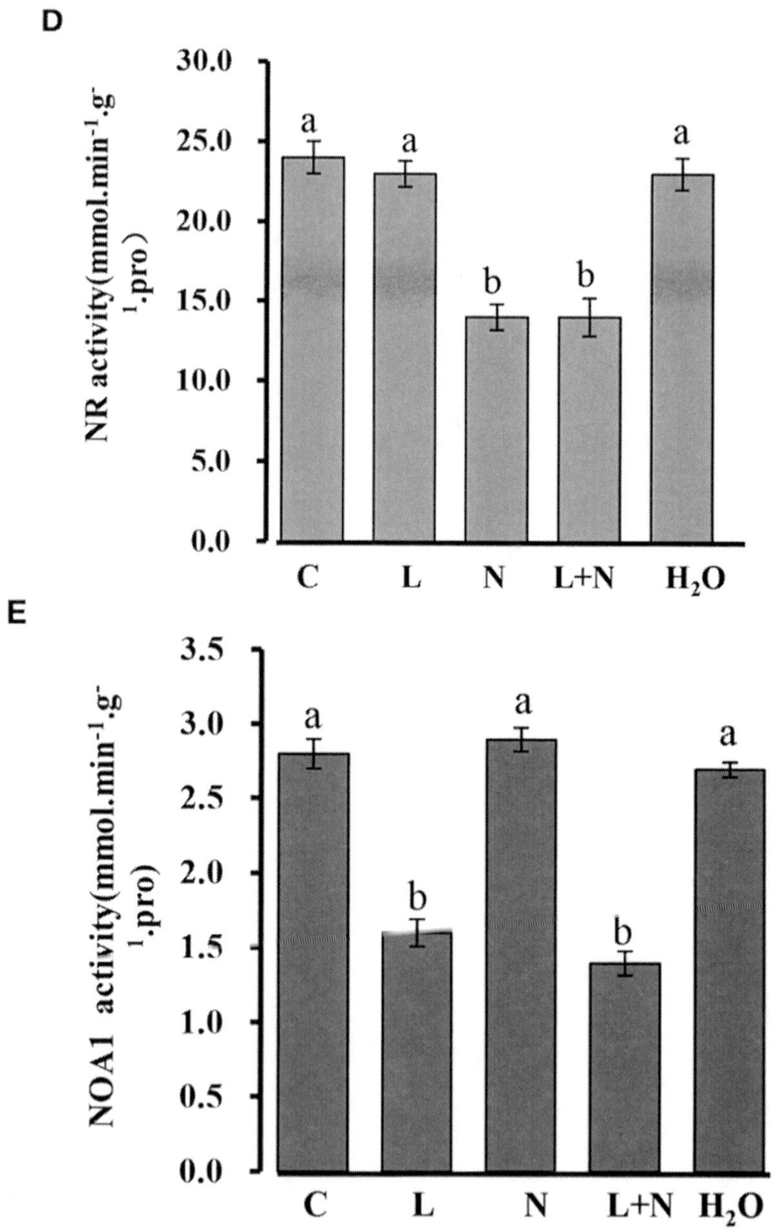

FIGURE 10. CHANGES IN ENDOGENOUS NO (NITRIC OXIDE) LEVELS IN CUCUMBER STOMATA UNDER NORMAL CONDITIONS. (A) NO levels in excised leaves were monitored by labeling with the NO-specific fluorescent probe DAF-FM DA and imaged by confocal microscopy. The right-hand image is a magnification of the red dotted square in the left image. Scale bar

= 100 μm. Two hundred micrometer NaN_3 and 200 μM L-NAME were applied to wild type cucumber plants grown under normal conditions. **(B)** NO levels in cucumber leaves under normal conditions. **(C)** NR involved in the NO production from nitrite plus NADH. Two ml of desalted cucumber extract were pre-incubated for 17 min with 5 mM AMP and 20 mM EDTA (open symbols), or with 5 mM ATP and 50 μM cantharidin (closed symbols) in a Petri dish placed in a reaction chamber (3.01–3.02 air min^{-1}, room temperature); NADH (250 μM) and KNO_2 (100 μM) were added as indicated. **(D,E)** Effect of inhibitors on NOA1 and NR activity under normal conditions. L, L-NAME; N, NaN_3; C, wild type and H_2O was used as a control. The mean values of three independent samples and standard errors are shown, and the same letter above the column indicates no significant differences at $P < 0.05$.

NAME treatment, green fluorescent spots were also observed in cucumber leaf sections pre-incubated with the NR activity inhibitor, sodium azide (NaN_3) (Figure 10A). Interestingly, NO-derived fluorescence was barely detectable in leaf sections pre-incubated with both L-NAME and NaN_3.

A quantitative assay was performed to detect endogenous NO levels (Figure 10B). Treatments with the inhibitors, L-NAME or NaN_3, significantly reduced NO levels in all leaf samples compared to those of the control, and when the two inhibitors were used in combination, endogenous NO levels decreased even more substantially, although they were not abolished entirely. In addition, we repeated the experiments using the inhibitors L-NNA (Rockel et al., 2002) for NOA1 and tungstate (Planchet et al., 2005) for NR, which gave similar results to those described above (Supplementary Figure S3A–D). Taken together these results suggest that NOA1 and NR act primarily in the endogenous NO pathway.

It is known that NO production can be modulated by preincubation with ATP, which indicates that NR activity (Rockel et al., 2002; Planchet et al., 2005). Desalted cucumber leaf extracts was used instead of purified NR in this study. It has been reported that auxiliary enzymes (NR kinase, P-NR phosphatase and 14-3-3 proteins) are required for this experiment (Rockel et al., 2002). Following the addition of NADH plus nitrite in 2 ml of reaction medium, NO was emitted into the gas phase above the solution. We observed that approximately 10 min after substrate addition, NO emission decreased again. In contrast, pre-incubation of the extract with ATP, $Mg^{2+}$and the protein phosphatase 2A inhibitor, cantharidine, caused an almost complete inhibition of nitrite-dependent NO production. The measured NR activity in an aliquot of the extract was 9.7 μmol nitrite g^{-1}FW $h^{-1}$for the AMP pretreated samples, but 1.2 μmol nitrite g^{-1}FW h^{-1} after preincubation with ATP plus cantharidine (Figure 10C). This result suggested that NR is involved in NO production in cucumber leaves.

NR and NOA1 activities were also measured in the presence of other inhibitors, and NOA1 activity was markedly reduced after adding its inhibitor L-NAME, but was not affected by NaN_3 (Figures 10D,E). Similarly, NR activity declined substantially after adding its inhibitor NaN_3, but was not affected by L-NAME. These results suggest that generation of endogenous NO in cucumber leaves may occurs largely independently in the NOA1 and NR pathways.

DISCUSSION

Cucumber is a warm-season horticultural crop with little, or no, frost tolerance (Liu et al., 2010), and while it is known that low temperature stress can be alleviated by NO in plants such as maize, rice and *A. thaliana* (Zhao et al., 2009; Liu et al., 2011; Wimalasekera et al., 2011; Yang et al., 2011b; Tan et al., 2013b), the molecular regulation of chilling tolerance by NO in cucumber is largely unknown.

Mitochondria Localized CsNOA1 Confers Chilling Stress Tolerance in Transgenic *A. thaliana*

Previous studies showed that NOA1 activity is linked to peroxisomes and chloroplasts (Corpas et al., 2001; Gould et al., 2003), while only a few reports describe mitochondria localized NO synthesis (Guo and Crawford, 2005; Hao et al., 2010). However, the Arabidopsis AtNOS1 gene or NOA1 gene, are not yet accepted as the genes involved in NO synthesis in plants. Initially AtNOS1 was considered as plant NOS (Guo et al., 2003) and is yet to be proven. Therefore, these disputed conclusions are not to be interpreted to reflect a standard NOS present in plants. In this current study, GFP-fused CsNOA1 was observed in the mitochondria, as fluorescence overlapped with Mito-tracker Red in the stomata, but not with Chl fluorescence (Figure 4). In further support of CsNOA1being localized to the mitochondria, we also observed that CsNOA1-GFP fluorescence overlapped with Mito-tracker Red in roots (Supplementary Figure S4).

We observed that ectopic expression of *CsNOA1* in the *A. thaliana* mutant *noa1*rescued the mutant phenotype and enhanced its tolerance to chilling stress (Figures 5A–F), which may be a consequence of increased levels of NO. Such an association with NO signaling has also observed in other plants. For example, studies using NR inhibitors, a NO scavenger, and a NO donor, showed that NR-dependent NO levels are positively correlated with freezing tolerance (Zhao et al., 2009). We observed that the degree of tolerance correlated with the level of *CsNOA1* expression (Figure 5), with a strong over-expression of *CsNOA1* resulting in a high degree of cold tolerance and high endogenous NO levels. This suggests that *CsNOA1* is important for chilling stress, and that the effect is dose dependent. We determined that *CsNOA1* mRNA levels in the L11 line were 3.5 fold higher than those in the L7 line, and that the chilling index (CI) and the expression of the cold tolerance marker gene *CBF3*was lower (Figure 5D). We also saw a higher Fv/Fm ratio in plants subjected to chilling stress, even though it was still lower than in plants grown under normal condition (Figure 5E).

Transgenic Cucumber CsNOA1 Overexpressing and RNAi Lines Exhibited Altered Chloroplast Ultrastructure and CBF3 Expression Patterns

Phenotypic observation and physiological indicators, such as Fv/Fm and *CBF3* expression, suggested a greater chilling tolerance in lines *CsNOA1* OE lines than in the *CsNOA1* RNAi lines (Figure 6). *NOA1* is required for chloroplast biogenesis (Flores-Pérez et al., 2008), and the ultrastructure of chloroplasts was therefore observed in all the different lines (Figures 7A–D). The chloroplasts of the OE lines were shown to be filled with swollen starch granules, unlike those of the non-transgenic control grown under the same conditions (Figure 7A), and we propose that an important function of *NOA1* is to ensure efficient chloroplast function during photosynthesis.

A major response of plants to chilling is the activation of the *CBF*-dependent pathway, which regulates the *CBF* regulon (Van Buskirk and Thomashow, 2006). In this study, we showed that the expression of *CBF3* was greatly induced in chilled *CsNOA1* OE lines compared with WT plants. In contrast, the level of *CBF3* expression was lower in the RNAi lines compared with WT plants under the same chilling stress conditions (Figure 6). These results indicate that *CsNOA1* affected the expression of *CBF3,* and link NO signaling with the *CBF*-dependent pathway. This is supported by a previous report suggesting that in the nia1nia2 NR double mutant, NO production is linked to the *CBF*-dependent pathway (Cantrel et al., 2011). We suggest that under chilling conditions, different sources of NO in different plants may be involved in the same cold/chilling signal pathway, or at least, in the *CBF*-dependent pathway.

CsNOA1 Affects Soluble Sugar Accumulation, but not Proline Levels, in Chill Stressed Transgenic Cucumber Lines

We observed that elevated NO levels can result in starch accumulation in cucumber leaves under normal conditions (Figures 7I,J), and that under chilling conditions, the starch content in *CsNOA1* OE cucumber lines was higher than in WT plants (Figures 8A–D). Therefore, the induction of starch accumulation may be a downstream effect of *CsNOA1*-induced NO production. To date, there have been few reports showing that NO can alter the expression of the starch biosynthesis related genes *SS1* and *SS3* or the starch degradation related gene SA1 (Figures 8A–F, Supplementary Tables 1–2). Our study provides evidence that NO signaling is linked to starch content (Supplementary Tables 1–2) and we detected an enhanced accumulation osf soluble sugars in the *CsNOA1* OE lines, which may partially account for the higher tolerance of these plants to chilling stress. The high transcript level of the SUG gene, which encodes an alkaline/neutral invertase, has previously been correlated with glucose accumulation (Maruyama et al., 2014), and a similar relationship between high transcript levels of this gene and enhanced levels of glucose has been reported in cold-treated *A. thaliana* plants (Kaplan et al., 2007).

SUG expression in the *CsNOA1* OE lines, RNAi lines, and WT cucumber seedlings in response to chilling stress were more than 6-fold higher in the OE lines in untreated plants (Supplementary Figure 2C), while the other three *SUG genes* showed no changes in all tested samples (data not shown). High expression of the *SUG* gene in the *CsNOA1* OE lines may effectively enhance osmoregulation capacity and improve water potential by the accumulation of sugars, thereby minimizing chilling induced damage.

Land plants accumulate free proline in response to a number of abiotic stresses, such as drought, salinity and freezing (Hare et al., 1999; Ashraf and Foolad, 2007; Hoque et al., 2007). We detected no changes in proline content in cucumber *CsNOA1* OE or RNAi lines when compared to WT before or after chilling stress (Figure 9B), as well as no differences in the expression of proline biosynthesis or transport related genes (Supplementary Figure S2A,B). In contrast, total soluble sugar levels increased substantially in the OE lines and were greatly reduced in the RNAi lines after exposure to chilling stress, when compared to WT plants (Figure 9A). This result is not congruent with an analysis of cold-stressed *A. thaliana* (Zhao et al., 2009), suggesting that the ameliorative effect of NO on chilling stress tolerance cannot be explained by osmoregulation mediated by proline content. It may also be that NO is not directly related to proline content or P5CS activity, and that the expression of P5CS is not directly regulated by NO.

NOA1 and NR, this Two Pathways Co-Exist in Cucumber Seedlings

Under normal condition, we took several classic inhibitors (Zhao et al., 2007a, 2009; Rümer et al., 2009) to perform and detect endogenous NO levels in wild type cucumber seedling (Figure 10; Supplementary Figures 3A–D). The results suggest that generation of endogenous NO in cucumber leaves occurs largely independently in the NOA1 and NR pathways. Under chilling condition, when the expression level of *NR* was analyzed by real time RT-PCR, no obvious change was observed between the transgenic cucumber lines and WT; however, *NR* transcript levels were generally higher under chilling conditions than under normal conditions (Figure 6E). We infer from these results that *CsNOA1* produced NO independent of NR expression This result is not consistent with studies of loquat fruit (Xu et al., 2012), where chilling induced NO generation was partially suppressed by the NR inhibitor tungstate, indicating that the NR pathway may be the main source of NO generation. Additionally, in A. thaliana, endogenous NO production is mainly generated by NR activity in WT leaves and chilling-induced NO release is reduced in the *NR* double mutant (Zhao et al., 2009; Cantrel et al., 2011). The results indicate that in cucumber, the two pathways can operate together under nomal condition, but under chilling stress, this stress signal, may trigger a certain pathway to produce NO and confer its tolerance to stress. Because of the lack of NOA1-and NR-related mutants in cucumber, inhibitors were chosen as a study tool, although those provide an indirect confirmation. Using new technology, for example, CRISPY-Cas9, to create CsNOA1 and NR null mutants would further verify this result. However, this new technology was technically immature in cucumber, and presently work is being performed in this area. Considering all the data, a model (Figure 11) is proposed in which *CsNOA1* transcription occurs independent of chilling stress. NO produced by *CsNOA1* directly activates the expression of *CBF3* and, meanwhile, NO activates the starch metabolic pathway, leading to increased expression of *SS1* and *SS3* and decreased expression of *SA1*. High expression of *CBF3* and accumulation of starch increases the stress tolerance of cucumber.

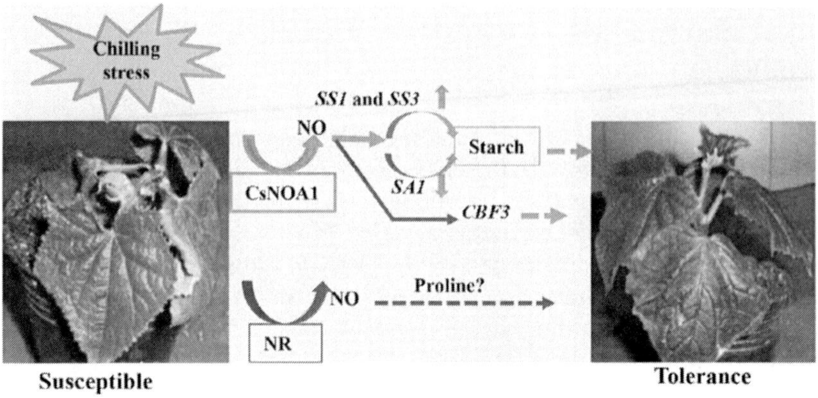

FIGURE 11. A PUTATIVE WORKING MODEL OF THE INVOLVEMENT OF CSNOA1 IN CHILLING STRESS. NO (nitric oxide) produced by the *CsNOA1* pathway activates the starch metabolic pathway, leading to an increase in starch content. *CBF3* expression is also affected by NO. *CBF3*-denpendent stress tolerance may be correlated with, or independent of, starch metabolism. Endogenous NO is generated independently of the NR pathway in cucumber.

Accession Numbers

Sequence data of NOA1 proteins in this study can be found in the cucumber Genome database, Arabidopsis Genome Initiative, Phytozome or Genbank/EMBL/Swiss-prot database and NCBI under the following accession numbers: *CsNOA1* (Csa5M168870), *Atnoa1* (At3G47450), *Zea mays* (NP-001168044.1), *Oryza sativa*(Os02g0104700), *Brassica juncea* (Bradi3g00690.1*)*, *Nicotiana benthamiana* (AB303300.1), *Ricinus communis* (XP002510962.1), *Medicago truncatula* (KEH36486.1), *Solanum tuberosum* (GU205181.1). For real-time PCR, sequence data of text genes in this study can be found in the cucumber Genome database under the following accession numbers: *CsP5CS1* (Csa3M073290), *CsSug1* (Csa7M308910), *CsSug2* (Csa2M034660), *CsSug3*(Csa3M168930), *CsSug4* (Csa5M615240), *CsCBF3* (Csa5M155570*)*, *CsSS1* (Csa6M497160), *CsProtran s*(Csa5M615830), *CsSS3* (Csa1M062920).

ACKNOWLEDGMENTS

This work was supported by The National Research and Development Project (2016YFD0101705), Beijing Agricultural Innovation Consortium Project (BAIC01-2016) and Beijing Agricultural Scientific and Technological

Project (20160415) to HR. We thank M Zhao at the institute of botany, the Chinese Academy of sciences (Beijing) for kindly providing the Atnoa1 mutant for this study.

REFERENCES

1. Anand, B., Surana, P., Bhogaraju, S., Pahari, S., and Prakash, B. (2009). Circularly permuted GTPase YqeH binds 30S ribosomal subunit: implications for its role in ribosome assembly. Biochem. Biophys. Res. Commun. 386, 602–606. doi: 10.1016/j.bbrc.2009.06.078
2. Ashraf, M., and Foolad, M. (2007). Roles of glycine betaine and proline in improving plant abiotic stress resistance. Environ. Exp. Bot. 59, 206–216. doi: 10.1016/j.envexpbot.2005.12.006
3. Beck, E. H., Heim, R., and Hansen, J. (2004). Plant resistance to cold stress: mechanisms and environmental signals triggering frost hardening and dehardening. J. Biosci. 29, 449–459. doi: 10.1007/BF02712118
4. Bright, J., Desikan, R., Hancock, J. T., Weir, I. S., and Neill, S. J. (2006). ABA-induced NO generation and stomatal closure in Arabidopsis are dependent on H_2O_2 synthesis. Plant J. 45, 113–122. doi: 10.1111/j.365-313X.2005.02615.x
5. Cantrel, C., Vazquez, T., Puyaubert, J., Rezé, N., Lesch, M., Kaiser, W. M., et al. (2011). Nitric oxide participates in cold-responsive phosphosphingolipid formation and gene expression in Arabidopsis thaliana. New Phytol. 189, 415–427. doi: 10.1111/j.1469-8137.2010.03500.x
6. Cheng, T., Chen, J., Allah, E. F., Wang, P., Wang, G., Hu, X., et al. (2015). Quantitative proteomics analysis reveals that S-nitrosoglutathione redutase (GSNOR) and nitric oxide signling enhance poplar defense against chilling stress. Planta 242, 1361–1390. doi: 10.1007/s00425-015-2374-5
7. Clough, S. J., and Bent, A. F. (1998). Floral dip: a simplified method for Agrobacterium-mediated transformation of Arabidopsis thaliana. Plant J. 16, 735–743. doi: 10.1046/j.1365-313x.1998.00343.x
8. Cook, D., Fowler, S., Fiehn, O., and Thomashow, M. F. (2004). A prominent role for the CBF cold response pathway in configuring the low-temperature metabolome of Arabidopsis. Proc. Natl. Acad. Sci. U.S.A. 101, 15243–15248. doi: 10.1073/pnas.0406069101
9. Corpas, F. J., Barroso, J. B., and del Río, L. A. (2001). Peroxisomes as a source of reactive oxygen species and nitric oxide signal molecules in plant cells. Trends Plant Sci. 6, 145–150. doi: 10.1016/S1360-1385(01)01898-2
10. Desikan, R., Griffiths, R., Hancock, J., and Neill, S. (2002). A new role for an old enzyme: nitrate reductase-mediated nitric oxide generation is required for abscisic acid-induced stomatal closure in Arabidopsis thaliana. Proc. Natl. Acad. Sci. U.S.A. 99, 16314–16318. doi: 10.1073/pnas.252461999
11. Flores-Pérez, Ú., Sauret-Güeto, S., Gas, E., Jarvis, P., and Rodríguez-Concepción, M. (2008). A mutant impaired in the production of plastome-encoded proteins uncovers a mechanism for the homeostasis of isoprenoid biosynthetic enzymes in Arabidopsis plastids. Plant Cell 20, 1303–1315. doi: 10.1105/tpc.108.058768
12. Foresi, N., Correa-Aragunde, N., Parisi, G., Caló, G., Salerno, G., and Lamattina, L. (2010). Characterization of a nitric oxide synthase from the plant kingdom: NO generation from the green alga Ostreococcus tauri is light irradiance and growth phase dependent. Plant Cell 22, 3816–3830. doi: 10.1105/tpc.109.073510
13. Guo, F. Q., Okamoto, M., and Crawford, N. M. (2003). Identification of a plant nitric oxide synthase gene involved in hormonal signaling. Science302, 100–103. doi: 10.1126/science.1086770

14. Gould, K. S., Lamotte, O., Klinguer, A., Pugin, A., and Wendehenne, D. (2003). Nitric oxide production in tobacco leaf cells: a generalized stress response? Plant Cell Environ. 26, 1851–1862. doi: 10.1016/S0014-5793(96)01232-X
15. Guillas, I., Zachowski, A., and Baudouin, E. (2011). A matter of fat: interaction between nitric oxide and sphingolipid signaling in plant cold response. Plant Signal. Behave. 6, 140–142. doi: 10.4161/psb.6.1.14280
16. Guo, F. Q., and Crawford, N. M. (2005). Arabidopsis Nitric oxide synthase1 is targeted to mitochondria and protects against oxidative damage and dark-induced senescence. Plant Cell 17, 3436–3450. doi: 10.1105/tpc.105.037770
17. Gupta, K. J., and Kaiser, W. M. (2010). Production and scavenging of nitric oxide by barley root mitochondria. Plant Cell Physiol. 51, 576–584. doi: 10.1093/pcp/pcq022
18. Hao, L., Yu, C., Li, B., and Wang, D. (2010). Molecular cloning and preliminary analysis of TaNOA in common wheat. Chin. J. Biotechnol. 26, 48–56.
19. Hare, P., Cress, W., and Van Staden, J. (1999). Proline synthesis and degradation: a model system for elucidating stress-related signal transduction. J. Exp. Bot. 50, 413–434. doi: 10.1093/jxb/50.333.413
20. Hoque, M. A., Okuma, E., Banu, N. A., Nakamura, Y., and Shimoishi, M. (2007). Exogenous proline mitigates the detrimental effects of salt stress more than exogenous betaine by increasing antioxidant enzyme activities. J. Plant Physiol. 164, 553–561. doi: 10.1016/j.jplph.2006.03.010
21. Huang, S., Li, R., Zhang, Z., Li, L., Gu, X., Fan, W., et al. (2009). The genome of the cucumber, Cucumis sativus L. Nat. Genet. 41, 1275–1281. doi: 10.1038/ng.475
22. Jefferson, R. A., Kavanagh, T. A., and Bevan, M. W. (1987). GUS fusions: beta-glucuronidase as a sensitive and versatile gene fusion marker in higher plants. EMBO J. 6, 3901–3907.
23. Kaplan, F., Kopka, J., Sung, D. Y., Zhao, W., Popp, M., Porat, R., et al. (2007). Transcript and metabolite profiling during cold acclimation of Arabidopsis reveals an intricate relationship of cold-regulated gene expression with modifications in metabolite content. Plant J. 50, 967–981. doi: 10.1111/j.1365-313X.2007.03100.x
24. Kuk Y. I. Shin J. S. (2007). Mechanisms of low-temperature tolerance in cucumber leaves of various ages. J. Am. Soc. Hortic. Sci. 132, 294–301. Available online at: http://journal.ashspublications.org/content/132/3/294.full
25. Kwan, Y. M., Meon, S., Ho, C. L., and Wong, M. Y. (2014). Cloning of nitric oxide associated 1 (NOA1) transcript from oil palm (Elaeis guineensis) and its expression during Ganoderma infection. J. Plant Physiol. 174, 131–136. doi: 10.1016/j.jplph.2014.10.003
26. Liu, B., Liu, X., Yang, S., Chen, C., Xue, S. D., Cai, Y., et al. (2016). Silencing of the gibberellin receptor homolog, CsGID1a, affects locule formation in cucumber (Cucumis sativus) fruit. New Phytol. 210, 551–561. doi: 10.1111/nph.13801
27. Liu, L., Duan, L., Zhang, J., Zhang, Z., Mi, G., and Ren, H. (2010). Cucumber over-expressing cold-induced transcriptome regulator ICE1 exhibits changed morphological characters and enhances chilling tolerance. Sci. Hortic. 124, 29–33. doi: 10.1016/j.scienta.2009.11.018
28. Liu, X., Wang, L., Liu, L., Guo, Y., and Ren, H. (2011). Alleviating effect of exogenous nitric oxide in cucumber seedling against chilling stress. Afr J Biotechnol. 10, 4380–4386. doi: 10.5897/AJB10.812
29. Maruyama, K., Takeda, M., Kidokoro, S., Yamada, K., Sakuma, Y., Urano, K., et al. (2009). Metabolic pathways involved in cold acclimation identified by integrated analysis of metabolites and transcripts regulated by DREB1A and DREB2A. Plant Physiol. 150, 1972–1980. doi: 10.1104/pp.109.135327
30. Maruyama, K., Urano, K., Yoshiwara, K., Morishita, Y., Sakurai, N., Suzuki, H., et al. (2014). Integrated analysis of the effects of cold and dehydration on rice metabolites, phytohormones, and gene transcripts. Plant Physiol. 164, 1759–1771. doi: 10.1104/pp.113.231720

31. Matre, P., Meyer, C., and Lillo, C. (2009). Diversity in subcellular targeting of the PP2A B′η subfamily members. Planta 230, 935–945. doi: 10.1007/s00425-009-0998-z
32. Mayer, B., and Hemmens, B. (1997). Biosynthesis and action of nitric oxide in mammalian cells. Trends Biochem. Sci. 22, 477–481. doi: 10.1016/S0968-0004(97)01147-X
33. Minami, A., Nagao, M., Ikegami, K., Koshiba, T., Arakawa, K., Fujikawa, S., et al. (2005). Cold acclimation in bryophytes: low temperature-induced freezing tolerance in Physcomitrella patens is associated with increases in expression levels of stress-related genes but not with increase in level of endogenous abscisic acid. Planta 220 414–423. doi: 10.1007/s00425-004-1361-z
34. Miura, K., Jin, J. B., Lee, J., Yoo, C. Y., Stirm, V., Miura, T., et al. (2007). SIZ1-mediated sumoylation of ICE1 controls CBF3/DREB1A expression and freezing tolerance in Arabidopsis. Plant Cell 19, 1403–1414. doi: 10.1105/tpc.106.048397
35. Moreau, M., Lee, G. I., Wang, Y., Crane, B. R., and Klessig, D. F. (2008). AtNOS/AtNOA1 is a functional Arabidopsis thaliana cGTPase and not a nitric-oxide synthase. J. Biol. Chem. 283, 32957–32967. doi: 10.1074/jbc.M804838200
36. Planchet, E., Jagadis Gupta, K., Sonoda, M., and Kaiser, W. M. (2005). Nitric oxide emission from tobacco leaves and cell suspensions: rate limiting factors and evidence for the involvement of mitochondrial electron transport. Plant J. 41, 732–743. doi: 10.1111/j.1365-313X.2005.02335.x
37. Rockel, P., Strube, F., Rockel, A., Wildt, J., and Kaiser, W. M. (2002). Regulation of nitric oxide (NO) production by plant nitrate reductase in vivo and in vitro. J. Exp. Bot. 53, 103–110. doi: 10.1093/jexbot/53.366.103
38. Ruelland, E., Vaultier, M. N., Zachowski, A., and Hurry, V. (2009). Cold signaling and cold acclimation in plants. Adv. Bot. Res. 49, 35–150. doi: 10.1016/s0065-2296(08)00602-2
39. Rümer, S., Gupta, K. J., and Kaiser, W. M. (2009). Plant cells oxidize hydroxylamines to NO. J. Exp. Bot. 60, 2065–2072. doi: 10.1039/jxb/erp077
40. Saitou, N., and Nei, M. (1987). The neighbor-joining method: a new method for reconstructing phylogenetic trees. Mol. Biol. Evol. 4, 406–425.
41. Sang, J., Zhang, A., Lin, F., Tan, M., and Jiang, M. (2008). Cross-talk between calcium-calmodulin and nitric oxide in abscisic acid signaling in leaves of maize plants. Cell Res. 18, 577–588. doi: 10.1038/cr.2008.39
42. Stöhr, C., and Stremlau, S. (2006). Formation and possible roles of nitric oxide in plant roots. J. Exp. Bot. 57, 463–470. doi: 10.1093/jxb/erj058
43. Sudhamsu, J., Lee, G. I., Klessig, D. F., and Crane, B. R. (2008). The structure of YqeH. An Atnoa1 ortholog that couples GTP hydrolysis to molecular recognition. J. Biol. Chem. 283, 32968–32976. doi: 10.1074/jbc.M804837200
44. Tan, J., Wang, C., Xiang, B., Han, R., and Guo, Z. (2013a). Hydrogen peroxide and nitric oxide mediated cold-and dehydration-induced myo-inositol phosphate synthase that confers multiple resistances to abiotic stresses. Plant Cell Environ. 36, 288–299. doi: 10.1111/j.1365-3040.2012.02573.x
45. Tan, J., Zhuo, C., and Guo, Z. (2013b). Nitric oxide mediates cold-and dehydration-induced expression of a novel MfHyPRP that confers tolerance to abiotic stress. Physiol. Plantarum. 149, 310–320. doi: 10.1111/ppl.12032
46. PubMed Abstract | CrossRef Full Text | Google Scholar
47. Theocharis, A., Clément, C., and Barka, E. A. (2012). Physiological and molecular changes in plants grown at low temperature. Planta 235, 1091–1105. doi: 10.1007/s00425-012-1641-y
48. CrossRef Full Text | Google Scholar
49. Van Buskirk, H. A., and Thomashow, M. F. (2006). Arabidopsis transcription factors regulating cold acclimation. Physiol. Plantarum. 126, 72–80. doi: 10.1111/j.1399-3054.2006.00625.x

50. Wendehenne, D., Pugin, A., Klessig, D. F., and Durner, J. (2001). Nitric oxide: comparative synthesis and signaling in animal and plant cells. Trends Plant Sci. 6, 177–183. doi: 10.1016/s1360-1385(01)01893-3
51. Wilson, I. D., Neill, S. J., and Hancock, J. T. (2008). Nitric oxide synthesis and signalling in plants. Plant Cell Environ. 31, 622–631. doi: 10.111/j.1365-3040.2007.01761.x
52. Wimalasekera, R., Tebartz, F., and Scherer, G. F. (2011). Polyamines, polyamine oxidases and nitric oxide in development, abiotic and biotic stresses. Plant Sci. 181, 593–603. doi: 10.1016/j.plantsci.2011.04.002
53. Xiong, L., and Zhu, J. K. (2001). Abiotic stress signla transduction in plants: molecular and genetic perspectives. Physiol. Plantarum. 112, 152–166. doi: 10.1034/j.1399-3054.2001.1120202.x
54. Xu, M., Dong, J., Zhang, M., Xu, X., and Sun, L. (2012). Cold-induced endogenous nitric oxide generation plays a role in chilling tolerance of loquat fruit during postharvest storage. Postharvest Biol. Technol. 65, 5–12. doi: 10.1016/j. postharvbio. 2011.10. 008
55. Yang, Q., He, H., Li, H., Tian, H., Zhang, J., Zhai, L., et al. (2011b). NOA1 functions in a temperature-dependent manner to regulate chlorophyll biosynthesis and Rubisco formation in rice. PLoS ONE 6:e20015. doi: 10.1371/journal.pone.0020015
56. Yang, H., Wu, F., and Cheng, J. (2011a). Reduced chilling injury in cucumber by nitric oxide and the antioxidant response. Food Chem. 127, 1237–1242. doi: 10.1016/j.foodchem.2011.02.011
57. Zhang, L. P., Mehta, S. K., Liu, Z. P., and Yang, Z. M. (2008). Copper-induced proline synthesis is associated with nitric oxide generation in Chlamydomonas reinhardtii. Plant Cell Physiol. 49, 411–419. doi: 10.1093/pcp/pcn017
58. Zhang, X., Zhou, Y., Ding, L., Wu, Z., Liu, R., and Meyerowitz, E. M. (2013). Transcription repressor HANABA TARANU controls flower development by integrating the actions of multiple hormones, floral organ specification genes, and GATA3 family genes in Arabidopsis. Plant Cell 25, 83–101. doi: 10.1105/tpc.112.107854
59. Zhang, Y., Zhang, X. L., Liu, B., Wang, W. J., Liu, X. W., Chen, C. H., et al. (2014). A GAMYB homologue CsGAMYB1 regulates sex expression of cucumber via an ethlyene-independent pathway. J. Exp. Bot. 65, 3201–3212. doi: 10.1093/jxb/eru176
60. Zhao, M. G., Chen, L., Zhang, L. L., and Zhang, W. H. (2009). Nitric reductase-dependent nitric oxide production is involved in cold acclimation and freezing tolerance in Arabidopsis. Plant Physiol. 151, 755–767. doi: 10.1104/pp.109.140996
61. Zhao, M. G., Tian, Q. Y., and Zhang, W. H. (2007b). Nitric oxide synthase-dependent nitric oxide production is associated with salt tolerance in Arabidopsis. Plant Physiol. 144, 206–217. doi: 10.1104/pp.107.096842
62. Zhao, M., Zhao, X., Wu, Y., and Zhang, L. (2007a). Enhanced sensitivity to oxidative stress in an Arabidopsis nitric oxide synthase mutant. J. Plant Physiol. 164, 737–745. doi: 10.1016/j.jplph.2006.03.002
63. Insights from the Cold Transcriptome and Metabolome of Dendrobium officinale: Global Reprogramming of Metabolic and Gene Regulation Networks during Cold Acclimation
64. Zhi-Gang Wu, Wu Jiang, Song-Lin Chen, Nitin Mantri, Zheng-Ming Taoand Cheng-Xi Jiang

CHAPTER 10

Guiding Strategies for Breeding Vegetable Cultivars

João Carlos da Silva Dias

Technical University of Lisbon, Instituto Superior de Agronomia, Tapada da Ajuda, Lisbon, Portugal

ABSTRACT

Vegetables are considered essential for wellbalanced diets. The production and marketing of vegetables crops are undergoing continuous change globally. This is mainly due to the growing demands of consumers for safe and healthy vegetables, increased urbanisation of societies, and the growth in scale and influence of supermarkets chains. Horticultural science can respond to many of these challenges through research, breeding and innovation that can seek to gain more efficient methods of crop production, refined post-harvest storage and handling methods, newer and higher value vegetable cultivars and demonstration of their health benefits. Vegetable breeding has to address and satisfy the needs of both the consumer and the producer. Innovation in vegetable breeding depends on specific knowledge, the development and application of new technologies, access to genetic resources, and capital to utilise them. The driving force behind this innovation is acquiring or increasing market share. Access to technology, as well as biodiversity, is essential for the development of new vegetable cultivars. A few multinational corporations, whose vast economic power has effectively marginalized the role of public sector breeding as well as local, small/medium-scale seed companies, dominate the global vegetable seed trade. For most vegetable crops, only a few multinational seed corporations are controlling large part of the world market. This situation makes a growing part of the global vegetable supply dependent on a few seed providers. The multinational seed corporations ensued from merging some small or medium vegetable breeding programs to reduce costs. There may be fewer vegetable breeders in the future and growers will rely on seeds with a narrow genetic base. In order to meet future needs of vegetable breeders it is important that educacional programs incorporate rapidly changing new technologies with classical content and methods. Active and positive connections between the private and public breeding sectors and large-scale gene banks are required to avoid a possible conflict involving breeders' rights, gene preservation and erosion. Horticulturists will need to develop cultural practices and vegetable breeders to breed vegetables for a multifunctional horticulture (diversity, health promotion, post-harvest, yearround suply, etc.) and to cope with harsher climate conditions and lower inputs than they have come to expect. Improved production systems that can cope with climate extremes must allow vegetables to produce under high temperatures, greater drought stress, increased soil salinity, and periodic flooding. This will involve a combination of improved vegetable cultivars and modified production systems.

Keywords: Vegetables; Horticulture; Health Benefits; Marketing; Trading; Breeding; Year-Round Production; Sustainability

1. VEGETABLES IN THE WORLD

1.1. Human Diets

Vegetables are grown worldwide in almost 200 countries and make up a major portion of the diet of humans in many parts of the world and play a significant role in human nutrition, especially as sources of vitamins (C, A, B1, B6, B9, E), minerals, dietary fiber and phytochemicals [1]. They can be also a major source of protein for the poor. Vegetables in the daily diet have been strongly associated with improvement of gastrointestinal health, good vision, and reduced risk of heart disease, stroke, chronic diseases such as diabetes, and some forms of cancer [2]. Low vegetable intake, in unbalanced diets, has been estimated to cause about 31% of ischaemic heart disease and 11% of stroke worldwide. The International Agency for Research on Cancer (IARC) estimates that the preventable percentage of cancer due to such diets ranges from 5% - 12% for all cancers, and 20% - 30% for upper gastrointestinal tract cancers. Some phytochemicals of vegetables are strong antioxidants and are thought to reduce the risk of chronic disease by protecting against free-radical damage, by modifying metabolic activation and detoxification of carcinogens, or even influencing processes that alter the course of tumor cells [2]. "Hidden hunger" or micronutrient and vitamin deficiencies is a pernicious problem around the world that is caused by a lack of vitamins and minerals such as vitamin A, iodine and iron in the human diet. It affects the health of between 2 and 3.5 billion people in the developing world and increases the risk of illness or death from infectious diseases and children do not develop to their full physical or mental potential [1,2]. In much of South Asia and sub-Saharan Africa, dietary deficiencies of iron and vitamin A are major health problems, resulting in millions of deaths each year. Iron deficiency anemia is one of the world's most prevalent dietary deficiencies. In a recent report on the world nutrition situation, the United Nations (UN) estimated that 4 to 5 billion people are affected by iron deficiency anemia, with the most severe incidence in South Asia and sub-Saharan Africa. Its effects on cognitive development, fatigue and mortality have significant economic impacts, estimated at 7.9% of GDP in the case of Bangladesh [3]. Vitamin A deficiency also is widespread and an estimated 190 million preschool children and 19 million pregnant women are affected. More than half of these are found in South and East Asia and under a third in Africa [4]. Vitamin A deficiency is a major cause of blindness and weakened resistance to disease, resulting in the deaths of up to 3 million children each year [1-3]. In Bangladesh, Cambodia, Nepal and the Philippines at least half of the preschool children and pregnant women are affected by micronutrient and vitamin deficiencies [5]. Small variation in maternal diets, particularly reduction in micronutrient content, can have a significant impact on fetal growth and development. The interplay of the different micronutrients and antioxidants found in vegetables has important health impacts, explaining for instance the higher birth weight of children in India, when mothers consumed higher rates of green leafy vegetables and fruits during pregnancy [6]. Diets rich in vegetables, in all their many forms to ensure an adequate intake of most micronutrients, dietary fibers, and phytochemicals can bring a much-needed measure of balance back to diets contributing to solve many of these nutrition problems [2,7].

Good health depends on dietary diversity, and as poverty increases, human diets become less diverse. In impoverished countries the poor have little choice and are forced to rely on the cheapest available staples (and eventually vegetables) and dietary diversity and health suffer. In richer countries, changes in the food systems have made poor quality processed foods high in carbohydrates and fats more affordable, available, and accessible; this has most impact on the diets of the poor [8]. A long term global focus on ensuring supplies of vegetable and staple crops has come at the expense of dietary diversity. The emphasis has been on ensuring supplies of macronutrients while putting less emphasis on major sources of micronutrients and vitamins, which means a focus on survival rather than on health. World vegetable production does not meet even the basic nutritional needs of most countries, and the resulting imbalanced diets are one of the world's most serious health problems. Continuing to focus on increasing the production of staples will only exacerbate the real food problem, which is one of imbalanced diets.

A world vegetable survey showed that 402 vegetable crops are cultivated worldwide, representing 69 families and 230 genera [9,10]. Vegetable crops, of which the leaves or young leafy shoots are consumed, were the most common group of vegetables utilized (53% of the total), followed by vegetable fruits (15%). Below ground crop vegetable organs ranked as follows in frequency of use: roots > tubers > rhizomes > corms > stolons and together comprised 17% of the total number. Many vegetable crops have more than one part used. Most of the vegetables are marketed fresh with only a small proportion processed because as most vegetables are perishable by nature, consumption shortly after harvest is the best guarantee for optimal quality. Of these marketed vegetables, only 67 (17%) have attracted great breeding attention by international seed companies, due to their large area of production and substantial consumption, 52 (13%) were considered minor, and other 87 (22%) were considered rare [11,12]. In 2010 the global vegetable seed market was estimated at US$4.1 billion, of which 36% were for solanaceous, 21% for cucurbits, 13% for roots and bulbs, 12% for large seed, 11% for brassicas, and 7% for leafy and others vegetables [13]. In the last 8 years global commercial vegetable seed sales had an annual growth rate of 5.8%. There are now 7.2 billion human beings inhabiting this planet, and it has been projected that world population growth may exceed 70 million annually over the next 40 years. It is expected to reach approximately 9.5 billion by 2050, when approximately 90% of the global population will reside in Asia, Africa, and Latin American countries. With the increase in world population and consumption, and the advent of a high degree of growth and added value through biotechnology, the global market of vegetable seeds is expected to expand in future years.

1.2. Consumption

The worldwide consumption and importance of vegetables in the diet is difficult to estimate owing to scant production statistics. Even where crop reporting services are an integral part of the agricultural infrastructure, information is available only for a small percentage of the vegetable crops grown. The consumption and caloric contribution of vegetables to the diet varies widely with geographical region, nationality, local customs, and cuisine. China is the largest consumer of vegetables in the world. Vegetables make up about 35% of per capita food consumption in China, a much higher share than the world average [14]. Besides India, other southern Asian countries such as Bangladesh, Cambodia, North and South Vietnam, Laos and Philippines are also high producers and great consumer of vegetables.

For example, vegetables comprise 40% of the Bangladeshi diet [15]. Many vegetables are consumed near where they are produced, especially in China, India and other Asian countries. The per capita consumption of vegetables in Asia has increased considerably from 41 kg to 141 kg between 1975 and 2003 [16], particularly in China where the per capita consumption has increased from 43 kg (1975) to 154 kg (2003). In Africa, the per capita consumption of vegetables lags far behind the world average, with less than the FAO/WFO (World Food Organization) recommended minimum uptake of 200 g of vegetables/day (73 kg/year).

Rapid growth in mean per capita incomes in developed countries during the 1990s enabled consumers to purchase a broader range of relatively expensive vegetable commodities such as off-season produce, relatively new or renewed vegetables and organic produce. Higher incomes of consumers in developed countries have also raised the demand for other attributes such as better quality vegetables and more variation in the daily menu. In developing countries, consumption and domestic vegetable markets are also expanding because of an emerging educated middle class with increasing incomes. China, India and Indonesia, countries with the largest populations, have large emerging middle classes, which impact the demand for high-value vegetables [17]. Although the overall downturn in the globally economy over the past decade has seen a decrease in the discretionary income of consumers in many countries. As a consequence the consumption of many high priced items such as fresh vegetables has been static or declining. This has also impacted on the rate of growth in exports from some countries.

Globally, an increasing proportion of the world's population is living in urban environments where their interest in and knowledge of farming, and therefore of food production, is becoming increasingly deficient. In 1950 approximately 71% of the world's population lived in rural locations while in 2010 this had declined to 50% (or as low as 10%, 18%, and 22% for the United Kingdom, the USA and France, respectively), and is projected to be as low as 30% globally by 2050 [18]. There is, in all developed countries, a growing expectation and dependence on the supply of a low-cost, year-round supply of premium quality vegetables. Paradoxically, these same urban-based consumers have become increasing vocal about issues such as carbon taxes, the use of pesticides and safety of vegetables, labour conditions for farm workers, the sustainability of production methods, and the environmental conditions under which it is produced. These are often driven as "matters of conscience" and are in isolation from the reality of current production methods or of the opportunity to realistically meet these consumer demands. This trend, in turn, has led to the increased importance of organic vegetables and labeled brands [19,20]. Most of the total organic arable land for vegetables (213,453 ha) is located in Europe

(74%), followed by North America (20%) and Latin America (3%), but represents only 0.04% of the total area of vegetables in the world [21]. Organic vegetable trade from developing to developed countries is currently growing at over 20% per year [22] but represents only 3% of total vegetable production [21].

Populations in developed countries are also shifting dramatically towards greater proportions of older people as a consequence of the "baby boom" of the 1940s and 1960s, and as a result of the lower birth rates in last recent decades. For example, in Japan and Europe (the two regions where severe population ageing will occur in the near future), within twenty years many countries will face a situation where the largest population cohort will be those over 65 and average age will be approaching 50. These shifts are impacting strongly on consumer consumption patterns as evidenced by smaller households, increased dining out, and increased demands for healthier vegetabes and foods.

Desire for year-round availability, increased diversity, and growing health awareness, have also been important reasons for increased consumption of vegetables in developed countries. For example, the dietary benefit of fresh produce, is the major reason for the 25% increase in fresh vegetable consumption in the United States during the 1977-1999 period [23]. On the other hand, factors such as increased participation by women in the labor market have created demand for processed, ready-to-eat convenience vegetable products.

1.3. Diversification

Consumers also want more vegetable diversification and a continuous supply. Vegetables are purchased partly based on eye appeal, which means that the development of desire to consume, increases market demand. Diversification also tends to increase consumption. Product differentiation, including new or renewed product introductions, is still a key strategy for expanding sales in vegetable markets. For example, the fresh tomato category has been differentiated to more than 10 offerings (beefsteak tomato, Roma type tomato, vine-ripe tomato, cocktail tomatoes on vine, tiny-plum tomatoes, mini-plum tomatoes, red cherry tomatoes on vine, attractive yellow and orange cherry tomatoes, mini San Marzano-type tomatoes, tear drop or pear shaped tomatoes, super or premium taste tomatoes). The introduction of specialty fresh baby leaf vegetable salads and fresh-cut products has opened new opportunities for domestic producers. The increased production of baby leaf vegetables in the world is intended to increase desire among elite consumers and is also an excellent way of supplying micronutrients. For example baby leaf curly kale, as well as other dark green leafy vegetables, is a rich source of lutein plus zeaxanthin carotenoids and when cooked contains seven times as much vitamin A as cooked broccoli and provides more calcium per 100 g than milk, yogurt, cooked broccoli or cooked spinach. Until recently kale was not a particularly popular vegetable in Europe (except Portugal), but as a baby leaf vegetable with 5 leaves, it is now accepted by many European consumers. An example of diversification in peas is the pea shoot. It is a nutritious leaf vegetable with high levels of vitamin C, folic acid and vitamin A. To exploit such opportunities, it is important to continue research and to disseminate information regarding the nutritional benefits of vegetables, develop new improved vegetable cultivars and lightly processed products, evaluate the economic opportunities and the market scope of these new products, and identify marketing trends and alternatives [12].

Increasingly, consumers also demand lightly processed, ready-to-eat vegetables and this is true not only for developed but also for developing countries. Lightly processed vegetables require appropriate packaging, such as plastic wrap, which is important to keep the cut vegetables fresh. New lightly processing and packaging technology enables fresh vegetables to be presented in a convenient format (e.g. increasing the shelf life of fresh pre-cut vegetables) while maintaining aspects of health, safety, and freshness. Providing customers with innovative combinations of products and services is the key to the lightly processed vegetables. The United States leads in the light processing industry, with estimated sales of USD$13 billion in 2008 in both retail and foodservice, which is double the amount in 2000. Brands play an important role in this segment. For example, the brands of Dole and Fresh Express together accounted for 88% of the United States packaged salad market in 2004. Vitacress is Europe's leading grower and packer of watercress, rocket, baby leaf salads, and other specialty vegetables. The growth rate in retail sales of lightly processed vegetables in several Asian countries over the past years has been tremendous. For example the total percentage of retail sales of processed vegetables during the 2001- 2005 period in Vietnam, China, Indonesia, South Korea, Thailand, India and Philippines was 11.2%, 9.6%, 7.8%, 6.8%, 6.1%, 6.0%, and 5.1% respectively [24].

2. VEGETABLE MARKETING AND SAFETY ISSUES

2.1. Vegetable Marketing

Vegetables are often highly perishable products, restricting the ability of producers to store them in order to cope with price flutuations. They are also of specific high value and they are usually sold through specialized markets.

The supply chain for vegetables produce has changed rapidly on a global scale over the past two or three decades. Huge supermarket chains have rapidly displaced smaller marketers as the main outlets for the year-round supply of vegetables. Furthermore, the number of major supermarket chains is currently fairly static and is predicted to reduce further, although expansion of supermarkets all over the world and specialy in Asia is very significant. Consequently the buying power is being consolidated into fewer supermarket chain retailers with significant consequences to both primary producers and to buyers .The following characteristics are common for the produce buyers in most if not all these huge supermarket chains: i) a need for year-round supply to allow consumers continuity of access to specific vegetable items; ii) a wide range of interesting products to challenge the existing product range and to entice and retain consumers; iii) the best cultivars; iv) highest possible quality with regard to flavour, appearance, and condition; and v) competitive pricing.

Although some consumers in many developed countries have become more interested in understanding the sources of the vegetables that they eat. This has led to a proliferation of farmer markets in many areas, including those focussed on vegetable crops produced using organic farming methods, which have become increasingly popular amongst consumers. Nonetheless, the proportion of produce that these markets are selling is still comparatively so small, on a global scale, and is usually (but not always) limited to provision of products during the current growing season (the exception is where local producers sometimes supplement their own products with imported ones).

Currently more than 65% of the vegetables produced in the world are sold by vegetable growers to wholesale dealers or huge supermarket chains, and relatively few growers sell their product at retail prices to consumers in farmer markets [12]. Globally the horticultural product markets are still dominated by a large number of wholesalers or middle-men, which means not only that the producers have a lower profit, but also the consumer often does not have access to lower-priced vegetables. Globally growers only receive 30% or less of the retail price. This situation is a serious problem for growers.

Domestic and international markets for vegetables are changing rapidly all over the world, partially fuelled by the spread of huge supermarkets chains. The top ten global supermarket chains in 2012 were: Wal-Mart (from USA), Carrefour (from France), Tesko (from UK), Metro (from Germany), Kroger (from USA), Schwarz (from Germany), Costco (from USA), Home Depot (from USA), Target (from USA) and Aldi (from Germany). Consumers increasingly purchase their vegetables and other foods in large convenience stores such as these super and hypermarkets chains. The proliferation of supermarkets in developed and developing countries creates both challenges and opportunities for vegetable promducers [12]. Indeed supermarkets may contribute to a higher demand of horticultural products and increase expectations for quality, safety, and presentation, while simultaneously excluding small growers from participating in procurement and contracts. The importance of supermarkets is growing especially fast in the southeast Asia region. Supermarkets are expanding in China 30% - 40% per year, 2 - 3 times faster than other developing countries [12]. According to USDA, China could support at least 9100 hypermarkets [25]. In China growth of modern retailing markets is spreading from the early growth areas of Guangzhou, Shanghai and Beijing to inner western regions. This is already having a profound effect on the way that vegetables are produced in the country. In Indonesia, Malaysia, and Thailand, supermarkets account for 33% of the market while in Taiwan and Philippines the figure is closer to 63% [26]. The 10 supermarket chains with the highest growth rates in last decade included two supermarket chains from China and one from South Korea [27,28]. And in Indonesia the local retail supermarket chain Ramayana is growing at 25% a year [29]. The share of fresh vegetables sold through supermarkets in some countries and metropolitan areas of Asia such as Malaysia, Bangkok (Thailand), Manila (Philippines), South Korea and China is 35%, 30%, 15%, 11% and 10%, respectively [30]. In Vietnam, where modernization of food retailing is only about 10 years old, and where wet markets are still the major source of food shopping, the supermarket's share of vegetable products increased from around 0.5% in 2000 to around 40% in 2007. The Metro supermarket chain has announced programs of assistance and consulting for 4000 growers and suppliers to help them to upgrade the quality, marketability, and competitiveness of their products [12].

Outside of southeast Asia, supermarket sales in Kenya are growing at 18% a year. They are expected to become the dominant urban food retailers by 2011 [31]. At Chile the supermaket Cencosud is growing at 30.5% a year [27,28].

The growing importance of supermarket outlets has implications of its own, regarding their methods of procurement and their quality standards. Supermarkets in the cities bring quality to the shelves. Vegetables are well packed and presented, providing scope for premium quality as well as novelty items. The difficulties that growers can experience is reflected in fairly rapid declines in the numbers of growers involved, as companies tend to de-list suppliers who do not meet expectations in terms of volume, quality, and timely delivery. In Malaysia, for example, the Giant supermarket chain had 200 vegetable suppliers in 2001, but just 30 in 2003. In Thailand, similar changes have been seen following the introduction of a distribution center for the TOPS supermarket chain [30]. Small-scale growers can find it difficult to produce and deliver according to the supermarket chain's standards [12]. In Brazil the Wal-Mart supermarket chain has about 1500 associated growers producing vegetables, of which 90% are large producers and only 10% are small producers. The standards of quality, safety and presentation make it difficult for the small producers to compete. In Portugal and in other European countries the situation is similar. The market is getting more refined in terms of quality and yield expectations, and there is a clear demand for excellent hybrid vegetable cultivars [12,32]. Success for vegetable growers will depend on their ability to access diverse markets and respond promptly to changes in market conditions. Growers grow vegetables for immediate marketing and their produce is subject to competition to decide the prices. Hence they cannot compromise the quality of the seeds they use for fear of rejection of their produce. They buy the best seeds, mainly hybrids, and frequently try new products to remain successful. The seed companies in turn get instant response and success if they develop new promising hybrids. Price of seeds is a more critical factor in marginal vegetable areas, where capital and spending input regimes are low, but less important where high yields can be obtained and the grower's produce can be sold profitably [12,32].

2.2. Safety Issues

Food safety legislation in the European Union and in the United States is introducing increasingly stricter standards. It is expected that the assurance of "safe" vegetable products will become increasingly important not only in these countries but all over the world. Human and environmental health risks of inapropriate vegetable production practices arise mainly from the following: i) innapropriate handling of pesticides by growers; ii) crop selection or location without due regard to the environment polution in the air, soil or water; iii) aplication of unsorted or insuficiently treated solid and liquid organic wastes to vulnerable crops; and iv) poor and unsanitary handling during marketing and distribution [12].

In general, vegetable production requires more plant protection products per hectare. Because of their cultivation intensity, vegetables suffer from many biological stresses including pests, diseases and weeds. Because of the diversity within and between plant families, their pest loads are far more varied and complex compared to field crops. Considerably fewer resources have been directed at improving their production and pest and disease management options, compared to field crops such as rice, wheat and maize [33]. Since vegetables are high value commodities with high "cosmetic" standards, the main method for controlling pathogens, pests, and weeds has been the use of pesticides. Vegetables are often consumed in fresh form, so pesticide residue and biological contamination in them is a serious issue. Vegetables account for a significant share of the global pesticide market. About 20% of the world's annual pesticides expenditures (amounting to US$ 45 billion in 2009) are spent for growing vegetables [34]. Almost 25 kg/ha of active pesticide substances are used on average in vegetable production in the European Union [35]. Vegetable production accounts for less than 1% of the United States crop area, but 14% of its total pesticide use [36]. Insecticides are regularly applied to control a complex of insect pests that cause damage by feeding directly on the plant or by transmitting pathogens, particularly viruses. Despite pesticide use, insects, pathogens, and weeds continue to cause a heavy toll on world vegetable production. Pre-harvest losses are globally estimated as 15% for insect pests, 13% for damage by pathogens, and about 12% for weeds [37].

Studies on pesticide levels in vegetables, in some developing and developed countries, found that the pesticide residues sometimes exceeded tolerance limits [38]. For example, in India a survey of pesticide residues in vegetable crops taken at the farm gate and in markets from 1999 to 2003 confirmed that of the 3,043 samples, two-thirds were found to have pesticide residues and 9% contained residues above the maximum residue levels (MRLs) [39]. The increase of residues in vegetables is a major concern to consumers and there are some concerns about the ability of small growers to meet exacting quality and safety standards of the commercial sector. The problem of residues is not restricted to developing countries. In Europe, a recent survey of pesticide residues in vegetable and fruit crops taken by the EU coordinate pesticide monitoring programme in 2008, carried out in 27 EU member states and two EFTA states (Norway and Iceland), and analysing 78 pesticides, confirmed that 21.2% of the samples have measurable residues and from the 58,784 of samples analised, 3.7% exceeded the MRLs [40]. In USA, about 35% of the food purchase by American consumers, have detectable levels of pesticide residues, and 1% to 3% of these foods have pesticide residue that are above the MRLs. Residue levels may be higher because the analytical methods now employed in the US detect only about one-third of the more than 800 pesticides in use on crops [41].

Pesticide residues are often attributed to the failure of growers to restrain application before harvesting and the use of prohibited pesticides. Residues affect the health of growers and consumers and contaminate the environment Also the presence of pesticide residues may restrict trade opportunities. In South Asia, pest and disease vectors of eggplant, tomato and legumes, notably eggplant fruit and shoot borer, cotton bollworm, root knot nematode, white fly, and legume pod-borer have been identified as the major targets of pesticide use and abuse with frequent and excessive applications of pesticide. However, eggplant fruit and shoot borer pheromone traps and net houses developed by AVRDC (The World Vegetable Center) have helped reduce pesticide application significantly.

The promotion of integrated and biological pest control is expanding worldwide. The use of improved vegetable cultivars with resistance or tolerance to diseases and pests can also contribute to the reduction of pesticide applications and pesticide residues [12].

Genetically Modified Crops (GMOs) are now widely used in a number of countries for the production of largescale agronomic crops, especially maize, soybean and canola (oil seed rape). This is, however, not the case for vegetables crops where concerns and debates about the safety and the wider justification of genetic modification prevail in many countries [42]. Of concern, is the lack of awareness about the role that conventional plant breeding plays in the production of new cultivars and the confusion that often prevails between conventional breeding and plants produced using gene transfer technologies. The changes outlined above are merely some of the factors that are impacting on current production practices for horticultural crops. They also provide a plethora of opportunities for horticultural science to solve many of the current problems facing producers and to offer solutions to those issues that are concerning consumers [42-44].

Along with potentially toxic pesticide residues, vegetables grown on polluted soils and in the vicinity of rail, roads and industrial areas can be contaminated with heavy metals (mostly lead), sulphur and nitrates. These toxicants affect the nervous, digestive and circulatory systems, particularly threatening the health of young children. Fundamental factors determining the presence of these metals and toxicants in vegetables are the distance of the production area from the source of pollution and the duration of exposure to the toxicants. Also the type of vegetable grown determines the potential contamination. In general, leafy vegetables and vegetables with longer growing period are the most sensitive. Several surveys on quality of soils and vegetable production show that that, all over the world, large areas of vegetable production are polluted.

Nitrate is formed from fertilizers, decaying plants, manure and other organic residues. Due to the increased use of synthetic nitrogen fertilizers and livestock manure in intensive vegetable production, vegetables and drinking water may contain higher concentrations of nitrate than in the past. The presence of nitrate in vegetables, as in water, is a serious threat to man's health. Nitrate per se is relatively non-toxic, but approximately 5% of all ingested nitrate is converted in saliva and the gastrointestinal tract to the more toxic nitrite. The only chronic toxic effects of nitrate are those resulting from the nitrite formed by its reduction by bacterial enzymes. Nitrite and N-nitroso compounds, which form when nitrite binds to other substances before or after ingestion (for example, the amines derived from proteins), are toxic and can lead to severe pathologies in humans. Thus, the assessment of the health risk of nitrate to humans should encompass the toxicity of both nitrite and N-nitroso compounds.

The accumulation of nitrates in the environment occurs also quite easily by vegetable production activities, mainly by the unpropriate use of nitrogen fertilizers, and affect particularly bodies of water such as groundwater. Excessive nitrogen applications are very commom in China due to rising population which requires increase vegetables supply, resulting in greater production of nitrate-rich runoff from vegetable crops grown. In Beijing, nitrogen fertilizer use has been steadily increasing from 263 kg/ha in 1991 to 344 Kg/ha in 2001. An extensive survey in Beijing during 1996-2000 period found: i) excessive nitrogen applications with up to five times required rates of nitrogen; ii) 78% of farms using 300-900 kg/ha on Chinese cabbage; iii) 35% of greenhouse tomatoes receiving more than 1000 Kg N/ha [45]. The consequences of these practices are that nearly 50% of nitrogen fertilizers leach into the soil and groundwater instead of being utilized by vegetable crops. As referred above it is known that high intake of nitrates from vegetables present potential adverse health implications. Vegetables grown and irrigated by water from these sources tend to accumulate a considerable amount of nitrates, mostly from fertilizers. Eighty percent of rural people and 50% of urban people in India rely on groundwater for drinking and for vegetable production. Nitrate concentration are over the permissible limit of 45 ppm in 11 Indian states covering 95 districts. A study on water quality in vegetable production areas in northern China showed that nitrate pollution in ground and drinking water has become a serious problem [46]. Nitrate contents exceeded the limit of 50 ppm for drinking water in over half of the 69 sites investigated, and in the worst case the nitrate content of groundwater reached up to 300 ppm. In all sites surveyed high amounts of nitrogen fertilizer were applied: recorded rates reached up to 1900 Kg/ha per year. Similar levels of nitrogen fertilizer input and emissions were observed in vegetable production in eastern and southern China [47]. Generally, nitrate accumulating vegetables are leafy and root Brassicaceae (Chinese cabbages, leafy oriental vegetables, cabbages, rocket, turnipgreens, cress, radish, kohlrabi), Chenopodiaceae (spinach, beetroot and Swiss chard), Asteraceae (lettuce, leaf chicory and escarola) and Apiaceae (parsley, celery, celeriac, carrot and fennel) [9,48]. Some of these vegetables are amongst the most highly consumed throughout the year in many regions of the world. Growers should then adhered to good practice of utilization of nitrogen fertilizers.

3. VEGETABLE TRADING

Globalization has also now become increasingly important among vegetable producers as indicated by the recent massive increases in international trade of fresh and lightly processed vegetable crops. As a result of the increased frequency of world travel, there is a substantially greater variety of vegetable crops available worldwide in many local markets. Worldwide, the total volume traded in vegetables has increased fivefold from 1965 to 2005 [24]. The European Union (EU), followed by North America and Japan, are the world's most important fresh vegetable import regions. A large share of this comes from Africa, Latin America and the Caribbean , and from China, each contributing roughly one-tenth to overall vegetable exports. While more than 75% of global vegetable production takes place in Asia and in the Oceania, only about 12% of the total value of exported horticultural products is generated in Asia and in the Oceania regions [24]. Because of the large share that Asia and the Oceania has in global population, domestic markets and their requirements are likely to continue to be more important in the near future than the attraction of export markets. In Asia, China, Thailand and the Philippines are the largest net exporters of vegetables, while Singapore, the Republic of Korea, and Malaysia are the largest net importers of vegetables [24]. In sub-Saharan Africa, Kenya and Côte d'Ivoire are two of the most important vegetable producers for export.

While many other agricultural commodities face stagnation and declining world prices, vegetable exports have grown dramatically. In many African countries, export of fresh vegetables has become a bright spot in an otherwise dim agrarian landscape [49,50]. There are several reasons for the rapid increase in vegetable trade over the past decade. In the 1990s, many countries, particularly in Africa and Latin America, changed their trade policies from protectionist to more openness, economic liberalization, and export diversification. This opened the way for several new trade agreements that facilitated trade in high-value agricultural products, such as vegetables. For example, liberalization under the North American Free Trade Agreement (NAFTA) underpinned the rapid growth in United States of agricultural imports from Canada and Mexico [51]. In the European Union, horticultural products have also benefited from preferential trade access. For instance, imports of fresh vegetables into the European Union rose by 150% during 1989-1997, with almost three-quarters of the value of these exports coming from sub-Saharan Africa. The majority of exports come from countries with preferred trade status under the Lomé Convention. For instance, in 2000, Kenya was the single largest supplier of green beans to the European Union, followed by Ethiopia, South Africa, and Switzerland. Kenya captured 53% of the total traded value. Côte d'Ivoire was the second largest supplier of green onions and shallots to the European Union, after New Zealand, capturing 16% of the total traded value [52]. Although within this same period, many world governments have increased the levels of protectionism against imported produce or at least have slowed down the opening of their markets to increased exposure to imports.

The increase in total volume of vegetables traded worldwide has been dramatic. Still, compared to overall exports of agricultural products, the importance of vegetable exports remains minor, comprising less than 10% of the total value. However, in recent years the share has been rising and it is projected to continue to rise faster than other agricultural products. During the 1990s, the value of fresh and processed vegetables imported by the European Union surpassed all other categories [50]. Growth in these commodities is also linked to changing trends in consumer preference and food retailing. In this situation many vegetable growers are eager to produce value-added horticultural crops as compared to field crops, and to obtain higher yields of high-quality products. International supermarket chains and large processors are becoming the main buyers of exported fresh vegetables [19] and small-scale growers worldwide need to be trained and organized to meet the challenge of supplying these international players. The major constraints to the participation of small-scale growers in international vegetable exports are the increased awareness that food quality and safety is receiving in the food trade and as well, an expansion in the number of non-tariff measures that developed countries apply to vegetable products [12,32].

Vegetables belong to the class of food items most frequently affected by sanitary and phytosanitary measures [53]. Sanitary issues refer to ensuring a safe food supply for consumers, while phytosanitary issues concern the protection of domestic crops from imported pests and diseases. The Sanitary and Phytosanitary Agreement (SPS) of the World Trade Organization specifies that countries can pursue their own levels of food safety standards. However SPS issues are sometimes used as a protectionist tool against imports since multilateral trade agreements have reduced the ability to protect domestic production with tariffs and quotas [54]. SPS regulations may be the most important barrier to international trade in fresh vegetables [53]. Thus exporters from less developed countries must be provided with training opportunities and information access on how to produce and supply safe products to developed countries. Traceability, phytosanitary, infrastructure, and productivity issues will continue to be a barrier for participation in the vegetable trade for most of the developing world. Application of agricultural chemicals is often poorly regulated, and industrial pollutants are common hazards in the soil, water, and air of developing countries. In the future, the inability of these countries to meet increasingly strict phytosanitary and traceability requirements for food products will constrict exports to developed countries [12]. Smallscale growers and processors in developing countries will thus have to learn to supply safe products with traceability labels, if their participation in global trade is to continue and to expand. Technologies for safe and environmentally friendly vegetable production as well as capacity building should therefore gain particular attention for training to enable small-scale growers to participate in vegetable production for international markets [12,32].

Food and nutritional security involves securing the whole food chain from production through to consumption. Horticulturists traditionally have focused on production, but issues such as equitable access to food and ensuring a balanced diet increasingly demand attention. Those with the least power and already substantially imbalanced diets are most affected by climate change and limited resources. Horticulturists need to build greater connections with health professionals who have more experience with aspects of consumption. We are all concerned by what we eat, and securing the food chain to provide a balanced diet (particularly for the most vulnerable), can point us toward an alternative way of viewing the role of horticulturists in a warming world.

4. MEETING CONSUMER PREFERENCES THROUGHOUT THE VALUE CHAIN WITH END-USER-DRIVEN VEGETABLE BREEDING

Vegetable breeding has to address and satisfy the needs of both the consumer and the grower. The general objectives for growers are good yield, disease and pest resistance, uniformity, and abiotic stress resistance. Objectives for consumers are quality, appearance, shelf life, taste, and nutritional value. Quality in vegetable crops, in contrast to field crops, is often more important than yield. For growers to survive, cultivars must be accepted by the market. Thus, color, appearance, taste, and shape are usually more important than productivity. For example, tomatoes to be used either fresh or in processing must have distinct quality characteristics. Fresh tomatoes must have acceptable flavor, color, texture, and other taste parameters to satisfy consumer demands and handling requirements. Processing tomatoes, on the other hand, must have intrinsic rheological characteristics that make them suitable for various processing applications, such as juice, ketchup, or sauce production. Traditional breeding requires the selection of a tomato genotype or a related wild species that has a desirable trait, such as early ripening or disease resistance, and crossing it with another tomato cultivar that has a good genetic background. The desired result is an earlier ripening tomato that makes it to the market sooner, or cultivars that resist pathogen attack. In this way, several thousands of tomato cultivars have been developed over the years. The final goal of vegetable breeding programs is then to release new cultivars having elite combinations of many desirable horticultural characteristics.

Consumers (and therefore supermarkets) are seeking new eating experiences, different quality attributes and improved convenience, and are prepared to pay a premium for such produce if their expectations are met. Plant breeding for improved taste, convenience, and consumer appeal has already contributed to increased per capita vegetable consumption with the development of products such as baby carrots, yellow and orange peppers, cherry and pear tomatoes, non-bitter cucumbers, mild tasting eggplants, seedless watermelons, and lettuces with different colors, textures and flavors for baby leaf and precut salads. In some instances, these premiums are considerable and bring benefit to the overall vegetable industry through improving the popularity of the vegetable crop overall. Further, the premiums achieved can exceed those that consumers are prepared to pay for the choice of consuming organic produce.

Other important objectives of vegetable breeding are disease and pest resistances. Since the early days of the 20th century, traditional breeding for disease resistance in vegetables has been a major method for controlling plant diseases. Cultivars that are resistant or tolerant to one or a few specific pathogens are already available for many vegetable crops. Resistant hybrids with multiple resistances to several pathogens exist and are currently used in vegetable production. For example, in tomato, the genetic control of pathogens is a very useful practice and most resistances are monogenic and dominant. So far, tomato breeding has resulted in cultivars with resistance to at least 15 pathogens, although with varying stability and level of expression . Tomato cultivars with some resistance to fungi or oomycetes (Alternaria alternata f. sp. lycopersici, Cladosporium fulvum, Fusarium oxysporum f. sp. lycopersici, Fusarium oxysporum f. sp. radicis-lycopersici, Phytophthora infestans, Pyrenochaeta lycopersici, Verticillium dahliae), bacteria (Corynebacterium michiganense, Pseudomonas solanacearum, Pseudomonas syringae pv. tomato), virus (beet curly top hybrigeminivirus, tomato mottle bigeminivirus, tomato spotted wilt tospovirus, and several variants of the tomato yellow leaf curl bigeminivirus), and nematodes (Meloidogyne spp.) are available . Many open-pollinated cultivars of tomato presently cultivated possess genetic resistance to three or four pathogens. With the increasing use of F1 hybrids it is possible to use cultivars combining from four up to six resistances .

Pest resistance is essential in vegetable production but is marginal in vegetable breeding research. There are few vegetable cultivars resistant to insects. Resistance may be unstable due to genetic variants of the insect that are able to overcome that source of resistance. Depending on the complexity of the interaction between the pest and the vegetable plant, plant resistance may break down rapidly or be long-lived. Insects, including aphids, whiteflies, thrips and leafhoppers, are also very important in vegetables because they vector many viruses. Viruses can substantially reduce production and quality and are becoming increasingly problematic worldwide due to the absence of virus resistant germplasm for many important vegetable crops. Aphid vectored viruses are particularly problematic because many are transmitted in a non-circulative and non-persistent manner [57,58]. This means that a very short time, i.e. a few seconds or minutes, is sufficient for aphids to acquire virus particles when probing on infected plants. A similarly short time period is enough for aphids to release virus particles when probing on healthy plants. The primary injury caused by aphid-vectored viruses arises not from direct feeding damage by the aphids, but from their ability to allow the virus to enter the plant and initiate the disease. A successful application of biotechnology has been the development of vegetable cultivars that resist insect transmitted viruses, as well as cultivars that directly resist insect feeding or development. Bt potato cultivars expressing resistance to Colorado potato beetle (Leptinotarsa decemlineata), and aphids associated with potato virus Y, and potato leaf roll virus, were approved for sale in the United States in 1995. These cultivars were marketed and sold under the trade name NewLeaf®, NewLeafY®, and NewLeafPlus®, until potato processors, concerned about consumer resistance and loss of market share in Europe and Japan, suspended contracts for Bt potatoes with growers in 2000 . Because of this consumer concern Bt potato cultivars were taken off the market in 2000. Two other Bt vegetable crops are under development: Bt eggplant targeted for control of eggplant fruit and shoot borer, and Bt crucifer vegetables targeted against diamondback moth [42,44].

The economic return of investment in breeding for disease and pest resistance may be low because it is dispersed among many different vegetable crop types. Also resistant cultivars compete directly with non–resistant ones that may still be used by growers with minimum problems. Therefore disease resistance is most important when the disease is a limiting factor in production, and is especially important for many virus diseases. The high interest in, and the increasing present demand for breeding for disease and pest resistance is related to a generalized interest in releasing "environmentally friendly" vegetable cultivars requiring sparse or no use of pesticides.

Breeding for postharvest traits, mainly transport quality, shelf life and cosmetic problems, is of increasing importance in vegetables. For example, in tomato, textural properties of fruits are important contributors to the overall quality for the fresh market and to the properties of products processed from tomatoes [61]. Because cell wall disassembly in ripening fruit contributes to fruit texture, modification of cell wall proteins and enzymatic activity during ripening can impact cell wall polysaccharide metabolism and influence texture. Lettuce and other leafy vegetables used for salads deteriorate rapidly following harvest, requiring a considerable investment of effort to maintain quality and shelf life of cut material. Harvesting increases respiration, stimulating deterioration, with increase in the synthesis of phenylalanine ammonia lyase and phenolic compounds, such as chlorogenic acid which cause tissue browning [62]. Consequently, delaying leaf senescence is an important target for breeding of leafy vegetables. Also in lettuce, breeding efforts have targeted tipburn, marginal browning, and rib discoloration, which detract from overall appearance [63]. Vegetable products with good transport quality, better shelf life and good appearance will be preferred by traders and also by consumers.

Since vegetables are rich in vitamins, minerals and other micronutrients, and therefore vital for health, breeding objectives should include improving their nutritional value. Historically vegetable breeders have applied selection pressure to traits related to agronomic performance, particularly yield and quality, because these are the traits important to the producer. Rarely have growers been paid for nutritional factors, so there have not been economic incentives to pay much attention to these traits. However, consumers are becoming more aware of these traits .

Vegetable breeding for nutritional quality was not mentioned as a primary goal in plant breeding text books through the mid-20th century (e.g. [64-66]). However vegetable breeding efforts targeting improved micronutrient content and composition had begun in the 1940s and 1950s with research describing the inheritance and development of tomato breeding stocks and lines high in provitamin A carotenoids and vitamin C [67-69]. Lincoln et al. [67] noted a fourfold variation in vitamin C among commercial cultivars and up to 1194 ppm in red-fruited tomato interspecific crosses with Solanum pimpinellifolium. Similar research leading to the development of darker orange, and consequently high provitamin A, carrots began in the 1970s [70]. Yellow core color occurs only in older open-pollinated carrot cultivars since uniform orange storage root color has been a trait of interest in carrot for over a century [71]. Similar studies were made in squash where rapid gains in carotenoid content have been made with phenotypic selection for orange color versus green and cream [72]. Genetic improvement to increase levels of specific micronutrients has been pursued in several other vegetables such as melon, spinach, sweetpotato, potato, lettuce, broccoli, pepper, watermelon, collard, kale, peas, and bean. This field of study is relatively new, and also complex because of mineral interactions with each other, and numerous other compounds in the soil and in the plant [73]. There is usually a large environmental effect, when the component is present in tiny amounts, such as for some micronutrients and phytochemicals. Success in vegetable breeding for higher vitamin and mineral content must consider not only substance concentration but also organic components in plants that can be abundant and either reduce or increase bioavailability [73].

With these numerous considerations, breeding vegetable plants for improved nutritional value is a complicated goal that needs expertise in many disciplines such as plant breeding, nutrition, and soil science. When a vegetable compound (micronutrient or phytochemical) is found to be important for human health, and growers, vegetable markets and seed companies can capitalize on the value of the compound, there may be an opportunity for vegetable breeders to increase the amount of this compound. Breeders can be successful in reaching this goal, if the vegetable crop contains genetic variability for the compound, if selection is effective without detrimental pleiotropic effects, and if there is an easy method to measure the compound.

Enhanced nutritional content would add value for poor, malnourished populations. Breeding for provitamin A carotenoids, iron, and zinc is of keen interest as a strategy to alleviate nutrient deficiencies in developing countries [7,74]. An example is the "golden tomato" which contains three to six times more provitamin A carotenoids than standard tomatoes. Developed at AVRDC with conventional breeding techniques, these improved nutritionally-rich tomato lines could help prevent many children of developing countries from going blind, since vegetarians and populations with limited access to animal products depend on provitamin A carotenoids for vitamin A. One "golden tomato" can provide a person's full daily vitamin A requirements. Tomato fruit and its processed products are the principal dietary sources of carotenoids such as lycopene. Lycopene is a potent antioxidant with the potential to prevent epithelial cancers and improve general human health. Therefore, there is considerable interest in elevating the levels of carotenoids in tomato fruit and thereby improve the nutritional quality of the crop [75,76]. The B gene from Solanum hirsutum shifts tomato carotenoid accumulation from lycopene almost entirely to b-carotene and results in orange fruit color [77]. This consequently dramatically increases the provitamin A carotenoid content. b-carotene content of commercial cultivars, as mentioned, is of interest [78], and several high b-carotene orange cherry tomato breeding lines have been bred [79]. Rainbow carrots, which are super sweet and crunchy, have multi-pigmented roots that naturally contain several antioxidants, such as lycopene, lutein, and anthocyanin. Similarly, yellow sweet potatoes are much more nutritious than white ones since they are high in provitamin A carotenoids. Unfortunately, the popularity of white fleshed sweet potato cultivars in many tropical regions may complicate the acceptance of more nutritious orange ones. Recent studies across a range of Andean potato cultivars show wide variation in calcium, iron and zinc content [80] as well as anthocyanins [81,82] due to the existence of red-, blueand purple-fleshed potatoes.

A vegetable, in order to have impact for its nutrient content, must be appealing to consumers. Sensory appeal, including color, is an attribute important to consumers when selecting many vegetables. Enhanced pigmentation of carrot, potato, tomato, and pepper, for example, is considered a quality factor [70,82]. In peppers carotenoid content of green, yellow, orange and particularly red peppers can be relatively high. Selection for high pigment is an important goal because these carotenoids are important for visual appeal in many markets but until now didn't have any specific nutritional impact [7,83]. In the past yellow or orange tomatoes could not compete with red tomatoes because they were unfamiliar to consumers, but now they are commercialized, and are challenging the market. This fact and recent development of orange colored cauliflower [84] and orange flesh cucumbers [85] reflect a new direction in vegetable breeding: nutritional quality. The white versions of these two vegetables lack carotenoids present in high enough concentration to alter their appearance. As nutritional quality becomes a more common breeding goal, and the novelty of unusual colors brings added value to seed companies and vegetable growers, unusual colors will quite certainly become more available and perhaps more widely consumed. But consumer requirements for quality: appearance, shelf life, and taste, must be met. Breeding to increase consumer appeal by improving convenience and the quality factors of a moderately nutritious crop often can be a more effective approach to increase intake of shortfall nutrients [86]. Nutritional quality identifiable by the consumer and available at a moderate price might induce increased consumption and thus confer an important marketing incentive for breeding activity.

Vegetables generally are sensitive to environmental extremes. High temperatures and moisture stress are already major causes of low yields in the tropics and these are likely to be exacerbated by climate change. Vegetable crops of predominantly temperate origin that are popular in the Western countries have come to dominate world production, aided by their spread to all continents during colonial times. Far less emphasis has been placed on tropical indigenous vegetables from Asia and Africa that in many cases may be better adapted to future warmer climate scenarios than the vegetable crops the world more commonly views. Most vegetables prefer cooler temperatures, and productivity is lowest in the hot and humid lowlands. Vegetable breeders need to rethink which vegetables a warmer world will need to rely on. High value vegetables preferentially receive more inputs than other vegetable crops to help them cope with unfavorable weather and nutritional deficiencies, but extremes of heat, drought, salinity, and flooding will be unavoidable in a warming world. Some vegetable cultural practices to cope with these stressors already exist, but the development of well adapted cultivars will take longer. Little work has been done to assess the genetic variability of most vegetables to cope with less than optimum growing conditions.

Improved production systems that can cope with climate extremes must allow vegetables to produce under high temperatures, greater drought stress, increased soil salinity, and periodic flooding. This will involve a combination of improved cultivars and modified production systems. Most research on the effect of environmental stress has been done on tomato and there is a great need for more research on other vegetables [87]. High temperatures decrease pollen productivity and viability [88], and inhibit photosynthesis in tomato [89] resulting in reduced fruit set, smaller and lower quality fruits, and ultimately a reduction in yield. Drought is the single biggest cause of crop failures worldwide [90] and vegetables are particularly susceptible as most are succulent plants consisting of generally more than 90% water [91]. Soil salinity is an increasing threat; high soil salinity afflicts about 20% of cultivated land and 33% of irrigated land worldwide [92]. Many globally important vegetables, including onion, are sensitive to soil salinity while cucumbers, eggplants, crucifers, peppers and tomatoes are only moderately sensitive [93]. Vegetable production in the tropics is often limited during the rainy season as most vegetables are highly susceptible to flooding. Genetic variation in this characteristic is limited, particularly in tomatoes, which rapidly wilt and die following a short period of flooding, particularly at high temperatures [94].

Breeding vegetables for heat tolerance has proved easier than breeding for drought, salinity, or flooding tolerance [87]. AVRDC has developed cultivars of tomato and Chinese cabbage with good adaptation to hot and humid tropical environments since the early 1970s. These improved lines initiated the successful cultivation of tomato and Chinese cabbage in the lowland tropics. The heat tolerant lines of tomato show no reduction in pollen viability from heat stress and produce a fruit set 15% higher than heat sensitive cultivars under the same conditions. More improved lines are needed for future climatic scenarios, and these must be able to match the yields of conventional cultivars under favorable growing conditions.

New cultivars must be developed from both existing tomato lines and from the relatively unexploited wild relatives of tomato. It is particularly challenging to breed for drought and salinity tolerance as this is controlled by multiple genes. It also varies with crop developmental stage [95]. Some wild relatives of tomato have good drought and salinity tolerance [96], but this is often genetically linked to undesirable fruit characteristics such as small fruit size and low fruit set and yield.

Developing vegetables that are more tolerant of environmental extremes should not be limited to only those of global importance. Many indigenous landrace vegetables from the tropics already are well adapted to the climatic conditions likely to be more widespread in the future. Many are highly nutritious and familiar to smallholder farmers, and can provide excellent opportunities to help farmers cope with climate change.

5. GUIDELINES AND PRIORITIES FOR RESEARCH ON BREEDING AND PRODUCTION OF VEGETABLES

5.1. Seed Breeding Industry

In the case of most vegetable crops, biodiversity and genetics are delivered in a marvelous package known as the seed. The special techniques of seed production, such as seed treatment for the control of planting diseases and viruses, and the combination of breeding improvement program such as development of hybrids and the incorporation of biotechnology, cannot be efficiently carried by individual growers. By these reasons in modern vegetable production the seed business is usually conducted as a distinct industry. High tech seed industry is a key part of modern horticulture that combines, seed production, genetic improvement, seed production, storage, and distribution. One of the main factors that determine success in vegetable production is biodiversity and genetic capacity. No practical breeding program can succeed without large numbers of lines (genotypes) to evaluate, select, recombine, and inbreed (fix genetically). This effort must be organized, so valid conclusions can be reached and decisions made. Scientists, breeders, support people, and facilities, budgets, and good management are requirements to assure success in the vegetable seed business. Science must be state of the art to maximize success in a competitive business environment. Since the continued need for fundamental breeding research is critical to sustain development of new technology and expansion of the knowledge base that supports cultivar development, competition among proprietary cultivars results in owner-companies striving to do the best possible research to develop their own products and to compete on genetic and physiological quality of vegetable seed in the marketplace. Reasonable profit margins are necessary to pay back the R&D costs to the owner and to fund future research on developing even better vegetable cultivars to stay competitive [97]. There is considerable genetic variation within the various vegetable species, which can be exploited in the development of superior proprietary cultivars. The consequences of this dynamic situation mean that relatively short-lived cultivars are replaced by either the owner of the cultivar or a competitor seed company. This intense competition means constantly improved and more sophisticated cultivars for the vegetable industry. Seed companies are in the business of manipulating genes to improve plant cultivar performance for a profit. The success of the research is judged by the success of the product in making a reasonable profit. The research must improve economic performance starting with the seed production costs and include the grower-shipper/processor and the end user. If any link in this sequence of events is weak or broken, the new cultivar likely will fail [97,98].

Private breeding companies are placing more and more emphasis on the development of hybrids to exploit heterosis, and to combine multiple disease and stress resistance, but also for economic purposes to ensure growers must purchase seed for each planting. Control of the parents prevents other seed companies from reproducing the hybrid.

Farmers pay all the breeding work and seed marketing costs when purchasing improved or hybrid vegetable seed. International seed companies are mainly interested in the breeding and production of vegetable seeds with a high commercial value. Traditional vegetable landraces have largely been neglected by seed companies, policymakers, and researchers. But while their production often takes place under low-input conditions, they contribute substantially to household food and livelihood security, particularly for small resource-poor farmers [99,100]. For example, in Africa traditional landraces constitute an important source of micronutrients, contributing between 30% and 50% of iron and vitamin A consumed, respectively, in poor households [100,101].

Breeding companies strived more and more to bring hybrid seeds onto the market. Worldwide the share of hybrid seed is increasing at a fast pace of 8% - 10% annually in most of the vegetables. More than two-thirds of the 5000 non-hybrid vegetable cultivars available in 1984 seed catalogs from North America were dropped during last 25 years [12,98].

Hybrid vegetable technology has made significant impact on most vegetable crops in developed countries, but a major limitation to vegetable production in many developing countries is the unavailability of high quality seeds. Hybrid seed production is a high level technology and cost intensive venture. Only well organized seed companies with good scientific manpower and wellequipped research facilities can afford hybrid seed production. The public sector in developing countries frequently does not have sufficient capacity to supply adequate quantities of good quality vegetable seed to poor farmers and at present, there are few private sector seed companies adapting cultivars to local environments, especially in the poorer countries [97,98,102]. Farmers themselves often produce seeds of locally preferred or traditional landraces, as the individual markets are too small and private companies have little interest in producing open pollinated cultivars [12,100]. Without proper seed production, processing technology, quality assurance, and management supervision, locally produced seeds are often contaminated by seed transmitted viruses and other disease organisms, and are genetically diverse. Lack of proper storage facilities and an effective monitoring mechanism often leads to low or uncertain seed viability and vigor. Moreover, low capital resources and poor market information discourage the development of seed-related agribusinesses. Seed quality and treatment are keys to product quality, and there is a need for upgrading quality control laboratories to meet international standards. By those reasons in modern vegetable production the seed business is usually conducted as a distinct industry [12,97,98].

The global seed trade is now dominated by international corporations whose vast economic power has effectively marginalized the roles of public sector plant breeding and local, small scale seed companies. Thirty years ago there were thousands of seed companies in the world, most of which were small and family owned. Today, the top six global seed companies control almost 50% of the commercial seed trade [97,98]. Some of these companies belong to worldwide corporations that are also involved with pesticides and biotechnology. The vegetables attracting the most breeding attention vary considerably between small and huge seed companies/ corporations. Small seed companies have a tendency to specialize in a few vegetable crops. In large international companies the breeding activity is more diverse, but is concentrated on the more economically important crops. In these companies, marker assisted selection (MAS) has become an integral component of many commercial vegetable breeding programs [103,104]. The initial aim of molecular breeding has been to supplement conventional methods with faster and more efficient breeding through MAS or marker-assisted backcrossing (MABC). Molecular markers that are closely linked to the trait of interest may be identified and applied in gene pyramiding, facilitating introgression of desirable traits into cultivars, early selection, etc. For more complex traits conferred by multi-genic traits, QTL analysis has been carried out. Markers bring additional value when they can be used to accelerate the development of new improved vegetable cultivars.

Developmental cycles of commercial hybrid cultivars range between 5 and 12 years. Besides breeding lines necessary for creating hybrid seed need to be refined for specific markets. All these high significant R&D investments make commercial vegetable breeding expensive.

A greater desire for year-round availability of vegetables has had a significant impact on seed companies, requiring full year-round production and consequently a global presence. Unlike for instance agrochemicals "where one size fits all", seed cultivars need to be adapted and differentiated to suit the agronomic needs of the respective region where the vegetable is grown. Active international trade and overseas vegetable seed production by contract is common in many countries. Each multinational company vies to provide better vegetable seeds to compete with domestic seed producers. In China, whose seed market is estimated to be valued at more than US$1.4 billion, the increased recognition of new and high-yielding hybrid cultivars has encouraged the local development of a large number of vegetable seed producers and distributors. Four types of vegetable seed producers were established: public seed companies, research institutes, foreign seed companies, and local seed companies. Private seed companies have been expanding rapidly in recent years and there are now thousands of small firms. Some companies have started to breed their own cultivars and establish marketing networks. They play a strong role in the Chinese vegetable seed industry. About 60 foreign seed companies have opened branch companies or stations in China. Most of them not only sell their vegetable seeds but also have established breeding stations. In other southeastern Asian countries such as India, Indonesia, Vietnam and Malaysia, the percentage of hybrid vegetable cultivars is lower than in China, and so a large expansion of seed companies has not yet occurred. While there has been rapid growth in the seed markets of developing countries due to a shift away from farm saved seed, the seed markets in developed countries, particularly those of Europe and Japan are stagnant. In Europe and the United States, the seed industry has been concentrated and is largely in the hands of large corporations and many small firms are closing [12,97,98].

Vegetable breeding strategy and targets are dependent on market trends. Successful breeders anticipate changes in the market by developing new cultivars that are ready to be released to the growers when their demand increases. Therefore, it will be interesting to see how breeding companies react to changes in vegetable consumption and to evaluate the potential influence that the vegetable market and growing systems may have on breeding targets and priorities.

5.2. Biodiversity and Germplasm Conservation

As stated biodiversity is the basis for plant breeding. Selection is impossible without diversity and new cultivars for farmers and growers cannot be developed without it. This makes access to this variation essential for breeders.

As referred above, about one half (52%) of the total number of vegetables cultivated in the world get commercial breeding attention by seed companies and, of those, only 17% are in large scale breeding programs, fostering a need for serious attention to maintenance of vegetable crop biodiversity. There has been a severe decline in the vegetable cultivar genetic base, as evidenced by the significant reduction, especially within the last 50 years, in the number and range of vegetable cultivars grown. During this period vegetable genetic diversity has been eroding all over the world and vegetable genetic resources are disappearing, on a global scale, at an unprecedented rate of 1.5% - 2% per annum [97,98,102]. Widespread adoption of simplified vegetable systems with low genetic diversity carries a variety of risks including food insecurity. In the short term, such systems risk potential crop failure. In the longer term, they encourage the reduction of the broad genetic base that contributes to high yields, quality traits, disease and pest resistance, etc. and thus compromise the future genetic health of vegetables.

Especially prominent among the "enemies" of genetic diversity are the commercial markets and economic social pressures that have influenced breeding practices that promote uniformity, encouraging extensive cultivation of preferred improved and hybrid vegetable cultivars with insufficient diversity. In addition, globalization has stimulated the consolidation of vegetable seed companies into huge corporations and the decline of small seed companies that serve local and regional markets. In consequence some vegetable breeding programs have been merged or eliminated to reduce costs. Thus, fewer and fewer companies/corporations are making critical decisions about the vegetable research agenda, and the future of vegetables worldwide. Inevitably, two things will happen. There will be fewer vegetable breeders in the future and growers will be dependent on a narrower genetic background that could contribute in the near future to food insecurity for poor growers and consumers. Also, with the advent of genetic engineering, these huge seed corporations are also assuming ownership of a vast array of living organisms and biological processes. Of equal concern are expanded uses of legal mechanisms, such as patents and plant breeder's rights that are removing vegetable plant germplasm from general public use [102, 105-107]. Intellectual property rights for plants was intended as a defensive mechanism to prevent the loss of invented cultivars to competitors. However, with the more stringent enforcement of plant breeding rights, and particularly with the application of the utility patent law in the United States to protect all forms of an innovation, this has become an offensive weapon to stifle competition and inhibit the flow of germplasm and information. This can have serious implications for the future conservation of vegetable genetic resources and for world food security [102,106,107].

Some landraces and old open-pollinated cultivars of vegetables have existed for long periods outside the commercial and professional plant breeding circles because they have been kept alive within communities by succeeding generations of seed savers. Unfortunately, active seed savers tend to be more and more a minority among the millions of vegetable growers. Due to the demand of commercial markets and the professionalization of the sector, many growers are no longer saving seeds. This is an additional threat to genetic diversity. So continued survival of landraces and open-pollinated cultivars of vegetables depends largely on popular interest and initiative as well as preservation in gene banks. We should be alerted and concerned about the loss of biodiversity in vegetables and about this impact on food security.

Vegetable growers have an important role in conserving and using vegetable biodiversity. The future of world food security depends not just on stored vegetable genes, but also on the people who use and maintain crop genetic diversity on a daily basis. In the long run, the conservation of plant genetic diversity depends not only on a small number of institutional plant breeders and seed banks, but also on the vast number of growers who select, improve, and use vegetable diversity, especially in marginal farming environments. That is why we should be also alerted and particularly alarmed by the current trend to use improved and hybrid vegetable cultivars exclusively.

Growers do not just save seeds, they are plant breeders who are constantly adapting their vegetable crops to specific farming conditions and needs. For over 300 vegetable generations, vegetable growers have been selecting seeds and adapting their plants for local use. This genetic biodiversity is the key to maintaining and improving the world's food security and nutrition. No plant breeder or genetic engineer starts from scratch when developing a new cultivar of tomato, pepper, cabbage or lettuce. They are building on the accumulated success of generations of growers, who have selected and improved vegetable seeds for thousands of years. If poor small-scale growers in marginal areas stop saving seeds, we will lose genetic diversity. Growers will lose the means to select and adapt vegetable crops to their unique farming conditions, which are characterized by low external inputs.

Hybrid seed technology is designed to prevent growers from saving seed from their harvest, thus forcing them to return to the commercial seed market every year. Hybrid vegetable seeds alone, and used globally, can be a dead-end to biodiversity. If growers abandon completely their traditional vegetable landraces in the process of adopting only hybrids, crop genetic diversity achieved over centuries will be lost forever. Many agronomic benefits will be lost to worldwide vegetable growers and thus to consumers. The exclusive adoption of hybrid cultivars in marginal areas may restrict the vegetable producing capacity of farmers, it will destroy biodiversity, and it may contribute in the long-term to food insecurity.

There is a growing awareness world-wide about the need to conserve plant germplasm for the use of future generations. Consequently, considerable media attention has been given to the creation of the Svalbard global seed vault and the related storage of seeds of many economically-important crops (see http://www.croptrust.org/main). However, the lower "status" of most of vegetable crops means that they are not treated with the same degree of priority and concern. Furthermore, the fact that many vegetable crops of comercial and horticultural significance are vegetatively propagated and not seed propagated restricts the options for their conservation other than through relatively expensive live plant collections. There are also considerable ongoing efforts by national governments and international organizations to preserve plant vegetable germplasm in gene banks. This is a valuable but static approach, as further evolutionary changes and improvements will not occur until the seeds are planted. It is also an activity that relies heavily on continued political stability and support, including sustained governmental funding. Active and positive connections between the private breeding sector and large-scale gene banks are required to avoid possible conflict involving breeders'rights and gene preservation.

The diversity of crop species will be promoted by the maintenance of crop gene banks by governments and non-governmental organizations, the continued use of diverse sources by plant breeders, especially in the public sector, and by the use of local cultivars and landraces by farmers.

5.3. Vegetable Breeders

Breeders play a key role in determining what we eat, since the plant cultivars they develop begin the dietary food chain. There are considerably fewer vegetable plant breeding positions than in field crops, and vegetable breeders are often required to be knowledgeable about a more diverse group of crops. Vegetable crops include many different crops morphotypes with great diversity within. So careful consideration must be given to the type of graduate education appropriate for preparing new vegetable breeders, as well as continuing education to keep them current and productive. Since many breeders will work on different vegetable crops during a career, how can they be prepared for inevitable change, the types of which are difficult to imagine? Who would have predicted in 1990 that a mere 20 years later, many plant genomes are sequenced and DNA-based selection methods linked with phenotypic selection and electronic data handling are becoming standard practice? Few of the vegetabe plant breeders beginning their careers two decades ago had much knowledge about and experience with DNA technology, but in order to be productive and competitive today they have had to learn about and integrate new technology and methods. There is no reason to suppose that changes will be any less dramatic for future vegetable breeders.

As with other courses of study, plant breeding graduate education must strike a balance between broad preparation and specialization. The number of vegetable breeders is small compared to other crops, therefore few if any graduate programs will be designated solely for vegetable. So too, limited number of university faculty working on vegetable crops will make it challenging to assemble a critical mass of faculty with research and teaching appointments and interest in a range of vegetable crop areas. Some pertinent questions for preparation of vegetable breeders should include: i) what types of new positions will be available in vegetable crop breeding; ii) what level of formal education will be appropriate; iii) how should preparation differ for global careers that may be in developed and developing countries/regions; iv) what skills will be needed for success; v) what vegetable crops should graduates be knowledgeable about; vi) are there educational knowledge, experience and skills that differ for horticultural vs. other agronomic crops; vii) what is needed to prepare vegetable breeders for career changes among different vegetable crops; and viii) what continuing education should be available for practicing breeders throughout their careers.

The first step is to attract outstanding students into vegetable breeding programs with modern, exciting and relevant curricula that provide clear paths for obtaining knowledge, experience and skills needed for success in attractive jobs that are currently available in plant breeding globally. For all countries, vegetable products are critical for good nutrition, healthy living and a dynamic business sector. Development of new, genetically improved vegetable cultivars that are profitable for growers and that meet the needs of consumers will contribute to a vital, dynamic vegetable industry world-wide.

The ready supply of a wide range of affordable vegetable crops in most developed countries has led to a marked complacency about the need for ongoing R & D vegetable breeding programs. Consequently, many governments world-wide have scaled back funding for vegetable breeding and production programs. Most research driven universities have now disbanded horticulture/ horticultural breeding departments (and often agronomy departments) and merged them into plant science/plant biology departments with a very strong focus on the molecular sciences. This erosion has been concentrated in what historically have been viewed as vital disciplines for vegetable breeding such as plant breeding, agronomy, pathology, and entomology.

Nonetheless, as indicated above, there are a basic need for research and breeding on vegetable crops given by the current consumer demands for affordability, safety and continuity of supply; increasing needs to achieve sustainable practices; and the need to deal with challenges arising from a more variable climate, the loss of productive soils through urban encroachment, and the loss of low-cost labour.

5.4. Vegetable Sustainable Production Practices and Technology Transfer

The concepts of sustainability of vegetable crops relate both to the use of resources, such as water and nutrients, in a manner which considers future needs, and the responsible use of compounds such as pesticides which will not compromise the quality of the environment (or the safety of the produce that is produced). Improved vegetable cultivars are also important since there are cultivars more apropriated for sustainable production. Vegetable breeding can be important for genetic enhancement of vegetables, and for an eco-friendly organic horticulture that will reduce human carbon footprints.

The most topical issue that is facing vegetable production in almost all areas of the world is the availability of water. Pressure on water resources for urban, industrial, recreational, conservation and other uses all appear to have higher priority within societies than the availability of water for vegetable crop production. Current demands on water from both river and subterranean sources are outstripping availability resulting in attention being given to water harvesting options with on-farm dams, the use of waste water and to the use of more efficient cultural practices. Research is urgently needed to resolve issues such as the development of drought tolerant vegetable crops, the management of crops under managed water deficits, dealing with increased salinity, and the use of low quality water.

Integrated crop production systems have been shown to be effective on a number of vegetable crops for reducing the number of pesticide applications and for shifting to new control compounds that have less persistence in the environment and which are of lower overall toxicity. Producers have behaved responsibly and with some urgency to ensure the adoption of new practices. GMOs cultivars can also be important for sustainable production [42,44]. Genetic engineering has the potential to address some of the most challenging biotic constraints faced by vegetables growers, which are not easily addressed through conventional plant breeding alone. Transgenic vegetable crops, enable also breeders to bring favorable genes, often inaccessible by conventional methods, into elite cultivars, improving their value considerably and offering unique opportunities for controlling insects and other pathogens [42,44].

Technology transfer from both university and government sources in many developed countries has markedly declined over the past two decades and have been replaced with a greater provision of services from the private sector. Paradoxically, the demand for information from public providers has nonetheless increased as providers of information in the private sector seek to stay abreast of the latest developments and as producers seek to secure the best advice possible. This can be very important for sustainable production and better knowledge of the vegetable producers about the cultivars available.

In developing countries, where often the provision of knowledge from both private and public sectors may be poorly developed, the demand for useful and relevant information is growing as vegetable enterprises become more sophisticated and often more focussed on exporting their vegetable produces. In parallel with such changes, there has been an unprecedented expansion of methods by which to transfer information-mobile phones, the internet, laptop computers, private and public networking, and so on. Equally, there is a growing expansion of the types of information that can be accessed including that from remote sensing, environmental monitoring, image analysis and others. The challenges relate not to the availability of information, but to the provision of relevant knowledge and the more effective means of transmitting the specific information that is required. Vegetable producers have rapidly adopted many of these technologies contributing for their success as farmers. The major challenge facing vegetable producers in the future will be to filter the information that is available into useful knowledge that will allow them to be more effective and efficient in their various enterprises.

5.5. Health Promotion

Regular consumption of a vegetable rich diet has undeniable positive effects on health since phytonutriceuticals of vegetables can protect the human body from several types of chronic diseases. The mechanism by which vegetables decrease risk of disease is complex and largely unknown. Various components of the whole food are likely to contribute to the overall health benefit. Various phytonutriceuticals with antioxidant properties may work directly by quenching free radicals or indirectly by participating in cell signaling pathways sensitive to redox balance [1,2]. Nutrients such as potassium contribute to blood pressure regulation. The dietary fiber content and type of different vegetables may also contribute to the overall health benefit, such as improving bowel transit, lowering cholesterol, helping manage blood glucose concentrations, and by transporting a significant amount of minerals and phytochemicals linked to the fibre matrix through the human gut [1]. Finaly, increaseing vegetables in the diet may reduce the intake of saturated fats, trans fats, and foods with higher caloric density, all of which may be related to a healthier overall diet [1].

A major activity over the past 2 - 3 decades has been the promotion of health benefits that can be gained through the consumption of various vegetables. This has been most evident from the various "5 a Day" and similar promotional programmes to promote fruits and vegetables consumption. These activities that have been a success have not just included fresh vegetable and fruit products but have also embraced products based on juice, dried products and extracts, many of which have extensive on-label claims. The reality is that many such claims have not in fact been clinically proven and regulators are increasingly requiring label claims to either be substantiated or removed. Clearly, there is a need for considerably more research in this area in order to fully specify the presumed health benefits in many vegetables [1,2]. This is an exciting area of endeavour where there is a convergence of horticultural science with medical and nutritional research. Notwithstanding the need to prove many of the specific health claims that are made for many vegetable crops, the consumption of vegetables per capita in markets such as the United States of America has increased over the past 2 - 3 decades. However, consumer preferences have changed markedly during that same period with a growing pattern of interest in convenience, year-round availability and with novelty. No doubt the perceived health benefits of some of these vegetable crops have also helped to enhance consumption.

Because each vegetable contains a unique combination of phytonutriceuticals (vitamins, minerals, dietary fiber and phytochemicals), a great diversity of vegetables should be eaten to ensure that individual's diet includes a combination of phytonutriceuticals and to get all the health benefits. The availability of a large diversity of vegetables year-round, allied to increase in mean per capita incomes in recent years and knowledge of vegetable health benefits, have enable consumers to include a variety of health promoting phytonutriceuticals in their diet [1,2].

Nutritional quality as understood by the consumers and available at a moderate price may encourage enhanced consumption, thereby conferring an important marketing incentive to vegetable plant breeding. Research on the health benefits of vegetables, from a horticultural and breeding perspective, needs to focus on key areas in the near future such as: i) to continue the evaluation of phytonutriceuticals content among older versus newer major cultivars; ii) to identify the genetic mecanisms that regulate the syntesis of their key phytochemicals, such as the glucosinolates, thiosulfides and flavonoids, in order to develop cultivars rich in a variety of phytochemicals and in order to ensure that a mixture of phytochemicals enters into the human diet; iii) to study the potencial change in the balance of these compounds (and their eventual synergisms or interactions); and iv) to identify the optimum conditions for mantaining these phytochemicals after harvest and processing since studies have shown that the biovailability of some of the phytochemicals increase dramatically after storage and processing and others de grade rapidly.

Genetic engineering enables vegetable breeders to incorporate desired transgenes into elite cultivars, thereby improving their value considerably. It further offers unique opportunities for improving nutritional quality and bringing other health benefits [42,43]. Many vegetable crops have been genetically modified to improve traits such as higher nutritional status or better flavour, and to reduce bitter-ness or anti-nutritional factors [42, 43]. Transgenic vegetables can be also used for vaccine delivery. Consumers could benefit further from eating more nutritious transgenic vegetables, e.g. an increase of crop carotenoids by metabolic sink manipulation through genetic engineering appears feasible in some vegetables [42,43]. Genetically engineering carrots containing increase Ca levels may boost Ca uptake, thereby reducing the incidence of Ca deficiencies such as osteoporosis. Fortified transgenic lettuce with zinc will overcome the deficiency of this micronutrient that severely impairs organ function. Folates deficiency, which is regarded as a global health problem, can also be overcome with transgenic tomatoes with folate levels that provide a complete adult daily requirement. Transgenic lettuce with improved tocopherol and resvera-trol composition may prevent coronary disease and arteriosclerosis and can contribute to cancer chemopreventative ac-tivity. Food safety and health benefits can also be enhanced through transgenic approaches, e.g. rural African re-source-poor consumers will benefit eating cyanide-free cultivars of cassava. Biotechnology-derived vegetable crops will succeed if clear advantages and safety are demonstrated to both growers and consumers.

5.6. Improved Post-Harvest and Supply Chain Methods

A major trend over the past decade, possibly as a consequence of smaller households and more dedication to leisure activities, has been the shift towards the consumption of fresh rather than frozen vegetables. Modern consumers do not store vegetable products for long periods within the household and are able to discern the difference in quality (taste, flavor, and texture) between vegetables that have been stored for short versus long periods.

The challenge for breeding and horticultural science is, therefore, to refine methods for shortterm storage so that premium quality is retained rather than to focus on longer-term storage for prolonged marketing. Interestingly, this potentially transforms the approaches that can be taken for chilling sensitive vegetables where long-term storage techniques have not been routinely available. The option to provide for new approaches that can be used for quality retention during short-term storage should be an absolute priority.

Maturity at harvest is the most important determinant of storage-life and of final fruit-vegetable quality. Many vegetables, in particular leafy vegetables, and immature fruit-vegetables (such as cucumbers, sweet corn, green beans, peas, and okras), attain optimum eating-quality prior to reaching full maturity. This often results in delayed harvest, and consequently in produce of low quality.

The method of harvesting (hand vs mechanical) can significantly impact upon the composition and postharvest quality of vegetables. Mechanical injuries (such as bruising, surface abrasions and cuts) can accelerate loss of water and vitamin C resulting in increased susceptibility to decay-causing pathogens. Most fresh vegetables are harvested by hand. Root crops (such as carrot, onion, potato, and sweet potato) and some commodities destined for processing (such as processing tomatoes) are mechanically harvested. Management of harvesting operations, whether manual or mechanical, can have a major impact on the quality of harvested vegetables. Proper management procedures include selection of optimum time to harvest in relation to product maturity and climatic conditions, training and supervision of workers, and proper implementation of effective quality control. Expedited and careful handling, immediate cooling after harvest, maintenance of optimum temperatures during transit and storage, and effective decay-control procedures are important factors in the successful post-harvest handling of vegetables. Attention must be paid to all of these factors, regardless of the method of harvesting used. These factors are nevertheless more critical in the case of mechanically harvested commodities.

It should be noted that any practice that reduces the number of produce handling steps will help minimize losses. Field packing (selection, sorting, trimming, and packaging) of produce at the time of harvest can greatly reduce the number of handling steps in preparation vegetables for marketing. Mobile field packing stations with adequate shading are used for vegetables (such as broccoli, cauliflower, and green beans) that do not require washing as part of their preparation for marketing.

Within each vegetable there is a range of genotypic variation in composition, quality, and post-harvest-life potential. Plant breeders have been successful in selecting carrot, sweet potato, and tomato cultivars with comparably high carotenoid levels and vitamin A content; onion and tomato cultivars with longer shelf-lives, sweet corn cultivars that maintain their sweetness longer after harvest; cantaloupe and watermelon cultivars with higher sugar content and firmer flesh, etc.. These are just a few examples of how genetic manipulation has contributed to improving the quality of vegetables and post-harvest potential.

Breeders have also developed vegetable cultivars with improved storage characteristics, including resistance to storage fungi and pests, as well as physiological disorders. More adequate and sustained funding for this work is required, as well as improved linkages among breeders, nutritionists, health sector experts, and policymakers. Low-cost safe storage structures for small and intermediate scale farmers need to be promoted.

Many opportunities exist for applying biotechnology to improve the post-harvest quality and safety of fresh vegetables. Priority goals in this regard, should, be focused on: i) attaining and maintaining good flavor and nutritional quality, so as to satisfy consumer demands; and ii) introducing resistance to physiological disorders and/or decay-causing pathogens and pests, so as to reduce the use of chemicals on vegetables. For example in tomato, the first transgenic vegetable crop, there are a lot of research conducted to manipulate fruit ripening, texture and nutritional quality using transgenic approaches. Many of the genes targeted include ethylene because of its role in fruit ripening [42,44]. Enzymes that regulate ethylene biosynthesis in plants are: S-adenosylmethionine (SAM) synthase, 1-aminocyclopropane-1-carboxylate (ACC) synthase, and ACC oxidase. The genes encoding these enzymes [108,109] as well as those that metabolize SAM or ACC [110,111] have been targeted in order to manipulate ethylene biosynthesis, thereby regulating fruit ripening. It has been clearly shown that modulation of ethylene biosynthesis using genetic engineering can yield tomato fruits with predictable ripening characteristics. Tomato fruit ripening manipulation has been however achieved by introducing anti-ripening genes (rin and nor) in heterozygous genotypes. These genes have been incorporated in many fresh and processing tomatoes [42,44].

5.7. Year-Round Supply

The year-round provision of safe, nutritious vegetable crops to consumers has, over the past 50 years, primarily been through a combination of locally-produced and imported vegetable crops that have been handled through various transport and storage regimes, and shipped to both local consumers and to opposite hemispheres using such methods. The high costs of energy for transport and for storage, changes in consumer consumption preferences, and changes in consumer concerns (such as computed carbon footprints and others schemes), increasingly mean that these options may no longer be viable for some markets that want vegetables produced under a reduced carbon footprints.

Many vegetables particularly in tropical and subtropical countries have now a gross over-supply during a very concentrated production season. The aim is to extend the production season, and consequently the period available for marketing. The use of different vegetable cultivars in different seasons, and photoperiodic and vernalization sensitive cultivars for changing the flowering and therefore the production season, togheter with the use of special fertiliser and water regimes for shifting the flowering and vernalization time are effective and proven approaches that can be used for year-round supply of vegetable markets. Temperate vegetable crops can also be developed for extended production seasons as evidenced by the change in day length dependency in strawberry. Improved production using improved landrace vegetables and production under greenhouses should be also considered.

6. OUTLOOK

The production and marketing of vegetables crops is undergoing continuous change globally. This is mainly due to the growing demands of consumers for safe and healthy vegetables, increased urbanisation of societies, and the growth in scale and influence of supermarkets chains. The changes outlined above are merely some of the factors that are impacting on current production practices for vegetable crops. They also provide a plethora of opportunities for vegetable breeding and horticultural science to solve many of the current problems facing producers and to offer solutions to those issues that are concerning consumers. Horticultural science can respond to many of these challenges through research, breeding and innovation that can seek to gain more efficient methods of crop production, refined post-harvest storage and handling methods, newer and higher value vegetable cultivars and demonstration of their health benefits.

Horticulturists will need to develop cultural practices and vegetable breeders to breed vegetables for a multifunctional horticulture (diversity, health promotion, postharvest, year-round supply, etc.) and to cope with harsher climate conditions and lower inputs than they have come to expect. Improved production systems that can cope with climate extremes must allow vegetables to produce under high temperatures, greater drought stress, increased soil salinity, and periodic flooding. This will involve a combination of improved vegetable cultivars and modified production systems.

The continued globalisation of the vegetable industry will see a rapid transfer and adoption of knowledge. This will include the transfer from developed to developing countries as major enterprises move to year-round production and in response to securing lower-cost land and labour. Many of these activities will, however, be carried out by the private sector where the protection of intellectual property, the development of unique and proprietary vegetable cultivars and the advocacy of brands will, undoubtedly, see a markedly diminished involvement of the public sector. Modern trends indicate that consumers are seeking increased variety, novelty, and healthy options in their eating choices.

Creation of vegetable hybrids is a key means towards the development of cultivars for modern vegetable production. Hybrid seed production is high technology and a cost intensive venture. Only well organized seed companies with good scientific manpower and well equipped research facilities can afford seed production. Due to globalization most vegetable breeding research and cultivar development in the world are presently conducted and funded in the private sector, mainly by huge multinational seed companies. Public vegetable breeders and cultivar development activities and research are disappearing worldwide. This means in general that there are fewer decision-making centers for vegetable breeding and cultivar development. This has also resulted in the focus on relatively few major vegetables produced worldwide, to the detriment of all the other cultivated vegetables. It is imperative that national governments and policymakers, as part of a social duty, invest in breeding research and cultivar development of traditional open-pollinated cultivars and in the minor and so-called "forgotten" vegetables. Developing vegetable cultivars that are more tolerant of environmental extremes should not be limited to only those of global importance. Many landrace vegetables from the tropics already are well adapted to the climatic conditions likely to be more widespread in the future. Many are highly nutritious and familiar to smallholder farmers, and can provide excellent opportunities to help farmers cope with climate change.

Smaller seed companies, which are usually specialize in few vegetable crops, must be supported, possibly through autonomous affiliation with the larger companies. More investments in this area will mean less expensive seed for growers to choose from, and an increased preservation of vegetable biodiversity. The accomplishment of this goal may require new approaches to vegetable breeding research and development by both the public and private sector.

Public plant breeding remains a key component of vegetable research systems worldwide, especially in developing countries. However, the increasing presence of private sector breeding and a decrease in national and international support make it difficult for the public sector to continue operating in the traditional manner. Declining funding for public vegetable breeding coupled with the rapid increase of vegetable production and consumption and an urbanizing population, has created a difficult situation. Public sector breeding must be strenghened. More public sector vegetable breeders are needed worldwide to select and to produce non-hybrid cultivars of the minor and "forgotten" vegetables. Breeding of vegetables and other minor crops must continue as a viable endeavor. Improving landrace vegetables is particularly important for the poorest. This will benefit small farmers, and will safeguard biodiversity and food security in developing countries.

Currently there are too few vegetable breeders to satisfy global demands and opportunities in the public and private sectors of developing and developed countries. In order to meet future needs, it is important that educational programs incorporate rapidly changing new technologies along with classical content and methods in order to meet professional needs for vegetable cultivar developers, researchers, teachers and support scientists.

João Carlos da Silva Dias Vegetable crop improvement powered through plant breeding is critical for sustainable production of vegetable crops that contribute to healthful diets and enhance quality of life for people around the world. Policymakers and investors have to turn their attention to enhanced funding for the vegetable and horticultural sector, allowing farmers to compete with their products on a world market increasingly determined by market quality standards and phytosanitary concerns and regulations. Many current vegetable breeding efforts remain under-funded and disorganized. There is a great need for a more focused, coordinated approach to efficiently utilize funding, share expertise, and continue progress in horticultural technologies and breeding programs. Only then will the silent vegetable and horticultural revolution currently underway benefit a significant portion of the world's poor nations, farmers, and landless laborers.

REFERENCES

1. Dias, J.S. (2012) Major classes of phytonutriceuticals in vegetables and health benefits: A review. Journal of Nutritional Therapeutics, 1, 31-62.
2. Dias, J.S. (2012) Nutritional quality and health benefits of vegetables: A review. Food and Nutrition Sciences, 3, 1354-1374. http://dx.doi.org/10.4236/fns.2012.310179

3. ACC/SCN. (2004) Fifth report on the world nutrition situation: Nutrition for improved development outcomes. United Nations ACC/SCN in collaboration with World Health Organization, Geneva.
4. WHO. (2009) Global health risks: Mortality and burden of disease attributable to selected major risks. World Health Organization, Geneva.
5. Hellen Keller International (HKI). (2010) Homestead food production model contributes to improved household food security, nutrition and female empowerment—Experience from scaling-up programs in Asia (Bangladesh, Cambodia, Nepal and Philippines). Nutrition Bulletin, 8, 1-8.
6. Rao, S., Yajnik, C.S., Kanade, A., Fall, C.H.D., Margetts, B.M. and Jackson, A.A. (2001) Intake of micronutrientrich foods in rural Indian mothers is associated with the size of their babies at birth: Pune maternal nutrition study. The Journal of Nutrition, 131, 1217-1224.
7. Dias, J.S. (2012) 1. Vegetable breeding for nutritional quality and health benefits. In: Carbone, K., Ed., Cultivar: Chemical Properties, Antioxidant Activities and Health Benefits. Nova Science Publishers, Inc., Hauppauge, New York, 1-81.
8. Friel, C. and Baker, P.I. (2009) Equity, food security and health equity in the Asia-Pacific region. Asia Pacific Journal of Clinical Nutrition, 18, 620-632.
9. Kays, S.J. and Dias, J.S. (1995) Common names of commercially cultivated vegetables of the world in 15 languages. Economic Botany, 49, 115-152. http://dx.doi .org/10.1007/BF02862917
10. Kays, S.J. (2011) Cultivated vegetables of the world: A multilingual onomasticon. Wageningen Academic Publishers, Wageningen. http://dx.doi.org/10.3920/978-90-8686-720-2
11. Dias, J.S. (2011) World importance, marketing and trading of vegetables. Acta Horticulturae, 921, 153-169.
12. Dias, J.S. and Ryder, E.J. (2011) World Vegetable Industry: Production, breeding, trends. Horticultural Reviews, 38, 299-356.
13. Monsanto. (2011) Supplemental toolkit for investors. Updated April 2011. Monsanto, St. Louis.
14. Gale, F. (2002) China's food and agriculture: Issues for the 21st century. Market and Trade Economics Division, Economic Research Service. USDA, Washington DC.
15. Rich, M. (2008) IPM program in Bangladesh wins award. Integrated Pest Management Collaborative Research Support Program.
16. FAO. (2009) FAOSTAT data. Food and Agriculture Organization of the United Nations, Rome. http://www.fao.org.
17. Senauer, B. and Goetz, L. (2003) The growing middle class in developing countries and the market for high-value food products. Economic Research Service. USDA, Washington DC.
18. United Nations. (2007) World urbanisation prospects: the 2007 revision population database. United Nations Population Division, UN, New York.
19. Reardon, T., Timmer, C.P., Barrett, C.B. and Berdegue, J. (2003) The rise of supermarkets in Africa, Asia, and Latin America. American Journal of Agricultural Economics, 85, 1140-1146. http://dx.doi.org/10.1111/j.0092-5853.2003.00520.x
20. Yussefi, M. and Willer, H. (2003) The world of organic agriculture 2003—Statistics and future prospects. International Federation of Organic Agriculture Movements (IFOAM), Tholey-Theley.
21. Willer, H. and Kilcher, L. (2009) The world of organic agriculture—Statistics and emerging trends 2009. FIBLIFOAM Report. IFOAM, Bonn, FiBL, Frick, Switzerland, ITC, Geneva.
22. Raynolds, L.T. (2004) The globalization of organic agrofood networks. World Development, 32, 725-743. http://dx.doi.org/10.1016/j.worlddev.2003.11.008
23. Regmi, A. and Gehlar, M. (2001) Consumer preferences and concerns shape global food trade. Global Food Trade, Food Review, 24, 2-8.
24. FAO. (2007) FAOSTAT data. Food and Agriculture Organization of the United Nations, Rome. http://www.fao.org

25. Bi, X., Dong, X., Huang, J., Hu, D. and Rozelle, S. (2004) Securing small producer participation in restructured national and regional agri-food systems—China Country Report. Regoverning Markets—Center for Chinese Agricultural Policy. Chinese Academy of Sciences, Beijing.
26. Weinberger, K. and Lumpkin, T.A. (2005) High value agricultural products in Asia and the Pacific for smallholder farmers: Trends, opportunities and research priorities. GFAR, Rome.
27. Deloitte. (2010) Emerging from downturn. Global powers of retailing. Deloitte.
28. Deloitte. (2011) Leaving home. Global powers of retailing 2011. Deloitte.
29. Manalili, N. and Tumlos, L.A.C. (2004) Regoverning market: Securing small producer participation in restructured national and regional agri-food system—Southeast Asia overview. Regoverning Markets. SEARCA, Los Banos.
30. Shepherd, A. (2005) The implications of supermarket development for horticultural farmers and traditional marketing systems in Asia. FAO, Rome.
31. Neven, D., Reardon, T., Chege, J. and Wang, H. (2006) Supermarkets and consumers in Africa: The case of Nairobi, Kenya. Journal of International Food & Agribusiness Marketing, 18, 103-123. http://dx.doi.org/10.1300/J047v18n01_06
32. Dias, J.S. and Ryder, E. (2012) Impact of plant breeding on the world vegetable industry. Acta Horticulturae, 935, 13-22.
33. Lumpkin, T., Weinberger, K. and Moore, S. (2005) Increasing income through fruit and vegetable production opportunities and challenges. Consultive group on International agricultural research science Forum. CGIAR Priorities: Science for the Poor. CGIAR, Marrakech.
34. Freedonia. (2010) World pesticide market. Freedonia Group Inc., Cleveland.
35. OECD. (1997) Impacts of fruit and vegetable production on the environment and policy responses: European Union community instruments relating to the environmental aspects of the fresh fruit and vegetable sector. OECD, Paris.
36. Osteen, C. (2003) Agricultural resources and environmental indicators: Pest management practices. In: Heimlich, R., Ed., Agricultural resources and environmental indicators, 2003. Agriculture Handbook AH722, Ch. 4.3. Washington DC.
37. Pimentel D. (1997) Techniques for reducing pesticide use. Economic and environmental benefits. Wiley, New York.
38. Mansour, S.A. (2004) Pesticide exposure—Egyptian scene. Toxicology, 198, 91-15.http://dx.doi.org/10.1016/j.tox.2004.01.036
39. Choudhary, B. and Gaur, K. (2009) The development and regulation of Bt brinjal in India. ISAAA Brief 38, International Service for the Acquisition of Agri-Biotech Applications (ISAAA).
40. European Food Safety Authority (EFSA). (2010) 2008 Annual report on pesticide residues according to article 32 of regulation (EC) no 396/2005. European Food Safety Authority Journal, 8, 1-442.
41. Pimentel, D. and Hart, K. (1999) Ethical, environmental, and public health implications of pesticide use. New perspectives in bioethics. Johns Hopkins University Press, Baltimore.
42. Dias, J.S. and Ortiz, R. (2012) Transgenic vegetable crops: Progress, potentials and prospects. Plant Breeding Reviews, 35, 151-246.
43. Dias, J.S. and Ortiz, R. (2012) Transgenic vegetable breeding for nutritional quality and health benefits. Food and Nutrition Sciences, 3, 1209-1219. http://dx.doi.org/10.4236/fns.2012.39159
44. Dias, J.S. and Ortiz, R. (2013) Transgenic vegetables for 21st century horticulture. Acta Horticulturae, 974, 15-30.
45. Wolf, J., Wijk, M., Cheng, X., Hu, Y., Diepen, C., Jongbloed, A., Keulen, H., Lu, C.H. and Roetter, R. (2003) Urban and peri-urban agricultural production in Beijing municipality and its impact on water quality. Environment and Urbanization, 15, 141-156.

46. Zhang, W.L., Tian, Z.X., Zhang, N. and Li, X.Q. (1996) Nitrate pollution of groundwater in northern China. Agriculture, Ecosystems & Environment, 59, 223-231. http://dx.doi.org/10.1016/0167-8809(96)01052-3
47. Sheldrick, W.F., Syers, J.K. and Lingard, J. (2003) Soil nutrient audits for China to estimate nutrient balances and output/input relationships. Agriculture, Ecosystems & Environment, 94, 341-354. http://dx.doi.org/10.1016/S0167-8809(02)00038-5
48. Santamaria, P. (2006) Nitrate in vegetables: Toxicity, content, intake and EC regulation. Journal of the Science of Food and Agriculture, 86, 10-17. http://dx.doi.org/10.1002/jsfa.2351
49. Dolan, C.S. and Humphrey, J. (2000) Governance and trade in fresh vegetables: The impact of UK supermarkets on the African horticulture industry. The Journal of Development Studies, 37, 147-176. http://dx.doi.org/10.1080/713600072
50. Dolan, C.S., Humphrey, J. and Harris-Pascal, C. (1999) Horticulture commodity chains: The impact of the UK market on the African fresh vegetable industry. IDS Working Paper 96. IDS, Brighton.
51. Putnam, J. and Allshouse, J. (2001) Imports' share of US diet rise in late 1990s. Food Review, 24, 15-21.
52. FAO (2005) FAOSTAT data. Food and Agriculture Organization of the United Nations, Rome. http://www.fao.org
53. Unnevehr, L.J. (2000) Food safety issues and fresh food product exports from LDCs. Agricultural Economics, 23, 231-240. http://dx.doi.org/10.1111/j.1574-0862.2000.tb00275.x
54. Cerrex. (2003) Study of the consequences of the application of sanitary and phytosanitary (SPS) measures on ACP countries. Report commissioned by CTA. CTA. Cerrex Ltd, UK. 1-108.
55. Grube, R.C., Radwanski, E.R. and Jahn, M. (2000) Comparative genetics of disease resistance within the Solanaceae. Genetics, 155, 873-887.
56. Lacerrot, H. (1996) Breeding strategies for disease resistance in tomato with emphasis to the tropics: Current status and research challenges. 1st International Symposium Tropical Tomato Disease, Recife, 126-132.
57. Zitter, T.A., Hopkins, D.L. and Thomas, C.E. (1996) Compendium of cucurbit diseases. Amer Phytopathological Society Press, St. Paul.
58. Gonsalves, D. (1998) Control of papaya ringspot virus in papaya: A case study. Annual Review of Phytopathology, 36, 415-437. http://dx.doi.org/10.1146/annurev.phyto.36.1.415
59. Thomas, P.E., Kaniewski, W.K. and Lawson, E.C. (1997) Reduced field spread of potato virus in potatoes transformed with the potato leafroll virus coat protein gene. Plant Disease, 81, 1447-1453. http://dx.doi.org/10.1094/PDIS.1997.81.12.1447
60. Grafius, E.J. and Douches, D.S. (2008) The present and future role of insect-resistant genetically modified potato cultivars in IPM. In: Romeis, J., Shelton, A.M. and Kennedy, G.G., Eds., Integration of insect-resistant genetically modified crops within IPM programs. Springer Science þ Business Media B.V., 195-221. http://dx.doi.org/10.1007/978-1-4020-8373-0_7
61. Barrett, D.M., Garcia, E. and Wayne, J.E. (1998) Textural modification of processing tomatoes. Critical Reviews in Food Science and Nutrition, 38, 173-258. http://dx.doi.org/10.1080/10408699891274192
62. Kang, H.M. and Saltveit, M.E. (2003) Wound-induced increases in phenolic content of fresh-cut lettuce are reduced by a short immersion in aqueous hypertonic solutions. Postharvest Biology and Technology, 29, 271-277. http://dx.doi.org/10.1016/S0925-5214(03)00043-7
63. Ryder, E.J. (1999) Lettuce, endive, and chicory. CAB Intl., Wallingford. [Citation Time(s):1]
64. Hayes, H.K. and Garber, R.J. (1927) Breeding crop plants. McGraw-Hill, New York.
65. Allard, R.W. (1960) Principles of plant breeding. Wiley, New York.
66. Fehr, W.R. (1987) Principles of cultivar development. Vol.1 Theory and Technique. Macmillan, New York.

67. Lincoln, R.E., Zscheile, F.P., Porter, J.W., Kohler, G.W. and Caldwell, R.M. (1943) Provitamin A and vitamin C in the genus Lycopersicon. Bot. Gaz, 105, 113-115. http://dx.doi.org/10.1086/335197
68. Lincoln, R.E. and Porter, J.W. (1950) Inheritance of Betacarotene in tomatoes. Genetics, 35, 206-211.
69. Tomes, M.L., Quackenbush, F.W., Nelson Jr., O.E. and North, B. (1953) The inheritance of carotenoid pigment systems in the tomato. Genetics, 38, 117-127.
70. Gabelman, W.H. and Peters, S. (1979) Genetical and plant breeding possibilities for improving the quality of vegetables. Acta Horticulturae, 93, 243-270.
71. Simon, P.W. (2000) Domestication, historical development, and modern breeding of carrot. Plant Breeding Reviews, 19, 157-190.
72. Sudhakar, P., Singh, J., Upadhyay, A.K. and Ram, D. (2002) Genetic variability for antioxidants and yield components in pumpkin (Cucurbita moschata Duch. Ex Poir.), Vegetable Science, 29, 123-126.
73. Frossard, E., Brucher, M., Mozafar, A. and Hurrell, R. (2000) Potential for increasing the content and bioavailability of Fe, Zn and Ca in plants for human nutrition. Journal of the Science of Food and Agriculture, 80, 861- 879. http://dx.doi.org/10.1002/(SICI)1097-0010(20000515)80:7<861::AID-JSFA601>3.0.CO;2-P
74. Hotz, C. and McClafferty, B. (2007) From harvest to health: Challenges for developing biofortified staple foods and determining their impact on micronutrient status. Food and Nutrition Bulletin, 28, S271-S279.
75. Fraser, P.D., Romer, S., Shipton, C.A., Mills, P.B., Kiano, K.W., Misawa, N., Drake, R.G., Schuch, W. and Bramley, P.M. (2002) Evaluation of transgenic tomato plants expressing an additional phytoene synthase in a fruit specific manner. Proceedings of the National Academy of Sciences of the United States of America, 99, 1092-1097.http://dx.doi.org/10.1073/pnas.241374598
76. Mehta, R.A., Cassol, T., Li, N., Ali, N., Handa, A.K. and Mattoo, A.K. (2002) Engineered polyamine accumulation in tomato enhances phytonutrient content, juice quality and vine life. Nature Biotechnology, 20, 613-618.
77. Premanchandra, B.R. (1986) Genetic regulation of carotene biosynthesis in selected tomato strains: Aspects of beta-carotene biosynthesis and B gene specificity. International Journal for Vitamin and Nutrition Research, 56, 35-43.
78. Markovic, Z., Zdravkovic, J., Mijatovic, M. and Damjanovic, M. (2002) Breeding potential of local tomato populations for beta-carotene and vitamin C. Acta Horticulturae, 579, 157-161.
79. Stommel, J.R., Abbott, J.A. and Saftner, R.A. (2005) USDA 02L1058 and 02L1059: Cherry tomato breeding lines with high fruit beta-carotene content. HortScience, 40, 1569-1570.
80. Andre, C.M., Ghislain, M., Berlin, P., Oufir, M., Herrera, M.R., Hoffman, L., Hausman, J.F., Larondelle, Y. and Evers, D. (2007) Andean potato cultivars (Solanum tuberosum L.) as a source of antioxidant and mineral micronutrients. Journal of Agricultural Food and Chemistry, 55, 366-378. http://dx.doi.org/10.1021/jf062740i
81. Brown, C.R., Culley, D., Yang, C.P., Durst, R. and Wrolstad, R. (2005) Variation of anthocyanin and carotenoid contents and associated antioxidant values in potato breeding lines. Journal of American Society Horticultural Science, 130, 174-180.
82. Reyes, L.F., MillerJr, J.C. and Zevallos, L.C. (2005) Antioxidant capacity, anthocyanins and total phenolics in purple and red-fleshed potato (Solanum tuberosum L.) genotypes. American Journal of Potato Research, 82, 271- 277. http://dx.doi.org/10.1007/BF02871956
83. Hanson, P.M., Yu, Y.R., Lin, S., Tsou, S.C.S., Lee, T.C., Wu, J., Jin, S., Gniffke, P. and Ledesma, D. (2004) Variation for antioxidant activity and antioxidants in subset of AVRDC—The World Vegetable Center Capsicum Core Collection. Plant Genetic Resources: Characterization and Utilization, 2, 153-166. http://dx.doi.org/10.1079/PGR200444
84. Dickson, M.H., Lee, C.Y. and Blamble, A.E. (1988) Orangecurd high carotene cauliflower inbreds, NY 156, NY 163, and 165. HortScience, 23, 778-779.

85. Simon, P.W. and Navazio, J.P. (1997) Early Orange Mass 400, Early Orange Mass 402, and Late Orange Mass, 404, High-carotene cucumber germplasm. HortScience, 32, 144- 145.
86. Simon, P.W., Pollak, L.M., Clevidence, B.A., Holden, J.M. and Haytowitz, D.B. (2009) Plant breeding for human nutritional quality. Plant Breeding Reviews, 31, 325-392.
87. de la Pena, R. and Hughes, J. (2007) Improving vegetable productivity in a variable and changing climate. International Crops Research Institute for the Semi-Arid Tropics, 4, 1-22.
88. Porch, T.G. and Jann, M. (2001) Effects of high-temperature stress on microsporogenesis in heat sensitive and heat tolerant genotype of Phaseolus vulgaris. Plant, Cell & Environment, 24, 723-731. http://dx.doi.org/10.1046/j.1365-3040.2001.00716.x
89. Hazra, P., Samsul, H.A., Sikder, D. and Peter, K.V. (2007) Breeding tomato (Lycopersicon esculentum Mill) resistant to high temperature stress. International Journal of Plant Breeding, 1, 31-40.
90. CGIAR. (2003) Applications of molecular biology and genomics to genetic enhancement of crop tolerance to abiotic stresses: A discussion document. Interim Science Council Secretariat, FAO, Rome.
91. AVRDC. (1990) Vegetable Production Training Manual. Asian Vegetable Research and Training Center, Shanhua.
92. Foolad, M.R. (2004) Recent advances in genetics of salt tolerance in tomato. Plant Cell, Tissue and Organ Culture, 76, 101-119. http:// dx.doi.org/10. 1023/B: TICU.0000007308.47608.88
93. Shannon, M.C. and Grieve, C.M. (1999) Tolerance of vegetable crops to salinity. Scientia Horticulturae, 78, 5-38. http://dx.doi.org/10.1016/S0304-4238(98)00189-7
94. Kuo, D.G., Tsay, J.S., Chen, B.W. and Lin, P.Y. (1982) Screening for flooding tolerance in the genus Lycopersicon. HortScience, 17, 76-78.
95. Foolad, M.R. (1999) Comparison of salt tolerance during seed germination and vegetative growth in tomato by QTL mapping. Genome, 42, 727-734. http://dx.doi.org/10.1139/g98-163
96. Rick, C.M. (1979) Potential improvement of tomato by controlled introgression of wild genes from wild species. Proc. Broadening the genetic base of crops. Wageningen, 3-7 July 1978, 167-173.
97. Dias, J.S. (2010) Impact of improved vegetable cultivars in overcoming food insecurity. Euphytica, 176, 125-136. http://dx.doi.org/10.1007/s10681-010-0237-5
98. Dias, J.S. (2010) 23.Impact of improved vegetable cultivars in overcoming food insecurity. In: Nath. P. and Gaddagimath, P.B., Eds., Horticulture and Livelihood Security. Scientific Publishers, Jodhpur, 303-339.
99. Cavendish, W. (2000) Empirical regularities in the povertyenvironment relationship of poor households: Evidence from Zimbabwe. World Development, 28, 1979-2003. http://dx.doi.org/10.1016/S0305-750X(00)00066-8
100. Weinberger, K. and Msuya, J. (2004) Indigenous vegetables in Tanzania: Significance and prospects. Technical Bulletin 31. AVRDC, Shanhua.
101. Gockowski, J., Mbazo'o, J., Mbah, G. and Moulende, T. (2003) African traditional leafy vegetables and the urban and peri-urban poor. Food Policy, 28, 221-235. http://dx.doi.org/10.1016/S0306-9192(03)00029-0
102. Dias, J.S. (2011) Biodiversity and vegetable breeding in the light of developments in intellectual property rights. In: Grillo, O. and Verona, G., Eds., Ecosystems Biodiversity, Chapter 17. INTECH Publ., Rijeka, 389-428.
103. Dias, J.S. (1989) The use of molecular markers in selection of vegetables. SECH, Actas de Horticultura, 3, 175- 181.
104. Dias, J.S. and Ortiz, R. (2013) New Strategies and Approaches for Improving Vegetables Cultivars. In: Nath, P., Ed., The Basics of Human Civilization-Food, Agriculture and Humanity, Volume-I-Present Scenario. Prem Nath Agricultural

Science Foundation (PNASF), Bangalore & New India Publishing Agency (NIPA), New Delhi, 349- 381.

105. Ryder, E.J. (2005) Intellectual property rights for plants. The case for a new law. Chronica Horticulturae, 45, 5-11.

106. Dias, J.S. (2012) Impact of the vegetable breeding industry and intellectual property rights in biodiversity and food security. In: Jones, A.M. and Hernandez, F.E., Eds., Food security: Quality, management, issues and economic implications. Nova Science Publishers Inc., Hauppauge, 57-86.

107. Dias, J.S. (2013) Impact of vegetable breeding industry and intellectual property rights in food security. In: Nath, P., Ed., The Basics of Human Civilization-Food, Agriculture and Humanity, Volume-I-Present Scenario. Prem Nath Agricultural Science Foundation (PNASF), Bangalore & New India Publishing Agency (NIPA), New Delhi, 173-198.

108. Hamilton, A., Lycett, G. and Grierson, D. (1990) Antisense gene that inhibits synthesis of the hormone ethylene in transgenic plants. Nature, 346, 284-287. http://dx.doi.org/10.1038/346284a0

109. Oeller P.W., Wong, L.M., Taylor, L.P., Pike, D.A. and Theologis, A. (1991) Reversible inhibition of tomato fruit senescence by antisense RNA. Science, 254, 437-439. http://dx.doi.org/10.1126/science.1925603

110. Klee, H.J., Hayford, M.B., Kretzmer, K.A., Barry, G.F. and Kishore, G.M. (1991) Control of ethylene synthesis by expression of a bacterial enzyme in transgenic tomato plants. Plant Cell, 3, 1187-1193.

111. Good, X., Kellogg, J.A., Wagoner, W., Langhoff, D., Matsumura, W. and Bestwick, R.K. (1994) Reduced ethylene synthesis by transgenic tomatoes expressing S-adenosylmethionine hydrolase. Plant Molecular Biology, 26, 781-790. http://dx.doi.org/10.1007/BF00028848

CHAPTER 11

A Preliminary Comparison of Antioxidants of Tomato Fruit Grown Under Organic and Conventional Systems

Apiradee Uthairatanakij[1,*], Sukanya Aiamla-or[1], Pongphen Jitareerat[1] and Ashariya Maneenoi[2]

[1]*Division of Postharvest Technology, King Mongkut's University of Technology Thonburi, Bangkok 10150, Thailand*

[2]*Lake Rajada Office, Adams Enterprises Ltd., Bangkok 10110, Thailand*

ABSTRACT

Organic farming is rapidly growing due to its perceived potential for producing higher nutritional quality. However, studies of organically- and conventionally-grown crops have not always shown differences between the systems. The objective of this research was to compare the antioxidant activities of organically-grown tomato to those from a conventional production system during postharvest cold storage. "Tub Tim Dang" tomato (*Solanum lycopersicum*L.) fruit were harvested at the breaker stage of maturity from both organic and conventional farms. Fruit were cold-stored at 10 °C for 20 days, and samples were collected at intervals to measure the activities of superoxide dismutase (SOD), catalase (CAT) and ascorbate peroxidase (APX) activities, and total antioxidant activity by the 2,2′-diphenyl-1-picrylhydrazyl (DPPH) and ferric reducing antioxidant power (FRAP) assays. The activities of SOD, CAT and APX of organic tomato fruit did not differ from those of conventional fruit during cold storage. In addition, there was no effect of production system on FRAP activity. In contrast, DPPH activity of organic tomato fruit was lower than conventional fruit through 10 days of cold storage, but it was higher at 15 and 20 days. These results indicated that organic production did not have a significant effects on these antioxidant traits of tomato.

Keywords: tomato; antioxidant enzyme; DPPH; FRAP

1. INTRODUCTION

Currently, many consumers are health conscious and aware of food quality, food safety and environmental protection, leading to an increasing demand for organic fresh products. Numerous studies on organic versus conventional crops have shown that organic fresh produce has significantly less chemical residue and greater nutritional content, including ascorbic acid, vitamin E, β-carotene and phenolics [1,2,3,4,5]. However, the differences in chemical composition between organic and conventional produce is dependent on cultural practices and environmental factors, such as time of harvest, fertilizer, water supply and soil properties [6,7,8]. Tomato (*Solanum lycopersicum* L.) fruit have high economic value and popularity in the consumer's diet, and are an important nutritional source of antioxidants [9]. Antioxidant content may change during the ripening of tomato, and this is reflected in changes in the antioxidant activity [10].

Reactive oxygen species (ROS) are generated during normal metabolism and also under stress conditions, and are eliminated by the antioxidant system [11]. Antioxidant enzymes, including superoxide dismutase (SOD), catalase (CAT) and ascorbate peroxidase (APX), and free radical scavengers provide defense mechanisms against ROS [12]. SOD is found in the chloroplast and cytosol, and is a key antioxidant enzyme that catalyzes conversion of superoxide radicals into dioxygen and hydrogen peroxide [13,14]. Afterwards, hydrogen peroxide is eliminated by CAT or APX. CAT is located in the peroxisome and removes hydrogen peroxide without reducing agents, while APX is located in the chloroplast and cytosol and requires ascorbate as an electron donor to eliminate hydrogen peroxide in the plant cell [15,16]. SOD, CAT and APX are therefore efficient antioxidant enzymes for detoxifying ROS because they are located in all compartments of the cell.

Some antioxidant levels are believed to be higher in organic produce that are activated by natural mechanisms of the plant defense systems against pests and diseases or other stress factors [17]. Therefore, the aim of this study was to compare total antioxidant activity and activity of the antioxidant enzymes SOD, CAT, and APX of organically- and conventionally-grown tomato fruit.

2. EXPERIMENTAL SECTION

2.1. Plant Preparation

"Tub Tim Dang" tomato seeds were planted in January 2013 under both organic and conventional management systems. Both farms were located in Phakthongchai district, Nakonratchasima province, Thailand (14°35′–15°00′ latitude, 101°45′–102°15′ longitude) approximately 10 km apart. The organic tomato plants were grown according to the standard protocol used by Adam Enterprises, Ltd. (certified by the United States Department of Agriculture, Bangkok, Thailand). Before transplanting the organic tomato seedlings, 40 MT·ha^{-1} of composted manure was applied to the soil. Then, fertilizer from fermented fish (100 mL·plant^{-1}) was applied three times a week until harvest. Conventional tomato fruit were grown on farms that followed Good Agricultural Practices. Before transplanting, 187.5 kg·ha^{-1} of NPK starter fertilizer (15-15-15) was applied to the soil. The plants were side-dressed with fertilizer 10 days after transplanting with 62.5 kg·hectare^{-1} of 46-0-0, with 187.5 kg·ha^{-1} of 15-15-15 NPK at 25 days, and with 187.5 kg·ha^{-1} of 13-13-21 NPK at 40 days. Irrigation was provided when it was necessary to avoid drought stress. Approximately 20 kg of fruit was harvested from both farms at the breaker stage, 55 days after transplanting. Fruit were washed with tap water, and placed in baskets before storage at 10 °C and 90% RH, where they started to ripen as indicated by fruit color change. At 0, 5, 10, 15 and 20 days of cold storage, about 30 fruit were randomly selected for a total approximate weight of 500 g and divided into three replicates. On the sampling day, tissues were frozen in liquid nitrogen and stored at −20 °C until analysis. From each replicate, 3 g of frozen sample was allocated for 2,2′-diphenyl-1-picrylhydrazyl (DPPH) and ferric reducing antioxidant power (FRAP) assays. For each analysis of enzyme activity, 5 g of frozen sample was used.

2.2. Total Antioxidant Activity

DPPH and FRAP assays followed the method of Brand-Williams et al. [18] and Benzie and Strain [19], respectively, using a spectrophotometer. A standard curve between 25 and 800 μM Trolox was linear, and results are expressed in Trolox equivalent antioxidant activity (TEAC)·g^{-1}·FW.

2.3. Enzyme Assay

For enzyme assays, 5 g of fruit tissue was homogenized in 10 mL cold extraction buffer (0.05 M sodium phosphate buffer, pH 7.8). The homogenate was filtered through a muslin cloth and centrifuged at 15,000 rpm for 20 min at 4 °C. The supernatant was used as a crude extract for SOD activity assays [20]. For the CAT and APX assays, 5 g of fruit tissue was homogenized in 10 mL of cold 0.1 M phosphate buffer at pH 7.0. The homogenate was filtered through a muslin cloth and centrifuged at 15,000 rpm at 4 °C for 20 min. The supernatant was used as a crude extract for CAT and APX activities [21].

The SOD activity was determined based on its capacity to inhibit the reduction of nitro-blue tetrazolium (NBT) by superoxide radicals generated by xanthine oxidase. One SOD unit was expressed as the amount of extract with percent inhibition of NBT reduction [22]. CAT activity was measured at 240 nm for decomposition of H_2O_2 (mM)·min^{-1} at 25 °C [23]. APX activity measured the decrease of ascorbic acid at 290 nm, and the result was expressed as mM ascorbic acid·min^{-1} at 28 °C [24]. Three replicates were performed for each main treatment, organic and conventional, and each reaction was repeated 3 times.

2.4. Statistical Analysis

Microsoft Excel (Redmond, WA, USA) was used to perform statistical analyses with a paired *t*-test to compare organic versus conventional means.

3. RESULTS

3.1. 2,2′-Diphenyl-L-Picrylhydrazyl (DPPH)

The DPPH assay indicated a significant difference between organic and conventional tomato fruit, with generally higher levels from the conventional system during the first 10 days of storage, while organic tomato fruit showed significantly higher levels than conventional fruit on days 15 and 20 (Figure 1A).

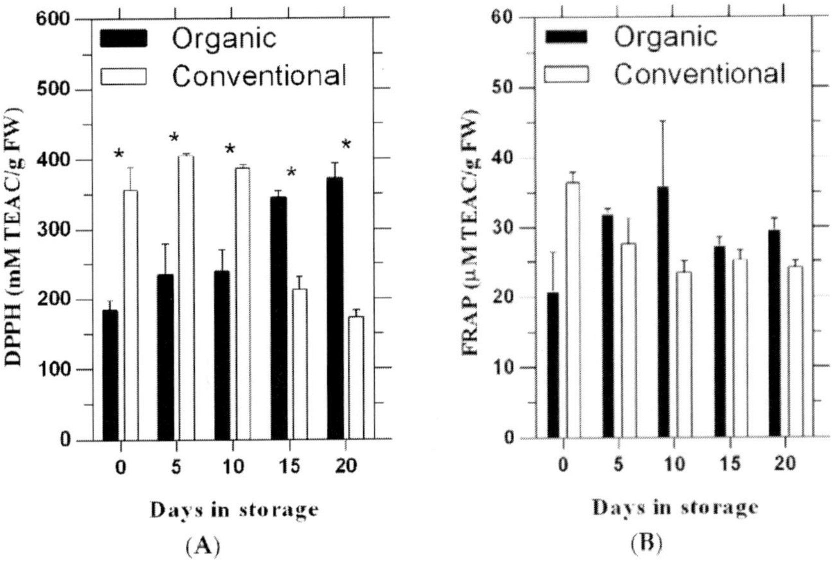

Figure 1. Changes in antioxidant activities by the DPPH (**A**) and FRAP (**B**) assays of tomato fruit during cold storage following production under organic and conventional production systems. Bars represent standard errors. * indicate the significant difference ($p \leq 0.05$).

3.2. Ferric Reducing Antioxidant Power (FRAP)

FRAP values from both systems varied from 21 to 36 $\mu M \cdot TEAC \cdot g^{-1} \cdot FW$ across storage days (Figure 1B), but the systems did not significantly differ.

3.3. The Activity of SOD, CAT and APX

SOD activity did not statistically differ between the two production systems across days of storage (Figure 2A). CAT activity did not differ between systems except at 15 days of storage when the organic system value was significantly greater than the conventional system (Figure 2B). No significant differences between systems were found for APX activity except that the organic value was significantly lower at 5 and higher at 10 days of storage than the conventional value (Figure 2C).

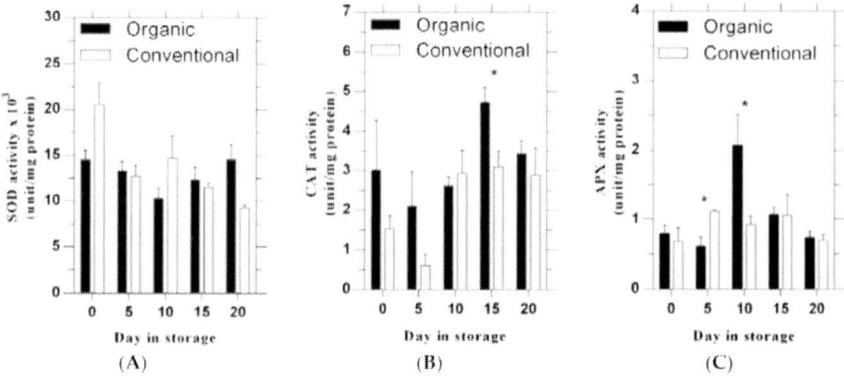

Figure 2. The activity of SOD (**A**), CAT (**B**) and APX (**C**) of tomato fruit during cold storage following production under organic and conventional production systems. Bars represent standard errors. * indicate the significant difference ($p \leq 0.05$).

4. DISCUSSION

Tomato fruit are a major source of phytochemicals and antioxidants such as lycopene, phenolics, flavonoids and vitamin C [24]. In this study, the antioxidant capacity of organic and conventional tomato during cold storage was compared using DPPH and FRAP radical scavenging capacity assays. These methods are based on electron transfer reactions which measure antioxidant reducing capacity [25]. Although DPPH activity was initially higher in conventional fruit through 10 days, it was lower than organic fruit at 15 and 20 days. In contrast, FRAP antioxidant activity did not differ between organic and conventional tomato fruit during storage. These two assays may have different sensitivities to the major antioxidants found in the fruit, as well as different reaction kinetics leading to the contrasting results [25]. Moreover, antioxidant effectiveness may depend on other factors such as polarity and solubility [26]. Some studies have reported significant correlations between the level of antioxidant activity and the phytochemical content of fruit [24,27]. Faller and Fialho [28] reported that organic tomato contained higher phytochemical contents than conventional tomato, but the assays in the present study gave conflicting results so it cannot be concluded that fruit from the two systems differed. No differences in DPPH assays were observed in other comparisons of organic versus conventional apple (*Malus x domestica*Borkh.), mango (*Mangifera indica* L.) or orange (*Citrus sinensis* L.) [28,29]. The antioxidant capacity in organic or conventional fruits can be affected by cultural practices and environmental factors

[4,5,9]. Riahi and Hdider [30] reported that organic fertilizer affected the antioxidant capacity in organic tomato. Similarly, Stracke et al. [31] demonstrated that the antioxidant capacity (FRAP, ORAC and TEAC assay) was not significant between organic and conventional apple.

SOD is a primary enzyme that eliminates superoxide radicles, converting them into dioxygen and hydrogen peroxide. Afterwards, hydrogen peroxide is catalyzed by CAT [12,13]. APX utilizes ascorbate as the electron donor reducing H_2O_2 to water, and prevents the accumulation of H_2O_2 to toxic levels under stress conditions [32]. In this study, the SOD activity did not significantly differ between fruit grown under the two cultural systems. In contrast, Oliveira et al. [33] reported that the activities of SOD and CAT were induced by organic production. In addition, no clear patterns in APX activity in organic or conventional tomato were evident.

5. CONCLUSIONS

The present study did not provide evidence that would indicate the superiority of nutritional composition of organically-grown tomato fruit in terms of antioxidant capacity by either the DPPH or FRAP assays, and also by antioxidant enzyme activities. Further work will be required to assess how differing maturities and harvest dates from each system affect these traits.

ACKNOWLEDGMENTS

The authors would like to thanks Adams Enterprises, Ltd. for providing organic tomatoes and Postharvest Innovation Center, Commission on higher Education, Bangkok, Thailand for scientific instruments.

REFERENCES

1. Zhao, X.; Carey, E.E.; Wang, W.; Rajashekar, C.B. Does organic production enhance phytochemical content of fruits and vegetables? Current knowledge and prospects for research. *Horttechnology* **2006**, *16*, 449–456.
2. Lombardi-Boccia, G.; Lucarini, M.; Lanzi, S.; Aguzzi, A.; Cappelloni, M. Nutrients and antioxidant molecules in yellow plums (*Prunus domestica* L.) from conventional and organic productions: A comparative study. *J. Agric. Food Chem.* **2004**, *52*, 90–94.

3. Parka, Y.S.; Imb, M.H.; Choic, J.H.; Yimc, S.H.; Leontowiczd, H.; Leontowiczd, M.; Suhaje, M.; Gorinstein, S. The effects of ethylene treatment on the bioactivity of conventional and organic growing "Hayward" kiwi fruits. *Sci. Hortic.* **2013**, *164*, 589–595.
4. Parka, Y.S.; Ham, K.S.; Kang, S.G.; Park, Y.K.; Namiesnik, J.; Leontowicz, H.; Leontowicz, M.; Ezra, A.; Trakhtenberg, S.; Gorinstein, S. Organic and conventional kiwi fruits, myths versus reality: Antioxidant, antiproliferative, and health effects. *J. Agric. Food Chem.* **2012**, *60*, 6984–6993.
5. Worthington, V. Nutritional quality of organic versus conventional fruits, vegetables, and grains. *J. Altern. Complement. Med.* **2001**, *7*, 161–173.
6. Winter, C.; Davis, S. Organic foods. *J. Food Sci.* **2006**, *71*, 117–124.
7. Mehdizadeh, M.; Darbandi, E.I.; Naseri-Rad, H.; Tobeh, A. Growth and yield of tomato (*Lycopersicon esculentum* Mill.) as influenced by different organic fertilizers. *Int. J. Agron. Plant Prod.* **2013**, *4*, 734–738.
8. Reganold, J.P.; Andrews, P.K.; Reeve, J.R.; Carpenter-Boggs, L.; Schadt, C.W.; Alldredge, J.R.; Ross, C.F.; Davies, N.M.; Zhou, J. Fruits and soil quality of organic and conventional strawberry agro ecosystems. *PLoS ONE* **2010**, *5*, e12346.
9. Toor, R.K.; Geoffrey, P.S.; Anuschka, H. Influence of different types of fertilizers on the major antioxidant components of tomatoes. *J. Food Compos. Anal.* **2006**, *19*, 20–27.
10. Cano, A.; Acosta, M.; Arnao, M.B. Hydrophilic and lipophilic antioxidant activity changes during on-vine ripening of tomatoes (*Lycopersicon esculentum* Mill.). *Postharvest Biol. Technol.* **2003**, *28*, 59–65.
11. Foyer, C.H.; Shigeoka, S. Understanding oxidative stress and antioxidant functions to enhance photosynthesis. *Plant Physiol.* **2011**, *155*, 93–100.
12. Rani, P.; Unni, M.; Karthikeyan, J. Evaluation of antioxidant properties of berries. *Indian J. Clin. Biochem.* **2004**, *19*, 103–110.
13. Bowler, C.; van Montagu, M.; Inze, D. Superoxide dismutase and stress tolerance. *Annu. Rev. Plant Biol.* **1992**, *43*, 83–116.
14. Abreu, I.A.; Cabelli, D.E. Superoxide dismutases—A review of the metal-associated mechanistic variations. *Biochim. Biophys. Acta* **2010**, *1804*, 263–274.
15. Willekens, H.; Chamnongpol, S.; Davey, M.; Schraudner, M.; Langebartels, C.; van Montagu, M.; Inzé, D.; van Camp, W. Catalase is a sink for H_2O_2 and is indispensable for stress defence in C_3 plants. *EMBO J.* **1997**, *16*, 4806–4816.
16. Kangasjärvi, S.; Lepistö, A.; Hännikäinen, K.; Piippo, M.; Luomala, E.M.; Aro, E.M.; Rintamäki, E. Diverse roles for chloroplast stromal and thylakoid-bound ascorbate peroxidases in plant stress responses. *Biochem. J.* **2008**, *412*, 275–285.
17. Brandt, K.; Mølgaard, J.P. Organic agriculture: Does it enhance or reduce the nutritional value of plant foods? *J. Sci. Food Agric.* **2001**, *81*, 924–931.
18. Brand-Williams, W.; Cuvelier, M.E.; Berset, C. Use of free radical method to evaluate antioxidant activity. *Lebensm. Wiss. Technol.* **1995**, *28*, 25–30.
19. Benzie, I.F.F.; Strain, J.J. The ferric reducing ability of plasma (FRAP) as a measure of "antioxidant power": The FRAP assay. *Anal. Biochem.* **1996**, *239*, 70–76.
20. Cakmak, I.; Marschner, H. Magnesium deficiency and high light intensity enhance activities of superoxide dismutase, ascorbate peroxidase, and glutathione reductase in bean leaves. *Plant Physiol.* **1992**, *98*, 1222–1227.
21. Nakano, Y.; Asada, K. Hydrogen peroxide is scavenged by ascorbate specific peroxidase in spinach chloroplast. *Plant Cell Physiol.* **1981**, *22*, 867–880.
22. Bartoli, C.G.; Simontacchi, M.; Guiamet, J.J.; Montadi, E.; Puntarulo, S. Antioxidant enzymes and lipid peroxidation during aging of *Chrysanthemum morifolium* RAM petals. *Plant Sci.* **1995**, *104*, 161–168.
23. Kato, M.; Shimizu, S. Chlorophyll metabolism in higher plants, VII. Chlorophyll degradation in senescence tobacco leaves, phenolic-dependent peroxidative degradation. *Can. J. Bot.* **1987**, *65*, 729–735.

24. Guil-Guerrero, J.L.; Rebolloso-Fuentes, M.M. Nutrient composition and antioxidant activity of eight tomato (*Lycopersicon esculentum*) varieties. *J. Food Compos. Anal.* **2009**, *22*, 123–129.
25. Huang, D.; Ou, B.; Prior, R.L. The chemistry behind antioxidant capacity assays. *J. Agric. Food Chem.* **2005**, *53*, 1841–1856.
26. Pokorny, J.; Yanishlieva, N.; Gordon, M. *Antioxidants in Food: Practical Application*; CRC Press: New York, NY, USA, 2001; p. 380.
27. Jin, P.; Wang, S.Y.; Wang, C.Y.; Zheng, Y. Effect of cultural system and storage temperature on antioxidant capacity and phenolic compounds in strawberries. *Food Chem.* **2011**, *124*, 262–270.
28. Faller, A.L.K.; Fialho, E. Polyphenol content and antioxidant capacity in organic and conventional plant food. *J. Food Compos. Anal.* **2010**, *23*, 561–568.
29. Bogs, J.; Bunning, M.; Stushnoff, C. Influence of biologically enhanced organic production on antioxidant and sensory qualities of (*Malus x domestica* Borkh. cv. Braeburn) apples. *Organ. Agric.* **2012**, *2*, 117–126.
30. Riahi, A.; Hdider, C. Bioactive compounds and antioxidant activity of organically grown tomato (*Solanum lycopersicum* L.) cultivars as affected by fertilization. *Sci. Hortic.* **2013**, *151*, 90–96.
31. Stracke Berenike, A.; Rüfer, C.E.; Bub, A.; Seifert, S.; Weibel, F.P.; Kunz, C.; Watzl, B. No effect of the farming system (organic/conventional) on the bioavailability of apple (*Malus domestica* Bork. cultivar Golden Delicious) polyphenols in healthy men: A comparative study. *Eur. J. Nutr.* **2010**, *49*, 301–310.
32. Salandanan, K.; Bunning, M.; Stonaker, F.; Kulen, O.; Kendall, P.; Stushnoff, C. Comparative analysis of antioxidant properties and fruit quality attributes of organically and conventionally grown melons (*Cucumis melo* L.). *Hortscience* **2009**, *44*, 1825–1832.
33. Oliveira, A.B.; Moura, C.F.H.; Gomes-Filho, E.; Marco, C.A.; Urban, L.; Miranda, M.R.A. The Impact of organic farming on quality of tomatoes is associated to increased oxidative stress during fruit development. *PLoS ONE* **2013**, *8*, e56354.

CHAPTER 12

Recent Progress in The Use of 'Omics Technologies in Brassicaceous Vegetables

Katja Witzel[1], Susanne Neugart[1], Silke Ruppel[1], Monika Schreiner[1], Melanie Wiesner[1]and Susanne Baldermann[1,2*]

[1]*Leibniz-Institute of Vegetable and Ornamental Crops Großbeeren/Erfurt e.V., Großbeeren, Germany*

[2]*Institute of Nutritional Science, University of Potsdam, Nuthetal, Germany*

ABSTRACT

Continuing advances in 'omics methodologies and instrumentation is enhancing the understanding of how plants cope with the dynamic nature of their growing environment. 'Omics platforms have been only recently extended to cover horticultural crop species. Many of the most widely cultivated vegetable crops belong to the genus *Brassica*: these include plants grown for their root (turnip, rutabaga/swede), their swollen stem base (kohlrabi), their leaves (cabbage, kale, pak choi) and their inflorescence (cauliflower, broccoli). Characterization at the genome, transcript, protein and metabolite levels has illustrated the complexity of the cellular response to a whole series of environmental stresses, including nutrient deficiency, pathogen attack, heavy metal toxicity, cold acclimation, and excessive and sub-optimal irradiation. This review covers recent applications of 'omics technologies to the brassicaceous vegetables, and discusses future scenarios in achieving improvements in crop end-use quality.

Keywords: genomics, transcriptomics, metabolomics, proteomics, crop, microbiomics

INTRODUCTION

Brassicaceous vegetables, which are cultivated worldwide, belong to the taxa *Brassica oleracea* (cabbage, broccoli, cauliflower, kale, Brussels sprouts, collard greens, savoy, kohlrabi, and Chinese kale) and *B. rapa*(turnip, mizuna, napa cabbage, cime di rapa, and turnip rape). Just as for crops generally, maintaining their productivity in the light of incipient climate change and the dynamically changing pest and pathogen community represents a major challenge for the biotechnologist and the plant breeder (Augustine et al., 2014). The importance of this class of vegetables lies not only in their contribution to the vitamin and mineral components of the human diet, but also in their beneficial effect on human health, which reflects the action of the glucosinolates, a group of secondary plant metabolites almost exclusively associated with this plant family (Higdon et al., 2007; Jeffery and Araya, 2009; Wu et al., 2013). The aliphatic glucosinolates (and their break down products) have attracted scientific attention (Reichelt et al., 2002; Halkier and Gershenzon, 2006; Sonderby et al., 2010; Wittstock and Burow, 2010); of particular note in this context is the anti-carcinogen isothiocyanate sulforaphane, the major break down product of glucoraphanin in broccoli (Fahey et al., 1997; Shapiro et al., 2001, 2006).

The use of 'omics technologies, which current gather information either at the DNA, RNA, protein or metabolite levels, can potentially provide a comprehensive picture of cellular physiology. DNA sequence (although not the epigenome) is largely independent of the growing environment, while the transcriptome, proteome and metabolome are all highly responsive (Figure 1). Tailoring a crop cultivar to a specific environment is the central challenge for the plant breeder, and reflects the reality that genotype on its own will not generally be sufficient to support a biotechnology-driven crop improvement program. Rather, a combination of one or more of the 'omics platforms is required to deliver reliable information. Such multiple 'omics data sets tend to be very large, as they represent a series of time point and/or treatment samplings; their analysis can only be addressed computationally (Bieda, 2012; Kohl et al., 2014; Schumacher et al., 2014). The demand for better data processing led to the release of new software packages freely available or provided by vendors (examples are given in Katajamaa and Oresic, 2007), however, the visualization of multi-omic data sets remains an important task for bioinformatics. Practical tools should create clear and meaningful visualizations without being overwhelmed by the complexity of the data sets (recent approaches are discussed by Gehlenborg et al., 2010.

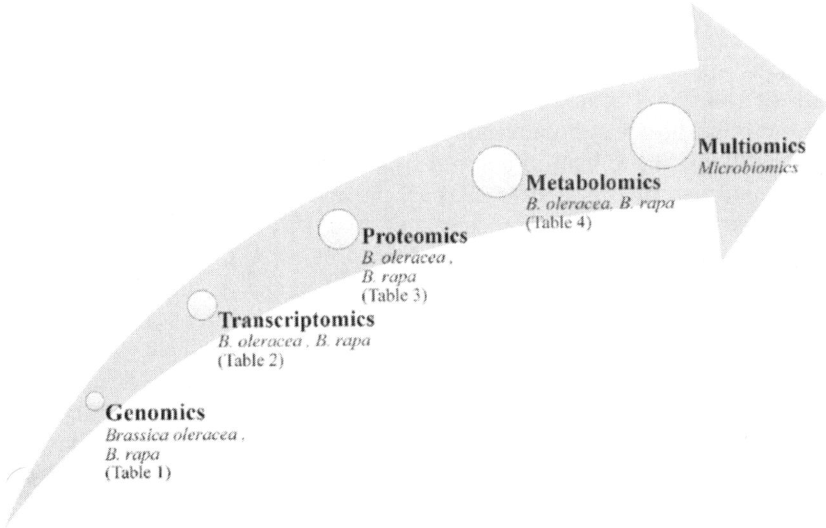

FIGURE 1. The application of 'omics technologies in brassicaceous vegetables. Arrow size reflect the complexity of the molecular data and the impact on phenotype.

The prior acquisition of a full genome sequence aids materially in the interpretation of such data. The genome sequence of the model plant *Arabidopsis thaliana*, a member of the *Brassicaceae* family, has been known for nearly 15 years (Kaul et al., 2000), and the full genome sequences of both *B. rapa* (Wang et al., 2011) and *B. oleracea* (Ayele et al., 2005; Liu et al., 2014b) have been published more recently. Substantial amounts of transcript-based data have been acquired for *B. oleracea* (Gao et al., 2014; Izzah et al., 2014; Kim et al., 2014).

Genomics

Genomic research has a great capability in speeding up breeding processes and several applications for crop improvement, through, e.g., marker-assisted selection and gene pyramiding. In case of brassicaceous vegetables, several populations have been generated to establish linkage maps using simple sequence repeat (SSR), amplified fragment length polymorphism (AFLP), nucleotide binding site or expressed sequence tag (EST) markers with the aim to genetically localize favorable traits by quantitative trait locus (QTL) analysis (summarized in Table 1). A survey of available online tools, covering mapping populations, linkage maps, gene sequences, and QTL, is presented by Li et al. (2013). Linkage maps were generated by crossing different genotypes of *B. oleracea* (Camargo et al., 1997; Iniguez-Luy et al., 2008) and *B. oleracea*var. *botrytis* (Gu et al., 2008). In the previous examples, a segregating offspring population was used to establish marker order and spacing. The same was shown for doubled haploid lines where offspring are homozygous (see Pink et al., 2008 for review). A doubled haploid population derived from crosses of two broccoli cultivars was genotyped by SSR and AFLP markers and QTL analysis identified loci for horticulturally important characteristics (Walley et al., 2012). Plant genetic resources have been characterized using AFLP markers to assess the huge genetic diversity present in gene banks, including Dutch and Italian *B. oleracea* (van Hintum et al., 2007; Maggioni et al., 2014) and Czech *B. oleracea* var. *capitata* accessions (Faltusova et al., 2011). A *B. rapa* collection consisting of 239 accessions was genotyped using SSR markers and subsequent association mapping identified two markers associated with flowering time (Zhao et al., 2010). Genetic basis of flowering time was also investigated in *B. rapa* by association mapping using natural variation and recombinant inbred lines (Lou et al., 2011). Phylogenetic relationships were established in *B. rapa* subspecies by AFLP markers (Takuno et al., 2007). The development of genetic maps from *Brassica* vegetables paved the way for QTL analysis and, together with accurate phenotyping, morphological, qualitative and yield traits were evaluated in different species. QTL mapping was conducted for eleven yield- and heading-related traits in doubled haploid Chinese cabbage lines and identified 46 main QTL (Liu et al., 2013b). In addition, 27 QTL were found for leaf and heading-related traits in a segregating population of Chinese cabbage (Ge et al., 2011). A segregating population of a cross between a Chinese cabbage with a turnip was used to identify loci related to morphological characteristics of the tap root (Lu et al., 2008). To investigate the genetic background of premature bolting under cold stress, a Chinese cabbage population was constructed based on crossing early and late bolting genotypes and QTL analysis identified 26 QTL (Wang et al., 2014c). Seven morphological traits were screened in a population derived from crossing a Chinese cabbage with a vegetable turnip and resulted in the detection of eight QTL (Kubo et al., 2010). A doubled haploid population of *B. oleracea* var. *capitata*served for detecting 13 QTL for heading-related traits (Lv et al., 2014). QTL analysis of a *B. oleracea* var. *italica* population, derived from a cross of a heat-sensitive and a heat-tolerant cultivar, identified AFLP markers correlated to floral development under heat stress (Lin et al., 2013). A high-density linkage map was established for a *B. oleracea* population segregating for carotenoid concentration in florets and three carotenoid QTL were found (Brown et al., 2014). Authors applied the *B. napus* SNP array and presented a 96% coverage of the *B. oleracea* genome. Localizing underlying factors affecting another major pigment, chlorophyll, in a Chinese cabbage population revealed five QTL for chlorophyll a and five QTL for chlorophyll b content (Ge et al., 2012).

Vegetables with high nutrient use efficiency are also developed in order to reduce fertilizer application. Association mapping of *B. oleracea* accessions identified some QTL related to potassium concentration in shoot and those were tested using substitution lines (White et al., 2010). The gain of knowledge on the genetic localization of favorable traits is transferred to breeding new lines through marker-assisted selection. Accumulating three major loci for clubroot (*Plasmodiophora brassicae*) resistance genes resulted in the development of Chinese cabbage lines with strengthened resistance (Matsumoto et al., 2012).

TABLE 1. Published genomics analyses in the brassicaceous vegetables.

Brassica species	Population	Plant material	Reference
B. oleracea	Segregating population	Cross of contrasting geno-types	Camargo et al. (1997)
B. oleracea	Segregating population	Cross of contrasting genotypes	Sebastian et al. (2000)
B. oleracea	Segregating population	Cross of contrasting genotypes	Gao et al. (2007)
B. oleracea	Gene bank accessions	–	van Hintum et al. (2007)
B. oleracea	Doubled-haploid	Cross of contrasting genotypes	Inguez-Luy et al. (2008)
B. oleracea	Gene bank accessions	DNA methylation	Salmon et al. (2008)
B. oleracea	TILLING	EMS mutagenesis	Himelblau et al. (2009)
B. oleracea	Gene bank accessions	–	White et al. (2010)
B. oleracea	Segregating population	Cross of contrasting genotypes	Brown et al. (2014)
B. oleracea		DNA methylation	Parkin et al. (2014)
B. oleracea var. *botrytis*	Segregating population	Cross of contrasting genotypes	Gu et al. (2008)
B. oleracea var. *capitata*	Gene bank accessions	–	Faltusova et al. (2011)
B. oleracea var. *capitata*	Doubled-haploid	Cross of contrasting genotypes	Wang et al. (2012)
B. oleracea var. *capitata*	Doubled-haploid	Cross of contrasting genotypes	Lv et al. (2014)
B. oleracea var. *italica*	Doubled-haploid	Cross of contrasting genotypes	Walley et al. (2012)
B. oleracea var. *italica*	Segregating population	Cross of contrasting genotypes	Lin et al. (2013)
B. oleracea, *B. rupestris*	Wild and cultivated populations, hybrid population		Maggioni et al. (2014)
B. rapa	Wild and cultivated populations	–	Takuno et al. (2007)
B. rapa	Segregating population	Cross of contrasting genotypes	Lu et al. (2008)
B. rapa	Segregating population	Cross of contrasting genotypes	Kubo et al. (2010)
B. rapa	Wild and cultivated populations	–	Takuno et al. (2007)
B. rapa	Segregating population	Cross of contrasting genotypes	Lu et al. (2008)
B. rapa	Segregating population	Cross of contrasting genotypes	Kubo et al. (2010)
B. rapa	TILLING	EMS mutagenesis	Stephenson et al. (2010)
B. rapa	Gene bank accessions	–	Zhao et al. (2010)
B. rapa	Segregating population, gene bank accessions	Cross of contrasting genotypes	Lou et al. (2011)
B. rapa	Segregating population	Cross of contrasting genotypes	Bagheri et al. (2012)
B. rapa	Hypermethylated population	DNA methylation	Amoah et al. (2012)
B. rapa	Segregating population	Cross of contrasting genotypes	Wang et al. (2014c)
B. rapa ssp. *pekinensis*	Segregating population	Cross of contrasting genotypes	Ge et al. (2011)
B. rapa ssp. *pekinensis*	Segregating population	Cross of contrasting genotypes	Ge et al. (2012)
B. rapa ssp. *pekinensis*	Marker-assisted selection	Gene pyramiding	Matsumoto et al. (2012)
B. rapa ssp. *pekinensis*	Doubled-haploid	Cross of contrasting genotypes	Liu et al. (2013b)

Advances in next generation sequencing technologies enabled surveying genotype-phenotype-relationships with the highest resolution to date (Wei et al., 2013; Varshney et al., 2014). A high-density linkage map was derived from the whole genome shotgun sequence of *B. oleracea* var. *capitata*, based on 1,227 genetic markers (Wang et al., 2012). The availability of *B. rapa* genome sequence aided in the generation of a linkage map for a recombinant inbred line descending from a vegetable leafy and an yellow sarson oilseed genotype of *B. rapa* (Bagheri et al., 2012). Recently, genomes of all three sequences *Brassicas* were compared to develop SSR markers and a total of 115,869, 185,662, and 356,522 primer pairs were designed from *B. rapa*, *B. oleracea*, and *B. napus*, respectively (Shi et al., 2014). EST markers were developed based on the sequenced transcriptomes of two cabbage lines susceptible or tolerant to black rot disease, demonstrating the feasibility especially for species without a reference genome sequence (Izzah et al., 2014).

One of the main strategies in reverse genetics is Targeting Induced Local Lesions IN Genomes (TILLING) and numerous applications are shown in functional genomics of model plants and crops (Kurowska et al., 2011; Chen et al., 2014). Random point mutations are generated by chemical mutagenesis and high-throughput screening for SNPs in the target gene isolates mutants with loss-of-function or gain-of-function phenotypes. Currently, two TILLING populations are established for brassicaceous vegetables, *B. oleracea*[1] (Himelblau et al., 2009) and *B. rapa*[2] (Stephenson et al., 2010) and both are freely accessible.

Next generation sequencing and AFLP mapping of methylation polymorphism allowed determining the DNA cytosine methylation status and, hence, enabled tackling the epigenome to understand gene expression variation (Niederhuth and Schmitz, 2014). The level of genome methylation was assessed in 30 *B. oleracea* populations and lines, and then related to phenotypic variability (Salmon et al., 2008). The *B. oleracea* genome was correlated to the leaf transcriptome and methylome and provided insights into polyplody events (Parkin et al., 2014). A chemically induced hypermethylated *B. rapa* population was developed recently and can serve for epiallele discovery (Amoah et al., 2012).

Transcriptomics

Transcriptomics is based on the characterization and quantification of RNAs present in a given plant, organ, tissue, or cell. A particular attraction of assaying transcription is that it presages gene expression and so has the potential to provide a link between genotype and phenotype. Rapid developments in the relevant technologies now allow for the design of very large-scaled experiments which can in principle capture and enumerate every transcript present in a given biological sample (summarized in Table 2). The early transcriptomic platforms were built around the concept of immobilizing transcripts in an array, which was used to capture the mRNA content of a sample on the basis of nucleotide sequence complementarity. A more sophisticated form of microarray replaced transcripts with 20–70 nt long oligonucleotides designed to target known gene sequences (Knudsen, 2004). In both cases, the protocol requires bathing the array in a solution of labeled RNA extracted from the test sample. Hybridization of the probe at a given site on the array is detected by the fluorescence emitted by the labeled probe. Typically, two parallel experiments need to be run, in which one sample is derived from a plant exposed to a certain treatment while the other is derived from a plant which has not been exposed (control). Alternatively, genetic contrasts (such as wild type *vs.* a mutant, wild type *vs.* a transgenic) can replace the treatment contrast. Many of such experiments have been reported in a range of plant species (for example, Rabbani et al. (2003) in rice, Yu and Setter (2003) in maize, Lee and Yun (2006) in pepper). The quantity of such data collected from the brassicaceous vegetable species is relatively limited, partly perhaps because their full genome sequences have only very recently been acquired. The close phylogenetic relationship between *Brassica* spp. and *A. thaliana* has encouraged a number of attempts to use an *A. thaliana*microarray to analyze *Brassica* spp. transcriptomes. With the recent acquisition of *Brassica* spp. genome sequences, a number of *Brassica*-specific microarrays are finally becoming available – these include the *B. rapa* 24 K oligo microarray and the Agilent *Brassica* microarray, which is able to detect the transcription of >27,000 unigenes in a range of *Brassica* spp. The genetic closeness of *Brassica* spp. and *A. thaliana* has allowed large numbers of *Brassica* unigenes to be assigned a function, but nevertheless, substantial numbers appear to encode either *Brassica*-specific proteins or represent non-coding RNA (Trick et al., 2009).

TABLE 2. Published transcriptomic analyses in the brassicaceous vegetables.

Brassica species	Plant organ/developmental stage	Study objective	Methodology	Reference
B. oleracea	Flowers	Male sterility	Microarray	Kang et al. (2008)
B. oleracea var. *italica*	Seeds and sprouts	Glucosinolate metabolism	RNA-seq analysis	Gao et al. (2014)
B. rapa	Flowers	Self-incompatibility	Microarray	Kwun et al. (2004)
B. rapa	plants	Abiotic stress	Microarray	Lee et al. (2008)
B. rapa	Seedlings, roots, petioles, leaves, flowers	Comparative analysis	RNA-seq	Kim et al. (2012)
B. rapa	Seedlings, roots, leaves, petiole	Abiotic stress	RNA-Seq	Devisetty et al. (2014)
B. rapa ssp. *pekinensis*	Seedlings, root tips	Cold stress	Microarray	Yang et al. (2005)
B. rapa ssp. *pekinensis*	Leaves	Heat shock transcription factor	Comparative genomic analysis	Song et al. (2014)
B. rapa ssp. *pekinensis*	Seedlings	Plant-microbe interactions	RNA-seq	Sun (2014)
B. rapa ssp. *rapa*	Seedlings	Etiolation	RNA-seq	Zhou et al. (2014)

A cDNA microarray constructed from a Chinese cabbage (*B. rapa* ssp. *pekinensis*) pistil-specific cDNA library approach was used by Kwun et al. (2004) to investigate the effect of CO_2 on self-incompatibility, and a similar strategy was followed by Yang et al. (2005) to identify genes up-regulated by low temperature stress. Meanwhile Lee et al. (2008) synthesized an oligo-based microarray from the sequences of 24,000 *B. rapa* unigenes to characterize the response to low temperature, salinity and drought. Based on the identification of candidate genes using a 2 × 104 k *Brassica* microarray to compare the gene expression of untreated and methyl jasmonate treated pak choi sprouts, Wiesner et al. (2013a, 2014) were able to identify the homologues genes involved in the synthesis of 1-methoxyindole-3-ylmethyl glucosinolate.

The major alternative to microarray technology relies on amplification rather than on hybridization. The most commonly used method, termed RNA-seq, is designed to generate a *de novo* assembly of the transcriptome. This has largely replaced the earlier approach called "serial analysis of gene expression" (SAGE). In general, a population of RNA (total or fractionated) is converted to a library of cDNA fragments with adaptors attached to one or both ends. Each molecule, with or without amplification, is then sequenced in a high-throughput manner to obtain short sequences from one end (single-end sequencing) or both ends (pair-end sequencing). The reads are typically 300–400 bp, depending on the DNA-sequencing technology used (Wang et al., 2009). In principle, any high-throughput sequencing technology can be used for RNA-Seq. The data generated from RNA-Seq experiments are in the form of an absolute frequency of each transcript identified, which obviates the need for using a reference sequence. For this reason, it can provide a greater level of insight and accuracy than microarray analysis can (Marioni et al., 2008; T'Hoen et al., 2008). In broccoli, the RNA-Seq approach has been taken to describe the transcription of genes involved in glucosinolate metabolism (Gao et al., 2014). Of the nearly 20 k unigenes recovered, more than 2,500 appeared to be differentially transcribed in a contrast between the seed and the seedling, and a large proportion of these were putative transcription factors. Curiously, the up-regulation of candidate glucosinolate synthesis genes was negatively correlated with the glucosinolate content of the germinating seedling; a possible explanation offered by the authors was that the high transcript abundance of an *TGG2* orthologue (encoding a myrosinase) had the effect of degrading aliphatic glucosinolates (Gao et al., 2014).

Small RNAs, which range in length from 19 to 25 nt (most are either 21 nt or 24 nt long), are ubiquitous in eucaryotic cells. Extensive transcriptome sequencing has revealed the presence of a large number of such RNAs, which cannot be translated into a protein, but rather act as an important class of regulators, especially in the context of plant development, signal transduction, metabolism and the response to biotic and abiotic stress (Ellendorff et al., 2009; Sun, 2012). The abundance of a particular species of a class of small RNAs, referred to as microRNAs (miRNAs), varies from plant species to plant species, from developmental stage to developmental stage and from tissue to tissue (Carrington and Ambros, 2003; Bartel, 2004; Kenan-Eichler et al., 2011). The transcriptional response of Chinese cabbage to infection by *Erwinia carotovora* ssp. *carotovora* included a marked alteration in the abundance of certain small RNAs (Sun, 2014). In turnip (*B. rapa* ssp. *rapa*), it appeared that various miRNAs are involved in the regulation of plant growth, development and differentiation in the absence of light (Zhou et al., 2014). A number of *Brassica* species are polyploids, the genomes of which have been markedly altered by the polyploidization event (Kenan-Eichler et al., 2011). Some of these changes have been associated with the activity of small RNAs. Kim et al. (2012) identified 412 distinct miRNAs in *B. rapa*, of which 216 were novel. The same study identified 29 novel miRNAs which were only found in the flower.

Proteomics

The acquisition of the complete genome sequence of a growing number of plant species[3] along with their transcriptomes continues apace. However, much of the physiology of the cell is determined by gene products (particularly, but not exclusively, proteins) rather than by nucleic acid. A proteomic analysis seeks to characterize the full protein complement present in a particular organism, organ, tissue, or cell. Initial attempts to derive this were based on two dimensional gel electrophoresis (2-DE), as described by O'Farrell (1975). Besides 2-DE, current proteomics platforms exploit more versatile LC (liquid chromatography)-based methods (Rabilloud et al., 2010; Matros et al., 2011), coupled with mass spectrometry (MS). Complicating the analysis, and unlike the nucleic acids, protein molecules are subject to a range of functional modification, such as phosphorylation, glycosylation, and acetylation (Cox and Mann, 2011). Nevertheless, proteomic approaches have been successfully applied to a number of plant species to study various developmental processes and environmental adaptation (see reviews by Barkla et al., 2013; Vanderschuren et al., 2013; Zhuang et al., 2014). Kehr and Buhtz (2011) have provided a summary of the literature regarding proteomic analyses in *Brassica* spp., a list dominated by experiments based on oilseed rape. In the following, we give emphasis to reports on how proteomic techniques are applied to catalog and characterize special, developmental-driven or environmentally induced alterations in the proteome of *brassicaceous* vegetables (summarized in Table 3).

Using a conventional 2-DE approach, Alam et al. (2013) succeeded in detecting about 1,300 distinct low abundance leaf proteins in Chinese cabbage. A characterization of the proteomic content of the xylem sap of broccoli and oilseed rape (and some non- brassicaceous) undertaken by Buhtz et al. (2004) showed little evidence of any species specificity,

TABLE 3. Published proteomic analyses in the brassicaceous vegetables.

Brassica species	Plant organ/developmental stage	Study objective	Methodology	Reference
Members of *Brassicaceae* family	Etiolated leaves of seedlings	Genotypic variation	2-DE	Marquès et al. (2001)
B. oleracea var. *capitata*, *B. rapa* var. *pekinensis*, synthesized *B. napus*	Seeds	Genotypic variation	2-DE and MS/MS	Hossain et al. (2014)
B. oleracea	Xylem saps	Mapping	1-DE and MS/MS	Buhtz et al. (2004)
B. oleracea	Vacuoles	Mapping	MS/MS	Schmidt et al. (2007)
B. oleracea	Xylem saps	Mapping, *N*-glycosylation	MS/MS	Ligat et al. (2011)
B. oleracea	Phloem tissues of stem	Mapping	MS/MS	Anstead et al. (2013)
B. oleracea var. *alboglabra*	Leaves, stems	Genotypic variation	2-DE	Albertin et al. (2005)
B. oleracea var. *alboglabra*	Leaves	Genotypic variation	2-DE and MS	Kong et al. (2010)
B. oleracea var. *botrytis italica*	Leaves, stems	Genotypic variation	2-DE	Albertin et al. (2005)
B. oleracea var. *botrytis*	Mitochondria	Mapping	2-DE and MS	Pawlowski et al. (2005)
B. oleracea var. *capitata*	Floral head	Effect of transgene	2-DE and MS/MS	Liu et al. (2011)
B. oleracea L. var *capitata*	Leaves	Cropping systems	2-DE and MS/MS	Nawrocki et al. (2011)
B. oleracea var. *capitata*	Stigma	Self-incompatibility	2-DE	Zeng et al. (2012)
B. oleracea var. *capitata*	Floral head	Effect of transgene	2-DE and MS/MS	Liu et al. (2013a)
B. oleracea var. *capitata*	Seeds	Genotypic variation	2-DE and MS/MS	Hossain et al. (2014)
B. oleracea var. *capitata*	Floral heads	Effect of transgene	2-DE and MS/MS	Liu et al. (2014a)
B. oleracea var. *italica*	Xylem sap	Salinity	2-DE and MS/MS	Fernandez-Garcia et al. (2011)
B. oleracea var. *italica*	Floral heads	Sodium selenate nutrition	2-DE and MS/MS	Sepúlveda et al. (2013)
B. rapa cvs. Natsumaki, Takamaru	Roots	Pathogen interaction	2-DE	Kaido et al. (2007)
B. rapa	Leaves	Genotypic variation	2-DE and MS	Kong et al. (2010)
B. rapa ssp. *chinensis*	Leaves	Effect of di-*n*-butyl phtalate	2-DE and MS/MS	Liao et al. (2006)
B. rapa var. *chinensis*	Leaves	Effect of di-*n*-butyl phtalate	2-DE and MS/MS	Liao et al. (2009)
B. rapa ssp. *chinensis*	Leaves	Mapping	2-DE	Alam et al. (2013)
B. rapa ssp. *chinensis*	Leaves	Pathogen interaction	2-DE and MS/MS	Sun et al. (2014)
B. rapa ssp. *chinensis*	Pistil	Self-incompatibility	2-DE and MS/MS	Wang et al. (2014a)
B. rapa ssp. *chinensis*	Leaves and roots	Nitrogen nutrition	2-DE and MS	Zhuang et al. (2014)
B. rapa ssp. *pekinensis*	Seedlings	Effect of atrazine	2-DE and MS	Li et al. (2008)
B. rapa var. *pekinensis*	Seeds	Genotypic variation	2-DE and MS/MS	Hossain et al. (2014)

while De Bernonville et al. (2014) were able to define the proteomic impact in the xylem sap induced by infection of *B. oleracea* with the pathogen *Xanthomonas campestris*pv. *campestris*. Ligat et al. (2011) expanded the *B. oleracea* data set by subjecting the xylem sap proteome to LC-MS/MS and the *N*-glycoproteome by prior enrichment with concanavalin A affinity chromatography and LC-MS/MS. Most of the ~200 proteins identified proved to be involved in cell wall-related carbohydrate metabolism, although a number of oxido-reductases and proteases were also revealed. A proteomic analysis of broccoli tissue enriched with phloem allowed the identification of 379 proteins, some of which were structural and others associated with the biotic and/or abiotic stress response (Anstead et al., 2013). In cauliflower, an analysis of isolated tonoplast membranes revealed 102 tonoplast integral and 214 peripheral proteins (Schmidt et al., 2007). A further study in this species was able to show that its mitochondrial proteome was highly similar to that of *A. thaliana*, and included at least 51 proteins involved in the generation of ATP, protein folding and protein transport (Pawlowski et al., 2005).

The host-pathogen interaction is important in brassicaceous vegetables, as in all crops. A root proteomic comparison between a pair of turnip cultivars showing a contrasting reaction to the causal agent of clubroot disease was able to demonstrate the presence of certain proteins in the resistant cultivar which were absent from the profile of the sensitive one (Kaido et al., 2007). Sun et al. (2014) tracked the proteomic response of Chinese cabbage resulting from infection with downy mildew (*Hyaloperonospora parasitica*), and were able to identify some 91 proteins which altered in their abundance; some of these were involved in ethylene signaling. When the abundances of specific transcripts and their gene products were compared, a correlation could be established for only about half of the 33 genes investigated; a finding which was interpreted as implying that many of the early events in the host's resistance response involved post-translational modification.

Abiotic stress is a further major production constraint. Analysis of the effect of the photosystem II inhibitor atrazine on the chloroplast proteome of Chinese cabbage has identified at least nine proteins as being differentially expressed; some of these were related to isoprene or protein synthesis (Li et al., 2008). Investigations of the impact of the industrial soil pollutant phthalate ester on either pak choi (Liao et al., 2006) or Chinese cabbage (Liao et al., 2009) have shown that the pollutant was responsible for a different set of up- or down-regulated proteins, even though the two species are so closely related to one another. Among the effects of salinity on broccoli are the down-regulation of several proteins involved in cell wall metabolism and the up-regulation of known plant defense-related proteins (Fernandez-Garcia et al., 2011). A third common source of abiotic stress is that imposed by nutrient supply. Wang et al. (2014b) have characterized the response of hydroponically grown pak choi to the supply of nitrogen in different forms. The plants responded to glycine supply by altering the expression of proteins involved in protein processing, amino acid metabolism and redox homeostasis. A similar experiment involving broccoli investigated the effect of supplying the plants with a non-limiting concentration of the essential trace element selenium (Sepúlveda et al., 2013), resulting in an enhanced accumulation in the inflorescence of a number of proteins associated with the general stress response, while that of pathogen defense-related proteins was reduced. The effect of an organic farming regime on the cabbage proteome has shown that the cropping system had a measurable effect on the accumulation of proteins involved in glycolysis, the Krebs cycle, amino acid metabolism and various detoxification processes (Nawrocki et al., 2011).

A plants protein composition is a key determinant of a species' and a cultivar's identity, and differences in specific plant features are better reflected by their proteome than by either their genome or transcriptome. Thus proteome variation has been suggested as providing a viable basis for the design of molecular markers able to accelerate breeding (Witzel et al., 2011). Marquès et al. (2001) made comparisons between the 2-DE profiles of various *Brassica* species in an attempt to establish the nature of the genetic relationship between them. PCR-based markers for the genomes A (*B. rapa*) and B (*B. oleracea*) were obtained by Kong et al. (2010) by aligning the proteomic profiles of young leaves. From some of those differentially expressed proteins, PCR-based markers were developed successfully. The effect of the polyploidization process on the proteome was assessed by Albertin et al. (2005), who made comparisons between the leaf and stem proteomes of haploid, diploid and tetraploid sprouting broccoli, and Chinese kale. The analysis concluded that the two different tissues were responsible for 40% of the proteomic variation observed, with only 10% ascribed to ploidy level differences. With the aim to identify proteins responsible for an improved vernalisation trait, the seed proteome of a somatic hybrid line was compared with those of its parent lines *B. oleracea* (cabbage) and *B. rapa* (Chinese cabbage) that showed low or high seed production during warm winter conditions, respectively (Hossain et al., 2014).

Understanding the basis of self-incompatibility has been a long-standing problem in the *Brassica* genus. A proteomic approach addressing this issue in Chinese cabbage was taken by Wang et al. (2014a), who analyzed the protein complement of the pistils from self-incompatible and compatible plants; in some cases, protein abundance was consistent with that of the relevant transcript, particularly with respect to those involved in energy metabolism, protein synthesis and the defense against stress. Recently, for a more cell-type specific proteomics approach related to self-incompatibility, a protocol for the analysis of cabbage stigma proteins was established (Zeng et al., 2012).

The market value of vegetables is greatly affected by their visual appearance and their post-harvest shelf life. As part of a transgenic-based attempt to prolong the marketability of broccoli, 2-DE was used to compare the proteomes of the inflorescences of wild type and transgenic lines produced to express delayed senescence (Liu et al., 2011). The conclusion was that a number of stress response related proteins and chaperones accumulated in the transgenic lines, while the build-up of 1-aminocyclopropane-1-carboxylate oxidase was only noted in the wild type. When the effect of exogenously supplying cytokinin (N^6-benzylaminopurine) was studied at the proteomic level, Liu et al. (2013a) detected only a minor degree of overlap with the transgenic contrast, an observation which was taken to suggest that the molecular basis of the delayed post-harvest senescence induced by exogenous cytokinin differed from that induced by endogenous cytokinin. Of the up-regulated stress response related proteins detected in the transgenic material, 17 proved to be heat-stable following cooking and were therefore considered to represent potential allergens (Liu et al., 2014a).

Metabolomics

Metabolomics, defined as the systematic study of the by-products of cellular processes, is an ambitious field, as the number of chemically distinct molecules involved in a typical plant sample is estimated to be at least 100,000 (Wink, 1988), each varying with respect to their abundance (De Vos et al., 2007). A number of these compounds are of physiological importance and/or represent dietary sources of antioxidants and other health enhancing compounds; most importantly with respect to vegetable crops, they make a major contribution to product color and flavor. Often these latter quality aspects rely not on individual compounds, but rather on specific mixtures (Ferguson and Schlothauer, 2012). The state-of-the-art analytical platforms employed for metabolomic analyses are a sophisticated combination of nuclear magnetic resonance spectroscopy (NMR), MS, gas chromatography (GC), LC, and capillary electrophoresis (CE). Changes in the primary and secondary metabolite pool have been characterized using a GC-MS approach (Hummel et al., 2013), while CE-MS is more suitable for characterizing the products associated with the central metabolism (Yang et al., 2012). NMR is appropriate for the targeting of phenolic compounds, carbohydrates, organic acids and amino acids, and of particular interest in the context of brassicaceous vegetables, glucosinolates (Table 4). Electrospray ionization (ESI), conducted either in positive (resulting in protonated species) or in negative (deprotonated species) mode is used for the analysis of semi-polar metabolites. Together, these technologies can identify and quantify a wide range of primary and secondary metabolites. For LC-MS and NMR applications, samples are typically extracted in alcohol and not derivatized. In contrast, GC-MS separations require a derivatization step (Kopka et al., 2005). Particularly high mass resolution can be achieved by the use of time-of-flight (TOF)-MS devices, and additional information can be obtained from the MS/MS fragmentation patterns and UV-VIS spectra. Recent software developments have improved the capacity to recognize different metabolites. Profiles based on mass, retention time and signal amplitude provide the data required for filtering biomarkers. Typically a data processing pipeline can be divided into two steps: data processing (filtering, feature detection, alignment, and normalization) and data analysis (algorithm selection, training, evaluation, and model examination; further explanation see Boccard et al., 2010).

Nevertheless, metabolite identification in non-targeted approaches in the absence of authentic reference compounds remains difficult. Various platforms and spectral databases are available online[4].

TABLE 4. Published non-targeted metabolomic analyses in the brassicaceous vegetables.

Brassica species	Plant organ/ developmental stage	Study objective	Methodology	Compounds	Reference
B. oleracea collection	Leaves	Factor identification for thermal degradation of glucosinolates	HPLC-MS LC-MS	Semi-polar	Hennig et al. (2014)
B. oleracea var. *botrytis* collection	Inflorescences	Genotypic variation of colored cultivars	GC-TOF-MS	Primary and secondary compounds*	Park et al. (2013)
B. oleracea var. *capitata*	Leaves	Discrimination of conventional and organic farming	LC-MS	Semi-polar	Mie et al. (2014)
B. oleracea var. *capitata*	Leaves	Genotypic variation	LC-MS	Phenolic compounds	Park et al. (2014)
B. oleracea var. *italic*	Inflorescences	Genotypic variation, selenium treatment, crop management	LC-MS	Phenolic compounds	Luthria et al. (2008)
B. rapa collection	Leaves	Response to pre-harvest bacterial contamination	^{1}H NMR	Polar	Jahangir et al. (2008b)
B. rapa collection	Leaves	Genotypic variation	^{1}H NMR	Polar	Abdel-Farid et al. (2007)
B. rapa collection	Leaves	Effect of metal-ion treatment	^{1}H NMR	Polar	Jahangir et al. (2008a)
B. rapa collection	Leaves	Response to pre-harvest fungal infection	^{1}H NMR	Polar	Abdel-Farid et al. (2009)
B. rapa collection	Leaves	Genotypic variation	LC-MS	Semi-polar	Del Carpio et al. (2011)
B. rapa var. *pekinensis*	Leaves	Discrimination by geographical areas	^{1}H NMR	Polar	Kim et al. (2013a)
B. rapa var. *rapa*, *Raphanus sativa*	Leaves	Ontogenetic variation	NMR, HPLC	Primary and secondary compounds*	Muhammad et al. (2014)

**Includes both a non-targeted and a targeted approach.*

The metabolome of *Brassica* spp. as that of all plants, is influenced by both genotype (Luthria et al., 2008) and developmental stage (Abdel-Farid et al., 2007), as well as by the plant's growing environment (Hall, 2006; Weimer and Slupsky, 2013). Metabolomic analyses in the brassicaceous vegetables have included both non-targeted (Table 4) and targeted (Table 5) approaches (Cevallos-Cevallos et al., 2009). The former have concern a range of both secondary (polyphenols and carotenoids) and primary metabolites (Hall, 2006), while the latter focused heavily on the identification and quantification of glucosinolates (Ramautar et al., 2006). The glucosinolates are a unique component of the brassicaceous vegetable metabolome (Verkerk et al., 2009). These compounds contain a sulfur-linked β-D-glucopyranose moiety and an amino acid-derived side chain, and fall into three definable classes: the aliphatic, indole and aromatic glucosinolates. They are structurally highly diverse, and within the *Brassicaceae* are mostly dominated by the aliphatic glucosinolates (Halkier and Gershenzon, 2006). Subsequently, aliphatic glucosinolates can be subdivided into straight- or branched-chain alk(en)yl glucosinolates with or without a hydroxy group and into a large group of those that contain an additional sulfur atom with various oxidation levels in the side chain (see for detailed structural overview, Hanschen et al., 2014). The glucosinolates are not known to be bioactive, but their hydrolysis products provide deterrence against feeding (Textor and Gershenzon, 2009), and their presence in the human diet has been associated with both health benefits (Vig et al., 2009) and taste (Beck et al., 2014).

TABLE 5. Published targeted metabolomic analyses in the brassicaceous vegetables, featuring glucosinolates and their break-down products.

Brassica species	Plant organ/Developmental stage	Study objective	Methodology	Reference
B. carinata	Leaves	(Onto)genetic variation, effect of water supply	HPLC-MS	Schreiner et al. (2009)
B. oleracea	Leaves	Genotypic variation	LC-MS	Rochfort et al. (2008)
B. oleracea var. *italica*	Inflorescences	Effect of nitrogen and sulfur supply	HPLC-MS	Schonhof et al. (2007a)
B. oleracea var. *italica*	Inflorescences	Effect of CO_2	HPLC-MS	Schonhof et al. (2007b)
B. oleracea var. *italica*	Sprouts	Effect of water supply and aphid treatment	HPLC-MS	Khan et al. (2011)
B. oleracea var. *italica*	Sprouts	Effect of UV-B	HPLC-MS	Mewis et al. (2012)
B. rapa	Roots	Effect of nitrogen and sulfur supply	HPLC-MS	Shumin et al. (2007)
B. rapa	Leaves	Effect of fungal infection	NMR	Abdel-Farid et al. (2010)
B. rapa	Leaves	Effect of fertilization	GC-MS	Okazaki et al. (2012)
B. rapa	Roots	Genotypic variation	LC-PDA-QTOF-MS	Lee et al. (2013)
B. rapa ssp. *chinensis*	Leaves, sprouts	Effect of methyl jasmonate	HPLC-MS	Wiesner et al. (2013a,b)
B. rapa ssp. *pekinensis*	Leaves, seeds	Organ differentiation	LC-ESI-MS, LC-UV	Hong et al. (2011)
B. rapa ssp. *rapa*	Leaves, roots, root exudates	Effect of methyl jasmonate, salicylic acid	HPLC-MS	Schreiner et al. (2011)

Gas chromatography/TOF-MS has been used to characterize the phytochemical diversity of the florets in cauliflower cultivars varying in inflorescence pigmentation (Park et al., 2013). A ^{1}H-NMR platform proved informative for discriminating between a set of Chinese cabbage cultivars from China and Korea: differences with respect to both primary and secondary metabolites were uncovered, although insufficient attention has yet to be given to the effect of the growing environment (Kim et al., 2013a). A non-targeted strategy was chosen by Mie et al. (2014) to reveal the impact of organic farming on the metabolome of cabbage, and led to the conclusion that this approach has potential for the authentication of organic products. The exposure of *B. carinata* seedlings to lithium ions induced a marked effect on the plant's lipid and phenolic content: sinapic acid esters and chloroplast lipids were replaced by benzoate derivatives, resveratrol and oxylipins (Li et al., 2009). When GC/TOF-MS was used to discriminate between cabbage cultivars varying with respect to their resistance to feeding by diamondback moth (*Plutella xylostella*) caterpillars, Kim et al. (2013b) were able to show that the levels of glycolic acid, quinic acid, inositol, fumaric acid, glyceric acid, trehalose, shikimic acid, and aspartic acid were all very different.

Non-targeted metabolic approaches have also been applied to the study of processing quality in brassicaceous vegetables. Changes in the content of certain flavonols have been associated with the thermal degradation of glucosinolates in *B. oleracea* (Hennig et al., 2014). Several analytical techniques have been used to assess the impact of industrial processing on the phytochemical composition of vegetable purées. In broccoli, most of the metabolites present in the purée were degraded by heating, including various health-related and flavor compounds, vitamins, carotenoids, flavonoids, glucosinolates, and oxylipins (Lopez-Sanchez et al., 2015). Post-harvest storage temperature also has a profound effect on the metabolome.

For example, in radish *(Raphanus sativus*), the content of phenylpropanoids, flavonoids, and glucosinolates decreased with storage time (Jahangir, 2010).

Microbiomics

Microbiomics refers to the application of 'omics technologies to the microbial community associated with the plant spermosphere, endosphere, rhizosphere, and phyllosphere. The plant microbiome has attracted the focus of several researchers in recent years (reviewed in Vorholt, 2012; Bakker et al., 2013; Bulgarelli et al., 2013), as it has an intimate interaction with the plant, affecting both its metabolism and physiology. Studies of the plant microbiome have relied to date either on the *in vitro* culture of the relevant micro-organisms or on the recognition of species identity from PCR amplicons derived from variable regions of their genome (such as the 16S rDNA locus) followed by sequencing, microarray analysis or various electrophoretic procedures (Rastogi et al., 2013; Turner et al., 2013; Muller and Ruppel, 2014). Metaproteomics approaches can complement such data sets to provide functional information relevant to the microbiome (Knief et al., 2012). As many vegetable are consumed as a fresh product, their associated microbiome will enter the human digestive system. Exemplarily, the lettuce (Hunter et al., 2010; Rastogi et al., 2012; Jackson et al., 2013; Williams and Marco, 2014) and spinach (Leff and Fierer, 2013; Lopez-Velasco et al., 2013) microbiomes are dependent on the host genotype, the cultivation conditions and the tissue sampled. The impact of commensal microbial diversity on crop disease has been investigated in a range of oilseed rape cultivars infected with the fungal pathogen *Verticillium longisporum* (Graner et al., 2003). Bacterial strains isolated from tolerant or susceptible plant cultivars showed biocontrol activity against the fungus and imply a possible application in crop protection. Meanwhile De Campos et al. (2013)used next generation sequencing to show how the components of the oilseed rape microbiome depended on the developmental stage of the plant. Links et al. (2014) demonstrated that the seeds of *B. juncea, B. napus,* and *B. rapa*, were associated with a core seed microbial population of geographically and ecologically different crops. The structure of the *Brassica* species -specific microbiome has been shown to be sensitive to the composition of the host's secondary metabolites (Ruppel et al., 2008). Hunter et al. (2014) established a close relationship between the composition of the root exudate and the structure of the rhizosphere microbiome. Certain microbial strains had a significant impact on the capacity of the rhizosphere to solubilize phosphorus (Schilling et al., 1998; Deubel et al., 2000). Determining the functional interaction between the microbiome's composition and plant metabolic activities is still at an early stage of development (Bakker et al., 2013).

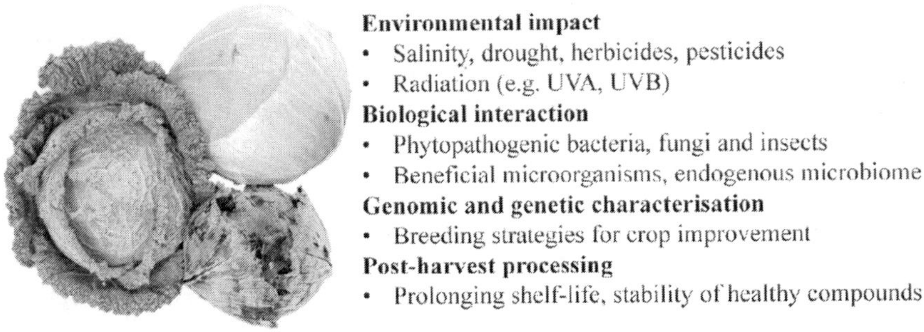

FIGURE 2. Potential applications of 'omics technologies for brassicaceous vegetables improvement.

Concluding Remarks

Many refinements in all the 'omics technologies can be expected to come on-stream over the next few years, and these will provide exciting new avenues of research in the brassicaceous vegetables. Some of the topics likely to feature in the near future are flux analysis, the identification of informative breeding markers for the selection of beneficial bioactives and enhanced responses to both abiotic and biotic stress (Figure 2). The increasing volume of transcriptomic data acquired continues to provide a wealth of information relevant to obtaining a detailed picture of key regulatory mechanisms and pathways active in the plant. Improving the capacity to identify the components of the proteome is a fast moving area of technology development. So far, proteomic analysis in the brassicaceous vegetables has relied almost exclusively on 2-DE separation, even though the more sophisticated chromatography and MS platforms have been quite widely used in other (particularly model) plant species; these technologies are of particular value in illuminating post-translational modifications. The acquisition of full genome sequences for some (or all) of the brassicaceous vegetables can only aid in facilitating proteomic analysis in this groups of crop plants. Gaining some control over the composition of metabolites will surely provide a powerful means of varietal improvement and the optimization of post-harvest technology.

The marriage of 'omics technology and plant breeding should trigger a major efficiency breakthrough in crop improvement, and also generate novel opportunities in the fields of nutrigenomics (systems approach to understand the relationship between diet and health) and microbiomics.

Although substantial progress has already been made both in the data acquisition and its interpretation, many technical and scientific challenges remain before the level of understanding of the workings of a complex biological system such as a plant rises above the current state. Global information of different hierarchies (genomics, transcriptomics, proteomics, and metabolomics) need to be integrated using mathematical and statistical methods to refine existing knowledge and make new discoveries.

There are successful studies were systems perspectives revealed correlations that had not been suggested by classical approaches in *Arabidopsis,* (Mochida and Shinozaki, 2011), tomato (Enfissi et al., 2010), maize (Casati et al., 2011), and others. Elucidating principles that govern relationships between biological instances will help to improve plant functions also in brassicaceous vegetables; i.e., stress-resistant plants and high- yield cultivars.

REFERENCES

1. Abdel-Farid, B. B., Kim, H. K., Choi, Y. H., and Verpoorte, R. (2007). Metabolic characterization of *Brassica rapa* leaves by NMR spectroscopy. *J. Agric. Food Chem.* 55, 7936–7943. doi: 10.1021/jf071294b
2. Abdel-Farid, I. B., Jahangir, M., Mustafa, N. R., Van Dam, N. M., Van Den Hondel, C., Kim, H. K., et al. (2010). Glucosinolate profiling of *Brassica rapa* cultivars after infection by *Leptosphaeria maculans* and *Fusarium oxysporum*. *Biochem. Syst. Ecol.* 38, 612–620. doi: 10.1016/j.bse.2010.07.008
3. Abdel-Farid, I. B., Jahangir, M., Van Den Hondel, C., Kim, H. K., Choi, Y. H., and Verpoorte, R. (2009). Fungal infection-induced metabolites in *Brassica rapa*. *Plant Sci.* 176, 608–615. doi: 10.1016/j.plantsci.2009.01.017
4. Alam, I., Sharmin, S. A., Kim, K. H., Kim, Y. G., Lee, J. J., and Lee, B. H. (2013). An improved plant leaf protein extraction method for high resolution two-dimensional polyacrylamide gel electrophoresis and comparative proteomics. *Biotech. Histochem.* 88, 61–75. doi: 10.3109/10520295.2012.729863
5. Albertin, W., Brabant, P., Catrice, O., Eber, F., Jenczewski, E., Chèvre, A.-M., et al. (2005). Autopolyploidy in cabbage (*Brassica oleracea*L.) does not alter significantly the proteomes of green tissues. *Proteomics* 5, 2131–2139. doi: 10.1002/pmic.200401092
6. Amoah, S., Kurup, S., Lopez, C. M. R., Welham, S. J., Powers, S. J., Hopkins, C. J., et al. (2012). A hypomethylated population of *Brassica rapa* for forward and reverse epi-genetics. *BMC Plant Biol.* 12:193. doi: 10.1186/1471-2229-12-193
7. Anstead, J., Hartson, S., and Thompson, G. (2013). The broccoli (*Brassica oleracea*) phloem tissue proteome. *BMC Genomics* 14:764. doi: 10.1186/1471-2164-14-764
8. Augustine, R., Arya, G. C., Nambiar, D. M., Kumar, R., and Bisht, N. C. (2014). Translational genomics in *Brassica* crops: challenges, progress, and future prospects. *Plant Biotechnol. Rep.* 8, 65–81. doi: 10.1007/s11816-013-0298-8
9. Ayele, M., Haas, B. J., Kumar, N., Wu, H., Xiao, Y. L., Van Aken, S., et al. (2005). Whole genome shotgun sequencing of *Brassica oleracea* and its application to gene discovery and annotation in *Arabidopsis*. *Genome Res.* 15, 487–495. doi: 10.1101/gr.3176505

10. Bagheri, H., El-Soda, M., Van Oorschot, I., Hanhart, C., Bonnema, G., Jansen-Van Den Bosch, T., et al. (2012). Genetic analysis of morphological traits in a new, versatile, rapid-cycling *Brassica rapa* recombinant inbred line population. *Front. Plant Sci.* 3:183. doi: 10.3389/fpls.2012.00183
11. Bakker, P. A. H. M., Berendsen, R. L., Doornbos, R. F., Wintermans, P. C. A., and Pieterse, C. M. J. (2013). The rhizosphere revisited: root microbiomics. *Front. Plant Sci.* 4:165. doi: 10.3389/fpls.2013.00165
12. Barkla, B. J., Castellanos-Cervantes, T., Diaz De León, J. L., Matros, A., Mock, H.-P., Perez-Alfocea, F., et al. (2013). Elucidation of salt stress defense and tolerance mechanisms of crop plants using proteomics - current achievements and perspectives. *Proteomics* 13, 1885–1900. doi: 10.1002/pmic.201200399
13. Bartel, D. P. (2004). MicroRNAs: genomics, biogenesis, mechanism, and function. *Cell* 116, 281–297. doi: 10.1016/S0092-8674(04)00045-5
14. Beck, T. K., Jensen, S., Bjoern, G. K., and Kidmose, U. (2014). The masking effect of sucrose on perception of bitter compounds in *Brassica* vegetables. *J. Sens. Stud.* 29, 190–200. doi: 10.1111/joss.12094
15. Bieda, M. (2012). Kepler for 'Omics Bioinformatics. *Procedia Comput. Sci.* 9, 1635–1638. doi: 10.1016/j.procs.2012.04.180
16. Boccard, J., Veuthey, J.-L., and Rudaz, S. (2010). Knowledge discovery in metabolomics: an overview of MS data handling. *J. Sep. Sci.*33, 290–304. doi: 10.1002/jssc.200900609
17. Brown, A. F., Yousef, G. G., Chebrolu, K. K., Byrd, R. W., Everhart, K. W., Thomas, A., et al. (2014). High-density single nucleotide polymorphism (SNP) array mapping in *Brassica oleracea*: identification of QTL associated with carotenoid variation in broccoli florets. *Theor. Appl. Genet.* 127, 2051–2064. doi: 10.1007/s00122-014-2360-5
18. Buhtz, A., Kolasa, A., Arlt, K., Walz, C., and Kehr, J. (2004). Xylem sap protein composition is conserved among different plant species. *Planta* 219, 610–618. doi: 10.1007/s00425-004-1259-9
19. Bulgarelli, D., Schlaeppi, K., Spaepen, S., Van Themaat, E. V. L., and Schulze-Lefert, P. (2013). Structure and functions of the bacterial microbiota of plants. *Annu. Rev. Plant Biol.* 64, 807–838. doi: 10.1146/annurev-arplant-050312-120106
20. Camargo, L. E. A., Savides, L., Jung, G., Nienhuis, J., and Osborn, T. C. (1997). Location of the self-incompatibility locus in an RFLP and RAPD map of *Brassica oleracea*. *J. Hered.* 88, 57–60. doi: 10.1093/oxfordjournals.jhered.a023057
21. Carrington, J. C., and Ambros, V. (2003). Role of microRNAs in plant and animal development. *Science* 301, 336–338. doi: 10.1126/science.1085242
22. Casati, P., Campi, M., Morrow, D. J., Fernandes, J. F., and Walbot, V. (2011). Transcriptomic, proteomic and metabolomic analysis of UV-B signaling in maize. *BMC Genomics* 12:321. doi: 10.1186/1471-2164-12-321
23. Cevallos-Cevallos, J. M., Reyes-De-Corcuera, J. I., Etxeberria, E., Danyluk, M. D., and Rodrick, G. E. (2009). Metabolomic analysis in food science: a review. *Trends Food Sci. Technol.* 20, 557–566. doi: 10.1016/j.tifs.2009.07.002
24. Chen, L., Hao, L. G., Parry, M. A. J., Phillips, A. L., and Hu, Y. G. (2014). Progress in TILLING as a tool for functional genomics and improvement of crops. *J. Integr. Plant Biol.* 56, 425–443. doi: 10.1111/jipb.12192
25. Cox, J., and Mann, M. (2011). "Quantitative, high-resolution proteomics for data-driven systems biology," in *Annual Review of Biochemistry*, Vol. 80, eds R. D. Kornberg, C. R. H. Raetz, J. E. Rothman, and J. W. Thorner (Palo Alto, CA: Annual Reviews), 273–299.
26. De Bernonville, T., Albenne, C., Arlat, M., Hoffmann, L., Lauber, E., and Jamet, E. (2014). "Xylem sap proteomics," in *Plant Proteomics*, eds J. V. Jorrin-Novo, S. Komatsu, W. Weckwerth, and S. Wienkoop (New York, NY: Humana Press), 391–405. doi: 10.1007/978-1-62703-631-3_28
27. De Campos, S. B., Youn, J. W., Farina, R., Jaenicke, S., Junemann, S., Szczepanowski, R., et al. (2013). Changes in root bacterial communities associated to two different

development stages of canola (*Brassica napus* L. var oleifera) evaluated through next-generation sequencing technology. *Microb. Ecol.* 65, 593–601. doi: 10.1007/s00248-012-0132-9

28. De Vos, R. C. H., Moco, S., Lommen, A., Keurentjes, J. J. B., Bino, R. J., and Hall, R. D. (2007). Untargeted large-scale plant metabolomics using liquid chromatography coupled to mass spectrometry. *Nat. Protoc.* 2, 778–791. doi: 10.1038/nprot.2007.95
29. Del Carpio, D. P., Basnet, R. K., De Vos, R. C. H., Maliepaard, C., Visser, R., and Bonnema, G. (2011). The patterns of population differentiation in a *Brassica rapa* core collection. *Theor. Appl. Genet.* 122, 1105–1118. doi: 10.1007/s00122-010-1516-1
30. Deubel, A., Gransee, A., and Merbach, W. (2000). Transformation of organic rhizodepositions by rhizosphere bacteria and its influence on the availability of tertiary calcium phosphate. *J. Plant Nutr. Soil Sci.* 163, 387–392. doi: 10.1002/1522-2624(200008)163:4<387::AID-JPLN387>3.0.CO;2-K
31. Devisetty, U. K., Covington, M. F., Tat, A. V., Lekkala, S., and Maloof, J. N. (2014). Polymorphism identification and improved genome annotation of *Brassica rapa* through deep RNA sequencing. *G3 (Bethesda)* 4, 2065–2078. doi: 10.1534/g3.114.012526
32. Ellendorff, U., Fradin, E. F., De Jonge, R., and Thomma, B. P. (2009). RNA silencing is required for *Arabidopsis* defence against *Verticillium* wilt disease. *J. Exp. Bot.* 60, 591–602. doi: 10.1093/jxb/ern306
33. Enfissi, E. M. A., Barneche, F., Ahmed, I., Lichtlé, C., Gerrish, C., Mcquinn, R. P., et al. (2010). Integrative transcript and metabolite analysis of nutritionally enhanced DE-ETIOLATED1 downregulated tomato fruit. *Plant Cell* 22, 1190–1215. doi: 10.1105/tpc.110.073866
34. Fahey, J. W., Zhang, Y. S., and Talalay, P. (1997). Broccoli sprouts: an exceptionally rich source of inducers of enzymes that protect against chemical carcinogens. *Proc. Natl. Acad. Sci. U.S.A.* 94, 10367–10372. doi: 10.1073/pnas.94.19.10367
35. Faltusova, Z., Kucera, L., and Ovesna, J. (2011). Genetic diversity of *Brassica oleracea* var. capitata gene bank accessions assessed by AFLP. *Electr. J. Biotechnol.* 14, 10. doi: 10.2225/vol14-issue3-fulltext-4
36. Ferguson, L. R., and Schlothauer, R. C. (2012). The potential role of nutritional genomics tools in validating high health foods for cancer control: broccoli as example. *Mol. Nutr. Food Res.* 56, 126–146. doi: 10.1002/mnfr.201100507
37. Fernandez-Garcia, N., Hernandez, M., Casado-Vela, J., Bru, R., Elortza, F., Hedden, P., et al. (2011). Changes to the proteome and targeted metabolites of xylem sap in *Brassica oleracea* in response to salt stress. *Plant Cell Environ.* 34, 821–836. doi: 10.1111/j.1365-3040.2011.02285.x
38. Gao, J. J., Yu, X. X., Ma, F. M., and Li, J. (2014). RNA-Seq analysis of transcriptome and glucosinolate metabolism in seeds and sprouts of broccoli (*Brassica oleracea* var. italic). *PLoS ONE* 9:e88804. doi: 10.1371/journal.pone.0088804
39. Gao, M. Q., Li, G. Y., Yang, B., Qiu, D., Farnham, M., and Quiros, C. (2007). High-density *Brassica oleracea* linkage map: identification of useful new linkages. *Theor. Appl. Genet.* 115, 277–287. doi: 10.1007/s00122-007-0568-3
40. Ge, Y., Ramchiary, N., Wang, T., Liang, C., Wang, N., Wang, Z., et al. (2011). Mapping quantitative trait loci for leaf and heading-related traits in Chinese cabbage (*Brassica rapa* L. ssp pekinesis). *Hortic. Environ. Biotechnol.* 52, 494–501. doi: 10.1007/s13580-011-0031-x
41. Ge, Y., Wang, T., Wang, N., Wang, Z., Liang, C., Ramchiary, N., et al. (2012). Genetic mapping and localization of quantitative trait loci for chlorophyll content in Chinese cabbage (*Brassica rapa* ssp pekinensis). *Sci. Hortic.* 147, 42–48. doi: 10.1016/j.scienta.2012.09.004
42. Gehlenborg, N., O'Donoghue, S. I., Baliga, N. S., Goesmann, A., Hibbs, M. A., Kitano, H., et al. (2010). Visualization of omics data for systems biology. *Nat. Methods* 7, S56–S68. doi: 10.1038/nmeth.1436

43. Graner, G., Persson, P., Meijer, J., and Alstrom, S. (2003). A study on microbial diversity in different cultivars of *Brassica napus* in relation to its wilt pathogen, *Verticillium longisporum*. *FEMS Microbiol. Lett.* 224, 269–276. doi: 10.1016/S0378-1097(03)00449-X
44. Gu, Y., Zhao, Q. C., Sun, D. L., and Song, W. Q. (2008). A genetic linkage map based on AFLP and NBS markers in cauliflower (*Brassica oleracea* var. botrytis). *Bot. Stud.* 49, 93–99.
45. Halkier, B. A., and Gershenzon, J. (2006). Biology and biochemistry of glucosinolates. *Annu. Rev. Plant Biol.* 57, 303–333. doi: 10.1146/annurev.arplant.57.032905.105228
46. Hall, R. D. (2006). Plant metabolomics: from holistic hope, to hype, to hot topic. *New Phytol.* 169, 453–468. doi: 10.1111/j.1469-8137.2005.01632.x
47. Hanschen, F. S., Lamy, E., Schreiner, M., and Rohn, S. (2014). Reactivity and stability of glucosinolates and their breakdown products in foods. *Angew. Chem. Int. Ed.* 53, 11430–11450. doi: 10.1002/anie.201402639
48. Hennig, K., De Vos, R. C. H., Maliepaard, C., Dekker, M., Verkerk, R., and Bonnema, G. (2014). A metabolomics approach to identify factors influencing glucosinolate thermal degradation rates in *Brassica* vegetables. *Food Chem.* 155, 287–297. doi: 10.1016/j.foodchem.2014.01.062
49. Higdon, J. V., Delage, B., Williams, D. E., and Dashwood, R. H. (2007). Cruciferous vegetables and human cancer risk: epidemiologic evidence and mechanistic basis. *Pharmacol. Res.* 55, 224–236. doi: 10.1016/j.phrs.2007.01.009
50. Himelblau, E., Gilchrist, E. J., Buono, K., Bizzell, C., Mentzer, L., Vogelzang, R., et al. (2009). Forward and reverse genetics of rapid-cycling *Brassica oleracea*. *Theor. Appl. Genet.* 118, 953–961. doi: 10.1007/s00122-008-0952-7
51. Hong, E., Kim, S. J., and Kim, G. H. (2011). Identification and quantitative determination of glucosinolates in seeds and edible parts of Korean Chinese cabbage. *Food Chem.* 128, 1115–1120. doi: 10.1016/j.foodchem.2010.11.102
52. Hossain, M. M., Li, X., Evans, I. H., and Rahman, M. A. (2014). A proteomic analysis of seed proteins expressed in a *Brassica* somatic hybrid and its two parental species. *J. Plant Tissue Cult. Biotechnol.* 24, 11–26. doi: 10.3329/ptcb.v24i1.19187
53. Hummel, J., Strehmel, N., Bölling, C., Schmidt, S., Walther, D., and Kopka, J. (2013). "Mass spectral search and analysis using the golm metabolome database," in *The Handbook of Plant Metabolomics*, eds W. Weckwerth and G. Kahl (Weinheim: Wiley-VCH Verlag GmbH & Co. KGaA), 321–343. doi: 10.1002/9783527669882.ch18
54. Hunter, P. J., Hand, P., Pink, D., Whipps, J. M., and Bending, G. D. (2010). Both leaf properties and microbe-microbe interactions influence within-species variation in bacterial population diversity and structure in the lettuce (*Lactuca* species) Phyllosphere. *Appl. Environ. Microbiol.* 76, 8117–8125. doi: 10.1128/AEM.01321-10
55. Hunter, P. J., Teakle, G., and Bending, G. D. (2014). Root traits and microbial community interactions in relation to phosphorus availability and acquisition, with particular reference to *Brassica*. *Front. Plant Sci.* 5:27. doi: 10.3389/fpls.2014.00027
56. Iniguez-Luy, F. L., Voort, A. V., and Osborn, T. C. (2008). Development of a set of public SSR markers derived from genomic sequence of a rapid cycling *Brassica oleracea* L. genotype. *Theor. Appl. Genet.* 117, 977–985. doi: 10.1007/s00122-008-0837-9
57. Izzah, N. K., Lee, J., Jayakodi, M., Perumal, S., Jin, M., Park, B. S., et al. (2014). Transcriptome sequencing of two parental lines of cabbage (*Brassica oleracea* L. var. capitata L.) and construction of an EST-based genetic map. *BMC Genomics* 15:149. doi: 10.1186/1471-2164-15-149
58. Jackson, C. R., Randolph, K. C., Osborn, S. L., and Tyler, H. L. (2013). Culture dependent and independent analysis of bacterial communities associated with commercial salad leaf vegetables. *BMC Microbiol.* 13:274. doi: 10.1186/1471-2180-13-274

59. Jahangir, M. (2010). *Stress Response and Health Affecting Compounds in Brassicaceae*. Ph.D. thesis, Department of Pharmacognosy and Metabolomics, Institute of Biology (IBL), Faculty of Science, Leiden University, Leiden.
60. Jahangir, M., Abdel-Farid, I. B., Choi, Y. H., and Verpoorte, R. (2008a). Metal ion-inducing metabolite accumulation in *Brassica rapa*. *J. Plant Physiol.* 165, 1429–1437. doi: 10.1016/j.jplph.2008.04.011
61. Jahangir, M., Kim, H. K., Choi, Y. H., and Verpoorte, R. (2008b). Metabolomic response of *Brassica rapa* submitted to pre-harvest bacterial contamination. *Food Chem.* 107, 362–368. doi: 10.1016/j.foodchem.2007.08.034
62. Jeffery, E. H., and Araya, M. (2009). Physiological effects of broccoli consumption. *Phytochem. Rev.* 8, 283–298. doi: 10.1007/s11101-008-9106-4
63. Kaido, M., Ishikawa, T., and Hori, H. (2007). Comparative proteomic analysis of the resistant response in *Brassica rapa* root culture to the clubroot disease agent *Plasmodiophora brassicae*. *Bull. Facul. Agric. Niigata Univ.* 60, 67–71.
64. Kang, J., Zhang, G., Bonnema, G., Fang, Z., and Wang, X. (2008). Global analysis of gene expression in flower buds of Ms-cd1 *Brassica oleracea* conferring male sterility by using an *Arabidopsis* microarray. *Plant Mol. Biol.* 66, 177–192. doi: 10.1007/s11103-007-9261-9
65. Katajamaa, M., and Oresic, M. (2007). Data processing for mass spectrometry-based metabolism. *J. Chromatogr. A* 1158, 318–328. doi: 10.1016/j.chroma.2007.04.021
66. Kaul, S., Koo, H. L., Jenkins, J., Rizzo, M., Rooney, T., Tallon, L. J., et al. (2000). Analysis of the genome sequence of the flowering plant *Arabidopsis thaliana*. *Nature* 408, 796–815. doi: 10.1038/35048692
67. Kehr, J., and Buhtz, A. (2011). "*Brassica* proteomics and metabolomics," in *Genetics, Genomics and Breeding of Oilseed Brassicas*, eds D. Edwards, J. Batley, I. Parkin, and C. Kole (Lebanon: Science Publishers, Inc), 174–193.
68. Kenan-Eichler, M., Leshkowitz, D., Tal, L., Noor, E., Melamed-Bessudo, C., Feldman, M., et al. (2011). Wheat hybridization and polyploidization results in deregulation of small RNAs. *Genetics* 188, 263–272. doi: 10.1534/genetics.111.128348
69. Khan, M. A. M., Ulrichs, C., and Mewis, I. (2011). Water stress alters aphid-induced glucosinolate response in *Brassica oleracea* var. italica differently. *Chemoecology* 21, 235–242. doi: 10.1007/s00049-011-0084-4
70. Kim, B., Yu, H. J., Park, S. G., Shin, J. Y., Oh, M., Kim, N., et al. (2012). Identification and profiling of novel microRNAs in the *Brassica rapa* genome based on small RNA deep sequencing. *BMC Plant Biol.* 12:218. doi: 10.1186/1471-2229-12-218
71. Kim, H. A., Lim, C. J., Kim, S., Choe, J. K., Jo, S. H., Baek, N., et al. (2014). High-throughput sequencing and de novo assembly of *Brassica oleracea* var. capitata L. for transcriptome analysis. *PLoS ONE* 9:e92087. doi: 10.1371/journal.pone.0092087
72. Kim, J., Jung, Y., Song, B., Bong, Y. S., Ryu, D. H., Lee, K. S., et al. (2013a). Discrimination of cabbage (*Brassica rapa* ssp pekinensis) cultivars grown in different geographical areas using H-1 NMR-based metabolomics. *Food Chem.* 137, 68–75. doi: 10.1016/j.foodchem.2012.10.012
73. Kim, J. K., Choi, S. R., Lee, J., Park, S.-Y., Song, S. Y., Na, J., et al. (2013b). Metabolic differentiation of diamondback moth (*Plutella xylostella* (L.)) resistance in cabbage (*Brassica oleracea* L. ssp. capitata). *J. Agric. Food Chem.* 61, 11222–11230. doi: 10.1021/jf403441t
74. Knief, C., Delmotte, N., Chaffron, S., Stark, M., Innerebner, G., Wassmann, R., et al. (2012). Metaproteogenomic analysis of microbial communities in the phyllosphere and rhizosphere of rice. *Isme J.* 6, 1378–1390. doi: 10.1038/ismej.2011.192
75. Knudsen, S. (2004). *Guide to Analysis of DNA Microarray Data*, 2nd Edn. Hoboken, N.J: Wiley-Liss. doi: 10.1002/0471670278
76. Kohl, M., Megger, D. A., Trippler, M., Meckel, H., Ahrens, M., Bracht, T., et al. (2014). A practical data processing workflow for multi-OMICS projects. *Biochim. Biophys. Acta* 1844, 52–62. doi: 10.1016/j.bbapap.2013.02.029

77. Kong, F., Ge, C., Fang, X., Snowdon, R. J., and Wang, Y. (2010). Characterization of seedling proteomes and development of markers to distinguish the *Brassica* A and C genomes. *J. Genet. Genomics* 37, 333–340. doi: 10.1016/S1673-8527(09)60051-5
78. Kopka, J., Schauer, N., Krueger, S., Birkemeyer, C., Usadel, B., Bergmüller, E., et al. (2005). GMD@CSB.DB: the Golm Metabolome Database. *Bioinformatics* 21, 1635–1638. doi: 10.1093/bioinformatics/bti236
79. Kubo, N., Saito, M., Tsukazaki, H., Kondo, T., Matsumoto, S., and Hirai, M. (2010). Detection of quantitative trait loci controlling morphological traits in *Brassica rapa* L. *Breed. Sci.* 60, 164–171. doi: 10.1270/jsbbs.60.164
80. Kurowska, M., Daszkowska-Golec, A., Gruszka, D., Marzec, M., Szurman, M., Szarejko, I., et al. (2011). TILLING - a shortcut in functional genomics. *J. Appl. Genet.* 52, 371–390. doi: 10.1007/s13353-011-0061-1
81. Kwun, M., Choi, Y., Yoon, H., Park, B. S., Kang, B. J., and Chung, Y. Y. (2004). Expression analysis of the pistil genes in controlling self-incompatibility of *Brassica* campestris by CO2 gas using microarray. *Mol. Cells* 18, 94–99.
82. Lee, J. G., Bonnema, G., Zhang, N. W., Kwak, J. H., De Vos, R. C. H., and Beekwilder, J. (2013). Evaluation of glucosinolate variation in a collection of turnip (*Brassica rapa*) germplasm by the analysis of intact and desulfo glucosinolates. *J. Agric. Food Chem.* 61, 3984–3993. doi: 10.1021/jf400890p
83. Lee, S. C., Lim, M. H., Kim, J. A., Lee, S. I., Kim, J. S., Jin, M., et al. (2008). Transcriptome analysis in *Brassica rapa* under the abiotic stresses using *Brassica* 24K oligo microarray. *Mol. Cells* 26, 595–605.
84. Lee, S., and Yun, S. C. (2006). The ozone stress transcriptome of pepper (*Capsicum annuum* L.). *Mol. Cells* 21, 197–205.
85. Leff, J. W., and Fierer, N. (2013). Bacterial communities associated with the surfaces of fresh fruits and vegetables. *PLoS ONE* 8:e59310. doi: 10.1371/journal.pone.0059310
86. Li, X., Gao, P., Gjetvaj, B., Westcott, N., and Gruber, M. Y. (2009). Analysis of the metabolome and transcriptome of *Brassica carinata*seedlings after lithium chloride exposure. *Plant Sci.* 177, 68–80. doi: 10.1016/j.plantsci.2009.03.013
87. Li, X., Ramchiary, N., Dhandapani, V., Choi, S. U., and Lim, Y. P. (2013). "Omics applications in *Brassica* species," in *OMICS Applications in Crop Science*, ed. D. Barh (Boca Raton, FL: CRC Press), 163–190. doi: 10.1201/b16352-7
88. Li, X., Zhang, W., Wang, Y., Wang, Z., and Peng, Y. (2008). Changes in chloroplast proteome of Chinese cabbage seedlings induced by PS II inhibiting herbicide Atrazine. *Acta Agron. Sin.* 34, 238–242. doi: 10.3724/SP.J.1006.2008.00238
89. Liao, C.-S., Yen, J.-H., and Wang, Y.-S. (2006). Effects of endocrine disruptor di-n-butyl phthalate on the growth of Bok choy (*Brassica rapa* subsp. chinensis). *Chemosphere* 65, 1715–1722. doi: 10.1016/j.chemosphere.2006.04.093
90. Liao, C.-S., Yen, J.-H., and Wang, Y.-S. (2009). Growth inhibition in Chinese cabbage (*Brassica rapa* var. chinensis) growth exposed to di-n-butyl phthalate. *J. Hazard. Mater.* 163, 625–631. doi: 10.1016/j.jhazmat.2008.07.025
91. Ligat, L., Lauber, E., Albenne, C., Clemente, H. S., Valot, B., Zivy, M., et al. (2011). Analysis of the xylem sap proteome of *Brassica oleracea* reveals a high content in secreted proteins. *Proteomics* 11, 1798–1813. doi: 10.1002/pmic.201000781
92. Lin, K. H., Chang, L. C., Lai, C. D., and Lo, H. F. (2013). AFLP mapping of quantitative trait loci influencing seven head-related traits in broccoli (*Brassica oleracea* var. italica). *J. Hortic. Sci. Biotechnol.* 88, 257–268.
93. Links, M. G., Demeke, T., Grafenhan, T., Hill, J. E., Hemmingsen, S. M., and Dumonceaux, T. J. (2014). Simultaneous profiling of seed-associated bacteria and fungi reveals antagonistic interactions between microorganisms within a shared epiphytic microbiome on triticum and *Brassica* seeds. *New Phytol.* 202, 542–553. doi: 10.1111/nph.12693
94. Liu, M.-S., Ko, M.-H., Li, H.-C., Tsai, S.-J., Lai, Y.-M., Chang, Y.-M., et al. (2014a). Compositional and proteomic analyses of genetically modified broccoli (*Brassica*

oleracea var. italica) harboring an agrobacterial gene. *Int. J. Mol. Sci.* 15, 15188–15209. doi: 10.3390/ijms150915188

95. Liu, S., Liu, Y., Yang, X., Tong, C., Edwards, D., Parkin, I. a. P., et al. (2014b). The *Brassica oleracea* genome reveals the asymmetrical evolution of polyploid genomes. *Nat. Commun.* 5, 3930. doi: 10.1038/ncomms4930

96. Liu, M.-S., Li, H.-C., Chang, Y.-M., Wu, M.-T., and Chen, L.-F. O. (2011). Proteomic analysis of stress-related proteins in transgenic broccoli harboring a gene for cytokinin production during postharvest senescence. *Plant Sci.* 181, 288–299. doi: 10.1016/j.plantsci.2011.06.005

97. Liu, M.-S., Li, H.-C., Lai, Y.-M., Lo, H.-F., and Chen, L.-F. O. (2013a). Proteomics and transcriptomics of broccoli subjected to exogenously supplied and transgenic senescence-induced cytokinin for amelioration of postharvest yellowing. *J. Proteom.* 93, 133–144. doi: 10.1016/j.jprot.2013.05.014

98. Liu, Y., Zhang, Y., Xing, J. Y., Liu, Z. Y., and Feng, H. (2013b). Mapping quantitative trait loci for yield-related traits in Chinese cabbage (*Brassica rapa* L. ssp pekinensis). *Euphytica* 193, 221–234. doi: 10.1007/s10681-013-0931-1

99. Lopez-Sanchez, P., De Vos, R. C. H., Jonker, H. H., Mumm, R., Hall, R. D., Bialek, L., et al. (2015). Comprehensive metabolomics to evaluate the impact of industrial processing on the phytochemical composition of vegetable purees. *Food Chem.* 168, 348–355. doi: 10.1016/j.foodchem.2014.07.076

100. Lopez-Velasco, G., Carder, P. A., Welbaum, G. E., and Ponder, M. A. (2013). Diversity of the spinach (*Spinacia oleracea*) spermosphere and phyllosphere bacterial communities. *FEMS Microbiol. Lett.* 346, 146–154. doi: 10.1111/1574-6968.12216

101. Lou, P., Xie, Q., Xu, X., Edwards, C. E., Brock, M. T., Weinig, C., et al. (2011). Genetic architecture of the circadian clock and flowering time in *Brassica rapa*. *Theor. Appl. Genet.* 123, 397–409. doi: 10.1007/s00122-011-1592-x

102. Lu, G., Cao, J. S., Yu, X. L., Xiang, X., and Chen, H. (2008). Mapping QTLs for root morphological traits in *Brassica rapa* L. based on AFLP and RAPD markers. *J. Appl. Genet.* 49, 23–31. doi: 10.1007/BF03195245

103. Luthria, D. L., Lin, L. Z., Robbins, R. J., Finley, J. W., Banuelos, G. S., and Harnly, J. M. (2008). Discriminating between cultivars and treatments of broccoli using mass spectral fingerprinting and analysis of variance-principal component analysis. *J. Agric. Food Chem.*56, 9819–9827. doi: 10.1021/jf801606x

104. Lv, H. H., Wang, Q. B., Zhang, Y. Y., Yang, L. M., Fang, Z. Y., Wang, X. W., et al. (2014). Linkage map construction using InDel and SSR markers and QTL analysis of heading traits in *Brassica oleracea* var. capitata L. *Mol. Breed.* 34, 87–98. doi: 10.1007/s11032-014-0019-1

105. Maggioni, L., Von Bothmer, R., Poulsen, G., Branca, F., and Jorgensen, R. B. (2014). Genetic diversity and population structure of leafy kale and *Brassica rupestris* Raf. in south Italy. *Hereditas* 151, 145–158. doi: 10.1111/hrd2.00058

106. Marioni, J. C., Mason, C. E., Mane, S. M., Stephens, M., and Gilad, Y. (2008). RNA-seq: an assessment of technical reproducibility and comparison with gene expression arrays. *Genome Res.* 18, 1509–1517. doi: 10.1101/gr.079558.108

107. Marquès, K., Sarazin, B., Chané-Favre, L., Zivy, M., and Thiellement, H. (2001). Comparative proteomics to establish genetic relationships in the Brassicaceae family. *Proteomics* 1, 1457–1462. doi: 10.1002/1615-9861(200111)1:11<1457::AID-PROT1457>3.0.CO;2-W

108. Matros, A., Kaspar, S., Witzel, K., and Mock, H.-P. (2011). Recent progress in liquid chromatography-based separation and label-free quantitative plant proteomics. *Phytochemistry* 72, 963–974. doi: 10.1016/j.phytochem.2010.11.009

109. Matsumoto, E., Ueno, H., Aruga, D., Sakamoto, K., and Hayashida, N. (2012). Accumulation of three clubroot resistance genes through marker-assisted selection in Chinese cabbage (*Brassica rapa* ssp pekinensis). *J. Jpn. Soc. Hort. Sci.* 81, 184–190. doi: 10.2503/jjshs1.81.184

110. Mewis, I., Schreiner, M., Nguyen, C. N., Krumbein, A., Ulrichs, C., Lohse, M., et al. (2012). UV-B irradiation changes specifically the secondary metabolite profile in

broccoli sprouts: induced signaling overlaps with defense response to biotic stressors. *Plant Cell Physiol.*53, 1546–1560. doi: 10.1093/pcp/pcs096

111. Mie, A., Laursen, K. H., Aberg, K. M., Forshed, J., Lindahl, A., Thorup-Kristensen, K., et al. (2014). Discrimination of conventional and organic white cabbage from a long-term field trial study using untargeted LC-MS-based metabolomics. *Anal. Bioanal. Chem.* 406, 2885–2897. doi: 10.1007/s00216-014-7704-0
112. Mochida, K., and Shinozaki, K. (2011). Advances in omics and bioinformatics tools for systems analyses of plant functions. *Plant Cell Physiol.* 52, 2017–2038. doi: 10.1093/pcp/pcr153
113. Muhammad, J., Abdel-Farid, I. B., Vos, R. C. H. D., Jonker, H. H., Choi, Y., and Verpoorte, R. (2014). Metabolomic variation of *Brassica rapa var. rapa* (var. raapstelen) and *Raphanus sativus* L. at different developmental stages. *Pak. J. Bot.* 46, 1445–1452.
114. Muller, T., and Ruppel, S. (2014). Progress in cultivation-independent phyllosphere microbiology. *FEMS Microbiol. Ecol.* 87, 2–17. doi: 10.1111/1574-6941.12198
115. Nawrocki, A., Thorup-Kristensen, K., and Jensen, O. N. (2011). Quantitative proteomics by 2DE and MALDI MS/MS uncover the effects of organic and conventional cropping methods on vegetable products. *J. Proteom.* 74, 2810–2825. doi: 10.1016/j.jprot.2011.06.021
116. Niederhuth, C. E., and Schmitz, R. J. (2014). Covering your bases: inheritance of DNA methylation in plant genomes. *Mol. Plant* 7, 472–480. doi: 10.1093/mp/sst165
117. O'Farrell, P. H. (1975). High-resolution 2-dimensional electrophoresis of proteins. *J. Biol. Chem.* 250, 4007–4021.
118. Okazaki, K., Shinano, T., Oka, N., and Takebe, M. (2012). Metabolite profiling of Komatsuna (*Brassica rapa* L.) field-grown under different soil organic amendment and fertilization regimes. *Soil Sci. Plant Nutr.* 58, 696–706. doi: 10.1080/00380768.2012.733924
119. Park, S., Arasu, M. V., Jiang, N., Choi, S. H., Lim, Y. P., Park, J. T., et al. (2014). Metabolite profiling of phenolics, anthocyanins and flavonols in cabbage (*Brassica oleracea* var. *capitata*). *Ind. Crops Prod.* 60, 8–14. doi: 10.1016/j.indcrop.2014.05.037
120. Park, S. Y., Lim, S. H., Ha, S. H., Yeo, Y., Park, W. T., Kwon, D. Y., et al. (2013). Metabolite profiling approach reveals the interface of primary and secondary metabolism in colored cauliflowers (*Brassica oleracea* L. ssp botrytis). *J. Agric. Food Chem.* 61, 6999–7007. doi: 10.1021/jf401330e
121. Parkin, I. A. P., Koh, C., Tang, H. B., Robinson, S. J., Kagale, S., Clarke, W. E., et al. (2014). Transcriptome and methylome profiling reveals relics of genome dominance in the mesopolyploid *Brassica oleracea*. *Genome Biol.* 15, 18. doi: 10.1186/gb-2014-15-6-r77
122. Pawlowski, T., Rurek, M., Janicka, S., Raczynska, K. D., and Augustyniak, H. (2005). Preliminary analysis of the cauliflower mitochondrial proteome. *Acta Physiol. Plant.* 27, 275–281. doi: 10.1007/s11738-005-0003-9
123. Pink, D., Bailey, L., Mcclement, S., Hand, P., Mathas, E., Buchanan-Wollaston, V., et al. (2008). Double haploids, markers and QTL analysis in vegetable *Brassicas*. *Euphytica* 164, 509–514. doi: 10.1007/s10681-008-9742-1
124. Rabbani, M. A., Maruyama, K., Abe, H., Khan, M. A., Katsura, K., Ito, Y., et al. (2003). Monitoring expression profiles of rice genes under cold, drought, and high-salinity stresses and abscisic acid application using cDNA microarray and RNA gel-blot analyses. *Plant Physiol.*133, 1755–1767. doi: 10.1104/pp.103.025742
125. Rabilloud, T., Chevallet, M., Luche, S., and Lelong, C. (2010). Two-dimensional gel electrophoresis in proteomics: past, present and future. *J. Proteom.* 73, 2064–2077. doi: 10.1016/j.jprot.2010.05.016
126. Ramautar, R., Demirci, A., and De Jong, G. J. (2006). Capillary electrophoresis in metabolomics. *Trends Analyt. Chem.* 25, 455–466. doi: 10.1016/j.trac.2006.02.004

127. Rastogi, G., Coaker, G. L., and Leveau, J. H. J. (2013). New insights into the structure and function of phyllosphere microbiota through high-throughput molecular approaches. *FEMS Microbiol. Lett.* 348, 1–10. doi: 10.1111/1574-6968.12225

128. Rastogi, G., Sbodio, A., Tech, J. J., Suslow, T. V., Coaker, G. L., and Leveau, J. H. J. (2012). Leaf microbiota in an agroecosystem: spatiotemporal variation in bacterial community composition on field-grown lettuce. *ISME J.* 6, 1812–1822. doi: 10.1038/ismej.2012.32

129. Reichelt, M., Brown, P. D., Schneider, B., Oldham, N. J., Stauber, E., Tokuhisa, J., et al. (2002). Benzoic acid glucosinolate esters and other glucosinolates from *Arabidopsis thaliana*. *Phytochemistry* 59, 663–671. doi: 10.1016/S0031-9422(02)00014-6

130. Rochfort, S. J., Trenerry, V. C., Imsic, M., Panozzo, J., and Jones, R. (2008). Class targeted metabolomics: ESI ion trap screening methods for glucosinolates based on MSn fragmentation. *Phytochemistry* 69, 1671–1679. doi: 10.1016/j. phytochem. 2008.02.010

131. Ruppel, S., Krumbein, A., and Schreiner, M. (2008). Composition of the phyllospheric microbial populations on vegetable plants with different glucosinolate and carotenoid compositions. *Microb. Ecol.* 56, 364–372. doi: 10.1007/s00248-007-9354-7

132. Salmon, A., Clotault, J., Jenczewski, E., Chable, V., and Manzanares-Dauleux, M. J. (2008). *Brassica oleracea* displays a high level of DNA methylation polymorphism. *Plant Sci.* 174, 61–70. doi: 10.1016/j.plantsci.2007.09.012

133. Schilling, G., Gransee, A., Deubel, A., Lezovic, G., and Ruppel, S. (1998). Phosphorus availability, root exudates, and microbial activity in the rhizosphere. *J. Plant Nutr. Soil Sci.* 161, 465–478. doi: 10.1002/jpln.1998.3581610413

134. Schmidt, U. G., Endler, A., Schelbert, S., Brunner, A., Schnell, M., Neuhaus, H. E., et al. (2007). Novel tonoplast transporters identified using a proteomic approach with vacuoles isolated from cauliflower buds. *Plant Physiol.* 145, 216–229. doi: 10.1104/pp.107.096917

135. Schonhof, I., Blankenburg, D., Muller, S., and Krumbein, A. (2007a). Sulfur and nitrogen supply influence growth, product appearance, and glucosinolate concentration of broccoli. *J. Plant Nutr. Soil Sci.* 170, 65–72. doi: 10.1002/jpln.200620639

136. Schonhof, I., Klaring, H. P., Krumbein, A., and Schreiner, M. (2007b). Interaction between atmospheric CO2 and glucosinolates in broccoli. *J. Chem. Ecol.* 33, 105–114. doi: 10.1007/s10886-006-9202-0

137. Schreiner, M., Beyene, B., Krumbein, A., and Stutzel, H. (2009). Ontogenetic changes of 2-propenyl and 3-indolylmethyl glucosinolates in *Brassica carinata* leaves as affected by water supply. *J. Agric. Food Chem.* 57, 7259–7263. doi: 10.1021/jf901076h

138. Schreiner, M., Krumbein, A., Knorr, D., and Smetanska, I. (2011). Enhanced glucosinolates in root exudates of *Brassica rapa* ssp. rapa mediated by salicylic acid and methyl jasmonate. *J. Agric. Food Chem.* 59, 1400–1405. doi: 10.1021/jf103585s

139. Schumacher, A., Rujan, T., and Hoefkens, J. (2014). A collaborative approach to develop a multi-omics data analytics platform for translational research. *Appl. Transl. Genomics* 3, 105–108. doi: 10.1016/j.atg.2014.09.010

140. Sebastian, R. L., Howell, E. C., King, G. J., Marshall, D. F., and Kearsey, M. J. (2000). An integrated AFLP and RFLP *Brassica oleracea*linkage map from two morphologically distinct doubled-haploid mapping populations. *Theor. Appl. Genet.* 100, 75–81. doi: 10.1007/s001220050011

141. Sepúlveda, I., Barrientos, H., Mahn, A., and Moenne, A. (2013). Changes in SeMSC, glucosinolates and sulforaphane levels, and in proteome profile in broccoli (*Brassica oleracea* var. Italica) fertilized with sodium selenate. *Molecules* 18, 5221–5234. doi: 10.3390/molecules18055221

142. Shapiro, T. A., Fahey, J. W., Dinkova-Kostova, A. T., Holtzclaw, W., Stephenson, K. K., Wade, K. L., et al. (2006). Safety, tolerance, and metabolism of broccoli sprout

glucosinolates and isothiocyanates: a clinical phase I study. *Nutr. Cancer* 55, 53–62. doi: 10.1207/s15327914nc5501_7

143. Shapiro, T. A., Fahey, J. W., Wade, K. L., Stephenson, K. K., and Talalay, P. (2001). Chemoprotective glucosinolates and isothiocyanates of broccoli sprouts: metabolism and excretion in humans. *Cancer Epidemiol. Biomarkers Prev.* 10, 501–508.
144. Shi, J. Q., Huang, S. M., Zhan, J. P., Yu, J. Y., Wang, X. F., Hua, W., et al. (2014). Genome-wide microsatellite characterization and marker development in the sequenced *Brassica* crop species. *DNA Res.* 21, 53–68. doi: 10.1093/dnares/dst040
145. Shumin, L., Schonhof, I., Krumbein, A., Long, L., Stuetzel, H., and Schreiner, M. (2007). 771 Glucosinolate concentration in turnip (*Brassica rapa* spp. *rapifera L.*) roots as 772 affected by nitrogen and sulfur supply. *J. Agric. Food Chem.* 55, 8452–8457. doi: 10.1021/jf070816k
146. Sonderby, I. E., Geu-Flores, F., and Halkier, B. A. (2010). Biosynthesis of glucosinolates - gene discovery and beyond. *Trends Plant Sci.*15, 283–290. doi: 10.1016/j.tplants.2010.02.005
147. Song, X., Liu, G., Duan, W., Liu, T., Huang, Z., Ren, J., et al. (2014). Genome-wide identification, classification and expression analysis of the heat shock transcription factor family in Chinese cabbage. *Mol. Genet. Genomics* 289, 541–551. doi: 10.1007/s00438-014-0833-5
148. Stephenson, P., Baker, D., Girin, T., Perez, A., Amoah, S., King, G. J., et al. (2010). A rich TILLING resource for studying gene function in *Brassica rapa*. *BMC Plant Biol.* 10:62. doi: 10.1186/1471-2229-10-62
149. Sun, C. B. (2014). Characterization of the small RNA transcriptome in plant-microbe (*Brassica/Erwinia*) interactions by high-throughput sequencing. *Biotechnol. Lett.* 36, 371–381. doi: 10.1007/s10529-013-1362-8
150. Sun, C., Wang, L., Hu, D., Riquicho, A. R. M., Liu, T., Hou, X., et al. (2014). Proteomic analysis of non-heading Chinese cabbage infected with *Hyaloperonospora parasitica*. *J. Proteom.* 98, 15–30. doi: 10.1016/j.jprot.2013.11.028
151. Sun, G. (2012). MicroRNAs and their diverse functions in plants. *Plant Mol. Biol.* 80, 17–36. doi: 10.1007/s11103-011-9817-6
152. Takuno, S., Kawahara, T., and Ohnishi, O. (2007). Phylogenetic relationships among cultivated types of *Brassica rapa* L. em. Metzg. as revealed by AFLP analysis. *Genet. Resour. Crop Evol.* 54, 279–285. doi: 10.1007/s10722-005-4260-7
153. Textor, S., and Gershenzon, J. (2009). Herbivore induction of the glucosinolate-myrosinase defense system: major trends, biochemical bases and ecological significance. *Phytochem. Rev.* 8, 149–170. doi: 10.1007/s11101-008-9117-1
154. T'Hoen, P. A., Ariyurek, Y., Thygesen, H. H., Vreugdenhil, E., Vossen, R. H., De Menezes, R. X., et al. (2008). Deep sequencing-based expression analysis shows major advances in robustness, resolution and inter-lab portability over five microarray platforms. *Nucleic Acids Res.* 36, e141. doi: 10.1093/nar/gkn705
155. Trick, M., Cheung, F., Drou, N., Fraser, F., Lobenhofer, E. K., Hurban, P., et al. (2009). A newly-developed community microarray resource for transcriptome profiling in *Brassica* species enables the confirmation of *Brassica*-specific expressed sequences. *BMC Plant Biol.* 9:50. doi: 10.1186/1471-2229-9-50
156. Turner, T. R., James, E. K., and Poole, P. S. (2013). The plant microbiome. *Genome Biol.* 14, 209. doi: 10.1186/gb-2013-14-6-209
157. Vanderschuren, H., Lentz, E., Zainuddin, I., and Gruissem, W. (2013). Proteomics of model and crop plant species: status, current limitations and strategic advances for crop improvement. *J. Proteom.* 93, 5–19. doi: 10.1016/j.jprot.2013.05.036
158. van Hintum, T. J. L., De Wiel, C., Visser, D. L., Van Treuren, R., and Vosman, B. (2007). The distribution of genetic diversity in a *Brassica oleracea* gene bank collection related to the effects on diversity of regeneration, as measured with AFLPs. *Theor. Appl. Genet.*114, 777–786. doi: 10.1007/s00122-006-0456-2

159. Varshney, R. K., Terauchi, R., and Mccouch, S. R. (2014). Harvesting the promising fruits of genomics: applying genome sequencing technologies to crop breeding. *PLoS Biol.* 12:e1001883. doi: 10.1371/journal.pbio.1001883
160. Verkerk, R., Schreiner, M., Krumbein, A., Ciska, E., Holst, B., Rowland, I., et al. (2009). Glucosinolates in *Brassica* vegetables: the influence of the food supply chain on intake, bioavailability and human health. *Mol. Nutr. Food Res.* 53, S219–S265. doi: 10.1002/mnfr.200800065
161. Vig, A. P., Rampal, G., Thind, T. S., and Arora, S. (2009). Bio-protective effects of glucosinolates - a review. *Lwt-Food Sci. Technol.* 42, 1561–1572. doi: 10.1016/j.lwt.2009.05.023
162. Vorholt, J. A. (2012). Microbial life in the phyllosphere. *Nat. Rev. Microbiol.* 10, 828–840. doi: 10.1038/nrmicro2910
163. Walley, P., Carder, J., Skipper, E., Mathas, E., Lynn, J., Pink, D., et al. (2012). A new broccoli × broccoli immortal mapping population and framework genetic map: tools for breeders and complex trait analysis. *Theor. Appl. Genet.* 124, 467–484. doi: 10.1007/s00122-011-1721-6
164. Wang, L., Peng, H., Ge, T., Liu, T., Hou, X., and Li, Y. (2014a). Identification of differentially accumulating pistil proteins associated with self-incompatibility of non-heading Chinese cabbage. *Plant Biol.* 16, 49–57. doi: 10.1111/plb.12016
165. Wang, X., Tang, D., and Huang, D. (2014b). Proteomic analysis of pakchoi leaves and roots under glycine–nitrogen conditions. *Plant Physiol. Biochem.* 75, 96–104. doi: 10.1016/j.plaphy.2013.12.012
166. Wang, Y. G., Zhang, L., Ji, X. H., Yan, J. F., Liu, Y. T., Lv, X. X., et al. (2014c). Mapping of quantitative trait loci for the bolting trait in *Brassica rapa* under vernalizing conditions. *Genet. Mol. Res.* 13, 3927–3939. doi: 10.4238/2014.May.23.3
167. Wang, W. X., Huang, S. M., Liu, Y. M., Fang, Z. Y., Yang, L. M., Hua, W., et al. (2012). Construction and analysis of a high-density genetic linkage map in cabbage (*Brassica oleracea* L. var. *capitata*). *BMC Genomics* 13:523. doi: 10.1186/1471-2164-13-523
168. Wang, X., Wang, H., Wang, J., Sun, R., Wu, J., Liu, S., et al. (2011). The genome of the mesopolyploid crop species *Brassica rapa*. *Nat. Genet.* 43, 1035–1039. doi: 10.1038/ng.919
169. Wang, Z., Gerstein, M., and Snyder, M. (2009). RNA-Seq: a revolutionary tool for transcriptomics. *Nat. Rev. Genet.* 10, 57–63. doi: 10.1083/nrg2484
170. Wei, L. J., Xiao, M. L., Hayward, A., and Fu, D. H. (2013). Applications and challenges of next-generation sequencing in *Brassica* species. *Planta* 238, 1005–1024. doi: 10.1007/s00425-013-1961-6
171. Weimer, B. C., and Slupsky, C. (2013). “Metabolomics in food and nutrition,” in *Metabolomics in Food and Nutrition*, eds B. C. Weimer and C. M. Slupsky (Cambridge: Woodhead), 237.
172. White, P. J., Hammond, J. P., King, G. J., Bowen, H. C., Hayden, R. M., Meacham, M. C., et al. (2010). Genetic analysis of potassium use efficiency in *Brassica oleracea*. *Ann. Bot.* 105, 1199–1210. doi: 10.1093/aob/mcp253
173. Wiesner, M., Hanschen, F. S., Schreiner, M., Glatt, H., and Zrenner, R. (2013a). Induced production of 1-methoxy-indol-3-ylmethyl glucosinolate by jasmonic acid and methyl jasmonate in sprouts and leaves of pak choi (*Brassica rapa* ssp. *chinensis*). *Int. J. Mol. Sci.* 14, 14996–15016. doi: 10.3390/ijms140714996
174. Wiesner, M., Zrenner, R., Krumbein, A., Glatt, H., and Schreiner, M. (2013b). Genotypic variation of the glucosinolate profile in pak choi (*Brassica rapa* ssp. *chinensis*). *J. Agric. Food Chem.* 61, 1943–1953. doi: 10.1021/jf303970k
175. Wiesner, M., Schreiner, M., and Zrenner, R. (2014). Functional identification of genes responsible for the biosynthesis of 1-methoxy-indol-3-ylmethyl-glucosinolate in *Brassica rapa* ssp. *chinensis*. *BMC Plant Biol.* 14:124. doi: 10.1186/1471-2229-14-124
176. Williams, T. R., and Marco, M. L. (2014). Phyllosphere microbiota composition and microbial community transplantation on lettuce plants grown indoors. *Mbio* 5, e01564–14. doi: 10.1128/mBio.01564-14

177. Wink, M. (1988). Plant breeding - importance of plant secondary metabolites for protection against pathogens and herbivores. *Theor. Appl. Genet.* 75, 225–233. doi: 10.1007/BF00303957

178. Wittstock, U., and Burow, M. (2010). Glucosinolate breakdown in *Arabidopsis*: mechanism, regulation and biological significance. *Arabidopsis Book* 8, e0134. doi: 10.1199/tab.0134

179. Witzel, K., Pietsch, C., Strickert, M., Matros, A., Röder, M., Weschke, W., et al. (2011). Mapping of quantitative trait loci associated with protein expression variation in barley grains. *Mol. Breed.* 27, 301–314. doi: 10.1007/s11032-010-9432-2

180. Wu, Q. J., Yang, Y., Vogtmann, E., Wang, J., Han, L. H., Li, H. L., et al. (2013). Cruciferous vegetables intake and the risk of colorectal cancer: a meta-analysis of observational studies. *Ann. Oncol.* 24, 1079–1087. doi: 10.1093/annonc/mds601

181. Yang, K. A., Lim, C. J., Hong, J. K., Jin, Z. L., Hong, J. C., Yun, D. J., et al. (2005). Identification of Chinese cabbage genes up-regulated by prolonged cold by using microarray analysis. *Plant Sci.* 168, 959–966. doi: 10.1016/j.plantsci.2004.11.011

182. Yang, Z., Kobayashi, E., Katsuno, T., Asanuma, T., Fujimori, T., Ishikawa, T., et al. (2012). Characterisation of volatile and non-volatile metabolites in etiolated leaves of tea (*Camellia sinensis*) plants in the dark. *Food Chem.* 135, 2268–2276. doi: 10.1016/j.foodchem.2012.07.066

183. Yu, L. X., and Setter, T. L. (2003). Comparative transcriptional profiling of placenta and endosperm in developing maize kernels in response to water deficit. *Plant Physiol.* 131, 568–582. doi: 10.1104/pp.014365

184. Zeng, J., Chen, S., Zhu, L., Gao, Q., Wang, X., and Yang, X. (2012). Establishment of two-dimensional electrophoresis system of *Brassica oleracea* L. var. *capitata* L. stigma protein. *China Vegetables* 1, 30–36.

185. Zhao, J. J., Artemyeva, A., Del Carpio, D. P., Basnet, R. K., Zhang, N. W., Gao, J., et al. (2010). Design of a *Brassica rapa* core collection for association mapping studies. *Genome* 53, 884–898.

186. Zhou, B., Fan, P., and Li, Y. (2014). High-throughput sequence analysis of small RNAs in skotomorphogenic seedlings of *Brassica rapa* ssp. *rapa*. *Gene* 548, 68–74. doi: 10.1016/j.gene.2014.07.010

187. Zhuang, J., Zhang, J., Hou, X. L., Wang, F., and Xiong, A. S. (2014). Transcriptomic, proteomic, metabolomic and functional genomic approaches for the study of abiotic stress in vegetable crops. *Crit. Rev. Plant Sci.* 33, 225–237. doi: 10.1080/07352689.2014.870420

FOOTNOTES

[1]http://www3.botany.ubc.ca/can-till

[2]http://revgenuk.jic.ac.uk

[3]phytozome.jgi.doe.gov/

[4]www.metabolomicsociety.org

CHAPTER 13

Assessing Vegetable Farmer Knowledge of Diseases and Insect Pests of Vegetable and Management Practices Under Tropical Conditions

A. F. Abang,[1] C. M. Kouamé,[2] M. Abang,[3] R. Hanna,[4] and A. K. Fotso[4]

[1]*AVRDC–The World Vegetable Center, Messa, Yaoundé, Cameroon*

[2]*World Agro-forestry Centre, Cote D'Ivoire, Abidjan*

[3]*International Centre of Tropical Agriculture, Kampala, Uganda*

[4]*International Institute of Tropical Agriculture, Messa, Yaoundé, Cameroon*

ABSTRACT

In order to understand management of vegetable pests under tropical conditions, farmer knowledge of pest etiology and epidemiology was investigated in Mar. 2009. About 78% of vegetable farmers were men aged 31–40, with 51% having little formal education. About 21% of farmers can identify diseases and 16% can identify insects. Over 80% grow susceptible varieties. Ninety-two percent of farmers used synthetics pesticides. About 92% were willing to accept resistant varieties. This willingness provides a basis for further collaboration to employ resistant varieties and promote integrated pest management.

Keywords: Cameroon, Chemicals, Crop production, Pesticide

A minimum consumption of 200 g/day/person, 73 kg/person/year, of vegetables is recommended (Food and Agriculture Organization, 2004). Presently, vegetable availability is 50 kg/person/year and in sub-Saharan

Africa (SSA), and vegetable consumption is only 43% of recommended rate (Food and Agriculture Organization, 2004). Vegetables are important sources of micronutrients and provide farmers with higher income·per hectare than cereal, root, and tuber crops (Asian Vegetable Research and Development Center [AVRDC], 2006).

Vegetables form part of a healthy diet because they provide vitamins and minerals. Biotic and abiotic factors are among the major constraints of vegetable production. Many tropical locations receive high rainfall per year that contributes to high disease incidence on vegetables (Bowen and Kratky, 1982). Rain, heavy dews, warm temperatures, and dry climates (mostly for insect infestation, which is affected by rain) have been reported as principal conditions that favor establishment of pests (Landston and Eaker, 2009). Iwuchukwu and Uzoho (2009) indicated that the most important financial constraints associated with vegetable production in Enugu State, Nigeria, were caused by the laborious nature of vegetable production and incidence of diseases and pests. Youdeowei (2002) also indicated that biotic constraints caused significant economic loss on vegetable in Ghana. In Cameroon, pests and diseases have been identified as major constraints to vegetable production (Ellis-Jones et al., 2008). Pests and diseases cause both economic and health problems for vegetable farmers.

Chemical control is practiced by farmers for higher gains (Gerken et al., 2001), but these pests can become resistant to chemical insecticides very quickly. Moreover, the misuse of chemical insecticides in terms of quantity applied or in dangerous combinations (Obeng-Ofori et al., 2002) has created problems including pest resistance, resurgence of pests, pesticide residues, destruction of beneficial insects, and environmental pollution (AVRDC, 2003). A survey of pesticide application in Cameroon conducted by Matthews et al. (2003) to manage crop pests and diseases reported illnesses caused by the use of pesticide. In a resource-poor farming system, strategies for pest management become more complicated. The sociological diversity in Africa contributes to disparate traits that farmers' value in varieties grown. It is necessary to understand why farmers continue to use certain crops and cropping practices before suggesting improvements in management of pests (Kiros-Meles and Abang, 2007).

This study was undertaken to determine knowledge of pests and diseases and their management and to determine socioeconomic demographics of producers. Specifically, the objectives were to determine farmers' experience in pest and disease identification, management, and information needs. The data varied due to differences in knowledge, understanding, sex, age, family status, and level of education (Caulkins and Hyatt, 1999).

MATERIALS AND METHODS

Sampling sites were in vegetable production regions and determined by use of multilocation random sampling. The sample size for each region was determined based on its share of total national vegetable production. The regions surveyed and sample size were as follows northwest (21 farmers); west (49 farmers), southwest (34 farmers), central (38 farmers), littoral (19 farmers), south (18 farmers), and east (15 farmers). The AVRDC crops used in the Vegetable Breeding and Seed System program were as follows: *Amaranthus*spp. (amaranth), cabbage (*Brassica oleracea* var. *capitata* L), pepper (*Capsicum* spp.), onion (*Allium cepa* L.), jute mallow (*Corchorus olitorius* L.); African nightshade or huckle berry (*Solanum scabrum* Miller), eggplant (*S. melongena* var. *esculentum* Nees), tomato (*Lycopersicon esculentum* L.), and okra (*Abelmoschus* spp.). Tools used to obtain information about farmers' perceptions and knowledge related to vegetable crop pests were open-ended questions administered in a semistructured questionnaire allowing data to be gathered in the farmers's cultural context (Björnsen-Gurung, 2003). The questionnaire contained 38 main questions. Ten farmers participated in the prestudy (pilot study) to insure farmer comprehension of typical questions and the ability of enumerators to administer it. Their responses were used to improve the final copy.

A codification manual was prepared and numbers ranging from 1 to 6 assigned to responses, depending on the number of responses expected per question. This codification was used to enter farmers' responses in Census and Survey Data Processing system software (Ver. 3.3, Census Bureau and ICF Macro, Washington, DC). The data in Census and Survey Data Processing system software were imported into Statistical Package for Social Sciences software (SPSS, Ver. 17.0.1, Chicago, IL) for editing and analysis, and Excel was used to produce figures. Descriptive statistics were used to determine frequencies of responses.

RESULTS

The total number of farmers interviewed was 194; 96% of them produced vegetables for market and 4% produced for both the market and home consumption. About 64% produced vegetables in rural areas, 32% in peri-urban areas, and 7% in urban areas. The categories of vegetable farmers varied between age, sex, family status, and level of education (Figure 1). Seventy-eight percent of vegetable farmers were men and 22% were women. Most (77%) were household heads.

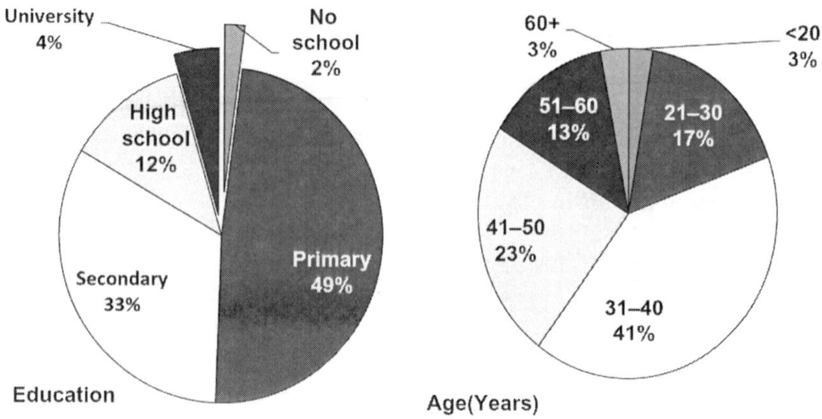

Figure 1 Age groups and level of education of vegetable farmers.

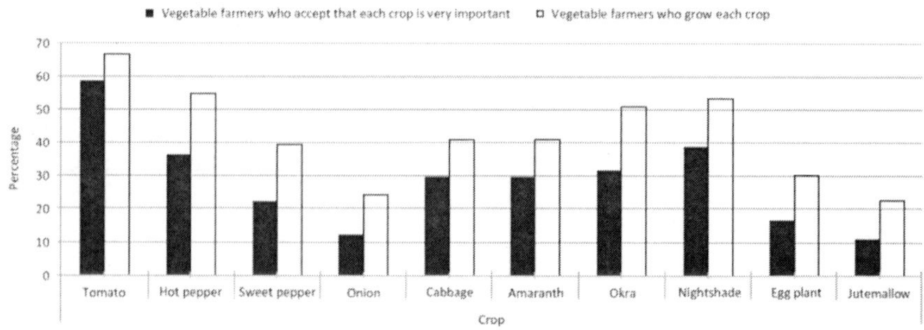

Figure 2 Comparison between farmers who consider each crop as very important and total percentage of farmers who grow each crop.

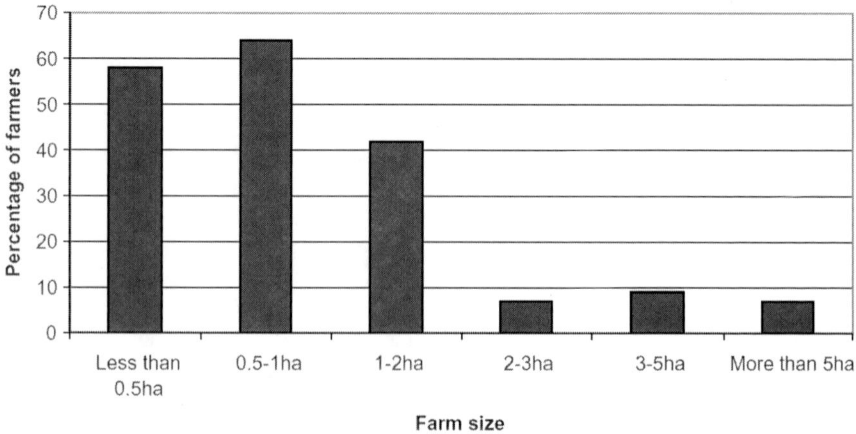

Figure 3 Farm size distribution used for vegetable production.

Experience in Crop Management

Tomato, hot pepper, okra, and cabbage are important regional crops in Cameroon (Figure 2) and are produced in different locations (Table 1). Producer experience in vegetable cultivation varied (Table 2). All land holdings were small (Figure 3).

Knowledge of Insect Pests and Diseases

All famers recognized the so-called mildiou (late blight) caused by *Phytophthora infestans* (Mont.) de Bary of tomato; the soil-borne bacterial wilt caused by *Ralstonia solanacearum* (Smith) on most solanaceous species; and aphids, which they can describe by symptoms on crops. The term "flies" was used as a general name for all insect pests that fly; they were able to differentiate between white flies and flea beetles from their color. Most defoliators and fruit borers were generally called caterpillars, which could only be differentiated from feeding habits described by the farmer. Only 18% of respondents were able to identify vegetable pests. When all pests were placed on the same scale, insect pests appeared to be more important than diseases (Table 3). According to respondents, eggplant and jute mallow did not have serious insect pest problems.

Table 1 Percentage of farmers who identified each crop as important by location

Region	Site	Tomato	Hot pepper	Cabbage	Okra	Nightshade	Sweet pepper	Onion	Amaranth	Eggplant	Jute mallow
Central	Bafia	100	60	88	71	79	67	33	0	67	71
	Obala	100	0	0	100	0	0	0	0	0	0
	Yaounde	94	0	50	0	100	92	0	100	67	0
East	Bertoua	100	33	0	75	60	67	0	50	0	0
	Abong-mbang	100	86	40	40	50	60	0	0	0	0
Littoral	Douala	20	43	33	67	86	0	0	0	0	0
	Njombe	33	89	100	67	100	0	0	0	0	0
Northwest	Santa	50	100	100	100	60	25	0	100	100	50
	Ndop	100	100	100	50	60	75	33	100	100	100
South	Ebolowa	91	40	33	36	30	50	0	20	33	0
	Ambam	83	50	0	0	100	33	0	0	0	0
Southwest	Buea	100	100	100	80	100	100	100	100	100	83
	Ekona	100	100	100	67	33	75	0	88	80	50
	Muyuka	33	40	88	57	100	33	0	0	71	33
West	Dschang	43	12	73	29	55	0	94	50	0	20
	Bangante	100	100	0	67	67	100	100	100	100	0
	Foumbot	100	88	67	92	89	82	100	73	100	0

Table 2 Farmer experience in market gardening

District	Years in agriculture	Years in market gardening
Bertoua (east)	17.4	14.9
Buea (southwest)	19.6	14.2
Douala (littoral)	20	10
Ebolowa (south)	10.9	8.3
Ekona (southwest)	11.8	7.8
Upper Nyong (east)	19.1	6.7
Lekie (central)	18.0	11.8
Mbam Ekim (central)	11.7	8.0
Menoua (west)	14.6	9.1
Mezam (northwest)	23.3	22.8
Mfoundi (central)	18.2	16.3
Moungo (littoral)	15.3	21.6
Muyuka (southwest)	15.9	14.0
Nde (west)	19.5	13.8
Ngoketunjia (northwest)	27.9	16.3
Noun (west)	15.7	13.9
Ntem valley (littoral)	9.0	5.7

Table 3 Percentage of growers who identified specific pests as a constraint in the production of vegetable crops

Pest type	Pest	Tomato	Hot pepper	Cabbage	Okra	Nightshade	Sweet pepper	Onion	Amaranth	Eggplant	Jute mallow
Insect	Unknown	66.9	87.6	92.2	89.2	86	93.7	99	93.8	98.8	98.3
	Ants	1.3	0.9		1.3	3.2			4		1
	Aphids	4	2.7	3	2.5	4.5	1.7		1	0.5	1
	Beetles				2.8	2	0.5		4.5		0.5
	Caterpillars	18.5	6.1	4.8	2.9	3.3	1.8	0.5	2	0.8	0.5
	Cricket	3.5			2		0.8	0.5	1.7	1	0.5
	White fly	9.3	2.7				2.9		1.8	0.8	
Disease	Unknown	70.7	93	97.9	9.6	95.2				99.5	99.5
	Blight	14.5	3.8	4	1.8	5.4	1.3	0.5	2	0.5	1
	Viruses	0.8	2		1.5	3	0.5				
	Bacteria Wilt	4.2	1		1.5	0.7					0.5

About equal numbers could or could not identify sources of pests and diseases on their farms (Table 4). Generally they could not predict disease outbreaks. They recognize disease problems when they can describe symptoms or when it is necessary to abandon fields for new sites. Generally they believed that all pests and disease problems in vegetable fields are not new. Most grow susceptible varieties and few stated that production of some vegetable crops and cultivars declined or were not able to be produced as a result of pest or disease problems. The varieties cultivated are not resistant. However, almost all respondents were willing to accept resistant varieties provided that the varieties have important traits (Figure 4).

Table 4 Sources of pests and disease identified by vegetable farmers

Source	Percentage
None	44.4
Climate change	8.6
Manure	0.5
Humidity	15.3
No treatment	7.5
Poor seed	2.0
Soil borne	12.1
Surrounding vegetation	7.1
Variety	1
Wind	1.5

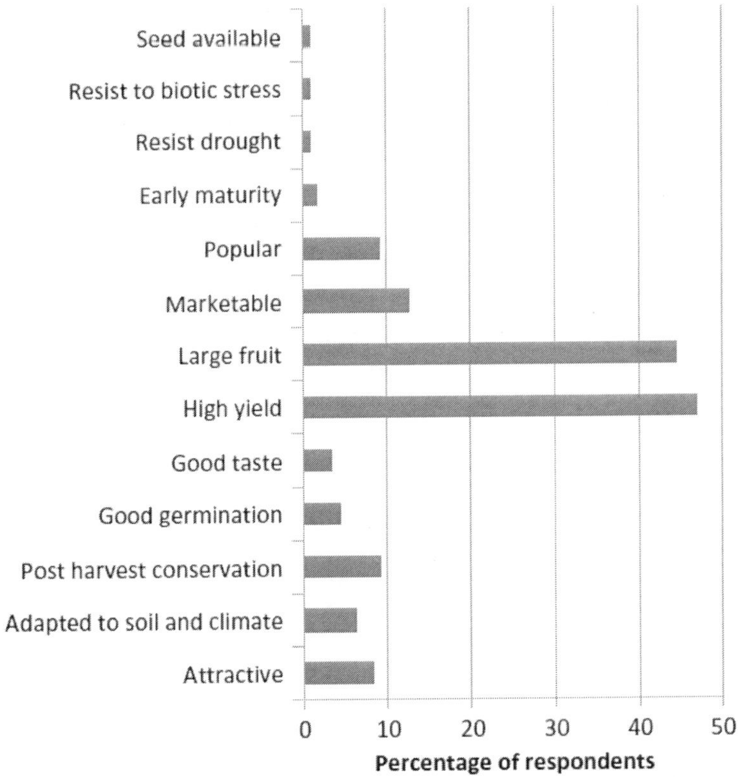

Figure 4 Acceptable traits of susceptible vegetable varieties. Precocity is early maturity of the crop.

Pest Management Practices

Most used synthetic pesticides against fungal diseases and insect pests (Figure 5). Producers recognized that a number of methods exist for control of insect pests and diseases other than chemical pesticides (Table 5). Sixteen percent of farmers know about natural enemies of insect pests. Insects identified by farmers as beneficial natural enemies were ants, bees, black ants, butterflies, crickets, ladybird beetles, praying mantis, snails, spiders, and wasps.

Table 5 Percentage of farmers who used methods other than chemical control (pesticides) for pest management

Alternative pest management	Percentage
Do not know	74.8
Use resistant varieties	0.5
Farming systems	6
Engine oil	0.5
Integrated pest management	0.5
Tobacco solution	4.5
Wood ash	10.2

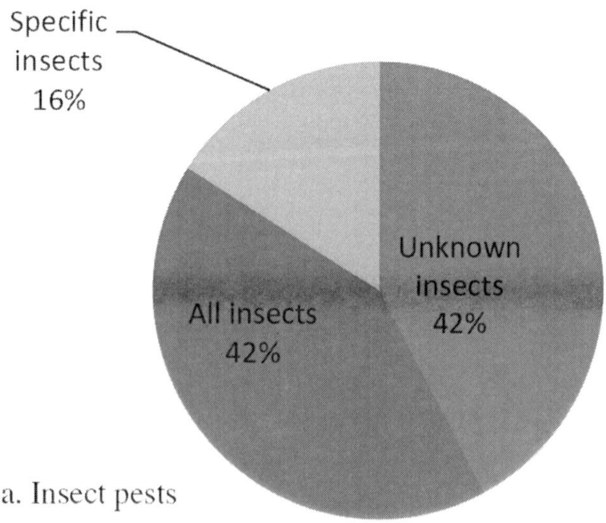

a. Insect pests

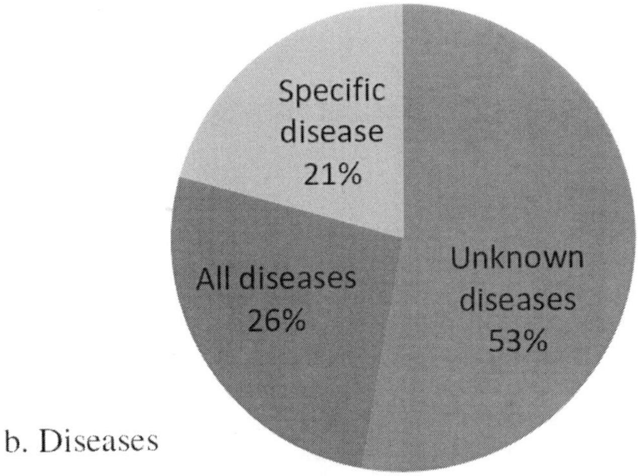

Figure 5 Farmer knowledge of control of a) insect pests and b) diseases.

Information Needs

Most respondents did not have access to information about integrated pest management, pesticide use and safety, or insect and disease identification (Table 6). The vegetable growers depend on the experience of others for advice on managing pests and diseases (Table 6).

Table 6 Source of vegetable pest management information in Cameroon

Source of information	Yes (%)	No (%)
Other farmers	73.1	26.9
Agro-dealers	20	80
Radio	5	95
Newspapers	10	90
Seed companies	16	84
Extension agents	38	62
TV	7.1	92.9
Other	11.7	88.3

With a high rate of pesticide use, 92% of farmers declared their willingness to adopt measures that would reduce pesticide use. Some measures proposed include use of wood ash, resistant and tolerant varieties, proper field management, and appropriate pesticides application. Other methods included good fertilization, use of biopesticides, and crop rotation. If seed of resistant or tolerant varieties were available to farmers, almost all (92%) would adopt their use.

DISCUSSION

Hovorka (2005) reported that urban agriculture is an important source of food and jobs in Botswana. This was not the case in this study as more farmers were in rural areas, although most inland valleys in cities were locations of vegetable farms. Most vegetables are produced in rural areas and peripheries of cities where resource-poor and less educated farmers are found. This could be the reason for the low level of farmer knowledge of vegetable pests and diseases and management methods. Most farmers were men, as was the case in Botswana (Obopile et al., 2008), where agriculture is not considered an activity for women. However, Ratta (1993) reported that women in some developing countries farm; men manage income-generating activities; and women manage vegetable production. Ratta (1993) further stated that farming is a viable alternative to waged labor for women who lack access to formal employment due to limited education, training, and other opportunities. Less than 50% of farmers completed at least primary school; fewer completed secondary school and level of education impacted awareness and effectiveness of technology information (Elizabeth and Zira, 2009). Socioeconomic characteristics of vegetable farmers in Nigeria indicated that 78% attended through primary school, corroborating our findings that most farmers are less educated. This confirms the conclusion by Ratta (1993) that farming is a viable alternative to waged labor for those who lack formal employment due to limited education.

All producers were small land holders, as was also determined by Elizabeth and Zira (2009). Autissier (1994) indicated that vegetable cultivation is production of annual plants (shrubby or herbaceous) in a delimited agrarian space, generally exploited intensively. The use of small vegetable farms could be a result of the intensive nature of vegetable production and the high cost of chemical inputs, which increases cost of production. In addition, lack of adequate knowledge of vegetable production and protection from biotic and maybe abiotic constraints could prevent vegetable farm size from increasing. Reduced size of the area enables resource-poor farmers to effectively manage their production.

Pests and diseases are important constraints to vegetable production in the tropics. Insect pests are more important in the dry season; diseases are more problematic in the rainy season. Humidity, soil-borne diseases, and climate change are reasons for pest and disease pressure (Mossler and Dunn, 2009). Despite the influence of moist, warm climates on vegetable pest severity, most farmers grow vegetables in all seasons; in the rainy season water is available and high income is generated from vegetable production.

Blight, mildew, and wilt were the most serious pests. Similarly, the causal agents for blight and mildew (Fontem and Schippers, 2004) and a range of viruses (Letts et al., 2009), misdirection of federal funds, equipment, and effort affect disease pressure. Aphids were the most recurrent insect pest on vegetable crops (Kekeunou et al., 2006; Praveen and Dhandapani, 2002). Most insect pests are vectors of disease organisms (Meyer, 2003). Viruses were identified by farmers among diseases in addition to blight and bacteria. Their ability to identify and talk about insect pests and diseases could indicate how economically important they are to farmers, since their identification knowledge is low.

Many farmers could identify factors that support pests and diseases. Landston and Eaker (2009) reported that rain, heavy dews, warm temperatures, and dry climates (mostly for insects, whose infestation is affected by rain) are principal conditions favoring pest establishment. Although cultural practices are recommended for pest management, the practices were not accepted over chemical pesticides to increase production (Holmer et al., 2001). Although all farmers used chemical pesticides, as also indicated by Obopile et al. (2007), they did not have a good knowledge of pesticide handling. Most commonly used insecticides were dimethoate, deltamethrin, and cypermethrine, and the most commonly used fungicides were maneb, metalaxyl, manizan, and mancozeb (Matthews et al., 2003). Here, and in the work of Elizabeth and Zira (2009), it was reported that most vegetable farmers received extension information from neighbors and had little or no contact with government departments. They also reported that almost all respondents were aware of extension services, and an equal percentage recognized the usefulness of extension services. However, most were never visited by the extension service, which likely resulted in farmers' inability to identify pests and diseases of vegetables, poor pest management skills, and lack of good knowledge of the use of chemical pesticides.

Vegetable production is changing from women to men. Agriculture is considered a job for the underprivileged and school dropouts. Vegetable production is a year-round activity in the tropics. Pests and diseases are important constraints to vegetable production. Knowledge of crop diseases constitutes a major obstacle in vegetable production systems. To increase availability of vegetables, their production needs to be intensified in urban areas, especially in developing countries (Hovorka, 2005). Vegetable breeding for resistance to major pests and diseases should consider consumer preference in addition to yield. Introduction of training programs for farmers on identification and management of pests and safe use of pesticides is necessary. This will improve farmer knowledge of diseases and insect pests and improve management practices, especially with the high illiteracy level among farmers and the aging farming population.

ACKNOWLEDGMENTS

The Asian Vegetable Research and Development Centre (AVRDC)–The World Vegetable Center acknowledges efforts of all institutions for their support with resources and scientists whose ideas have been included in this report. We appreciate financial, equipment, and human resource contributions of the entomology team of the International Institute of Tropical Agriculture (IITA)–Cameroon and the vegetable breeding and seed system project of AVRDC–Cameroon. We are grateful for contributions of Dr. Gaston Tsopmbeng of the University of Dschang and Dr. Maurice Tindo from IITA for their professional input and data collection. We acknowledge the resource personnel and vegetable research partners of the Catholic Rural Animation Center Bertoua, Ashu A. Tambe of the Center for Assistance to Sustainable Development at Ebolowa, Thiery Tsou Fematouo of the Centre Internationale de Promotion et de Recuperation in Yaounde and Institut de la Recherche Agronomique pour le Development in Foumbot and Ekona, who facilitated the work by making contacts with farmers and creating a conducive atmosphere for the study.

REFERENCES

1. Asian Vegetable Research and Development Center. 2003. Effects of natural and synthetic pesticides in controlling Diamondback moth in cabbage. AVRDC–The World Vegetable Centre, Shanhua, Taiwan.
2. Asian Vegetable Research and Development Center. 2006. Vegetables matter. (http://www.avrdc.org).
3. Autissier, V. 1994. Jardins des villes, Jardins des champs: Maraîchage en Afrique de l'Ouest, du diagnostic à l'Intervention, collection le point. Gret, Paris, France.
4. Björnsen-Gurung, A. 2003. Insects—A mistake in God's creation? Tharu farmers' perception and knowledge of insects: A case study of Gobardiha village development committee, Dang-Deukhuri, Nepal. Agr. Human Values 20:337–370.
5. Bowen, J.E. and B.A. Kratky. 1982. Greenhouses for the tropics. World Farming 24:6.
6. Caulkins, D. and S.B. Hyatt. 1999. Using consensus analysis to measure cultural diversity in organizations and social movements. Field Methods 11:5–26.
7. Elizabeth, S. and D. Zira. 2009. Awareness and effectiveness of vegetable technology information packages by vegetable farmers in Adamawa State, Nigeria. J. Agr. Res. 4(2):65–70.
8. Ellis-Jones, J., J. Stenhouse, H. Gridley, J. Hellaand, and M. Onim. 2008. Baseline study on vegetable production and marketing. Vegetable breeding and seed systems for poverty reduction in Africa. (aciar.gov.au/aifsc/.../vegie_s coping_study_report_05_july_2012.pdf).

9. Fontem, D.A. and R.R. Schippers. 2004. *Solanum scabrum* Mill, pp. 493–498. In: G.J.H. Grubben and O.A. Denton (eds.). Plant resources of tropical Africa. PROTA Foundation, Wageningen, Netherlands/Backhungo Publishers, Leiden, The Netherlands CTA.
10. Food and Agriculture Organization. 2004. FAOSTAT. Food and Agr. Organization, Rome, Italy. (http://www.fao.org).
11. Gerken, A., J.V. Suglo, and M. Braun. 2001. Crop protection policy in Ghana. Pesticide Policy Project produced by Plant Protection and Regulatory Services Directorate of Ministry of Food and Agriculture, Accra, Ghana.
12. Holmer, R.J., L.G. Janubas, G.E. Potutan, and W.H. Schnitzler. 2001. Pest management strategies of urban and periurban vegetable growers in Cagayan de Oro City. 13th NOMCARRD Regional Symp. on R&D Highlights, Farmers/Investors Forum and Media Conf., Dalwangan, Malaybalay City, Philippines, 9–11 Aug. 2001.
13. Hovorka, A.J. 2005. Gender, commercial urban agriculture and food supply in greater Gaberone, Botswana, p. 126–138. In: L.J.A. Mougeot (ed.). Agropolis: The social, political and environmental dimensions of urban agriculture. Earthscan, London.
14. Iwuchukwu, J.C. and U.C. Uzoho. 2009. Constraints to vegetable production among women in Enugu North Agricultural Zone of Enugu State. J. Agr. Ext. 13(1):16–23.
15. Kekeunou, S., J. Messi, S. Weise, and M. Tindo. 2006. Insect pests' incidence and variations due to forest landscape degradation in the humid forest zone of southern Cameroon: Farmers' perception and need for adopting an integrated pest management strategy. African J. Biotechnol. 5(7):555–562.
16. Kiros-Meles, A. and M.M. Abang. 2007. Farmers' knowledge of crop diseases and control strategies in the regional state of Tigrai, northern Ethiopia: Implications for farmer-researcher collaboration in disease management. Tigrai Agr. Res. Inst., Mekele, Tigrai, Ethiopia.
17. Letts, J., P. Lefeuvre, L. Couston, M. Hoareau, T. Magali, B. Reynaud, D.P. Martin, and V. Arvind. 2009. Complete genomic sequences of Tomato yellow leaf curl Mali virus isolates infecting tomato and pepper from the North Province of Cameroon. Arch. of Virol. 154(3):535–540.
18. Landston, D. and T. Eaker. 2009. Disease control in the home vegetable garden. Cooperative extension, University of Georgia, Athens, Ga.
19. Matthews, G., T. Wiles, and P. Baleguel. 2003. A survey of pesticide application in Cameroon. Crop Protection 22(5):707–714.
20. Meyer, J.R. 2003. Pest control tactics, ENT425 Homepage, Department of Entomology, North Carolina State University. Available at http://www.cals.ncsu.edu/course/ent425/
21. Mossler, M.A. and E. Dunn. 2009. Florida crop/pest management profile: Okra. Univ. of Florida, IFAS Ext. Publication, Gainesville, Fla.
22. Obeng-Ofori, D., E.O. Owusu, and E.T. Kaiwa. 2002. Variation in the level of carboxylesterase activity as an indicator of insecticide resistance in populations of the diamondback moth, *Plutella xylostella* (L.) attacking cabbage in Ghana. J. Ghana Sci. Assoc. 4(2):52–62.
23. Obopile, M., D.C. Munthali, and B. Matilo. 2007. Farmers' knowledge, perceptions and management of vegetable pests and diseases in Botswana. Dept. of Crop Sci. and Production, Botswana College of Agr., Gaborone, Botswana.
24. Praveen, P.M. and N. Dhandapani. 2002. Eco-friendly management of major pests of okra [*Abelmoschus esculentus* (L.) Moench]. J. Veg. Crop Production 7(2):3–11.
25. Ratta, A. 1993. City women farm for food and cash. Intl. Ag-Sieve 6(2):1–2.
26. Youdeowei, A. 2002. Integrated pest management practices for the production of vegetables. IPM Extension Guides, Vol. 4. Ministry of Food and Agr. (MoFA) Plant Protection and Regulatory Services Directorate (PPRSD), Ghana with German Dev. Cooperation, Deutsche Gesellschaft fur Technische Zuzammenarbeit, Accra, Ghana.

CHAPTER 14

Breeding Trends of Fruit and Vegetable Crops for Organic Production in Lithuania

Audrius Sasnauskas*, Vidmantas Bendokas, Rasa Karklelienė, Danguolė Juškevičienė, Tadeušas Šikšnianas, Dalia Gelvonauskienė, Rytis Rugienius, Danas Baniulis, Sidona Sikorskaitė-Gudžiūnienė, Ingrida Mažeikienė, Audrius Radzevičius, Nijolė Maročkienė, Eugenijus Dambrauskas and Vidmantas Stanys

Institute of Horticulture, Lithuanian Research Centre of Agriculture and Forestry, Babtai LT-54333, Lithuania;

ABSTRACT

At the Institute of Horticulture, Lithuanian Research Centre of Agriculture, Babtai, Lithuania, 27 fruit cultivars and 25 vegetable cultivars suitable for organic production were included in the National Plant Variety List and into the European Union Common Catalogue of Agricultural Plant and Vegetable Varieties. The cultivar development programs have focused on improving cold hardiness, resistance to diseases and pests, and yield, as well as on shipping, shelf life qualities, plant development, and biotechnological applications.

Keywords: breeding; cultivar; orchard plants; vegetables; biotechnology; genetics

1. INTRODUCTION

Climatic conditions and the socio-economic environment have given rise to unique horticultural traditions characteristic to Lithuania. Plant breeding for organic production has been an important activity at the Institute of Horticulture, Lithuanian Research Centre of Agriculture and Forestry (IH-LRCAF) in Babtai, Lithuania. The cabbage breeding program was started in 1924, onion in 1926, legume crops in 1929, pear in 1940, blackcurrant in 1946, root crops in 1948, apples and stone fruits in 1952, and strawberry in 1953. The efficiency of horticultural plant breeding depends on scientific progress in the following areas: speeding up generation times, creation of screening methods for valuable characters in the earliest plant development stages, investigation of genetic control mechanisms of plant characters and valuable properties, and development of methods for creation of initial breeding material. In breeding efforts directed toward interspecific crossing, selection of acceptable hybrids, male sterility and creation of stable lines are important issues [1]. The breeding program is focused on pest and disease resistance, quality traits, productivity, biochemical composition, tolerance of unfavorable environmental factors, and adaptability to local agro-climatic conditions [2,3,4,5,6]. Successful cultivar development depends on the availability of genetic material, knowledge of important plant traits, and methods of breeding. Biotechnological methods, based on the newest achievements in genetics, physiology and biochemistry provide new possibilities for breeding. Biotechnology and molecular techniques may guarantee production of high quality fruit and vegetable cultivars [7,8].

The aim of this paper was to review fruit and vegetable crop breeding, investigations of gene pools, and creation of new cultivars and hybrids at IH-LRCAF in Lithuania.

2. BREEDING EFFORTS

Crossbreeding and individual or family group selection have been used for fruit and vegetable breeding. Mass selection has been employed in primary seed production for maintenance of valuable family features as long as possible. Primary breeding material and first generation hybrids have been obtained. A description of the cultivars that were created is reported below, including 7 blackcurrant (*Ribes nigrum* L.), 2 strawberry (*Fragaria × ananassa* Duch.), 3 apple (*Malus domestica* Borkh.), 3 pear (*Pyrus communis* L.), 4 sweet cherry (*Prunus avium* L.), 2 sour cherry

(*Prunus cerasus* L.), 5 plum (*Prunus domestica* L.), 5 carrot (*Daucus carota* L.), 3 red beet (*Beta vulgaris* L.), 2 radish (*Raphanus sativus* L.), 3 cucumber (*Cucumis sativus* L.), 4 tomato (*Solanum lycopersicum* L.), 2 sweet pepper (*Capsicum annuum* L.), and 2 garlic (*Allium sativum*L.) cultivars, and 1 each white cabbage (*Brassica oleracea* L.), onion (*Allium cepa* L.), chive (*Allium schoenoprasum* L.), and bean (*Phaseolus vulgaris* L.) cultivar. The study was carried out in the experimental fields for organic and conventional horticultural plant production at the IH-LRCAF.

3. RESULTS

3.1. Fruit Crops

3.1.1. Pome Fruits

The genetic origin of resistance to apple scab, one of the most damaging diseases of apple worldwide, caused by the fungal pathogen *Venturia inaequalis*, was tested at the embryonic apple development stage using isolated cotyledons. Application of the method allows a 92%–100% recovery of progeny and provides a valuable tool for efficient assessment of disease resistance [9]. Screening methods of scab-resistant genotypes by using cotyledons in vitro were employed for estimation of donor characteristics and potential for breeding and preliminary estimation of interaction of nuclear and cytoplasmic genes. Donors of monogenic V_f and V_m resistance to *Venturia inaequalis* were introduced into the breeding program. Marker-assisted selection (MAS) has been extensively used in apple breeding to identify inherited resistance genes and to reduce the time required for selection of resistant seedlings, and it is also an important tool for searching for resistance donors. The most commonly used markers were based on SSRs and SNPs. This led to new scab-immune apple cultivars. Each of these cultivars was suitable in yield, storage life and fruit quality. DNA analysis using a *Vfa*1 gene-like sequence-specific marker was performed.

Our results showed that the cultivars “Aldas”, “Skaistis” and “Rudenis” were immune to *Venturia ineaqualis* races 1 to 5, and had a DNA fragment of 500 bp specific for the *Vfa*1 gene. The evolving virulence of the apple scab pathogen stimulated efforts to adapt new resistance breeding approaches, such pyramidization of resistance genes. A total of 159 apple hybrids with pyramidic resistance were obtained: 133 seedlings carried *Rvi5* and *Rvi6* genes, and 26 hybrids inherited *Rvi5* and *Rvi10* genes [10,11].

Fruit production and quality and scab resistance are important directions of the pear breeding program. Crossing of local with introduced cultivars has resulted in new pear cultivars, such as "Lukna", "Gaisra" and "Liepona" [12].

3.1.2. Stone Fruits

Crossing local, cold-resistant and disease-resistant cultivars developed abroad has led to the creation of the sweet cherry cultivars "*Jurgita*", "*Mindaugė*", "*Jurga*", "*Meda*" [13], the sour cherry cultivars "*Vytėnų žvaigždė*" and "*Notė*" [14], and the plum cultivars "*Gynė*", "*Jūrė*", "*Rausvė*", "*Kauno vengrinė*", and "*Aleksona*" [15]. A testing protocol for differentiation of resistance to brown rot (*Monilinia fructicola*) in sweet and sour cherry genotypes was developed, and is being applied for development of new sweet and sour cherry cultivars resistant to the disease.

3.1.3. Small Fruits

Organic blackcurrant production requires cultivars resistant to fungal diseases and pests. The limiting factor is lack of resistance in *R. nigrum*, as plants of other *Ribes* species are often more resistant to key pests and diseases. Therefore, interspecific hybridisation has been used to introduce resistance to pests and diseases into commercially-acceptable blackcurrant cultivars [16]. An investigation of inheritance of disease resistance in interspecific hybrids of *R. americanum*, *R. aureum*, *R. janczewskii*, *R. nigrum*, *R. pauciflorum*, *R. petraeum*, *R. sanguineum*, *R. ussuriense* and *R. uva-crispa* resulted in a large number of the hybrids displaying resistance to powdery mildew, a small percentage showing resistance to *Septoria* leaf spot, and no plants demonstrating any resistance to anthracnose.

Large scale hybridizations of the progeny were made with hybrids from the *R. nigrum* spp. scandinavica line. Breeding with this line for resistance to fungal diseases led to the development and release of a number of powdery mildew-resistant blackcurrants. An improvement in blackcurrant quality achieved during past decades has included an increase in average fruit size from 0.9 to 1.5 g, and a 5 to 7 days advancement of flowering time [17]. The new cultivars include:

"*Gagatai*" ("*Minaj Shmyriov*" × "*Öjebyn*"). The fruit are large (1.3 g), and very rich in ascorbic acid (197 mg/100 g) and other chemical compounds. Plants are highly resistant to frost, powdery mildew, and gall mite (*Cecidophyopsis ribis*).

"*Pilėnai*"("*Minaj Shmyriov*" × "*Öjebyn*"). The fruit are large (1.0 g), and contain 153.1 mg/100 g ascorbic acid. The cultivar is highly resistant to powdery mildew, of medium resistance to leaf spot, and slightly affected by gall mite.

"Tauriai" (*"Titania"* × *"Beloruskaja Sladkaja"*). The fruit are large (1.5 g), and contain 161 mg/100 g ascorbic acid. Plants are highly resistant to powdery mildew and gall mite.

"Blizgiai" (*"Öjebyn"* × *"Minaj Shmyriov"*). The flowering and ripening seasons are earlier than of the cv. *"Minaj Shmyriov"* (1–4 days). The fruit are large (1.5 g), and contain 176 mg/100 g ascorbic acid. The cultivar is resistant to powdery mildew and gall mite in blackcurrant cultivars in Lithuania started in the 1960"s [18]. To overcome the gall mite problem, a new direction in development of genetic lines was initiated by hybridization with the Siberian blackcurrant lineage and selection for field resistance to gall mite. This led to the release of the complex gall mite resistant cultivars *"Vakariai"* and "Dainiai". The main purpose of the breeding program for resistance to gall mite was to introduce the *P* gene from *R. nigrum*spp. *sibiricum* into *R. nigrum*. At the beginning, the agronomic traits of hybrids resistant to gall mite were poor, but these hybrids were backcrossed, and later three commercial cultivars, *"Dainiai"*, *"Kriviai"* and *"Kupoliniai"* [19,20], were released. *"Dainiai"*, released in 2001, is a cultivar with high fruit quality and complete resistance to gall mite, and was also used as a donor of resistance to gall mite. Progeny of *"Dainiai"* were used to map the *P* gene. A 107 bp AFLP marker was identified and its potential in screening progeny for resistance to gall mite hybrids was evaluated [21].

The breeding program for strawberry involved hybridization with European cultivars. PCR-based markers for the *Rpf1*gene were developed and used for screening strawberries for red stele resistance. Two cultivars, *"Dangė"* and *"Saulenė"*, were created, distinguished by their good resistance to fungal diseases, good size and taste, and high yield [22]. A concentrated and high yield with excellent tasting fruits are characteristic of *"Saulene"* (*"Shuksan"* × *"Senga Sengana"*), suitable for fresh use. The mid- to late-season cultivar *"Dange"* (*"Venta"* × *"Redgauntlet"*) distinguishes itself by good fruit size, attractiveness, and stable, high yield every year. New wild strawberry cultivars were developed and submitted for DUS testing (Distinctness, Uniformity and Stability) in Poland.

3.2. Vegetable Crops

3.2.1. Root Vegetables

Investigations of carrots and red beets under different production techniques showed that carrot production was less dependent on technology. The total yield of all cultivars of carrots was higher in a conventional system. The greatest differences were obtained with the cultivar "Šatrija BS", with yield of 48.3 t/ha grown conventionally and 41.0 t/ha grown organically (Table 1). The remaining varieties were less variable. Marketable production was also less varied.

Table 1. Carrots and red beet productivity in Lithuania, Babtai, 2010–2014.

Cultivars	Organic Production		Conventional Production	
	Total Yield (t/ha)	Marketable Production (%)	Total Yield (t/ha)	Marketable Production (%)
		Carrot		
"Ieva" H	65.5	88.9	70.0	90.0
"Rokita" H	50.7	88.0	65.3	89.8
"Svalia" H	53.2	87.1	57.7	88.4
"Garduoles"	57.0	87.5	60.0	88.5
"Šatrija BS"	41.0	84.2	48.3	85.4
LSD_{05}	3.186		4.224	
		Red beet		
"Joniai"	34.1	75.9	47.2	79.9
"Rikiai"	41.3	82.8	55.5	86.8
"Ilgiai"	40.0	79.4	51.0	80.3
LSD_{05}	7.272		2.469	

Carrot cultivars of the Nantes type are the most popular in Lithuania. Therefore, attention was given to root shape. Roots of the new varieties ("Ieva" H, "Rokita" H, "Svalia" H, "Garduolės", "Šatrija BS") are medium in size and diameter, with a cylindrical, pointed shape. Carrots in Lithuania grow well in a light soil, but the hybrid "Rokita" grows well in a heavier soil. Carotene accumulation ranges from 20–22.5 mg/100 g, dry soluble material from 11%–12.5%, and total sugar from 7.5%–8.5%. Carrots are suitable for harvesting in autumn and for storing [23,24].

Red beet cultivars of different types were created at the Institute [25,26]. The cultivar "Joniai" is distinguished by its small foliage, round and oval-round shape of the root, and not clearly concentrated rings [27]. "Ilgiai" is of medium size with emerald green leaves, with medium-long cylindrical roots. "Rikiai" is a medium early two-seeded (75%) cultivar. The colour of the foliage is dark green with a light yellow tone. The shape of the root is elliptical to round-elliptical with a thin main root with regular skin. "Rikiai" are tolerant to the drought [28]. All new cultivars are suitable for fresh use, storing, and processing.

Cultivars of radish were distinguished by their earliness. Extra earliness is characteristic of "Babtų žara". Its root is round to oval-round, the colour is bright red, the flesh is white, juicy, and delicate in taste. The radish "Liliai" represented medium earliness and has a longer growing cycle. They are suitable for growing in the open field and in greenhouses. Sowing time is possible early in spring or the second and third weeks of August. Roots are round, massive, and red in colour with a purple tone.

3.2.2. Fruit Vegetables

The new cucumber hybrids "Gintai" and "Krukiai BS" are suitable for growing in the open field and in greenhouses [29,30] "Gintai" is a mid-season heterotic hybrid with high yield. It has external attractiveness, is 7–10 cm long, and is light green in colour. "Krukiai BS" is a mid—to late-season hybrid. The fruit is green with white stripes, and does not turn yellow for a long time. The fruit is 8–10 cm long with a diameter of 3.5–4.0 cm. Bees are necessary for pollination. "Daugiai" is distinguished with early ripening and intensive yield. The fruit is 10–13 cm long, dark green, and does not turn yellow for a long time. They are suitable for the fresh market and pickling. Production is optimal in spring and summer in greenhouses.

The new tomato cultivars "Svara", "Viltis", "Balčiai" are suitable for growing in unheated greenhouses and in the open field. "Svara" is a late-season cultivar distinguished with high yield and fungal disease resistance. Fruit are red and weigh approximately 70 g. The fruit are round, and taste somewhat acidic. The cultivar "Viltis" is early ripening. Plants are of medium height and have a short duration of production. Fruit weight at first harvest reach 120–150 g, and later weights decrease to 80–90 g. Fruit are red with a sweet taste, and are flat-round. "Balčiai" is a mid-season cultivar with fruit having red colour, good taste, medium size (75–95 g), with a flat-round shape. "Skariai" is a mid-season cultivar suitable for growing in unheated greenhouses. Fruit are tasty, red in colour, thin skinned, are oblong-oval, cylindrical, and massive (90–150 g). It is sensitive to lack of fertilizer. "Rutuliai" is a mid-season, tall cultivar with red, round, mid-sized fruit. They are suitable for growing in the unheated greenhouses.

Cultivars of sweet pepper "Alanta" and "Reda" are mid-season with a plant height of 0.75–0.80 m, and are self-pollinating [31]. The cylindrical fruit of "Alanta" are distinguished with massive, orange, tasty fruit. This variety is suitable for growing in unheated greenhouses. The shape "Reda" is cut-cylindrical, with external attractiveness, and dark red colour. It grows well in unheated greenhouses, under covers of plastic film, and in the open field.

3.2.3. Onions

In *Allium* L. breeding, the new cultivars "Žiemiai" and "Vasariai" of garlic were created. A flat-round bulb shape is typical for "Žiemiai", the outer skin of the bulb is white, and the cloves are big with 6–9 pieces in a bulb. Bulbils are small, about 150–170 in an inflorescence. The cultivar is disease and pest resistant. "Vasariai" is a mid-season cultivar suitable for planting in spring. The outer skin of the bulb is white, and the bulb consists of 10–16 cloves that are arranged and concentrated into two rounds. The bulb size is 21–27 g.

Homozygotic lines of onions were created by applying a gynogenesis method [32,33]. The cultivar "Babtų didieji" is a mid-season type, with bulbs suitable for fresh use, storage, and processing. The shape of the bulb is round, it weighs 95–116 g, the skin is yellow-orange, and the fresh is white and mildly sweet.

The chive "Aliai" is a perennial plant grown as a bunch. Leaves of "Aliai" start growth early in spring, and the first yield can be harvested 20 days later. Leaves are suitable for fresh use until flowering. The leaves of this cultivar are thin, small and tasty.

3.2.4. Cabbage

The white cabbage "Bagočiai" is a mid- to late-season cultivar. The shape of the head is oval or flat-oval. The colour of the outer leaves is grey-green. "Bagočiai" is distinguished with a good biochemical composition. They are suitable for pickling. Fresh heads can be stored until March.

3.2.5. Legumes

The bean "Baltija" belongs to the bush pod group. The cultivar has a mid-early season. The pod length may reach 11–13 cm. Pods are straight or straight-curved. Seeds are white and oblong. The weight of 1000 seeds is 300–350 g.

4. CONCLUSIONS

Cultivars deemed best suited for organic production include: blackcurrant ("Gagatai", "Pilėnai","Tauriai", "Blizgiai", "Vakariai", "Dainiai", "Kriviai", "Kupoliniai"), strawberry ("Dangė" and "Saulenė"), apple ("Aldas", "Skaistis", "Rudenis"), pear ("Lukna", "Gaisra", "Liepona"), sweet cherry ("Jurgita", "Mindaugė", "Jurga", "Meda"), sour cherry ("Vytėnų žvaigždė", "Notė"), plum ("Gynė", "Jūrė", "Rausvė", "Kauno vengrinė", "Aleksona"), carrot ("Ieva" H, "Rokita" H, "Svalia" H, "Garduolės", "Šatrija BS"), red beet ("Joniai", "Rikiai", "Ilgiai"), radish ("Babtų žara", "Liliai"), cucumber ("Gintai", "Krukiai BS", "Daugiai"), tomato ("Svara", "Viltis", "Balčiai", "Skariai"), sweet pepper ("Alanta",

"Reda"), garlic ("Žiemiai", "Vasariai"), white cabbage ("Bagočiai"), onion ("Babtų didieji"), chive ("Aliai"), and bean ("Baltija"). In addition, the new vegetable cultivars were included in the EU Common catalogue of varieties of vegetable species in 2015.

ACKNOWLEDGMENTS

This work was carried out within the framework of the long-term research program "Horticulture: agro-biological basics and technologies" implemented by the Lithuanian Research Centre for Agriculture and Forestry.

REFERENCES

1. Liekis, A. Augalų Selekcija (Mokslo Straipsnių Rinkinys). Available online: http://elibrary.lt/inf_res4.phtml?id=69103(accessed on 10 March 2016). (In Lithuanian)
2. Armolaitienė, J. Morkų veislė "Garduolės 2". *Sod. ir darž.* **1997**, *16*, 47–51. (In Lithuanian)
3. Gaučienė, O. Morkų hibridas "Svalia" F_1. *Sod. ir darž.* **1997**, *16*, 57–62. (In Lithuanian)
4. Petronienė, D. "Ilgiai"—Nauja raudonųjų burokėlių veislė. *Sod. ir darž.* **2001**, *20*, 42–47. (In Lithuanian)
5. Petronienė, D.; Viškelis, P. Biochemical composition and preservation of various red beet cultivars. *Sod. ir darž.* **2004**, *23*, 89–97.
6. Karklelienė, R. Morkų ir burokėlių lietuviškų veislių bei hibridų ypatumai ekologinėje ir intensyvioje daržininkystėje. *Sod. ir darž.* **2006**, *25*, 193–200. (In Lithuanian)
7. Baniulis, D.; Gelvonauskienė, D.; Rugienius, R.; Sasnauskas, A.; Stanienė, G.; Stepulaitienė, I.; Frercks, B.; Sikorskaitė, S.; Lukoševičiūtė, V.; Mažeikienė, I.; et al. Orchard plant breeding, genetics, and biotechnology research at the Institute of Horticulture. LRCAF. *Sod. ir darž.* **2013**, *32*, 21–48.
8. Karklelienė, R.; Radzevičius, A.; Juškevičienė, D.; Maročkienė, N.; Bobinas, Č. Daržo augalų selekcijos tyrimų apžvalga. *Sod. ir darž.* **2013**, *32*, 69–86. (In Lithuanian)
9. Gelvonauskiene, D.; Stanys, V. Screening of scab resistant apple seedlings using isolated cotyledons. *Fruit Sci.* **2000**, *207*, 56–60.
10. Sikorskaite, S.; Gelvonauskiene, D.; Stanys, V.; Baniulis, D. Characterization of microsatellite loci in apple (*Malus* × *domestica* Borkh.) cultivars. *Zemd. Agric.* **2012**, *99*, 131–138.
11. Sikorskaite, S.; Gelvonauskiene, D.; Bendokas, V.; Stanys, V.; Baniulis, D. *Malus* sp.-*V. inaequalis* interaction characteristics among local apple cultivars in Lithuania. *Acta Hortic.* **2013**, *976*, 567–572.
12. Lukoševičius, A. *Kriaušių veislės (katalogas)*; Lietuvos Sodininkystės ir Daržininkystės Institutas: Babtai, Lithuania, 2003. (In Lithuanian)
13. Stanys, V.; Frercks, B.; Šikšnianienė, J.B.; Stepulaitienė, I.; Gelvonauskienė, D.; Stanienė, G.; Bobinas, Č. Identification of sweet cherry (*Prunus avium* L.) cultivars using AFLP and SSR markers. *Zemd. Agric.* **2012**, *99*, 437–444.

14. Stepulaitienė, I.; Žebrauskienė, A.; Stanys, V. Frost resistance is associated with development of sour cherry (*Prunus cerasus* L.) generative buds. *Zemdirb. Agric.* **2013**, *100*, 175–178.
15. Lukoševičius, A. *Slyvų Veislės (Katalogas)*; Lietuvos Sodininkystės ir Daržininkystės Institutas: Babtai, Lithuania, 2002; p. 110. (In Lithuanian)
16. Stanys, V.; Staniene, G.; Shikshnianas, T.; Bobinas, C. Interspecific hybridization in *Ribes* genus. *Acta Hortic.* **2004**, *663*, 861–864.
17. Sasnauskas, A.; Siksnianas, T.; Stanys, V.; Bobinas, C. Evaluation of agronomical characters of blackcurrant cultivars and selections in Lithuania. *Acta Hortic.* **2012**, *946*, 189–194.
18. Bendokas, V.; Mazeikiene, I.; Baniulis, D.; Stanys, V.; Siksnianas, T. Application of *P* gene donors in breeding of black currant resistant to gall mite. *Acta Hortic.* **2013**, *976*, 523–527.
19. Siksnianas, T.; Sasnauskas, A. Black currant cultivars Gagatai, Kriviai, Kupoliniai. *Sod. ir darž.* **1998**, *17*, 23–33.
20. Siksnianas, T. Development of productive, resistant to fungal diseases and gall mite black currant cultivars. *Sod. ir darž.* **2005**, *24*, 16–24.
21. Mazeikiene, I.; Bendokas, V.; Stanys, V.; Siksnianas, T. Molecular markers linked to resistance to the gall mite in blackcurrant. *Plant Breed.* **2012**, *131*, 762–766.
22. Rugienius, R.; Sasnauskas, A.; Shikshnianas, T. "Saulene" and "Dange"—Two recent Lithuanian strawberry cultivars. *Acta Hortic.* **2004**, *649*, 73–76.
23. Karklelienė, R.; Radzevičius, A.; Bobinas, Č. Productivity and root-crop quality of Lithuanian carrot (*Daucus Sativus*Röhl.) breeder lines. *Proc. Latv. Acad. Sci.* **2009**, *63*, 63–65.
24. Karklelienė, R.; Radzevičius, A.; Dambrauskienė, E.; Survilienė, E.; Bobinas, Č.; Duchovskienė, L.; Kavaliauskaitė, D.; Bundinienė, O. Root yield, quality and plant resistance to diseases of organically grown carrot hybrids and cultivars. *Zemdirb. Agric.* **2012**, *99*, 393–398.
25. Petronienė, O.D.; Viškelis, P. Įvairių veislių tipų ir grupių raudonųjų burokėlių (*Beta vulgaris* L.) biocheminė sudėtis. *Maisto Chemija ir Technologija* **2005**, *38*, 42–47. (In Lithuanian)
26. Karklelienė, R.; Viškelis, P.; Radzevičius, A.; Duchovskienė, L. Evaluation of productivity and biochemical composition of perspective red beet breeding number. *Proc. Fourth Balkan Symp. Vegetables Potatoes.* **2009**, *1*, 255–260.
27. Petronienė, D. "Joniai"—Nauja burokėlių veislė. *Sod. ir darž.* **2000**, *19*, 81–86. (In Lithuanian)
28. Karklelienė, R.; Radzevičius, A.; Maročkienė, N.; Juškevičienė, D.; Dambrauskienė, E. *Raudonojo burokėlio* (*Beta vulgaris* L. subsp. *vulgaris* convar. *vulgaris* var. *vulgaris*) veislė "Rikiai". *Sod. ir darž.* **2013**, *32*, 49–54. (In Lithuanian)
29. Dambrauskas, E. *Agurkų hibridai* "Krukiai" ir "Žalsviai". *Sod. ir darž.* **2001**, *20*, 76–82. (In Lithuanian)
30. Dambrauskas, E. *Trumpavaisiai partenokarpiniai* agurkai "Pūkiai" F_1, "Troliai" F_1 ir "Ulonai" F_1. *Sod. ir darž.* **2001**, *20*, 32–41. (In Lithuanian)
31. Maročkienė, N.; Karklelienė, R.; Bobinas, Č. *Saldžiosios paprikos* veislės "Alanta" biologinių ūkinių savybių įvertinimas. *Sod. ir darž.* **2009**, *28*, 127–134. (In Lithuanian)
32. Juškevičienė, D.; Stanys, V.; Bobinas, Č. *Gynogenesis pecularities* of *Allium* L. vegetables grown in Lithuania. *Biologija***2005**, *3*, 6–9.
33. Juškevičienė, D.; Stanys, V. *Valgomojo svogūno* (*Allium cepa* L.) ginogenezė ir homozigotinių linijų kūrimas. *Sod. ir darž.***2007**, *20*, 180–187. (In Lithuanian)

CHAPTER 15

The Ionomic Study of Vegetable Crops

Toshihiro Watanabe[1*], Eriko Maejima[1], Tomoko Yoshimura[1], Masaru Urayama[1], Aiko Yamauchi[1], Masako Owadano[1], Ryosuke Okada[1], Mitsuru Osaki[1], Yoshinori Kanayama[2], Takuro Shinano[3]

[1]*Research Faculty of Agriculture, Hokkaido University, Kita-9, Nishi-9, Kitaku, Sapporo, 0608589, Japan,*

[2]*Graduate School of Agricultural Science, Tohoku University, Sendai, 9818555, Japan,*

[3]*Agricultural Radiation Research Center, NARO Tohoku Agricultural Research Center, 50 Aza Harajyukuminami, Arai, Fukushima, 9602156, Japan*

ABSTRACT

Soil contains various essential and nonessential elements, all of which can be absorbed by plants. Plant ionomics is the study of the accumulation of these elements (the ionome) in plants. The ionomic profile of a plant is affected by various factors, including species, variety, organ, and environment. In this study, we cultivated various vegetable crop species and cultivars under the same field conditions and analyzed the level of accumulation of each element in the edible and nonedible parts using ionomic techniques. The concentration of each element in the edible parts differed between species, which could be partly explained by differences in the types of edible organs (root, leaf, seed, and fruit). For example, the calcium concentration was lower in seeds and fruit than in other organs because of the higher dependency of calcium accumulation on xylem transfer. The concentration of several essential microelements and nonessential elements in the edible parts also varied greatly between cultivars of the same species, knowledge of which will help in the breeding of vegetables that are biofortified or contain lower concentrations of toxic elements. Comparison of the ionomes of the fruit and leaves of tomato (*Solanum lycopersicum*) and eggplant (*S. melongena*) indicated that cadmium and boron had higher levels of accumulation in eggplant fruit, likely because of their effective transport in the phloem. We also found that homologous elements that have been reported to share the same uptake/transport system often showed significant correlation only in a few families and that the slopes of these relationships differed between families. Therefore, these differences in the characteristics of mineral accumulation are likely to affect the ionomic profiles of different families.

INTRODUCTION

At least 17 elements are known to be essential nutrients for plants; however, plants also absorb and accumulate various nonessential elements [1, 2]. In contrast, humans need at least 25 elements [3], a major source of which is crops. Mineral malnutrition is a widespread public health issue in both developing and developed countries, with up to two-thirds of the world's population being considered at risk of inadequate intakes of one or more essential mineral elements [3]. Iron (Fe) and zinc (Zn) deficiencies are common in humans, particularly in developing countries [4]. One possible way of addressing this issue is through biofortification, whereby crops are developed that contain essential mineral elements, increasing their nutritional value for humans [5].

Some crops also contain toxic elements, which can represent a public health threat. Crops that are cultivated on arable land that is polluted with heavy metals and metalloids often contain a high concentration of these elements in their edible parts. In some Asian countries, for example, cadmium (Cd) contamination in rice (*Oryza sativa*) grains poses a serious threat to public health; therefore, many studies have been conducted to address this issue [6, 7]. In addition, crops grown in soil that contains radioactive elements such as cesium (Cs) or strontium (Sr) as a result of a nuclear accident can pose a risk to human health. Therefore, comprehensive information about the potential for different species and cultivars of crops to accumulate each type of element in its edible parts is very important for human health.

Ionomics is the study of the accumulation of all metals, metalloids, and nonmetals in living organisms using high-throughput elemental analysis technologies (e.g. Inductively Coupled Plasma-Mass Spectrometry (ICP-MS) and ICP-Atomic Emission Spectrometry (ICP-AES)) [8] and can be applied to various types of plant science studies. Ionomics has the advantage in revealing network among various mineral elements in organism. For example, ionomic analyses have been conducted to isolate the genes that are responsible for mineral transport and homeostasis in plants [9, 10]. Furthermore, ionomics can also be applied to investigate phylogenetic and environmental influences on plant mineral accumulation [11–13]. Thus, ionomics is a powerful tool to study mineral dynamics in plants. Recently, ionomics have also been applied to seeds of staple crops, such as cereals and beans, and are expected to contribute to improving the malnutrition and food safety [14, 15]. However, there are very few reports of application of this technology to the edible parts of non-staple crops, particularly in vegetable crops.

In the present study, therefore, we used ionomics to comprehensively characterize mineral accumulation in the edible parts of various crop species. We cultivated a number of different vegetable crop species, including various cultivars, under the same field conditions and analyzed the ionome of their edible parts. Moreover, we also analyzed the ionome in nonedible leaves and compared it with that in edible fruits to understand the ionome distribution between source and sink organs. The findings of this study will be useful for enhancing the nutritional value and safety of these crops in the future.

MATERIALS AND METHODS

Cultivation

In 2011, 24 vegetable crop species, including two rootstocks, were cultivated in the experimental field of Hokkaido University, Japan. The crop species that were used in this study are shown in Table 1. The experimental field had a Gleyic Fluvisol soil type, the general chemical properties (exchangeable elements, Truog P, and soil pH) of which before cultivation are described in S1 Table. The soil analysis was conducted by the method of Watanabe et al. [2]. For all crop species, nitrogen (N), phosphorus (P), and potassium (K) fertilizers were uniformly applied as ammonium sulfate (140 kg N ha^{-1}), superphosphate (130 kg P_2O_5 ha^{-1}), and potassium sulfate (100 kg K_2O ha^{-1}). Seeds of carrot (*Daucus carota* subsp. *sativus*), radish (*Raphanus sativus* var. *longipinnatus*), turnip (*Brassica rapa* var. *rapa*), komatsuna (*B. rapa* var. *perviridis*), bok-choy (*B. rapa* var. *chinensis*), nabana (*B. rapa* var. *nippo-oleifera*), napa cabbage (*B. rapa* var. *pekinensis*), garland chrysanthemum (*Glebionis coronaria*), podded pea (*Pisum sativum*), green pea (*P. sativum*), and kidney bean (*Phaseolus vulgaris*) were directly sown in the field. Seeds of the remaining species were sown in the 300 mL pot containing soils, grown in a greenhouse at Hokkaido University, and transplanted to the field. The sowing and transplanting details are provided in Table 1. All species used in this study are usually cultivated in this prefecture at this season.

Table 1. List of the vegetable species used in this study and details about the sowing, transplanting, and sampling procedures.

Species number	Family	Common name[a]	Scientific name	Major edible part	Analyzed part	Intrarow × row (cm)	Sowing date	Transplanting date	Sampling date	No. samples
1	Malvaceae	Okra (2)	*Abelmoschus esculentus* (L.) Moench	Fruit	Fruit	50 × 60	May-5	Jun-17	Aug-10	1 plant/ replicate
2	Apiaceae	Carrot (6)	*Daucus carota* subsp. *sativus* (Hoffm.) Arcang.	Root	Root	13 × 13	May-24	-	Sep-13	1 plant/ replicate
3	Brassicaceae	Radish (17)	*Raphanus sativus* L. var. *longipinnatus* L.H.Bailey	Root	Root	30 × 90	May-24	-	Jul-11	1 plant/ replicate
4	Brassicaceae	Turnip (14)	*Brassica rapa* L. var. *rapa*	Root	Root	30 × 90	May-24	-	Jul-7	1 plant/ replicate
5	Brassicaceae	Broccoli (10)	*Brassica oleracea* L. var. *italica*	Flower bud	Flower bud	50 × 90	Apr-26	Jun-9	13-Jul or 4-Aug	1 plant/ replicate
6	Brassicaceae	Cauliflower (3)	*Brassica oleracea* L. var. *botrytis*	Flower bud	Flower bud	50 × 90	Apr-26	Jun-9	Jul-19	1 plant/ replicate
7	Brassicaceae	Cabbage (18)	*Brassica oleracea* L. var. *capitata*	Leaf	Leaf	50 × 90	Apr-26	May-25	Jul-26	1 plant/ replicate
8	Brassicaceae	Komatsuna (8)	*Brassica rapa* L. var. *perviridis*	Leaf	Leaf	13 × 13	May-24	-	Jun-27	2 plants/ replicate
9	Brassicaceae	Bok-choy (7)	*Brassica rapa* L. var. *chinensis*	Leaf	Leaf	13 × 13	May-24	-	Jul-6	2 plants/ replicate
10	Brassicaceae	Nabana (2)	*Brassica rapa* L. var. *nippo-oleifera*	Leaf	Leaf	13 × 13	May-24	-	Jun-27	2 plants/ replicate
11	Brassicaceae	Napa cabbage (14)	*Brassica rapa* L. var. *pekinensis* Rupr.	Leaf	Leaf	50 × 90	May-26	-	Aug-4	1 plant/ replicate
12	Asteraceae	Garland chrysanthemum (2)	*Glebionis coronaria* (L.) Cass. ex Spach	Leaf	Leaf	13 × 13	May-24	-	Jul-11	2 plants/ replicate
13	Asteraceae	Lettuce (24)	*Lactuca sativa* L.	Leaf	Leaf	50 × 90	Apr-20	May-25	Jul-6	1 plant/ replicate
14	Cucurbitaceae	Cucumber (14)	*Cucumis sativus* L.	Fruit	Fruit	50 × 60	Apr-27	Jun-7	Jul-19	2 fruits from 2 plants/ replicate
15	Solanaceae	Tomato (16)	*Solanum lycopersicum* L.	Fruit	Fruit, Leaf	50 × 90	Apr-20	Jun-8	Aug-17	1 fruit from 1 plant/ replicate, 5 fully expanded leaves from 1 plant/ replicate

16	Solanaceae	Grape tomato (2)	*Solanum lycopersicum* L.	Fruit	Fruit, Leaf	50 × 90	Apr-20	Jun-8	Aug-17	5 fruits from 1 plant/ replicate, 5 fully expanded leaves from 1 plant/ replicate
17	Solanaceae	Tomato rootstock (10)	*Solanum lycopersicum* L.	Fruit	Fruit, Leaf	50 × 90	Apr-20	Jun-8	Aug-17	3 fruits from 1 plant/ replicate, 5 fully expanded leaves from 1 plant/ replicate
18	Solanaceae	Eggplant (7)	*Solanum melongena* L.	Fruit	Fruit, Leaf	50 × 90	Apr-20	Jun-22	Jul-25	1 fruit from 1 plant/ replicate, 2 fully expanded leaves from 1 plant/ replicate
19	Solanaceae	Eggplant rootstock (5)	*Solanum melongena* L., *S. torvum* Sw., *S. integrifolium* Poir.	-	Leaf	50 × 90	Apr-20	Jun-22	Jul-25	2 fully expanded leaves from 1 plant/ replicate
20	Fabaceae	Podded pea (6)	*Pisum sativum* L.	Pod	Pod with seed	30 × 90	Jun-4	-	Jul-28	10 pods from 3 plants/ replicate
21	Fabaceae	Green pea (3)	*Pisum sativum* L.	Seed	Seed, Pod with seed	30 × 90	Jun-4	-	Aug-8	10 pods from 3 plants/ replicate
22	Fabaceae	Kidney bean (8)	*Phaseolus vulgaris* L.	Pod	Pod with seed	30 × 90	Jun-4	-	Jul-28	10 pods from 3 plants/ replicate
23	Amaryllidaceae	Chinese chive (6)	*Allium tuberosum* Rottler ex Spreng.	Leaf	Leaf	50 × 90	Apr-20	Jun-29	Oct-14	1 plant/ replicate
24	Amaryllidaceae	Green onion (25)	*Allium fistulosum* L.	Leaf	Leaf	5 × 100	Apr-20	Jun-22	Oct-14	1 plant/ replicate

[a] The number of cultivars is indicated in parentheses. The cultivar names are not included in this paper to avoid any conflicts of interest.

Each cultivar was cultivated in at least four rows with sufficient row length. The edible part of each cultivar was randomly collected avoiding the borders of each cultivation plot, with three replicates. In addition, both the fruit and the leaves were sampled from tomato (*Solanum lycopersicum*) (both grape tomato and rootstock) and eggplant (*S. melongena*), whereas only the leaves were sampled from the rootstock of eggplant. Leaf sampling involved collecting the largest fully expanded leaf from each plant. The sampling details are provided in Table 1. Sampling for crops harvested multiple times, such as cucumber, tomato, eggplant, and beans, was conducted at the mid-harvest stage. All samples were dried in an oven at 70°C for 7 days, following which they were weighed and then ground for mineral analysis.

Mineral analysis

Plant samples were digested in 2 mL of 61% (w/v) HNO_3 (EL grade; Kanto Chemical, Tokyo, Japan) at 110°C in a DigiPREP apparatus (SCP Science, QC, Canada) for approximately 2 h until the solution had almost disappeared. When the samples had cooled, 0.5 mL H_2O_2(semiconductor grade; Santoku Chemical, Tokyo, Japan) was added and the samples were heated at 110°C for a further 20 min. Once digestion was complete, the tubes were cooled and made up to a volume of 10 mL by adding 2% (w/v) HNO_3 in Milli-Q water. The concentrations of the following elements were measured using an ICP-MS (ELAN DRC-e; Perkin Elmer, Waltham, MA, USA): lithium (Li), boron (B), sodium (Na), K, chromium (Cr), cobalt (Co), nickel (Ni), arsenic (As), molybdenum (Mo), Cd, magnesium (Mg), aluminum (Al), P, calcium (Ca), manganese (Mn), Fe, copper (Cu), Zn, Sr, barium (Ba), and Cs.

Statistical analyses

All statistical analyses were performed using Sigmaplot 11.0 (Systat Software, Inc., San Jose, CA, USA) and Minitab 14 (Minitab Inc., State College, PA, USA).

RESULTS AND DISCUSSION

Ionome of the edible parts

First, to examine the reliability of cultivation and sampling methods of this study, cultivar Komatsuna4 was cultivated in 2012 at a different site in the same field under the same cultivation conditions as in 2011. The results showed that the mineral concentrations in leaves of Komatsuna4 in 2012 were almost the same as those in 2011 (S2 Table), implying that the methods in this study are reliable. The concentrations of each mineral in the edible parts of the sampled crops were shown as boxplots in Figs 1–6 and all data were shown in S3 Table. The concentrations in the nonedible leaves were also shown for comparison (Figs 1–3). A boxplot is a simple and convenient graphical tool that is widely used to visualize the distribution of continuous univariate data [16]. Both interspecific and intraspecific variations in the concentrations of elements in edible parts were observed though the nature of the edible parts differed between crop species (Figs 1–6). In edible parts, the concentration of K was lower in both the seeds and the pods with seeds of legumes than in the other species (Figs 1 and 4). The concentration of Ca was higher in the leaves of komatsuna, bok-choy, and

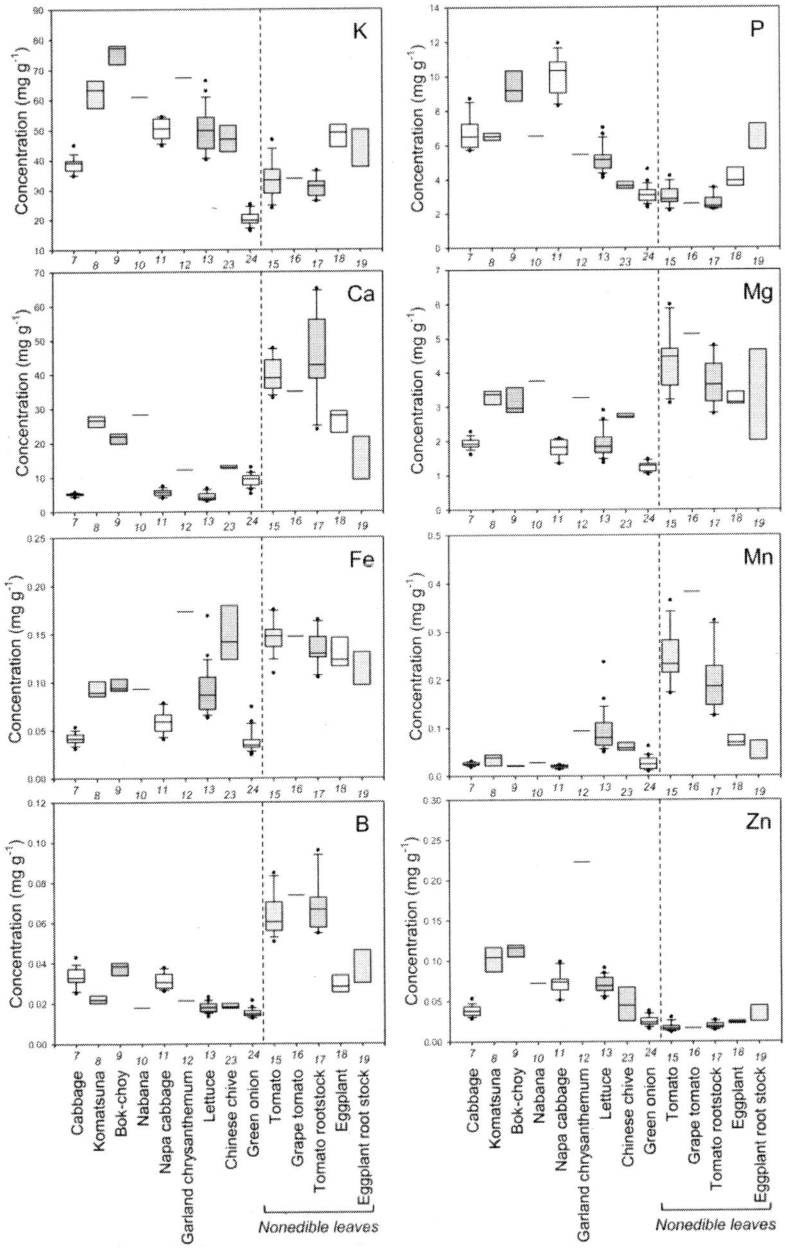

Figure 1. Boxplots showing the concentration of K, P, Ca, Mg, Fe, Mn, B, and Zn in leaves of various crop species. Concentrations in nonedible leaves are also shown. Numbers in x-axis indicate the species number in Table 1.

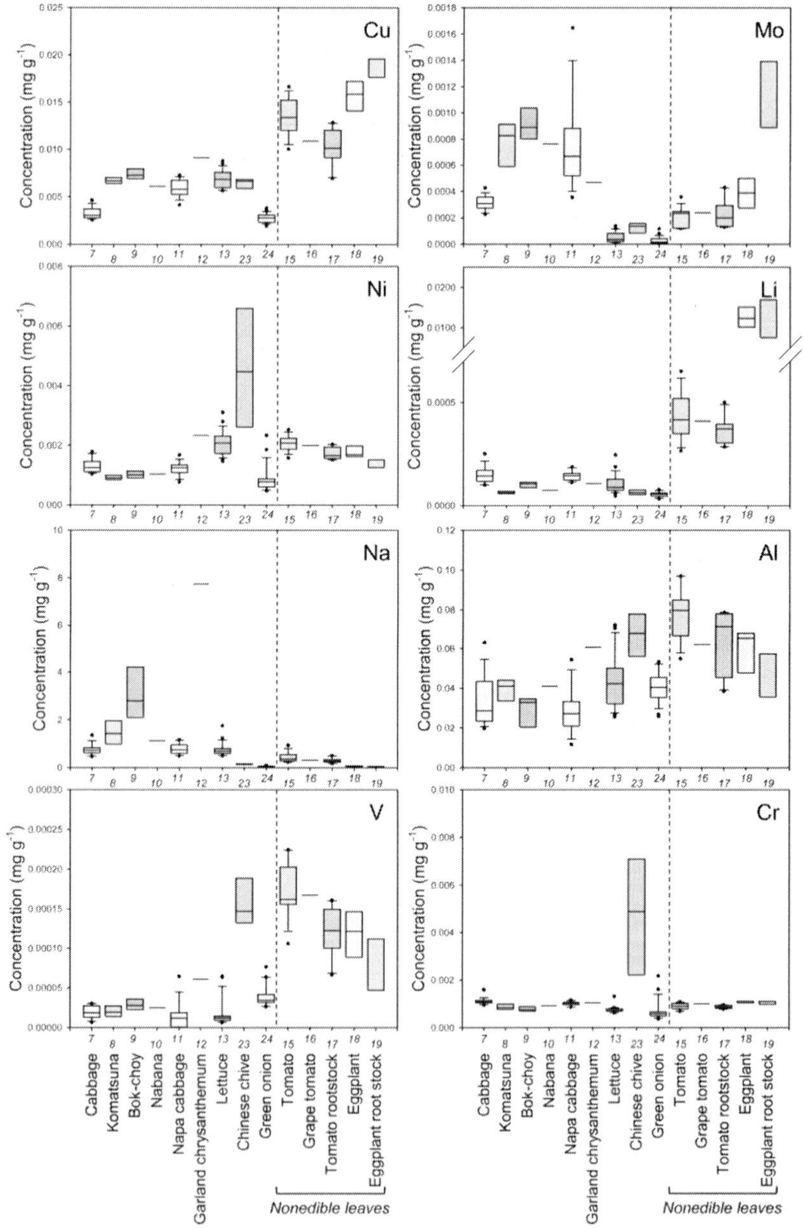

Figure 2. Boxplots showing the concentration of Cu, Mo, Ni, Li, Na, Al, V, and Cr in leaves of various crop species. Concentrations in nonedible leaves are also shown. Numbers in x-axis indicate the species number in Table 1.

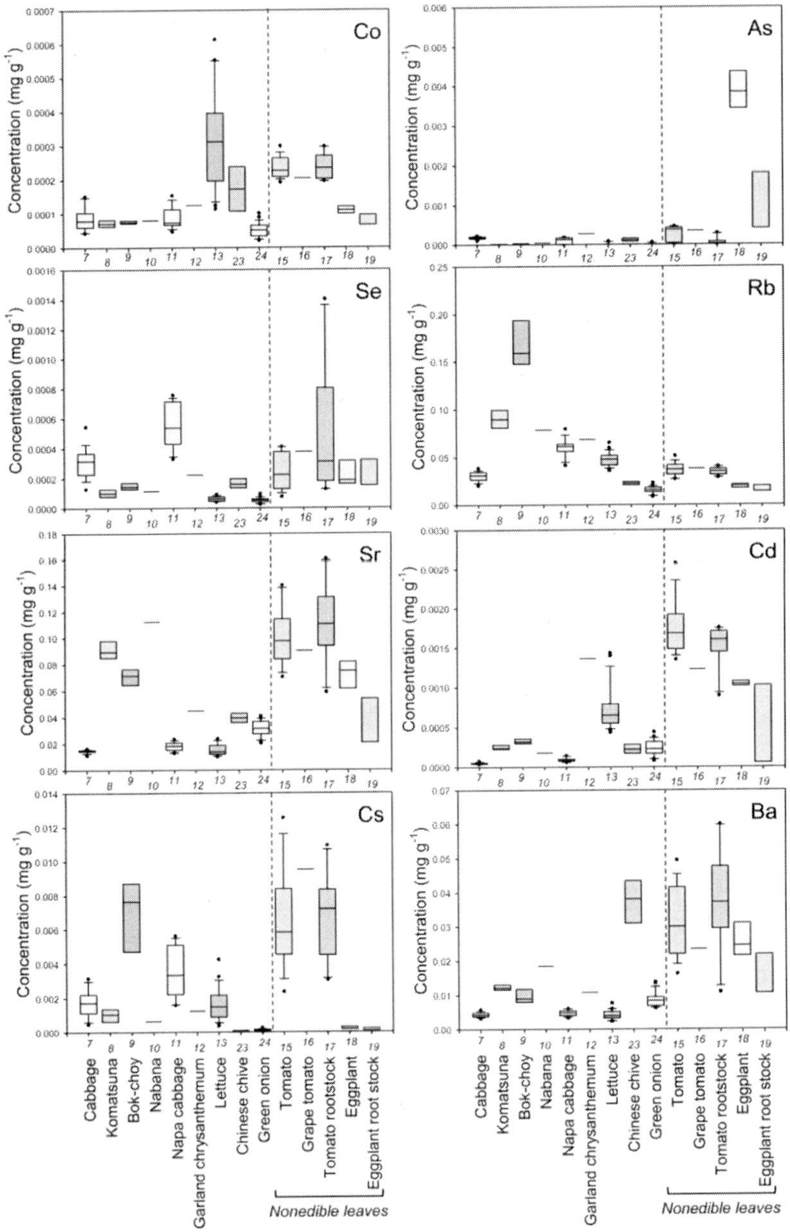

Figure 3. Boxplots showing the concentration of Co, As, Se, Rb, Sr, Cd, Cs, and Ba in leaves of various crop species. Concentrations in nonedible leaves are also shown. Numbers in x-axis indicate the species number in Table 1.

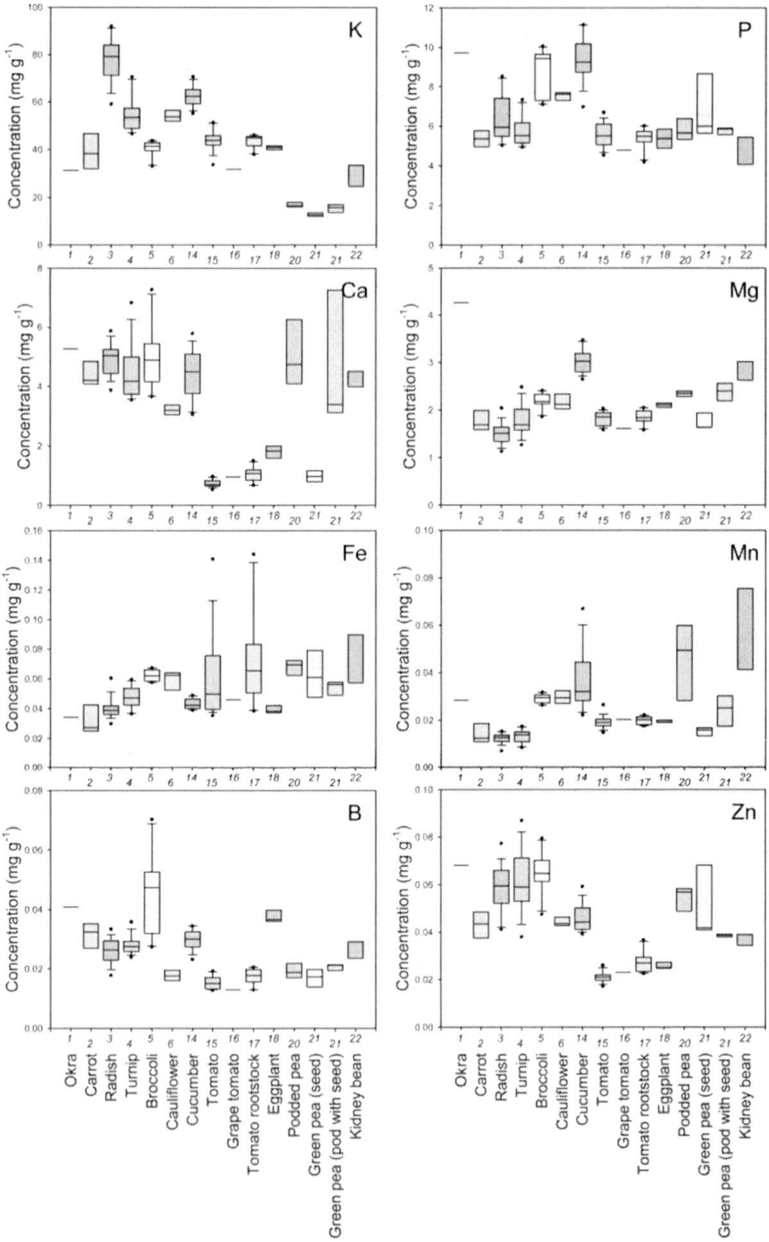

Figure 4. Boxplots showing the concentration of K, P, Ca, Mg, Fe, Mn, B, and Zn in non-leaf edible parts of various crop species. Numbers in x-axis indicate the species number in Table 1.

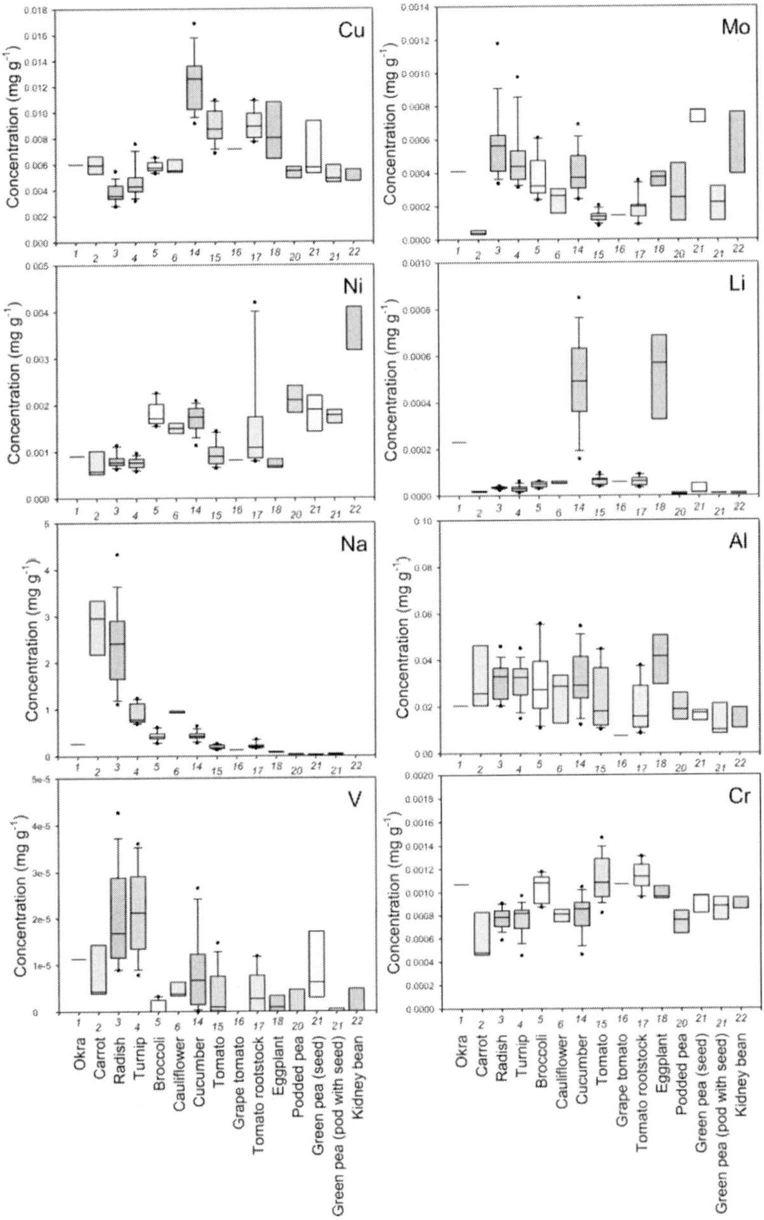

Figure 5. Boxplots showing the concentration of Cu, Mo, Ni, Li, Na, Al, V, and Cr in non-leaf edible parts of various crop species. Numbers in x-axis indicate the species number in Table 1.

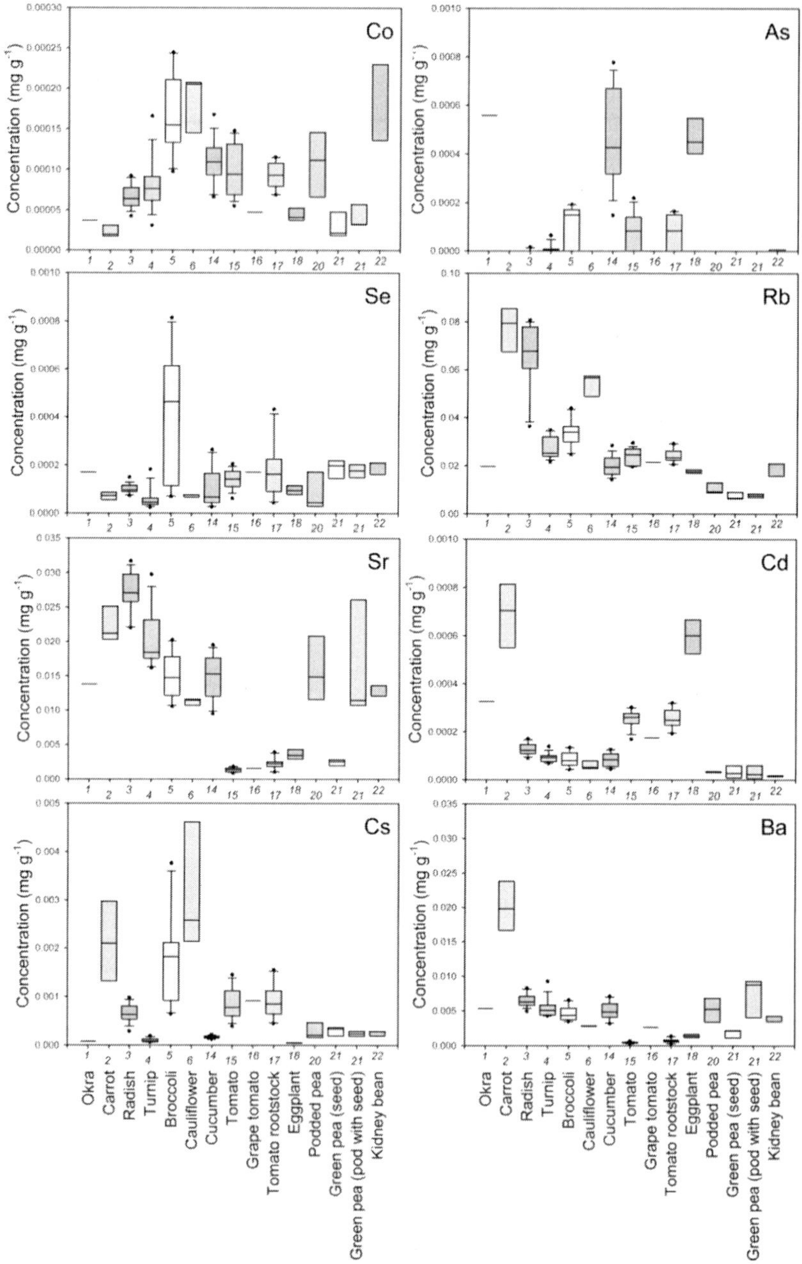

Figure 6. Boxplots showing the concentration of Co, As, Se, Rb, Sr, Cd, Cs, and Ba in non-leaf edible parts of various crop species. Numbers in x-axis indicate the species number in Table 1.

nabana, and lower in the fruit of Solanaceae species and the seed in green pea (Fig 1). The lower accumulation of Ca in fruit and seeds could be due to the distribution of Ca within the plant largely depending on xylem transfer [17], whereas fruit and seeds mainly use phloem transfer to accumulate minerals [18]. The large difference in Ca concentration between leaf and fruit (Figs 1 and 4) supports this prediction.

In general, there was little variation in the concentration of macroelements between different cultivars of each crop species. However, there tended to be a higher level of variation in the concentration of essential microelements in the edible parts of different crops and, in some cases, cultivars (Figs 1, 2, 4 and 5). For example, garland chrysanthemum and Chinese chive leaves contained a higher concentration of Fe, and a larger variation in Fe concentrations was observed between cultivars of lettuce (leaves), tomato (fruit), and eggplant (fruit) (Figs 1 and 4). A large variation was also observed in the concentration of the following microelements in the edible parts of different cultivars: Mn in lettuce, B in broccoli, Cu in cucumber, Mo in radish, turnip, and napa cabbage, and Ni in tomato rootstock and Chinese chive (Figs 1, 2, 4 and 5). The concentration of Zn was higher in the leaves of komatsuna and bok-choy, and lower in the fruits of Solanaceae species, with little variation between different cultivars of each species (Figs 1 and 4).

There also tended to be higher variation in the concentration of nonessential elements than essential macroelements, with greater levels of variation between cultivars in some cases (Figs 2, 3, 5 and 6). For example, the leaves of Chinese chive showed a remarkably higher concentration of V, Cr, and Ba as well as the essential microelements Fe and Ni (Figs 1–3).

Among the elements analyzed in this study, Fe, Zn, Cu, Ca, Mg, and Se are considered to be particularly important elements for biofortification [5]. Therefore, the information obtained from this study will be of use for selecting or breeding biofortified crops or cultivars. For example, we found that even within the genus *Brassica*, the average concentration and/or the level of variation in the concentration of each of these elements in the leaves differed between species, varieties, or cultivars (Figs 1 and 2).

From a food safety perspective, Cd, As, Cs, and Sr pose a particularly important risk to human health [19, 20]. We found that the leaves of some cultivars of garland chrysanthemum and lettuce exhibited higher concentrations of Cd (Fig 3). It has previously been shown that Cd is homologous to Zn, with the two elements exhibiting similar behaviors during uptake and transport in some plant species [21, 22]. However, although the leaves of garland chrysanthemum contained a higher concentration of Zn as well as Cd, all cultivars of lettuce had lower concentrations of Zn than other those of species (Fig 1), suggesting that the specificity of the uptake and transport systems for Cd and Zn varies even within the same family. Among the nonleaf vegetables, the roots of carrot and the fruit of eggplant contained a higher concentration of Cd, with little variation between the cultivars (Fig 6). It has previously been reported that eggplant often accumulates high concentrations of Cd in its fruit due to the effective xylem loading of this element [23].

Cucumber and eggplant fruits tended to contain higher concentrations of As (Fig 6). Interestingly, variations in the concentration of Sr among different crop species and cultivars were almost equal to variations in Ca, while the concentrations of Mg and Ba showed different trends (Figs 1, 3, 4 and 6). Sr is homologous to Ca, and it has previously been reported that the concentrations of these two elements are highly correlated in the leaves of 138 plant families growing under different environmental conditions [12]. Therefore, this finding implies that we must be careful when cultivating crops that have been selected or bred for Ca-biofortification in areas that are contaminated with radioactive Sr.

There was a high level of variation in the concentration of Cs in the edible parts of different species and cultivars, with a lower concentration tending to be found in fruits or seeds (Figs 3and 6). The leaves of bok-choy showed the highest average concentration of Cs, but there was also considerable variation between cultivars. The variation in the concentration of Cs that was observed between different crop species was not consistent with that seen for K and Na, which are homologous elements, but was similar to that of Rb (Figs 1–6).

Ionome of the leaves

It is generally difficult to examine phylogenetic influences on plant ionomes because they are affected by the growth environment, such as the soil nutrient status [2]. However, in the present study, we analyzed the ionome of the leaves of various crop species, including tomato, grape tomato, eggplant, and their rootstocks, which were cultivated under the same field conditions, allowing us to compare these in the absence of any environmental effects.

First, we compared the ionome of the leaves among species using principal component analysis (PCA), including the concentration of all elements (Fig 7A), essential elements (Fig 7B), or nonessential elements (Fig 7C) as the variable. The PCA scores were then presented on the basis of combinations of PC1 and PC2. When conducting the PCA with all elements, the score plots were loosely grouped according to family, although lettuce and garland chrysanthemum (both in Asteraceae) were plotted separately along PC2 (Fig 7A). Brassicaceae showed smaller variations in PC1 and larger variations in PC2, whereas Solanaceae and Amaryllidaceae exhibited larger variations in PC1 and smaller variations in PC2. In the loading plot, elements were loosely classified into two groups; groups 1 and 2 mainly contributed to PC1 and PC2, respectively (Fig 7B). Interestingly, Zn was plotted in group 2, separately from its homologous elements in group 1, and showed similar direction to P. The interaction between Zn and P in their uptake and transport [1] may be related to this result. The results of the PCA that only included the essential elements showed nearly the same trends as the results of the PCA that included all the elements (Fig 7C). In contrast, in the PCA that only included the nonessential elements, tomato and eggplant (both including rootstock) were separated by PC2 (Fig 7C), indicating that they differ in the uptake and/or translocation of nonessential elements despite belonging to the same family.

We then examined the correlation between the concentration of each pair of elements (Cs-Na, Cs-K, Cs-Rb, Sr-Ca, Cd-Fe, and Cd-Zn), which have been reported to partly share the same uptake/transport systems in plants [12, 24–27]. The concentration of Cs was only significantly correlated with that of Na in Brassicaceae, and Cs and K were only significantly correlated in Solanaceae (Fig 8). In contrast, a significant correlation between Cs and Rb was observed in all plant families, although the slope of this relationship differed between families. This suggests that these two elements at least partly share the same uptake/transport systems, but their affinities differ between plant families. Sr was highly correlated with Ca, which may be due to their similar behavior in both plants and the soil [28], and corresponds to the earlier findings for the edible parts (Figs 1, 3, 4 and 6). However, there were slight differences in the slope of this relationship between Solanaceae and other families (Fig 8), implying that the uptake/transport systems in Solanaceae have a slightly higher affinity for Ca.

The concentration of Cd was significantly correlated with the concentration of Fe in Brassicaceae and Asteraceae, and Zn in both these families plus Amaryllidaceae (Fig 8). The slope of this relationship differed greatly between families, however. The concentration of Cd was not significantly correlated with the concentrations of Fe and Zn in Solanaceae, which is particularly interesting since the accumulation of Cd in eggplant fruit often poses a risk to human health. It has been suggested that the xylem-loading citrate transporter gene *FRD3* is responsible for the transport of Cd from the roots to the shoots in *Solanum* spp. [29] and is involved in loading citrate into the xylem, ensuring the delivery of Fe from the roots to the shoots [30]. Therefore, the lack of a significant correlation between Cd and Fe concentrations in the leaves of Solanaceae suggests that Cd uptake and transport mechanisms are very complicated in this family.

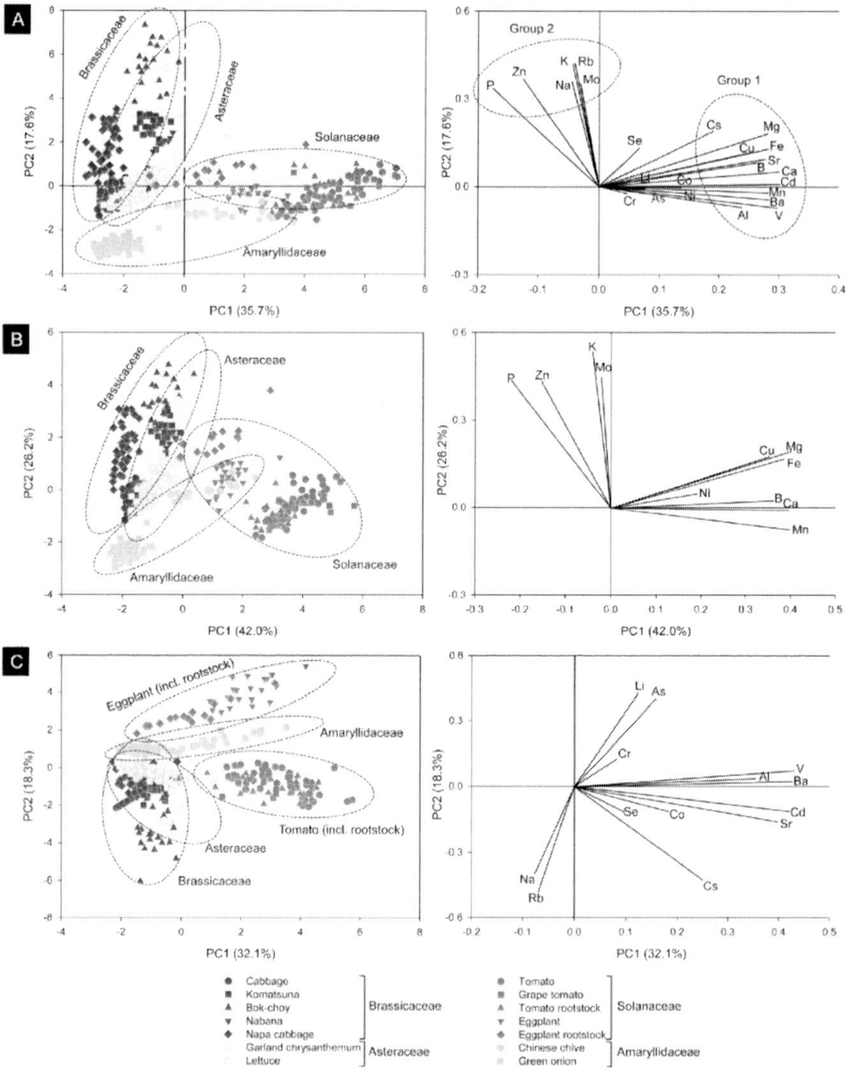

Figure 7. Principal component analysis (PCA) of elements in the leaves of plants analyzed in this study. A: Scores on PC1 and PC2 with all elements included and the corresponding loading plot. B: Scores on PC1 and PC2 with only the essential elements included and the corresponding loading plot. C: Scores on PC1 and PC2 with only the nonessential elements included and the corresponding loading plot.

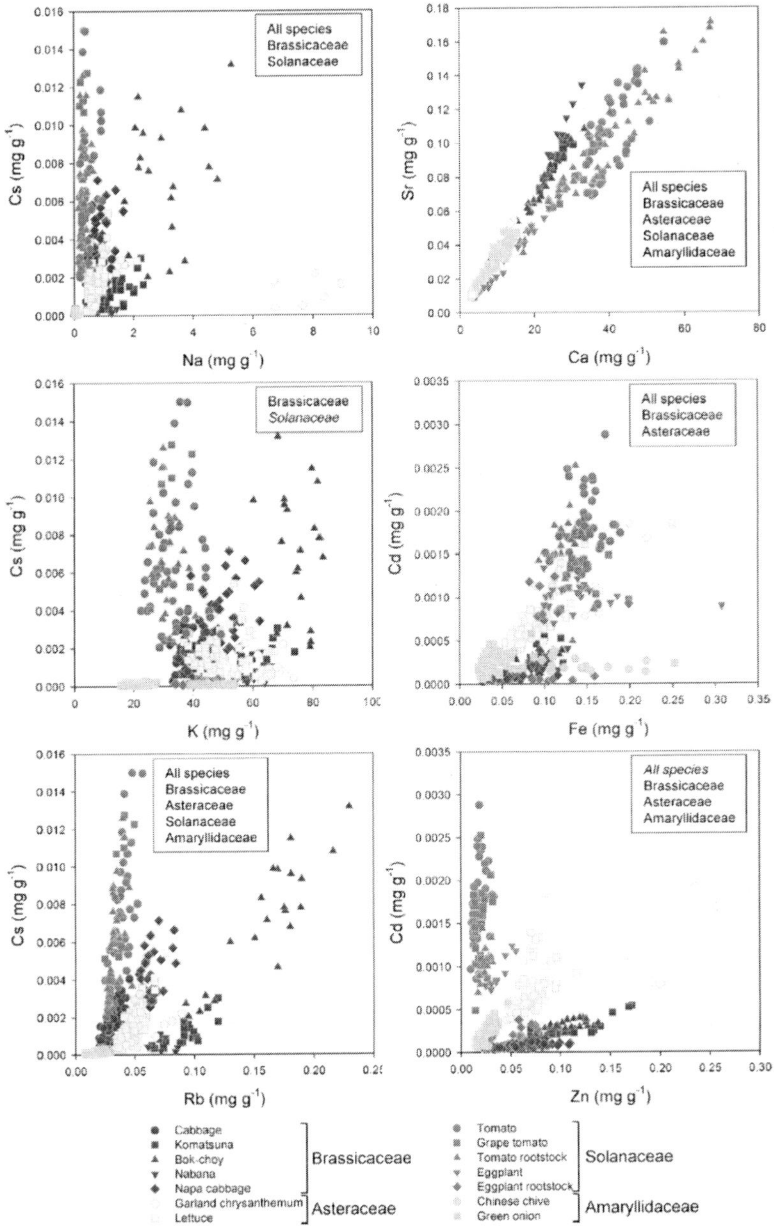

Figure 8. Correlation between the concentrations of pairs of homologous elements that possibly share the same uptake/transport system. Families with a significant correlation ($p < 0.01$) are shown in a box on each graph (italics = negative correlation).

Ionomic interactions between the leaves and fruit

We examined the correlation between the concentrations of each element in the leaves and fruit of tomato and eggplant (Table 2). Correlation coefficient was calculated using the average concentration in each cultivar. The relationship between these was found to differ between these two species, despite both belonging to family Solanaceae. In eggplant, there was no significant correlation between the leaves and the fruit for any element. In contrast, in tomato, there was a significant positive correlation between the concentration of several elements in the leaves and the fruit, particularly for nonessential elements (Table 2). Although the fact that many elements showed no such relationship suggests that it would be difficult to predict the fruit ionome based on the leaf ionome, re-examination with sufficient sample size is necessary in the future study.

Table 2. Correlation coefficients and ratio of the concentrations of each element in the leaves and fruit of different cultivars of tomato and eggplant.

		Correlation coefficient		Fruit/leaf ratio (±SE)	
		Tomato (n = 16)	Eggplant (n = 7)	Tomato (n = 16)	Eggplant (n = 7)
Essential	K	0.478	–0.331	1.34 ± 0.05	0.87 ± 0.03
	P	0.456	–0.431	1.88 ± 0.07	1.32 ± 0.10
	Ca	0.521*	0.451	0.02 ± 0.00	0.07 ± 0.00
	Mg	0.429	0.104	0.43 ± 0.02	0.66 ± 0.02
	Fe	0.495	–0.157	0.41 ± 0.04	0.30 ± 0.03
	Mn	0.119	0.685†	0.08 ± 0.00	0.27 ± 0.01
	B	0.373	–0.016	0.25 ± 0.01	1.31 ± 0.08
	Zn	0.243	0.491	1.28 ± 0.07	1.07 ± 0.04
	Cu	0.429	0.401	0.67 ± 0.02	0.54 ± 0.05
	Mo	0.787**	0.300	0.71 ± 0.04	1.03 ± 0.11
	Ni	0.611*	–0.081	0.47 ± 0.02	0.41 ± 0.03
Nonessential	Li	0.502*	0.146	0.16 ± 0.01	0.04 ± 0.01
	Na	0.600*	0.570	0.49 ± 0.03	1.55 ± 0.12
	Al	0.443	–0.252	0.30 ± 0.04	0.76 ± 0.11
	V	0.347	0.410	0.02 ± 0.01	0.02 ± 0.01
	Cr	0.608*	–0.386	1.26 ± 0.04	0.92 ± 0.03
	Co	0.100	0.593	0.42 ± 0.03	0.41 ± 0.03
	As	0.673**	–0.473	0.41 ± 0.22	0.12 ± 0.01
	Se	–0.019	0.503	0.72 ± 0.11	0.43 ± 0.09
	Rb	0.557*	0.730†	0.66 ± 0.02	0.90 ± 0.04
	Sr	0.176	0.439	0.01 ± 0.00	0.05 ± 0.00
	Cd	0.401	0.700†	0.15 ± 0.01	0.57 ± 0.02
	Cs	0.679**	0.661	0.14 ± 0.01	0.22 ± 0.02
	Ba	0.329	0.034	0.01 ± 0.00	0.06 ± 0.01

† (only for eggplant), *, and ** indicate statistical significance at $P < 0.1$, 0.05, and 0.01, respectively.

Although the transport and distribution of inorganic elements are mediated by both the xylem and phloem transportation systems, different elements have different levels of dependency on each of these [17]. In the phloem, K, Mg, P, and Na usually have a high mobility, Fe, Zn, Cu, B, and Mo have an intermediate mobility, and Ca and Mn have a low mobility [1]. The fruit/leaf ratio for the concentration of each element may reflect its mobility in the phloem and the efficiency of xylem transport of the element into the fruit (Table 2). The trends in these ratios were nearly the same in tomato and eggplant, with the exception of Ca, Mn, B, Na, and Cd, which had a higher ratio in eggplant than tomato, implying more efficient transport in the phloem and/or higher contribution of xylem transport [23] of these elements resulting in their higher accumulation in the fruit of this species (Figs 1–6).

CONCLUSION

This study applied ionomics to various vegetable species and cultivars that were cultivated under nearly identical conditions in large agricultural fields, allowing differences in the concentration of various essential and nonessential elements in the edible parts of crops to be compared with minimal environmental effects. This provides new insights into phylogenetic influences on the ionome which will aid our understanding of the mineral uptake and transport mechanisms in plants and may contribute to improving human health.

ACKNOWLEDGMENTS

This study was supported financially by Grants-in-Aid for Scientific Research (No. 24580088 (TW), No. 26660019 (YK, TW), and No. 16H02534 (YK, TW)) from the Japan Society for the Promotion of Science.

REFERENCES

1. Marschner P. Mineral Nutrition of Higher Plants, 3rd ed. New York: Academic Press; 2012.
2. Watanabe T, Urayama M, Shinano T, Okada R, Osaki M. Application of ionomics to plant and soil in fields under long-term fertilizer trials. SpringerPlus. 2015; 4: 781. pmid:26702370
3. White PJ, Brown PH. Plant nutrition for sustainable development and global health. Ann Bot. 2010; 105: 1073–1080. pmid:20430785
4. Younis K, Ahmad S, Badpa A. Malnutrition: causes and strategies. J Food Process Technol. 2015; 6: 434.
5. White PJ, Broadley MR. Biofortification of crops with seven mineral elements often lacking in human diets–iron, zinc, copper, calcium, magnesium, selenium and iodine. New Phytol. 2009; 182: 49–84. pmid:19192191
6. Ishikawa S, Abe T, Kuramata M, Yamaguchi M, Ando T, Yamamoto T, et al. A major quantitative trait locus for increasing cadmium-specific concentration in rice grain

is located on the short arm of chromosome 7. J Exp Bot. 2010; 61: 923–934. pmid:20022924
7. Watanabe T, Murata Y, Nakamura T, Sakai Y, Osaki M. Effect of zero-valent iron application on cadmium uptake in rice plants grown in cadmium-contaminated soils. J Plant Nutr. 2009; 32: 1164–1172.
8. Salt DE, Baxter I, Lahner B. Ionomics and the study of the plant ionome Ann Rev Plant Biol. 2008; 59: 709–733.
9. Baxter IR, Vitek O, Lahner B, Muthukumar B, Borghi M, Morrissey J, et al. The leaf ionome as a multivariable system to detect a plant's physiological status. Proc Natl Acad Sci USA. 2008; 105: 12081–12086. pmid:18697928
10. Chao D-Y, Gable K, Chen M, Baxter I, Dietrich CR, Cahoon EB, et al. Sphingolipids in the root play an important role in regulating the leaf ionome in *Arabidopsis thaliana*. Plant Cell. 2011; 23: 1061–1081. pmid:21421810
11. Sha Z, Oka N, Watanabe T, Tampubolon BD, Okazaki K, Osaki M, et al. Ionome of soybean seed affected by previous cropping with mycorrhizal plant and manure application. J Agric Food Chem. 2012; 60: 9543–9552 pmid:22950648
12. Watanabe T, Broadley MR, Jansen S, White PJ, Takada J, Satake K, et al. Evolutionary control of leaf element composition in plants. New Phytol. 2007; 174: 516–523. pmid:17447908
13. White PJ, Broadley MR, Thompson JA, McNicol JW, Crawley MJ, Poulton PR, et al. Testing the distinctness of shoot ionomes of angiosperm families using the Rothamsted Park Grass Continuous Hay Experiment. New Phytol. 2012; 196: 101–109. pmid:22803633
14. Sha Z, Oka N, Watanabe T, Tampubolon BD, Okazaki K, Osaki M, et al. Ionome of soybean seed affected by previous cropping with mycorrhizal plant and manure application. J Agric Food Chem. 2012; 60: 9543–9552. pmid:22950648
15. Shakoor N, Ziegler G, Dilkes BP, Brenton Z, Boyles R, Connolly EL, et al. 2016: Integration of experiments across diverse environments identifies the genetic determinants of variation in *Sorghum bicolor* seed element composition. Plant Physiol. 2016; 170: 1989–1998. pmid:26896393
16. Reese RA. Boxplots. Significance. 2005; 2: 134–135
17. Karley AJ, White PJ. Moving cationic minerals to edible tissues: potassium, magnesium, calcium. Curr Opin Plant Biol. 2009; 12: 291–298. pmid:19481494
18. Patrick JW, Offler CE. Compartmentation of transport and transfer events in developing seeds. J Exp Bot. 2001; 52: 551–564. pmid:11373304
19. Järup L. Hazards of heavy metal contamination. British Medical Bulletin. 2003; 68: 167–182. pmid:14757716
20. Oestling O, Kopp P, Burkart W Foliar uptake of cesium, iodine and strontium and their transfer to the edible parts of beans, potatoes and radishes. Int J Radiat Appl Instrum, Part C. 1989; 33: 551–554.
21. Liu F, Tang Y, Du R, Yang H, Wu Q, Qiu R. Root foraging for zinc and cadmium requirement in the Zn/Cd hyperaccumulator plant *Sedum alfredii*. Plant Soil. 2010; 327: 365–375.
22. Ma JF, Information C, Ueno D, Zhao F-J, McGrath SP. Subcellular localisation of Cd and Zn in the leaves of a Cd-hyperaccumulating ecotype of *Thlaspi caerulescens*. Planta. 2005; 220: 731–736. pmid:15517354
23. Yamaguchi N, Mori S, Baba K, Kaburagi-Yada S, Arao T, Kitajima N, et al. Cadmium distribution in the root tissues of solanaceous plants with contrasting root-to-shoot Cd translocation efficiencies. Environ Exp Bot. 2011; 71: 198–206.
24. Bækgaard L, Mikkelsen MD, Sørensen DM, Hegelund JN, Persson DP, Mills RF, et al. A combined zinc/cadmium sensor and zinc/cadmium export regulator in a heavy metal pump. J Biol Chem. 2010; 285: 31243–31252. pmid:20650903
25. Benito B, Haro R, Amtmann A, Cuin TA, Dreyer I (2014) The twins K^+ and Na^+ in plants. J Plant Physiol. 2014; 171: 723–731. pmid:24810769
26. Qi Z, Hampton CR, Shin R, Barkla BJ, White PJ, Schachtman DP. The high affinity K^+transporter AtHAK5 plays a physiological role in planta at very low

K^+ concentrations and provides a caesium uptake pathway in Arabidopsis. J Exp Bot. 2008; 59: 595–607. pmid:18281719

27. Nakanishi H, Ogawa I, Ishimaru Y, Mori S, Nishizawa NK. Iron deficiency enhances cadmium uptake and translocation mediated by the Fe^{2+} transporters OsIRT1 and OsIRT2 in rice. Soil Sci Plant Nutr. 2006; 52: 464–469.
28. Drouet T, Herbauts J. Evaluation of the mobility and discrimination of Ca, Sr and Ba in forest ecosystems: consequence on the use of alkaline-earth element ratios as tracers of Ca. Plant Soil. 2008; 302: 105–124.
29. Yamaguchi H, Fukuoka H, Arao T, Ohyama A, Nunome T, Miyatake K, et al. Gene expression analysis in cadmium-stressed roots of a low cadmium-accumulating solanaceous plant, *Solanum torvum*. J Exp Bot. 2010; 61: 423–437. pmid:19837731
30. Rogers EE, Guerinot ML. FRD3, a member of the multidrug and toxin efflux family, controls iron deficiency responses in Arabidopsis. Plant Cell. 2002; 14: 1787–1799. pmid:12172022

CHAPTER 16

Progress in High-Amylose Cereal Crops through Inactivation of Starch Branching Enzymes

Juan Wang[1,2], Pan Hu[1,2], Zichun Chen[1,2], Qiaoquan Liu[1,2] and Cunxu Wei[1,2*]

[1]*Jiangsu Key Laboratory of Crop Genetics and Physiology/Co-Innovation Center for Modern Production Technology of Grain Crops, Yangzhou University, Yangzhou, China*

[2]*Joint International Research Laboratory of Agriculture and Agri-Product Safety, Yangzhou University, Yangzhou, China*

ABSTRACT

High-amylose cereal starches provide many health benefits for humans. The inhibition or mutation of starch branching enzyme (SBE) genes is an effective method to develop high-amylose cereal crops. This review summarizes the development of high-amylose cereal crops through the inactivation of one or more SBE isoforms or combination with other genes. This review also reveals the causes of increase in amylose content in high-amylose crops. A series of changes, including amylopectin structure, crystalline structure, thermal properties, and hydrolysis properties, occurs as amylose content increases. The different morphological starch granules nominated as heterogeneous starch granules or differently stained starch granules are detected in high-amylose cereal crops. Detailed studies on four heterogeneous starch granules in high-amylose rice, which is developed by antisense RNA inhibition of SBEI/IIb, indicate that granules with different morphologies possess various molecular structures and physicochemical and functional properties. This variation diversifies their applications in food and non-food industries. However, current knowledge regarding how these heterogeneous starch granules form and why they exhibit regional distribution in endosperm remain largely unknown.

Keywords: high-amylose cereal crop, starch branching enzyme, starch molecular structure, starch property, heterogeneous starch granule

INTRODUCTION

High-amylose starches, rich in resistant starch (RS), have been extensively investigated because of their many potential health benefits. RS is a portion of starch that can evade degradation in the upper gastrointestinal tract and function as a substrate for the bacterial fermentation in the large intestine (Nugent, 2005). RS-enriched food can decrease glycemic and insulin responses and reduce the risk of developing type II diabetes mellitus, obesity, and cardiovascular diseases (Granfeldt et al., 1995; Regina et al., 2006; Zhu et al., 2012). The proportion of amylose is positively related to RS content (Sang et al., 2008; Cai et al., 2015; Lin et al., 2016b; Zhou et al., 2016). High-amylose starches exhibit various physicochemical properties, including high gelling strength, excellent film-forming ability, and ease of retrogradation; with these properties, industrial applications, such as adhesive, paper, and biodegradable plastic production, have been developed (Avella et al., 2002). Therefore, considerable research has focused on high-amylose starches.

Starch is composed of approximately 15–25% linear amylose. The remaining portion of starch consists of highly branched and organized amylopectin (Smith et al., 2002). Amylose in the endosperm can only be synthesized by granule bound starch synthase I (GBSSI), which is tightly bound to starch granules to extend α-1,4 linkages of glucose polymers (Jeon et al., 2010). Amylopectin synthesis is mainly attributed to three biosynthetic enzymes, namely, starch synthases (SSs), starch branching enzymes (SBEs), and debranching enzymes (DBEs). SSs and SBEs are responsible for the elongation of glucan chains and the formation of branch points, respectively (Jeon et al., 2010; Tetlow and Emes, 2014). The irregular glucan structure produced by SSs and SBEs is trimmed by DBEs to derive an ordered amylopectin branched chains (Nakamura, 2002). Plastidial phosphorylase (Pho1) also participates in starch synthesis; however, its precise role remains unclear (Yu et al., 2001; Satoh et al., 2008).

The amylose content (AC) in cereal endosperm can be effectively increased by applying two methods, that is, enhancing GBSSI expression or eliminating SBEs, SSIIa, or other enzymes involved in amylopectin synthesis (Umemoto and Terashima, 2002; Itoh et al., 2003; Crofts et al., 2012; Zhou et al., 2016). Once a certain level is reached, AC no longer increases in GBSSI-enhanced lines possibly because of the limited non-reducing ends in amylose and the substrate competition between amylose and amylopectin (Tsai et al., 1970; Smith et al., 2002; Sestili et al., 2012). The promotion of AC in SSIIa nulls varies widely in different species; for instance, the elimination of SSIIa unlikely change AC in rice *alk*(Umemoto et al., 2004, 2008), moderately increases AC (40–50%) in maize *sugary2* (*su2*) and wheat *sgp-1* (Yamamori et al., 2000; Tziotis et al., 2004; Zhang et al., 2004), and yields about 50% AC in barley *sex6* (Morell et al., 2003). Compared with the effectiveness of SBE inactivation for high-amylose cereal crops, which have remarkable increase in AC (Li et al., 2008; Carciofi et al., 2012; Zhu et al., 2012; Regina et al., 2015), the enhanced AC in SSIIa nulls is moderate (Yamamori et al., 2000; Tziotis et al., 2004; Zhang et al., 2004). Therefore, AC can be promoted effectively by suppressing or eliminating one or more SBE activities in cereal crops (Table 1).

Table 1. The inactivated gene, amylose content in starch, and starch crystallinity of high-amylose cereal crops.

Species	The original AC in starch	The inactivated genes		Name of high AC line	AC in starch[b]	Starch crystallinity	References
		Mutation	Down regulation[a]				
Maize	25–30%	SBEIIb		H99ae,OH43ae, B89ae, B84ae	61.7–67.7% (I)	B	Li et al., 2008
		SBEIIb (HAM)		GEMS-0067	88.2% (G)	B	Jiang et al., 2010b
Rice	*Japonica* 15.4–19.6%	SBEIIb		EM10	26.5% (I)	B	Nishi et al., 2001
			SBEIIb (~80%)	ami-BEIIb	41.2% (I)	B	Butardo et al., 2011
			SBEIIb (~50%)	hp-BEIIb	34% (I)	A or C	Butardo et al., 2011
			unknown	Goami 2(G2)	33.96% (I)	B	Kang et al., 2003
		SSIIIa		*ss3a*	24.8% (G)	A	Fujita et al., 2007
		SBEIIb, SSIIIa		*ss3a/be2b*	45% (G)	B	Asai et al., 2014
	Indica about 25%	SSIIIa		*b10*	~34% (I)	A	Zhou et al., 2016
			SBEI (~100%), SBEIIb (>90%)	TRS	64.8% (I)	C	Zhu et al., 2012
Wheat	22.9–32.3%	SBEIIa (Tilling)		SBEIIa Mutant	55.7% (K)	–	Slade et al., 2012
			SBEIIa (most), SBEIIb (most)	hp-SBEIIa	74.4% (G)	–	Regina et al., 2006
		SBEIIa, SBEIIb		CS2-F11	86.6% (G)	–	Regina et al., 2015
Barley	29.9–31.6%		SBEIIa (>80%)	SBEIIa⁻	38% (G)	–	Regina et al., 2010
			SBEIIa (>90%), SBEIIb (>80%)	SBEIIa⁻ / SBE IIb↓	67.2% (G)	–	Regina et al., 2010
			SBEIIa (>90%), SBEIIb (>96%)	SBEIIa⁻ / SBEIIb⁻	76.2% (G)	–	Regina et al., 2010
			SBEI (87%), SBEIIa (73%), SBEIIb (74%)	SBE RNAi4.1	99.1% (G)	B	Carciofi et al., 2012

[a] *The content of parenthesis means the reduced expression of targeted genes.*

[b] *The AC determined by iodine colorimetry, GPC, and K-AMYL kit method is labeled as (I), (G), and (K), respectively.*

This review summarizes the development of high-amylose cereal crops through the inactivation of SBE(s) and a series of starch changes caused by the increased amylose percentage.

Development of High-Amylose Cereal Crops Through SBE Inactivation

Cereal crops have three SBE classes, namely, SBEI, SBEIIa, and SBEIIb. Biochemical observations indicate that SBEI preferentially branches longer chains, whereas SBEII isoforms have a higher capacity for transferring of short chains (Guan and Preiss, 1993; Takeda et al., 1993; Guan et al., 1997; Nishi et al., 2001; Nakamura, 2002; Nakamura et al., 2010; Tetlow and Emes, 2014). Among these observations, studies on eliminating SBEI activity show no effect on the starch composition and kernel morphology (Blauth et al., 2002; Satoh et al., 2003b; Regina et al., 2004); however, in rice and maize, minor changes in the fine structure of amylopectin have been reported (Fujita et al., 2006; Xia et al., 2011). However, AC resulting from the inactivation of either one of two SBEII isoforms and the combination of two or three SBE isoforms is significantly increased in different species (Table 1).

Amylose Increase Through SBEIIa or SBEIIb Deficiency

In maize and rice, SBEIIb inactivation is required to obtain a high-amylose phenotype, whereas no significant changes including AC and amylopectin fine structure are observed in SBEIIa-defective endosperms (Blauth et al., 2001; Nakamura, 2002; Satoh et al., 2003a). Mutations of *SBEIIb* in maize and rice are commonly referred to as *amylose-extender* (*ae*) mutants. Normal maize occupies about 25–30% AC, whereas the AC in H99*ae*, OH43*ae*, B89*ae*, and B84*ae* inbred *ae*-lines has been reported to reach at least 60%, ranging from 61.7 to 67.7% (Li et al., 2008). In rice containing Wx^b, which mutates at the 5′ end of intron one, *japonica* type contains low GBSSI mRNA and protein levels, leading to reduced AC in contrast to *indica* type, which has a highly expressed Wx^a allele (Isshiki et al., 2000). The *ae* mutants from these two backgrounds have higher AC than their wild-type parents. However, AC increase extent is not more than 15%, which is significantly lower than that in analogous maize mutants, which show at least 35% promotion (Nishi et al., 2001; Shannon et al., 2009). Another *ae*-like mutant Goami 2 in rice, also known as Suweon 464, which is derived from high-quality temperate *japonica* rice variety (Iipumbyeo), shows about 2-fold increase in AC (to around 34.0%, see Kang et al., 2003). However, the specific mutation responsible for high-amylose phenotype of Goami 2 remains to be proved.

On the contrary, an almost complete *SBEIIb* inhibition in wheat and barley causes a minor change in AC, whereas *SBEIIa* inhibition can evidently increase AC (Regina et al., 2006, 2010). By targeting induced local lesions in genomes (TILLING), wheat *SBEIIa* mutant increases AC from 23 to 55% (Slade et al., 2012). In barley transgenic lines, by RNA-mediated silencing technology, a high-amylose phenotype (>38%) is observed when SBEIIa expression is reduced by >80% (Regina et al., 2010).

In summary, SBEIIb in rice and maize plays a distinct role in starch synthesis. By contrast, SBEIIa in wheat and barley is more significant than SBEIIb. High-amylose phenotype generation is determined by the difference in SBEIIa and SBEIIb expression levels in cereals. In maize, SBEIIb, which is at least 50 times the abundance of SBEIIa, is the predominant isoform in endosperm (Gao et al., 1997). On the other hand, in rice, the ratio of SBEIIb and SBEIIa is closer to 5:1, which partly elucidates the higher AC promotion in maize *ae* mutants. Furthermore, SBEIIa is expressed at much higher levels than SBEIIb counterpart in both wheat and barley kernels (Regina et al., 2006, 2010). Therefore, SBEIIa inhibition is more effective for developing high-amylose species in the two crops.

Higher Amylose Increase through Inactivation of Two or More SBEs

Combining reduced SBE isoforms with the base of single SBEIIa or SBEIIb inactivation is an efficient way to acquire high-amylose crops, especially for wheat and barley (Table 1). A transgenic resistant starch rice line (TRS) with 60% AC is developed by simultaneously declining the expressions of SBEIIb and SBEI in an *indica* rice variety (Te-qing, TQ). The AC of TRS is significantly higher than that of TQ-derived rice line by declining the expression of SBEIIb alone (Wei et al., 2010c; Zhu et al., 2012). In wheat, transformation with a hairpin construct targeting *SBEIIa*alone not only reduces transcription to <10%, but also leads to an almost complete loss of SBEIIb in the protein levels, which yields an AC of ~75% in transgenic wheat (Regina et al., 2006). Furthermore, a genetic strategy to combine deletion and single nucleotide polymorphism (SNP) generates wheat genotypes with the complete absence of SBEIIa from all three genomes and the absence of SBEIIb from one of the genomes, which elevates the AC to an unprecedented ~85% (Regina et al., 2015). In barley, a combination of SBEIIa and SBEIIb inhibition leads to an amylose increase up to 76.2% at maximum, which is much higher than the suppression by SBEIIa alone (Regina et al., 2010, 2012). In addition, an "amylose-only barley" is derived through concerted repression of SBEI, SBEIIa, and SBEIIb using a chimeric RNAi hairpin (Carciofi et al., 2012). Through this method, the highest known AC-contained lines in rice, wheat, and barley are derived (Table 1).

Higher Amylose Increase through SBEII Isoform Deficiency with Other Genes

By introducing a modifier gene or by coupling with isozyme deficiencies related to starch biosynthesis on the base of inactivation of SBEII, additive effects show a more profound meaning on AC in cereal crops (Wu et al., 2009; Jiang et al., 2010b; Asai et al., 2014).

When an unknown number of high-amylose modifier (HAM) gene is introduced into a homozygous *ae*-mutant background in maize, AC is further increased (Wu et al., 2009). The commercial maize GEMS-0067, which is the highest known AC-contained line (>85%), is derived from the predigree of [GUAT209:S13 × (OH43ae × H99ae)] (Li et al., 2008; Wu et al., 2009). GUAT209:S13 is a cross-breed serving as HAM gene. Jiang et al. (2010b) found that HAM gene dosage has an additive effect on AC; amylose percentage ranges from 68.9 to 88.2%, when the HAM gene-dosage levels are 0, 16.7, 33.3, 50, 66.7, 83.3, and 100%.

Single SBEII mutation coupled with other high-amylose mutation is an effective method in producing high-amylose starches. SSIIIa deficiency in rice produces 30.7% AC phenotype. However, when this mutation is introduced into SBEIIb-defective background, the AC is further promoted into 45% in *ss3a/sbe2b* double mutant (Asai et al., 2014).

Factors Affecting Amylose Increase in High-Amylose Cereal Crops

Three factors are proposed to be responsible for amylose enrichment caused by SBE isoform lesions in endosperm starch. First, amylopectin synthesis reduction emphasizes the amylose percentage in these high-amylose lines compared with the normal ones. Second, the increase protein amount of GBSSI or higher GBSSI activity results in an elevated amylose synthesis. Third, the extra-long chains (ELCs) resulting from amylopectin, which can bind iodine to develop a dark-blue color, significantly enriches AC value in high-amylose varieties, (Jane et al., 1999). Factors responsible for the amylose increase are different across various species.

In rice, different researchers have various opinions on amylose enrichment. An increase in amylose is attributed to the first two factors. In rice *ss3a/be2b* mutant, a significantly higher AGPase activity resulting from SSIIIa and SBEIIb deficiency leads to a high concentration of ADP-glucose. GBSSI yields a higher Km for ADP-glucose than other soluble SS isozymes. Coupled with the fact that the GBSSI amounts in *ss3a/sbe2b*and *sbe2b* are significantly higher than those in wild type varieties; these results strongly explain why amylose synthesis is remarkably enriched in the endosperm of *sbe2b* and *ss3a/sbe2b* compared with wild-type varieties (Nishi et al., 2001; Asai et al., 2014). In addition, amylopectin synthesis stops at later starch development stage when dehydration begins in *sbe2b*-related mutants, emphasizing the AC in *sbe2b* and *ss3a/sbe2b* (Asai et al., 2014).

For other researchers, the third factor is the main one responsible for amylose increase in rice. The *waxy ae* double mutant, which eliminates SBEIIb in an amylose-free background, still contains amylose, and AC increase extent is almost similar to that in its counterpart in *ae* single mutant (Nishi et al., 2001). Thus, amylose increase in the double mutant is not due to the true amylose increase but to modified amylopectin. The maximum wavelength of the absorbance of the starch-iodine complex (λmax) of starch from *waxy ae* is 32 nm higher than that from the *waxy*mutant. Although ELC is not observed directly, this result strongly suggests that the AC increase is caused predominantly by the abnormal structure of amylopectin. Except genetic evidence, structural analyses on debranched starch from two *SBEIIb* down-regulated lines, amiRNA and hpRNA demonstrate directly that AC doubling in these two transgenic lines is not due to an increase in the relative proportion of amylose but to the elevated proportion of ELC, which ranges from degree of polymerization (DP) 120–1,000 (Butardo et al., 2011).

For maize, the amylose increase mainly results from ELC. Although starch contains no amylose, similar to the *waxy ae* double mutant of rice, the *waxy ae* mutant of maize displays an apparent AC of 34.5% (Jane et al., 1999). On the other hand, detailed structural analyses have demonstrated that those long chains do not originate from the real amylopectin of high-amylose starches, but from intermediate components (IC), which is a fraction consisting of branched molecules with molecular weights smaller than amylopectin but similar to amylose (Wang et al., 1993; Kasemsuwan et al., 1995) and is capable of escaping 1-butanol precipitation with amylopectin (Jane and Chen, 1992; Li et al., 2008; Peymanpour et al., 2016). When removing amylose from high-amylose starches, a mixture containing both IC and amylopectin is further separated using gel permeation chromatography (GPC) (Li et al., 2008). Peymanpour et al. (2016) divided this mixture according to their distribution of molecular structure into high molecular weight fraction (HMF) and low molecular weight fraction (LMF). HMF is the typical amylopectin, whereas LMF corresponds to the IC fraction, which increases with AC and appears to have substantially more of long chains than HMF. Li et al. (2008) further divided IC fraction according to the blue-values into large-M_w IC and low-M_w IC. Chain-length distribution results indicate that large-M_w IC has a similar branch structure to amylopectin but with smaller molecular-weight. On the other hand, low-M_w IC, which is the main source of fluctuating apparent AC in starch, has longer chains compared with amylopectin and large-M_w IC (Li et al., 2008).

Structure Changes of Amylopectin in High-Amylose Cereal Crops

In those high-amylose lines, reduced or completely lost SBEs actually does not only result in a simple promotion of AC within the granules, but also complicates the starch structure, specifically the amylopectin structure. Amylopectin is mainly composed of short- to mediated-length chains in normal and waxy cultivars. However, starch from SBEs-deficient lines has an abnormal amylopectin structure, which is enriched with long branch-chains but depleted of short ones (Nishi et al., 2001; Yao et al., 2004). For example, in rice *ae* mutants with the *DP* ≤ 17, chains with DP 8–12 are remarkably reduced. On the other hand, the long branch-chains with *DP* > 24, which connect the clusters of amylopectin, are more abundant compared with those in the wild type varieties (Nishi et al., 2001; Asai et al., 2014). Furthermore, through GPC analyses of a series of different AC-contained starches, Lin et al. (2016c) found that the ratio of long amylopectin branch-chains in high-amylose species is positively correlated with the AC.

Effect of different SBEIIb dosages on the amylopectin in rice is widely researched. First, due to the triploid characteristic of endosperm, Nishi et al. (2001) performed reciprocal crosses between null *ae* mutant (*aeaeae*) and wild type (*AeAeAe*) to generate F1 seeds with two different doses of the SBEIIb allele, duplex (*AeAeae*) and simplex (*Aeaeae*), which shows 74 and 26% protein levels relative to the wild type, respectively. This result implies that the protein level of SBEIIb appears to increase almost linearly with the increase in the number of dominant *Ae* alleles; a similar trend is observed in gene dosage effect on SBEIIb activity. The proportion of short branch-chains with *DP* ≤ 17 is 74% in duplex (*AeAeae*), 72% in simplex (*Aeaeae*), and 66% in nulliplex (*aeaeae*), indicating that *DP* ≤ 17 increases with the increasing number of Ae allele; however, the increase is not linear. Second, when SBEIIb gene is introduced into SBEIIb-defective mutant, resulting transgenic rice plants show a wide range of SBEIIb activities (Tanaka et al., 2004). As the SBEIIb activity increases, short branch-chains with *DP* ≤ 13 become more frequent, whereas the number of long branch-chains with *DP* = 15–30 and *DP* ≥ 40 decline. Although structural and physicochemical properties caused by different SBEIIb dosages are not remarkably increased with the increase of Ae allele or SBEIIb expression level, the above results reveal that starch composition and molecular structure are entirely dependent on the level of SBEIIb activity. In addition, *SBEIIb* overexpression results in a severely shrunken phenotype of the kernel, which is caused by the accumulation of excessive branched, water-soluble polysaccharides instead of amylose and amylopectin (Tanaka et al., 2004).

Starch Properties of High-Amylose Cereal Crops

Increased amylose, especially the enriched percentage of amylopectin long branch-chains dramatically affected starch properties including crystalline structure (Cheetham and Tao, 1998; Nishi et al., 2001; Li et al., 2008; Jiang et al., 2010b; Wei et al., 2010b,d; Butardo et al., 2011; Man et al., 2013a; Asai et al., 2014; Cai et al., 2014c; Lin et al., 2016c), thermal properties (Jiang et al., 2010c; Regina et al., 2010; Wei et al., 2011; Man et al., 2014; Pan et al., 2017), and hydrolysis properties (Jiang et al., 2010d; Qin et al., 2011; Man et al., 2012a,b, 2013b,c; Cai J. et al., 2014; Huang et al., 2015; Lin et al., 2016a). These properties determine the quality of starch and the application of high-amylose crops.

Crystalline Structure of High-Amylose Starch

Starch is stored as granules with alternating semi-crystalline and amorphous growth rings. The semicrystalline ring consists of the lamellar structure of alternating crystalline and amorphous regions. Amylopectin branch-chains form double helices and are laterally packed to form crystalline regions (Blazek and Gilbert, 2011). Starch crystallinity has types A and B according to X-ray diffraction pattern. A-type crystallinity is formed by amylopectin with short branch-chains and closed branching points, whereas B-type crystallinity is formed by amylopectin with long branch-chains and distant branching points. Usually, waxy and normal cereal crops contain A-type starch (A-type crystallinity), tuberous crops consist of B-type starch (B-type crystallinity), and some legumes and rhizomes possess C-type starch, which is a mixture of both A- and B-type crystallinities (Cheetham and Tao, 1998; Blazek and Gilbert, 2011).

For high-amylose crops with SBEs inactivation, their amylopectin branch-chain length increases and branching degree decreases with increasing AC, leading to the change of starch crystallinity (Table 1). Most high-amylose crop starches display B-type crystallinity (Nishi et al., 2001; Li et al., 2008; Jiang et al., 2010b; Butardo et al., 2011; Asai et al., 2014; Lin et al., 2016c), which is easily understood through the change in amylopectin structure. However, some high-amylose crops have C-type starch (Cheetham and Tao, 1998; Wei et al., 2010a,d; Butardo et al., 2011; Cai et al., 2014c; Huang et al., 2015; Lin et al., 2016c). Cheetham and Tao (1998) proposed that starch changes from A-type to B-type via C-type as AC increases in maize. The transition occurs at approximately 40% AC, a value capable of maintaining C-type starch. Starch with AC lower than 40% has A-type crystallinity, whereas AC higher than 40% has B-type crystallinity. Cai et al. (2014c) thought that high-amylose maize starch with AC 35.6% and apparent AC 56% contain differently sized granules and large and small granules exhibit A- and B-type crystallinities, respectively. However, how and where these different-sized starch granules are synthesized in maize endosperm remains unclear. High-amylose rice TRS with 60% AC is composed of polygonal, aggregate, elongated, and hollow granules distributed in specific endosperm regions (Wei et al., 2010c; Cai et al., 2014b). Polygonal and elongated granule has A- and B-type crystallinity, respectively. However, the aggregate granule contains both A- and B-type crystallinities within the same granule (Wei et al., 2010a,b; Man et al., 2014). TRS is a good material to investigate the synthesis of different crystallinities in the future.

Thermal Properties of High-Amylose Starch

For the applications of starch in food industry, heating starch in water is necessary. During heating, starch granules absorb water and swell, crystalline structure is destroyed, and birefringence is lost. This process is defined as starch gelatinization and can be detected by hot-stage microscopy and differential scanning calorimetry (Cai et al., 2014a). AC and amylopectin structure in starch granule significantly affect starch gelatinization (Kaur et al., 2007; Wei et al., 2011; Qin et al., 2012; Cai et al., 2014d). Amylose can maintain the integrity of swollen granule and restrain its swelling (Tester and Morrison, 1992). Granule swelling is positively correlated to amylopectin short branch-chains, whereas negatively to amylopectin long branch-chains (Salman et al., 2009; Lin et al., 2016a). In addition, lipid content in starch granule is positively correlated with AC, and lipid-complexed amylose chains also restrict granular swelling (Tester and Morrison, 1992). Therefore, high-amylose starches have lower granule swelling.

For gelatinization temperature, normal starches usually show a clear gelatinization transition peak due to the dissociation of short-chain double-helical crystallites of amylopectin molecules (Jiang et al., 2010c). However, high-amylose starches yield two thermal transition peaks that correspond to the melting of amylopectin double-helical crystallite and amylose-lipid complex (Jiang et al., 2010c; Qin et al., 2012; Man et al., 2014). In high-amylose starch with relatively low AC increase, the amylopectin has longer branch-chains and its double-helical crystallite requires a higher gelatinization temperature than normal starch does; on the other hand, high AC causes the simultaneous formation of an amylose-lipid complex resulting in the second thermal transition peak appearance (Jiang et al., 2010c; Qin et al., 2012; Man et al., 2014). As AC continually increases, the molecular structure of amylopectin is severely disordered; thus, intensity of the first peak gradually becomes weak, whereas the second peak becomes strong (Regina et al., 2010; Man et al., 2014). When AC becomes the dominant component in the starch granules, the first peak disappears and only the second one is preserved (Jiang et al., 2010c; Regina et al., 2010; Man et al., 2014).

Hydrolysis of High-Amylose Starch

Starch hydrolysis by amylase is involved in many biological and industrial processes (Tawil et al., 2011). Usually, amylase hydrolyzes starch in two ways as follows: Amyloglucosidase erodes the outer surface of the granule (exocorrosion), and α-amylase creates channels leading to the granule center and consequently leads to its breakdown from the inside out (endocorrosion) (Li et al., 2004). The susceptibility of starch to amylase is influenced by many factors, such as granule morphology, size, integrity and porosity, AC, and crystalline structure (Blazek and Gilbert, 2010). Generally, large starch granule with low relative surface area is slowly hydrolyzed compared with a small granule (Kim et al., 2008). Surface pores and internal channels of starch granule can increase enzyme diffusion into the interior of granule (Naguleswaran et al., 2011, 2012). B-type starch has higher resistance to amylase than A-type starch; amylose inhibits starch hydrolysis (Jiang et al., 2010d; Man et al., 2013b,c; Lin et al., 2016a,b). Proportion of the elongated and aggregate starch granules in high-amylose maize and rice increases with the increase in AC. Elongated and aggregate granules, consisting of two or many fused sub-granules, have larger size than normal starch, show smooth surface without porosity and internal channel, and exhibit high amylose in the outer layer of the granule (Jiang et al., 2010a,d; Wei et al., 2010c; Cai et al., 2014d). Therefore, B-type crystallinity, large size, and smooth surface increase the resistance of high-amylose starches to amylase hydrolysis, and the outer region of the granule is highly resistant to hydrolysis (Jiang et al., 2010d; Man et al., 2013b,c).

Heterogeneity of Starch Granules in High-Amylose Cereal Crops

Recently, a highly diverse population of differently stained and different morphological starch granules are detected in the endosperm of high-amylose cereal crop (Jiang et al., 2010a; Wei et al., 2010c; Butardo et al., 2011; Wellner et al., 2011; Carciofi et al., 2012; Liu et al., 2013; Cai et al., 2014d; Man et al., 2014). More interestingly, different morphological granules have an obviously regional distribution in mature kernels (Wellner et al., 2011; Liu et al., 2013; Cai et al., 2014b).

Occurrence of Differently Stained Starch Granules

Isolated starch granules from high-amylose maize and rice grains show different colors under normal and polarized light when being stained with iodine (Cai et al., 2014b,d). More interesting, in the kernel of maize *ae* mutants, a narrow band of sub-aleurone is occupied by blue-staining granules, whereas the central endosperm region is dominated by pink-staining granules. Between the differently stained regions, a transition region with a range of biphasic granules containing blue- and pink-stained colors simultaneously are detected (Liu et al., 2013). This observation indicates that the three differently stained starch granules exhibit a spatial distribution and that the heterogeneity of starch granules increases slightly from the inner to outer layer. Raman microscopy reveals that the pink regions contain starch molecules with reduced levels of branching compared with those blue-stained regions, which are ordered (crystalline) but not radially oriented (Wellner et al., 2011). Thus far, the physicochemical and functional properties of the three differently stained granules have yet to be extensively investigated.

Occurrence of Different Morphological Starch Granules

Recently, several reports state that high-amylose crops have different morphological starch granule distinguished from the normal ones. Normal maize contains angular or spherical granules, whereas in the single mutant *ae*, a small percentage (~1.7%) of elongated granules is observed. Furthermore, this kind of granules has a positive relationship with AC. The percentage of elongated granules is increased to 32% in GEMS-0067 endosperm with 85% AC (Jiang et al., 2010a). Some aggregate starch granules are also reported in high-amylose maize (Cai et al., 2014d). Wheat and barley grains have large lenticular starch granules and small spherical ones; however, the large granules from their high-amylose counterparts become sickle-shaped (Regina et al., 2006, 2010). In barley amylose-only line, multi-lobed starch granules with elongated, rough, and globular morphology dominates the whole kernel (Carciofi et al., 2012). Rice grain is filled with homogeneous compound granules, which can be easily separated to individual polygonal ones with various angles during starch isolation process. In different source of high-amylose rice, including *ae* and Goami 2 mutants and *SBEIIb*-inhibited lines, polygonal granule is mixed with large voluminous, nonangular rounded bodies (Nishi et al., 2001; Kang et al., 2003; Butardo et al., 2011). In high-amylose rice TRS, mature kernel has four types of different morphological starches including polygonal, aggregate, elongated, and hollow granules (Wei et al., 2010c; Cai et al., 2014b).

Even though different morphological starch granules are popularly detected in high-amylose cereals, their physicochemical properties still remain unknown due to the fact that they are always studied as a mixture with the normal ones. Therefore, all the results from high-amylose starch are averaged values. Actually, not only do starch granules with different morphologies have the different structures, but also they contain different physicochemical properties across identical and different species. For example, in wheat and barley, the large granule has higher AC, gelatinization enthalpy and pasting viscosity, and lower amylopectin short branch-chains and swelling power than its small counterpart, which results in different end uses (Salman et al., 2009; Naguleswaran et al., 2011; Li et al., 2013). Small starches are predominantly applied as fat substitute, paper coating, and as carrier material in cosmetics, whereas large starches are used to manufacture biodegradable plastic film, carbonless copy paper, and brewing beer (Lindeboom et al., 2004). Thus, separation and purification of different morphological starch granules are essential for further applications in food and non-food industries.

Research on Purified Heterogeneous Starch Granules from High-Amylose Rice TRS

Contrary to the distribution of different morphological starch granules in the grain of *ae* maize mutants (Liu et al., 2013), the heterogeneous starch granules with polygonal, aggregate, elongated, and hollow shapes in high-amylose rice TRS grain have obvious regional distribution from the inner region to the outer region. The polygonal starch is located in the innermost region, the aggregate, and elongated starches exist in the center region, and the hollow starch is distributed in the outmost region (Cai et al., 2014b). To investigate their individual structure and physicochemical properties, the four granule types are separated and purified on the basis of their regionally distributed characteristics (Man et al., 2014). GPC analysis shows that from the polygonal to hollow starch granules, their AC displays a gradually elevated trend, from 35.9 to 75.8%. The four granule types are essentially different in relative crystallinity, short-range ordered structure, relative proportions of single helix, double helix, and amorphous conformation; these differences imply that their structures are completely divergent (Man et al., 2014). The different structures of TRS heterogeneous starch granules yield significantly different XRD patterns, thermal properties, and digestion properties. Polygonal starch exhibits A-type crystallinity, aggregate and elongated starches show C-type crystallinity, whereas hollow starch has no crystallinity (Man et al., 2014). Cai et al. (2014b) found that polygonal and elongated starches have the lowest and highest gelatinization temperatures, respectively, while the gelatinization temperature of hollow starch is undetected due to its extremely weak or no birefringence. The *in vitro* digestion properties of native starches from the four granule types are investigated, and results reveal that polygonal and aggregate starches possess monophasic digestograms, whereas elongated and hollow starches exhibit biphasic digestograms from 0 to 8 h mainly because of their different structural compositions (Huang et al., 2016).

CONCLUSIONS AND FUTURE PERSPECTIVES

In maize and rice, mutating or suppressing *SBEIIb* is necessary to obtain high-amylose lines. However, the efficiency for an AC increase in rice is much lower than that in maize. When HAM genes are introduced to maize *ae* mutants and SBEI and SBEIIb are simultaneously inhibited in *indica* rice, AC is further increased and thus reaches 88.2 and 60%, respectively. In wheat and barley, high-amylose lines with AC (>85%) are derived by simultaneously suppressing *SBEIIb* and *SBEIIa*. The enhanced amylose synthesis and the suppressed amylopectin synthesis and the ELC from amylopectin/IC are responsible for AC increase in high-amylose species. Along with the AC increase, fewer longer branch-chains are detected in amylopectin. The significantly altered structure of high-amylose starches causes a series of other important changes, such as crystallinity, thermal properties, and hydrolysis properties.

Different morphological starches and differently stained granules are detected in high-amylose rice and maize. They are also regularly distributed in the grains from the inner to outer regions and occupy different molecular structures. Four heterogeneous granules in high-amylose rice TRS possess different physicochemical and digestion properties, which can be used for further applications in food and non-food industries. Future biochemical experiments should be performed to elucidate how heterogeneous starch granules are formed in one kernel. The different morphological starches are also detected in other high-amylose cereal crops, but whether these heterogeneous starches exhibit regional distribution in kernels and have different physicochemical properties are unknown.

Studies have also yet to identify the causes of regional distribution of heterogeneous starches in kernels. Grain filling originally begins in the core and then spreads to the outer region during endosperm development in rice and maize. Combined with regionally distributed phenotypes in high-amylose grains, various regulated mechanisms may be observed in different regions of grains. For example, white-core mutants, such as *flo4, flo5,* and *rsr1* (Kang et al., 2005; Ryoo et al., 2007; Fu and Xue, 2010), and aberrant-periphery mutants, including *flo7* (Zhang et al., 2016), play essential roles in the early stage and late stages of endosperm development in rice, respectively. However, we have yet to reveal how starch synthesis-related enzymes are regulated in grain during endosperm development from the inner parts to the outer parts.

CONFLICT OF INTEREST STATEMENT

The authors declare that the research was conducted in the absence of any commercial or financial relationships that could be construed as a potential conflict of interest.

ACKNOWLEDGMENTS

This work is supported by grants from the National Key Research and Development Program of China (2016YFD0101002), the Natural Science Foundation of Jiangsu Province (BK20160457), the China Postdoctoral Science Foundation (2016590509), the Qing Lan Project of Jiangsu Province, the Talent Project of Yangzhou University, and the Priority Academic Program Development of Jiangsu Higher Education Institutions.

REFERENCES

1. Asai, H., Abe, N., Matsushima, R., Crofts, N., Oitome, N. F., Nakamura, Y., et al. (2014). Deficiencies in both starch synthase IIIa and branching enzyme IIb lead to a significant increase in amylose in SSIIa inactive japonica rice seeds. *J. Exp. Bot.* 65, 5497–5507. doi: 10.1093/jxb/eru310
2. Avella, M., Errico, M. E., Rimedio, R., and Sadocco, P. (2002). Preparation of biodegradable polyesters/high-amylose-starch composites by reactive blending and their characterization. *J. Appl. Polym. Sci.* 83, 1432–1442. doi: 10.1002/app.2304
3. Blauth, S. L., Kim, K., Klucinec, J., Shannon, J. C., Thompson, D., and Guiltinan, M. (2002). Identification of *Mutator* insertional mutants of starch-branching enzyme 1 (*sbe1*) in *Zea mays* L. *Plant Mol. Bio.* 48, 287–297. doi: 10.1023/A:1013335217744
4. Blauth, S. L., Yao, Y., Klucinec, J. D., Shannon, J. C., Thompson, D. B., and Guiltinan, M. J. (2001). Identification of mutator insertional mutants of starch-branching enzyme 2a in corn. *Plant Physiol.* 125, 1396–1405. doi: 10.1104/pp.125.3.1396
5. Blazek, J., and Gilbert, E. P. (2010). Effect of enzymatic hydrolysis on native starch granule structure. *Biomacromolecules* 11, 3275–3289. doi: 10.1021/bm101124t
6. Blazek, J., and Gilbert, E. P. (2011). Application of small-angle X-ray and neutron scattering techniques to the characterisation of starch structure: a review. *Carbohydr. Polym.* 85, 281–293. doi: 10.1016/j.carbpol.2011.02.041
7. Butardo, V. M., Fitzgerald, M. A., Bird, A. R., Gidley, M. J., Flanagan, B. M., Larroque, O., et al. (2011). Impact of down-regulation of starch branching enzyme IIb in rice by artificial microRNA- and hairpin RNA-mediated RNA silencing. *J. Exp. Bot.* 62, 4927–4941. doi: 10.1093/jxb/err188
8. Cai, C., Cai, J., Zhao, L., and Wei, C. (2014a). *In situ* gelatinization of starch using hot stage microscopy. *Food Sci. Biotechnol.* 23, 15–22. doi: 10.1007/s10068-014-0003-x

9. Cai, C., Huang, J., Zhao, L., Liu, Q., Zhang, C., and Wei, C. (2014b). Heterogeneous structure and spatial distribution in endosperm of high-amylose rice starch granules with different morphologies. *J. Agric. Food Chem.* 62, 10143–10152. doi: 10.1021/jf502341q
10. Cai, C., Lin, L., Man, J., Zhao, L., Wang, Z., and Wei, C. (2014c). Different structural properties of high-amylose maize starch fractions varying in granule size. *J. Agric. Food Chem.* 62, 11711–11721. doi: 10.1021/jf503865e
11. Cai, C., Zhao, L., Huang, J., Chen, Y., and Wei, C. (2014d). Morphology, structure and gelatinization properties of heterogeneous starch granules from high-amylose maize. *Carbohydr. Polym.* 102, 606–614. doi: 10.1016/j.carbpol.2013.12.010
12. Cai, J., Man, J., Huang, J., Liu, Q., Wei, W., and Wei, C. (2015). Relationship between structure and functional properties of normal rice starches with different amylose contents. *Carbohydr. Polym.* 125, 35–44. doi: 10.1016/j.carbpol.2015.02.067
13. Cai, J., Yang, Y., Man, J., Huang, J., Wang, Z., Zhang, C., et al. (2014). Structural and functional properties of alkali-treated high-amylose rice starch. *Food Chem.* 145, 245–253. doi: 10.1016/j.foodchem.2013.08.059
14. Carciofi, M., Blennow, A., Jensen, S. L., Shaik, S. S., Henriksen, A., Buleon, A., et al. (2012). Concerted suppression of all starch branching enzyme genes in barley produces amylose-only starch granules. *BMC Plant Biol.* 12:223. doi: 10.1186/1471-2229-12-223
15. Cheetham, N. W. H., and Tao, L. (1998). Variation in crystalline type with amylose content in maize starch granules: an X-ray powder diffraction study. *Carbohydr. Polym.* 36, 277–284. doi: 10.1016/S0144-8617(98)00007-1
16. Crofts, N., Abe, K., Aihara, S., Itoh, R., Nakamura, Y., Itoh, K., et al. (2012). Lack of starch synthase IIIa and high expression of granule-bound starch synthase I synergistically increase the apparent amylose content in rice endosperm. *Plant Sci.* 193–194, 62–69. doi: 10.1016/j.plantsci.2012.05.006
17. Fu, F. F., and Xue, H. W. (2010). Coexpression analysis identifies rice starch regulator1, a rice AP2/EREBP family transcription factor, as a novel rice starch biosynthesis regulator. *Plant Physiol.* 154, 927–938. doi: 10.1104/pp.110.159517
18. Fujita, N., Yoshida, M., Asakura, N., Ohdan, T., Miyao, A., Hirochika, H., et al. (2006). Function and characterization of starch synthase I using mutants in rice. *Plant Physiol.* 140, 1070–1084. doi: 10.1104/pp.105.071845
19. Fujita, N., Yoshida, M., Kondo, T., Saito, K., Utsumi, Y., Tokunaga, T., et al. (2007). Characterization of SSIIIa-deficient mutants of rice: the function of SSIIIa and pleiotropic effects by SSIIIa deficiency in the rice endosperm. *Plant Physiol.* 144, 2009–2023. doi: 10.1104/pp.107.102533
20. Gao, M., Fisher, D. K., Kim, K. N., Shannon, J. C., and Guiltinan, M. J. (1997). Independent genetic control of maize starch-branching enzymes IIa and IIb. *Plant Physiol.* 114, 69–78. doi: 10.1104/pp.114.1.69
21. Granfeldt, Y., Drews, A., and Björck, I. (1995). Arepas made from high amylose corn flour produce favorably Low glucose and insulin responses in healthy humans. *J. Nutr.* 125, 459–465.
22. Guan, H., Li, P., Imparl-Radosevich, J., Preiss, J., and Keeling, P. (1997). Comparing the properties of *Escherichia coli* branching enzyme and maize branching enzyme. *Arch. Biochem. Biophys.* 342, 92–98. doi: 10.1006/abbi.1997.0115
23. Guan, H. P., and Preiss, J. (1993). Differentiation of the properties of the branching isozymes from maize (*Zea mays*). *Plant Physiol.* 102, 1269–1273. doi: 10.1104/pp.102.4.1269
24. Huang, J., Lin, L., Wang, J., Wang, Z., Liu, Q., and Wei, C. (2016). *In vitro* digestion properties of heterogeneous starch granules from high-amylose rice. *Food Hydrocolloids* 54, 10–22. doi: 10.1016/j.foodhyd.2015.09.016
25. Huang, J., Shang, Z., Man, J., Liu, Q., Zhu, C., and Wei, C. (2015). Comparison of molecular structures and functional properties of high-amylose starches from rice transgenic line and commercial maize. *Food Hydrocolloids* 46, 172–179. doi: 10.1016/j.foodhyd.2014.12.019

26. Isshiki, M., Nakajima, M., Satoh, H., and Shimamoto, K. (2000). Dull: rice mutants with tissue-specific effects on the splicing of the waxy pre-mRNA. *Plant J.* 23, 451–460. doi: 10.1046/j.1365-313x.2000.00803.x
27. Itoh, K., Ozaki, H., Okada, K., Hori, H., Takeda, Y., and Mitsui, T. (2003). Introduction of *Wx* transgene into rice *wx* mutants leads to both high- and low-amylose rice. *Plant Cell Physiol.*44, 473–480. doi: 10.1093/pcp/pcg068
28. Jane, J., Chen, Y. Y., Lee, L. F., McPherson, A. E., Wong, K. S., Radosavljevic, M., et al. (1999). Effects of amylopectin branch chain length and amylose content on the gelatinization and pasting properties of starch. *Cereal Chem.* 76, 629–637. doi: 10.1094/CCHEM.1999.76.5.629
29. Jane, J. L., and Chen, J. F. (1992). Effect of amylose molecular size and amylopectin branch chain length on paste properties of starch. *Cereal Chem.* 69, 60–65.
30. Jeon, J. S., Ryoo, N., Hahn, T. R., Walia, H., and Nakamura, Y. (2010). Starch biosynthesis in cereal endosperm. *Plant Physiol. Biochem.* 48, 383–392. doi: 10.1016/j.plaphy.2010.03.006
31. Jiang, H., Campbell, M., Blanco, M., and Jane, J. L. (2010d). Characterization of maize amylose-extender (ae) mutant starches: part II. structures and properties of starch residues remaining after enzymatic hydrolysis at boiling-water temperature. *Carbohydr. Polym.* 80, 1–12. doi: 10.1016/j.carbpol.2009.10.060
32. Jiang, H., Horner, H. T., Pepper, T. M., Blanco, M., Campbell, M., and Jane, J. L. (2010a). Formation of elongated starch granules in high-amylose maize. *Carbohyd. Polym.* 80, 533–538. doi: 10.1016/j.carbpol.2009.12.016
33. Jiang, H., Jane, J. L., Acevedo, D., Green, A., Shinn, G., Schrenker, D., et al. (2010b). Variations in starch physicochemical properties from a generation-means analysis study using amylomaize V and VII parents. *J. Agric. Food Chem.* 58, 5633–5639. doi: 10.1021/jf904531d
34. Jiang, H., Lio, J., Blanco, M., Campbell, M., and Jane, J. L. (2010c). Resistant-starch formation in high-amylose maize starch during kernel development. *J. Agric. Food Chem.* 58, 8043–8047. doi: 10.1021/jf101056y
35. Kang, H. G., Park, S., Matsuoka, M., and An, G. (2005). White-core endosperm *floury endosperm-4* in rice is generated by knockout mutations in the C_4-type pyruvate orthophosphate dikinase gene (*OsPPDKB*). *Plant J.* 42, 901–911. doi: 10.1111/j.1365-313X.2005.02423.x
36. Kang, H. J., Hwang, I. K., Kim, K. S., and Choi, H. C. (2003). Comparative structure and physicochemical properties of Ilpumbyeo, a high-quality japonica rice, and its mutant, Suweon 464. *J. Agric. Food Chem.* 51, 6598–6603. doi: 10.1021/jf0344946
37. Kasemsuwan, T., Jane, J., Schnable, P., Stinard, P., and Robertson, D. (1995). Characterization of the dominant mutant amylose-extender (*Ae1-5180*) maize starch. *Cereal Chem.* 72, 457–464.
38. Kaur, L., Singh, J., McCarthy, O. J., and Singh, H. (2007). Physico-chemical, rheological and structural properties of fractionated potato starches. *J. Food Eng.* 82, 383–394. doi: 10.1016/j.jfoodeng.2007.02.059
39. Kim, J. C., Kong, B. W., Kim, M. J., and Lee, S. H. (2008). Amylolytic hydrolysis of native starch granules affected by granule surface area. *J. Food Sci.* 73, C621–C624. doi: 10.1111/j.1750-3841.2008.00944.x
40. Li, J. H., Vasanthan, T., Hoover, R., and Rossnagel, B. G. (2004). Starch from hull-less barley: V. *in-vitro* susceptibility of waxy, normal, and high-amylose starches towards hydrolysis by alpha-amylases and amyloglucosidase. *Food Chem.* 84, 621–632. doi: 10.1016/S0308-8146(03)00287-5
41. Li, L., Jiang, H., Campbell, M., Blanco, M., and Jane, J. L. (2008). Characterization of maize amylose-extender (*ae*) mutant starches. Part I: relationship between resistant starch contents and molecular structures. *Carbohydr. Polym.* 74, 396–404. doi: 10.1016/j.carbpol.2008.03.012
42. Li, W., Shan, Y., Xiao, X., Luo, Q., Zheng, J., Ouyang, S., et al. (2013). Physicochemical properties of A- and B-starch granules isolated from hard red and soft red winter wheat. *J. Agric. Food Chem.* 61, 6477–6484. doi: 10.1021/jf400943h

43. Lin, L., Cai, C., Gibert, R. G., Li, E., Wang, J., and Wei, C. (2016a). Relationships between amylopectin molecular structures and functional properties of different-sized fractions of normal and high-amylose maize starches. *Food Hydrocolloids* 52, 359–368. doi: 10.1016/j.foodhyd.2015.07.019
44. Lin, L., Guo, D., Huang, J., Zhang, X., Zhang, L., and Wei, C. (2016b). Molecular structure and enzymatic hydrolysis properties of starches from high-amylose maize inbred lines and their hybrids. *Food Hydrocolloids* 58, 246–254. doi: 10.1016/j.foodhyd.2016.03.001
45. Lin, L., Guo, D., Zhao, L., Zhang, X., Wang, J., Zhang, F., et al. (2016c). Comparative structure of starches from high-amylose maize inbred lines and their hybrids. *Food Hydrocolloids* 52, 19–28. doi: 10.1016/j.foodhyd.2015.06.008
46. Lindeboom, N., Chang, P. R., and Tyler, R. T. (2004). Analytical biochemical and physicochemical aspects of starch granule size, with emphasis on small granule starch: a review. *Starch*56, 89–99. doi: 10.1002/star.200300218
47. Liu, D., Parker, M. L., Wellner, N., Kirbyb, A. R., Cross, K., Morris, V. J., et al. (2013). Structural variability between starch granules in wild type and in *ae* high-amylose mutant maize kernels. *Carbohydr. Polym.* 97, 458–468. doi: 10.1016/j.carbpol.2013.05.013
48. Man, J., Lin, L., Wang, Z., Wang, Y., Liu, Q., and Wei, C. (2014). Different structures of heterogeneous starch granules from high-amylose rice. *J. Agric. Food Chem.* 62, 11254–11263. doi: 10.1021/jf503999r
49. Man, J., Qin, F., Zhu, L., Shi, Y., Gu, M., Liu, Q., et al. (2012a). Ordered structure and thermal property of acid-modified high-amylose rice starch. *Food Chem.* 134, 2242–2248. doi: 10.1016/j.foodchem.2012.04.100
50. Man, J., Yang, Y., Huang, J., Zhang, C., Chen, Y., Wang, Y., et al. (2013a). Effect of simultaneous inhibition of starch branching enzymes I and IIb on the crystalline structure of rice starches with different amylose content. *J. Agric. Food Chem.* 61, 9930–9937. doi: 10.1021/jf4030773
51. Man, J., Yang, Y., Huang, J., Zhang, C., Zhang, F., Wang, Y., et al. (2013b). Morphology and structural properties of high-amylose rice starch residues hydrolyzed by amyloglucosidase. *Food Chem.* 138, 2089–2098. doi: 10.1016/j.foodchem.2012.12.009
52. Man, J., Yang, Y., Zhang, C., Zhang, F., Wang, Y., Liu, Q., et al. (2013c). Morphology and structural characterization of high-amylose rice starch residues hydrolyzed by porcine pancreatic α-amylase. *Food Hydrocolloids* 31, 195–203. doi: 10.1016/j.foodhyd.2012.11.003
53. Man, J., Yang, Y., Zhang, C., Zhou, X., Dong, Y., Zhang, F., et al. (2012b). Structural changes of high-amylose rice starch residues following *in vitro* and *in vivo* digestion. *J. Agric. Food Chem.* 60, 9332–9341. doi: 10.1021/jf302966f
54. Morell, M. K., Kosar-Hashemi, B., Cmiel, M., Samuel, M. S., Chandler, P., Rahman, S., et al. (2003). Barley *sex6* mutants lack starch synthase IIa activity and contain a starch with novel properties. *Plant J.* 34, 173–185. doi: 10.1046/j.1365-313X.2003.01712.x
55. Naguleswaran, S., Li, J., Vasanthan, T., and Bressler, D. (2011). Distribution of granule channels, protein, and phospholipid in triticale and corn starches as revealed by confocal laser scanning microscopy. *Cereal Chem.* 88, 87–94. doi: 10.1094/CCHEM-04-10-0062
56. Naguleswaran, S., Li, J., Vasanthan, T., Bressler, D., and Hoover, R. (2012). Amylolysis of large and small granules of native triticale, wheat and corn starches using a mixture of α-amylase and glucoamylase. *Carbohydr. Polym.* 88, 864–874. doi: 10.1016/j.carbpol.2012.01.027
57. Nakamura, Y. (2002). Towards a better understanding of the metabolic system for amylopectin biosynthesis in plants: rice endosperm as a model tissue. *Plant Cell Physiol.* 43, 718–725. doi: 10.1093/pcp/pcf091
58. Nakamura, Y., Utsumi, Y., Sawada, T., Aihara, S., Utsumi, C., Yoshida, M., et al. (2010). Characterization of the reactions of starch branching enzymes from rice endosperm. *Plant Cell Physiol.* 51, 776–794. doi: 10.1093/pcp/pcq035

59. Nishi, A., Nakamura, Y., Tanaka, N., and Satoh, H. (2001). Biochemical and genetic analysis of the effects of *amylose-extender* mutation in rice endosperm. *Plant Physiol.* 127, 459–472. doi: 10.1104/pp.010127
60. Nugent, A. P. (2005). Health properties of resistant starch. *Nutr. Bull.* 30, 27–54. doi: 10.1111/j.1467-3010.2005.00481.x
61. Pan, T., Zhao, L., Lin, L., Wang, J., Liu, Q., and Wei, C. (2017). Changes in kernel morphology and starch properties of high-amylose brown rice during the cooking process. *Food Hydrocolloids* 66, 227–236. doi: 10.1016/j.foodhyd.2016.11.035
62. Peymanpour, G., Marcone, M., Ragaee, S., Tetlow, I., Lane, C. C., Seetharaman, K., et al. (2016). On the molecular structure of the amylopectin fraction isolated from "high-amylose" *ae*maize starches. *Int. J. Biol. Macromol.* 91, 768–777. doi: 10.1016/j.ijbiomac.2016.06.029
63. Qin, F., Man, J., Cai, C., Xu, B., Gu, M., Zhu, L., et al. (2012). Physicochemical properties of high-amylose rice starches during kernel development. *Carbohydr. Polym.* 88, 690–698. doi: 10.1016/j.carbpol.2012.01.013
64. Qin, F., Man, J., Xu, B., Hu, M., Gu, M., Liu, Q., et al. (2011). Structural properties of hydrolyzed high-amylose rice starch by α-amylase from *Bacillus licheniformis*. *J. Agric. Food Chem.*59, 12667–12673. doi: 10.1021/jf203167f
65. Regina, A., Berbezy, P., Kosar-Hashemi, B., Li, S., Cmiel, M., Larroque, O., et al. (2015). A genetic strategy generating wheat with very high amylose content. *Plant Biotechnol. J.* 13, 1276–1286. doi: 10.1111/pbi.12345
66. Regina, A., Bird, A., Topping, D., Bowden, S., Freeman, J., Barsby, T., et al. (2006). High-amylose wheat generated by RNA interference improves indices of large-bowel health in rats. *Proc. Natl. Acad. Sci. U.S.A.* 103, 3546–3551. doi: 10.1073/pnas.0510737103
67. Regina, A., Blazek, J., Gilbert, E., Flanagan, B. M., Gidley, M. J., Cavanagh, C., et al. (2012). Differential effects of genetically distinct mechanisms of elevating amylose on barley starch characteristics. *Carbohydr. Polym.* 89, 979–991. doi: 10.1016/j.carbpol.2012.04.054
68. Regina, A., Kosar-Hashemi, B., Ling, S., Li, Z., Rahman, S., and Morell, M. (2010). Control of starch branching in barley defined through differential RNAi suppression of starch branching enzyme IIa and IIb. *J. Exp. Bot.* 61, 1469–1482. doi: 10.1093/jxb/erq011
69. Regina, A., Kosar-Hashemi, B., Li, Z., Rampling, L., Cmiel, M., Gianibelli, M. C., et al. (2004). Multiple isoforms of starch branching enzyme-I in wheat: lack of the major SBE-I isoform does not alter starch phenotype. *Funct. Plant Biol.* 31, 591–601. doi: 10.1071/FP03193
70. Ryoo, N., Yu, C., Park, C. S., Baik, M. Y., Park, I. M., Cho, M. H., et al. (2007). Knockout of a starch synthase gene *OsSSIIIa/Flo5* causes white-core floury endosperm in rice (*Oryza sativa*L.). *Plant Cell Rep.* 26, 1083–1095. doi: 10.1007/s00299-007-0309-8
71. Salman, H., Blazek, J., Lopez-Rubio, A., Gilbert, E. P., Hanley, T., and Copeland, L. (2009). Structure-function relationships in A and B granules from wheat starches of similar amylose content. *Carbohydr. Polym.* 75, 420–427. doi: 10.1016/j.carbpol.2008.08.001
72. Sang, Y., Bean, S., Seib, P. A., Pedersen, J., and Shi, Y. C. (2008). Structure and functional properties of sorghum starches differing in amylose content. *J. Agric. Food Chem.* 56, 6680–6685. doi: 10.1021/jf800577x
73. Satoh, H., Nishi, A., Fujita, N., Kubo, A., Nakamura, Y., Kawasaki, T., et al. (2003a). Isolation and characterization of starch mutants in rice. *J. Appl. Glycosci.* 50, 225–230. doi: 10.5458/jag.50.225
74. Satoh, H., Nishi, A., Yamashita, K., Takemoto, Y., Tanaka, Y., Hosaka, Y., et al. (2003b). Starch-branching enzyme I-deficient mutation specifically affects the structure and properties of starch in rice endosperm. *Plant Physiol.* 133, 1111–1121. doi: 10.1104/pp.103.021527
75. Satoh, H., Shibahara, K., Tokunaga, T., Nishi, A., Tasaki, M., Hwang, S. K., et al. (2008). Mutation of the plastidial α-glucan phosphorylase gene in rice affects the synthesis

and structure of starch in the endosperm. *Plant Cell* 20, 1833–1849. doi: 10.1105/tpc.107.054007
76. Sestili, F., Botticella, E., Proietti, G., Janni, M., D'Ovidio, R., and Lafiandra, D. (2012). Amylose content is not affected by overexpression of the *Wx-B1* gene in durum wheat. *Plant Breed.*131, 700–706. doi: 10.1111/j.1439-0523.2012.02004.x
77. Shannon, J. C., Garwood, D. L., and Boyer, C. D. (2009). "Genetics and physiology of starch developmen," in *Starch, Chemistry and Technology*, eds J. BeMiller and R. L. Whistler (New York, NY: Academic Press), 23–82.
78. Slade, A. J., McGuire, C., Loeffler, D., Mullenberg, J., Skinner, W., Fazio, G., et al. (2012). Development of high amylose wheat through TILLING. *BMC Plant Biol.* 12:69. doi: 10.1186/1471-2229-12-69
79. Smith, A. M., Zeeman, S. C., and Denyer, K. (2002). "The synthesis of amylose," in *Starch, Advance in Structure and Function*, eds T. L. Barsby, A. M. Donald, and P. J. Frazier (Cambridge: Royal Society of Chemistry), 150–163.
80. Takeda, Y., Guan, H. P., and Preiss, J. (1993). Branching of amylose by the branching isoenzymes of maize endosperm. *Carbohydr. Res.* 240, 253–263. doi: 10.1016/0008-6215(93)84188-C
81. Tanaka, N., Fujita, N., Nishi, A., Satoh, H., Hosaka, Y., Ugaki, M., et al. (2004). The structure of starch can be manipulated by changing the expression levels of starch branching enzyme llb in rice endosperm. *Plant Biotech. J.* 2, 507–516. doi: 10.1111/j.1467-7652.2004.00097.x
82. Tawil, G., Viksø-Nielsen, A., Rolland-Sabaté, A., Colonna, P., and Buléon, A. (2011). In depth study of a new highly efficient raw starch hydrolyzing a-amylase from *Rhizomucor sp. Biomacromolecules* 12, 34–42. doi: 10.1021/bm100913z
83. Tester, R. F., and Morrison, W. R. (1992). Swelling and gelatinization of cereal starches. III. some properties of waxy and normal nonwaxy barley starches. *Cereal Chem.* 69, 654–658.
84. Tetlow, I. J., and Emes, M. J. (2014). A review of starch-branching enzymes and their role in amylopectin biosynthesis. *IUBMB Life* 66, 546–558. doi: 10.1002/iub.1297
85. Tsai, C. Y., Salamini, F., and Nelson, O. E. (1970). Enzymes of carbohydrate metabolism in the developing endosperm of maize. *Plant Physiol.* 46, 299–306. doi: 10.1104/pp.46.2.299
86. Tziotis, A., Seetharaman, K., Wong, K. S., Klucinec, J. D., Jane, J. L., and White, P. J. (2004). Structural properties of starch fractions isolated from normal and mutant corn genotypes using different methods. *Cereal Chem.* 81, 611–620. doi: 10.1094/CCHEM.2004.81.5.611
87. Umemoto, T., Aoki, N., Lin, H., Nakamura, Y., Inouchi, N., Sato, Y., et al. (2004). Natural variation in rice *starch synthase IIa* affects enzyme and starch properties. *Funct. Plant Biol.* 31, 671–684. doi: 10.1071/FP04009
88. Umemoto, T., Horibata, T., Aoki, N., Hiratsuka, M., Yano, M., and Inouchi, N. (2008). Effects of variations in starch synthase on starch properties and eating quality of rice. *Plant Prod. Sci.* 11, 472–480. doi: 10.1626/pps.11.472
89. Umemoto, T., and Terashima, K. (2002). Activity of granule-bound starch synthase is an important determinant of amylose content in rice endosperm. *Funct. Plant Biol.* 29, 1121–1124. doi: 10.1071/PP01145
90. Wang, Y. J., White, P., Pollak, L., and Jane, J. (1993). Amylopectin and intermediate materials in starches from mutant genotypes of the Oh43 inbred line. *Cereal Chem.* 70, 521–525.
91. Wei, C., Qin, F., Zhou, W., Chen, Y., Xu, B., Wang, Y., et al. (2010a). Formation of semi-compound C-type starch granule in high-amylose rice developed by antisense RNA inhibition of starch branching enzyme. *J. Agric. Food Chem.* 58, 11097–11104. doi: 10.1021/jf1024533
92. Wei, C., Qin, F., Zhou, W., Xu, B., Chen, C., Chen, Y., et al. (2011). Comparison of the crystalline properties and structural changes of starches from high-amylose transgenic rice and its wild type during heating. *Food Chem.* 128, 645–652. doi: 10.1016/j.foodchem.2011.03.080

93. Wei, C., Qin, F., Zhou, W., Yu, H., Xu, B., Chen, C., et al. (2010b). Granule structure and distribution of allomorphs in C-type high-amylose rice starch granule modified by antisense RNA inhibition of starch branching enzyme. *J. Agric. Food Chem.* 58, 11946–11854. doi: 10.1021/jf103412d

94. Wei, C., Qin, F., Zhu, L., Zhou, W., Chen, Y., Wang, Y., et al. (2010c). Microstructure and ultrastructure of high-amylose rice resistant starch granules modified by antisense RNA inhibition of starch branching enzyme. *J. Agric. Food Chem.* 58, 1224–1232. doi: 10.1021/jf9031316

95. Wei, C., Xu, B., Qin, F., Yu, H., Chen, C., Meng, X., et al. (2010d). C-type starch from high-amylose rice resistant starch granules modified by antisense RNA inhibition of starch branching enzyme. *J. Agric. Food Chem.* 58, 7383–7388. doi: 10.1021/jf100385m

96. Wellner, N., Georget, D. M. R., Parker, M. L., and Morris, V. J. (2011). *In situ* Raman microscopy of starch granule structures in wild type and *ae* mutant maize kernels. *Starch* 63, 128–138. doi: 10.1002/star.201000107

97. Wu, Y., Campbell, M., Yang, Y., Lii, Z. W., and Ibrahim, A. M. H. (2009). Genetic analysis of high amylose content in maize (*Zea mays* L.) using a triploid endosperm model. *Euphytica*166, 155–164. doi: 10.1007/s10681-008-9798-y

98. Xia, H., Yandeau-Nelson, M., Thompson, D. B., and Guiltinan, M. J. (2011). Deficiency of maize starch-branching enzyme i results in altered starch fine structure, decreased digestibility and reduced coleoptile growth during germination. *BMC Plant Biol.* 11:95. doi: 10.1186/1471-2229-11-95

99. Yamamori, M., Fujita, S., Hayakawa, K., Matsuki, J., and Yasui, T. (2000). Genetic elimination of a starch granule protein, SGP-1, of wheat generates an altered starch with apparent high amylose. *Theor. Appl. Genet.* 101, 21–29. doi: 10.1007/s001220051444

100. Yao, Y., Thompson, D. B., and Guiltinan, M. J. (2004). Maize starch-branching enzyme isoforms and amylopectin structure. In the absence of starch-branching enzyme IIb, the further absence of starch-branching enzyme Ia leads to increased branching. *Plant Physiol.* 136, 3515–3523. doi: 10.1104/pp.104.043315

101. Yu, Y., Mu, H. H., Wassermanm, B. P., and Carman, G. M. (2001). Identification of the maize amyloplast stromal 112-kD protein as a plastidic starch phosphorylase. *Plant Physiol.* 125, 351–359. doi: 10.1104/pp.125.1.351

102. Zhang, L., Ren, Y., Lu, B., Yang, C., Feng, Z., Liu, Z., et al. (2016). FLOURY ENDOSPERM7 encodes a regulator of starch synthesis and amyloplast development essential for peripheral endosperm development in rice. *J. Exp. Bot.* 67, 633–647. doi: 10.1093/jxb/erv469

103. Zhang, X., Colleoni, C., Ratushna, V., Sirghle-Colleoni, M., James, M. G., and Myers, A. M. (2004). Molecular characterization demonstrates that the *Zea mays* gene *sugary 2* codes for the starch synthase isoform SSIIa. *Plant Mol. Biol.* 54, 865–879. doi: 10.1007/s11103-004-0312-1

104. Zhou, H., Wang, L., Liu, G., Meng, X., Jing, Y., Shu, X., et al. (2016). Critical roles of soluble starch synthase SSIIIa and granule-bound starch synthase Waxy in synthesizing resistant starch in rice. *Proc. Natl. Acad. Sci. U.S.A.* 2016, 12844–12849. doi: 10.1073/pnas.1615104113

105. Zhu, L., Gu, M., Meng, X., Cheung, S. C. K., Yu, H., Huang, J., et al. (2012). High-amylose rice improves indices of animal health in normal and diabetic rats. *Plant Biotechnol. J.* 10, 353–362. doi: 10.1111/j.1467-7652.2011.00667.x

CHAPTER 17

Genetic Transformation and Genomic Resources for Next-Generation Precise Genome Engineering in Vegetable Crops

Teodoro Cardi*, Nunzio D'Agostino and Pasquale Tripodi

Consiglio per la ricerca in agricoltura e l'analisi dell'economia agraria (CREA), Centro di ricerca per l'orticoltura, Pontecagnano Faiano, Italy

ABSTRACT

In the frame of modern agriculture facing the predicted increase of population and general environmental changes, the securement of high quality food remains a major challenge to deal with. Vegetable crops include a large number of species, characterized by multiple geographical origins, large genetic variability and diverse reproductive features. Due to their nutritional value, they have an important place in human diet. In recent years, many crop genomes have been sequenced permitting the identification of genes and superior alleles associated with desirable traits. Furthermore, innovative biotechnological approaches allow to take a step forward towards the development of new improved cultivars harboring precise genome modifications. Sequence-based knowledge coupled with advanced biotechnologies is supporting the widespread application of new plant breeding techniques to enhance the success in modification and transfer of useful alleles into target varieties. Clustered regularly interspaced short palindromic repeats (CRISPR)/CRISPR-associated protein 9 system, zinc-finger nucleases, and transcription activator-like effector nucleases represent the main methods available for plant genome engineering through targeted modifications. Such technologies, however, require efficient transformation protocols as well as extensive genomic resources and accurate knowledge before they can be efficiently exploited in practical breeding programs. In this review, we revise the state of the art in relation to availability of such scientific and technological resources in various groups of vegetables, describe genome editing results obtained so far and discuss the implications for future applications.

Keywords: vegetable crops, genome editing, in vitro regeneration, genetic transformation, whole genome sequences, genomics, breeding

INTRODUCTION

Vegetable crops include a large number of species belonging to various families, characterized by multiple geographical origins, large genetic variability and diverse reproductive features. As result of natural and artificial selection, various vegetables are differentially used worldwide for many purposes, either as fresh or processed products. Due to their nutritional value, vegetables have an important place in human diet, providing, in combination with freshness and taste, protection against various non-transmissible diseases and reduction of the so-called "hidden hunger" especially in developing countries (Desjardins, 2014).

Fast changes in climate and consumers' needs as well as the appearance of emerging and re-emerging plant pests and diseases require the continuous development of novel genotypes adapted to varying horticultural systems (Anderson et al., 2004). The availability of genetic resources plays an important role in such a process. Indeed, a large number of crop wild relatives and local traditional varieties is conserved *in situ* and *ex situ* by private or public institutions. In addition, local vegetable varieties (landraces) are still grown and reproduced by farmers, based on their adaptation to specific environments and on the presence of valuable traits. New genomics and phenomics tools allow the throughout characterization of such resources and should facilitate their better use in breeding. Nonetheless, the overall level of biodiversity in horticultural systems is decreasing, with threats to their economic, social, and environmental sustainability (Maggioni, 2004; Silva Dias, 2014).

New plant breeding techniques, in particular gene transfer based on cisgenic approaches and next-generation precision genome engineering relying on genome editing technologies, can play a key role in accessing genetic resources and using them in functional studies and streamlined breeding strategies (Cardi, 2016) (Figure 1). Such approaches, however, require efficient transformation protocols as well as extensive genomic resources and accurate knowledge before they can be efficiently exploited in practical breeding programs. In this review, we revise the state of the art in relation to availability of such scientific and technological resources in various groups of vegetables, describe genome editing results obtained so far and discuss the implications for future applications.

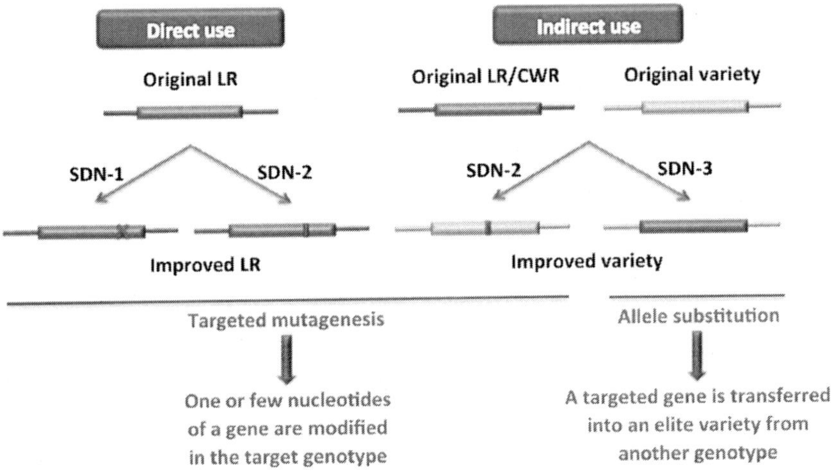

FIGURE 1. Novel plant breeding techniques applied to the exploitation of genetic resources. LR, landraces; CWR, crop wild relatives. SDN-1, 2, 3, site-directed nuclease usage for gene knockout, gene editing or gene replacement/stacking, respectively. If the gene replaced by the latter approach derives from a sexually compatible genetic resource, it can be dubbed also as "cisgenesis at the same locus."

Regeneration and Transformation

Available plant transformation methods include indirect (i.e., requiring an intermediate biological vector, usually the bacterium *Agrobacterium tumefaciens*) and direct methods (electroporation or PEG-mediated transformation of protoplasts, biolistics, etc.; Finer, 2010). Successful transformation, however, relies on various phases, being the introduction and integration of DNA into the plant genome as well as the selection and regeneration of transformed cells the most important and, sometimes, limiting factors (Wędzony et al., 2014; Altpeter et al., 2016). Regeneration of transformed plants has been extensively pursued in vegetable crops, although large differences in efficiency among families, species, and cultivars have been reported (Klocke et al., 2010). Plant regeneration is generally achieved via *in vitro* culture systems, using a range of explants and following two alternative pathways: *de novo* shoot organogenesis (DNSO) or somatic embryogenesis (SE; Wędzony et al., 2014).

Within the Solanaceae family, eggplant (*Solanum melongena*) and, to a lesser extent, tomato (*S. lycopersicum*) have been successfully subjected to genetic transformation with various purposes using different approaches (Klocke et al., 2010). Regeneration in eggplant has been obtained either through SE or DNSO starting from protoplasts, tissue explants, microspores (Rotino et al., 2014). In tomato, Agrobacterium-mediated transformation of cotyledon explants has been most frequently achieved, albeit protoplast protocols are also available. Genetic variability of response to both *in vitro* systems is well known (Kurtz and Lineberger, 1983; Derks, 1992). The dominant *Rg1* gene involved in regeneration capacity has been transferred from *S. peruvianum* into *S. lycopersicum* and subsequently mapped on chromosome 3 (Koornneef et al., 1993). Recently, its effect has been characterized in a different genetic background also in combination with other relevant genes (Pino et al., 2010; Lombardi-Crestana et al., 2012). By contrast, although some positive results have been reported (Klocke et al., 2010), pepper (*Capsicum* spp.) is considered a recalcitrant species due to several issues that jeopardize the *in vitro*response (Kothari et al., 2010). Some interesting results, however, based on the over-expression of two heterologous transcription factors involved in the regeneration process (i.e., WUSCHEL and BABY BOOM), have been recently published in sweet and chili pepper (Solís-Ramos et al., 2009; Heidmann et al., 2011).

Brassica oleracea and *B. rapa* include the majority of vegetables belonging to Brassicaceae. Transformation, usually achieved by inoculation of seedling explants with Agrobacterium (direct gene transfer using protoplasts has been also accomplished) followed by regeneration through organogenesis, is strongly dependent on the genotype and various other factors, being *B. rapa* genotypes more difficult to transform than *B. oleracea* ones, and cabbage (*B. oleracea* var. *capitata*) the most difficult type within the latter. A strong genetic component with significant additive effects both for *A. tumefaciens* susceptibility and regeneration ability from cotiledonary explants as well as from leaf protoplasts has been found in *B. oleracea* [(Sparrow et al., 2004) and references cited therein], suggesting the possibility to incorporate those traits in recalcitrant genotypes by transferring the relevant genes. *Raphanus sativus*, another vegetable of the same family, is considered recalcitrant and only few reports are available. Advances in regeneration and transformation in this group of crops have been reviewed elsewhere (Vinterhalter et al., 2007; Klocke et al., 2010; Sparrow et al., 2011; Park et al., 2012; Kumar and Srivastava, 2016). With the aim to avoid tissue culture systems, similarly to the related species *Arabidopsis thaliana, in planta*transformation systems based on floral dipping or flower bud microinjection have been also attempted in various genotypes (Curtis and Nam, 2001; Sparrow et al., 2011).

Within the Cucurbitaceae family, transformation has been accomplished in all the three genera including vegetable crops: *Cucumis* (cucumber and melon), *Cucurbita* (squash, pumpkin and zucchini), and *Citrullus* (watermelon). Various (young) explants have been generally inoculated with Agrobacterium followed by regeneration through shoot organogenesis, but transformation efficiency varied largely with the genotype used (Klocke et al., 2010; Manamohan et al., 2011). Some improvements in the protocol have been lately reported for melon (Bezirganoglu et al., 2014; Zhang et al., 2014), cucumber (Wang et al., 2015), *Cucurbita* spp. (Nanasato et al., 2013), and watermelon (Liu et al., 2016). In the latter species, an efficient regeneration procedure *via* SE from embryogenic calli derived from leaf explants has been also recently published (Vinoth and Ravindhran, 2015). Attempts to develop *in planta* methods (either *via*pollen tube or microinjection of the shoot apical meristem) have been made in watermelon and cucumber (Chen et al., 1998; Baskaran et al., 2016). As far as genetic and molecular aspects of shoot regeneration are concerned, a simple dominant control of regeneration ability from leaf explants has been found in cucumber (Nadolska-Orczyk and Malepszy, 1989), while distinct expression profiles of WUSCHEL-related homeobox (WOX) genes have been associated with different regeneration abilities in watermelon (Zhang N. et al., 2015).

Among Asteraceae, lettuce (*Lactuca sativa*) and chicory (*Cichorium intybus*) have been largely and successfully used in many transformation experiments for a variety of purposes, using either Agrobacterium inoculation of various explants (mainly cotyledons and true leaves) or direct gene transfer (electroporation/PEG treatment of protoplasts or particle bombardment of tissue explants) (Davey et al., 2007; Klocke et al., 2010; Song et al., 2014; Matvieieva, 2015). Shoot regeneration from hairy roots has also been accomplished. Regeneration normally proceeds by organogenesis, but SE has also been reported. Recently, the "surface response" method has been employed in lettuce to optimize plant regeneration (Gómez-Montes et al., 2015). While a mature regeneration/transformation technology is available for both above-mentioned species, *de novo* shoot regeneration and cell transformation have been observed only sporadically and with very low efficiency in another important vegetable of this family, i.e., globe artichoke (*Cynara cardunculus* var. *scolymus*; Menin et al., 2012).

Carrot (*Daucus carota*), a member of the Apiaceae family, is considered a model species for *in vitro* SE. Factors affecting transformation efficiency have been defined in the late 1980 – early 1990. Subsequently, genetic transformation has been pursued for various objectives using either indirect (*A. tumefaciens* and *A. rhizogenes*) or direct gene transfer methods (electroporation/PEG-mediated transformation of protoplasts and biolistics) (Punja et al., 2007; Baranski, 2008; Klocke et al., 2010). Much less work has been performed in other vegetables of the same family, such as fennel (*Foeniculum vulgare*), celery (*Apium graveolens*), and parsley (*Petroselinum crispum*; Baranski, 2008; Klocke et al., 2010). Fennel is able to regenerate *in vitro* by either organogenesis or embryogenesis, albeit with a marked genotypic effect (Anzidei et al., 2000; Jakhar and Choudhary, 2012; Saxena et al., 2012), but no transgenic plants have been generated. A reproducible protocol has been developed for rapid and efficient production of transgenic celery plants *via* somatic embryo regeneration from *A. tumefaciens*-inoculated leaf sections, cotyledons, and hypocotyls (Song et al., 2007). Only transgenic calli have been obtained in parsley (Baranski, 2008).

Successful Agrobacterium-mediated transformation in the leafy vegetable *Spinacia oleracea*(Chenopodiaceae) has been first published in 1995 and later applied for transferring agronomically relevant genes (Klocke et al., 2010). In the meantime, efforts have been made to improve transformation and regeneration procedures. Although the genotype has generally shown a significant effect, regeneration has been achieved either through organogenesis or embryogenesis depending mainly on the explant type and auxin/cytokinin ratio: cotyledon explants and low ratios facilitate the former, whereas root explants and high ratios favor the latter. In addition, low temperature (14°C), photoperiod, light intensity, and GA_3content had substantial effects on shoot regeneration (Geekiyanage et al., 2006; Chin et al., 2009; Nguyen et al., 2013b), while low levels of hygromycin (0.5 mg l^{-1}) have been found to promote SE (Milojević et al., 2012). An optimized regeneration protocol has been recently published (Nguyen et al., 2013a).

Vegetable legumes (harvested as green immature pods and seeds or, e.g., in cowpea, also as leaves) and pulses (harvested for the dry seed) have long been considered as recalcitrant to *in vitro* transformation and regeneration, although significant progress has been shown in the recent past (Somers et al., 2003; Klocke et al., 2010; Dewir et al., 2016; Gatti et al., 2016; Nguyen et al., 2016). Difficulties in achieving organogenesis and SE from differentiated tissues have prompted the development of regeneration/transformation protocols from meristematic tissues, but this has made difficult the selection of uniform (non-chimeric) transformed shoots. In several cases, the release of phenol oxidation products from explants cultured *in vitro* inhibits cell division and provokes tissue darkening and death, while rooting of regenerated shoots can be an additional critical step. All those problems have been encountered in applying genetic transformation to common bean (*Phaseolus vulgaris*), that, compared to related species (e.g., *P. acutifolius, P. coccineus, P. polyanthus*), exhibits a lower regeneration potential. Nevertheless, various direct and indirect transformation methods have been attempted in this species, showing some interesting results (Veltcheva et al., 2005; Hnatuszko-Konka et al., 2014). The improvement of the *in vitro* selection step through the use of a systemic herbicide has resulted in a significant increase of transformation efficiency (Aragão and Campos, 2007). Recently, it has been shown that the endogenous hormonal content in embryogenic calli could be altered and SE increased by down-regulating the expression of the *PvTRX1h* gene, encoding for a histone methyltransferase (Barraza et al., 2015). Genetic engineering problems and efforts similar to those described for *P. vulgaris* have been recently reviewed in the related species *Vigna unguiculata*, for which electroporation, biolistics and Agrobacterium-based methods have all been used, the latter being the most common (Citadin et al., 2011). Due to the involvement of meristematic tissues in the regeneration process, also for this species it has been necessary to develop alternative selection regimes for increasing the recovery of transformed shoots (Aragão and Campos, 2007; Bakshi et al., 2012; Bakshi and Sahoo, 2013). Similarly to other legumes, also in garden pea (*Pisum sativum*), first transformed in the late 1980, meristems have been the explants of choice, but the transformation protocols have been gradually improved and successfully employed for various purposes, although a strong genotypic effect has been shown (Klocke et al., 2010; Mikschofsky and Broer, 2012). The broad bean (*Vicia faba*) is probably one of the most difficult legume species to regenerate and transform, requiring particular efforts to solve the problem of tissue blackening *in vitro* (Klenotičová et al., 2013), and only few successful experiments have been reported using Agrobacterium-mediated transformation of meristematic cells or stem segments [reviewed in (O'Sullivan and Angra, 2016)]. Due to the intrinsic difficulties of legume regeneration systems, a range of methods not requiring tissue culture of explants have been proposed (Somers et al., 2003), including the electroporation of nodal axillary buds in pea and cowpea (Chowrira et al., 1996), the Agrobacterium inoculation of germinating seeds in pea and bean (Liu et al., 2005; Svabova et al., 2005), the inoculation of flower buds in cowpea (Ilori and Pellegrineschi, 2011).

Onion, shallot, garlic, leek (Amaryllidaceae), and asparagus (Asparagaceae), as other Monocotyledons, have been initially considered recalcitrant to *in vitro* regeneration and not amenable to Agrobacterium-mediated genetic transformation. Subsequently, however, several approaches have been developed [reviewed in (Gantait et al., 2010; Klocke et al., 2010)]. In onion and shallot (*Allium cepa*), either immature zygotic embryos or young calli derived from mature embryos have been successfully used as explants in Agrobacterium-mediated transformation experiments, but the former approach can rely on a higher availability of explants during the year (Eady et al., 2000; Zheng et al., 2001). Using the zygotic embryo-derived calli, shallot genotypes gave better results than onion ones. Furthermore, to overcome some limits of both embryo-based protocols, embryogenic calli initiated from seedling radicles have been used both with Agrobacterium and biolistic systems, showing comparable results (Aswath et al., 2006). Agrobacterium and biolistic approaches have been applied for garlic (*A. sativum*) as well. Either immature zygotic embryos, similarly to onion, or calli from root segments have been transformed *via* Agrobacterium, the former approach being successful also with leek (*A. porrum*; Zheng et al., 2004; Eady et al., 2005). However, the difficulty to get seed-derived tissues has recently prompted the development of alternative methods based on immature leaf tissues (Kenel et al., 2010) or roots from *in vitro* plantlets (Ahn et al., 2013). Finally, transformation in *Asparagus officinalis* has been demonstrated as early as in the '90 with a range of approaches, including Agrobacterium, protoplast electroporation, and biolistics (Klocke et al., 2010), but regeneration and transformation protocols have not been developed further.

The availability of *in vitro* culture and genetic transformation protocols for various vegetables is summarized in Table 1.

Genomic Resources

Next-Generation Sequencing technologies have determined a significant advancement in data generation. Indeed, large datasets are now being generated across various model and non-model plant species by sequencing whole genomes and/or by applying genome reduction strategies (i.e., RNA-sequencing, hybridization-based enrichment, restriction enzyme-based enrichment, etc.). The list of vegetable crops with publicly available complete or draft genome sequences is getting rich very quickly (Michael and VanBuren, 2015).

Within the Solanaceae family, complete genome sequences have been assembled, annotated, and published in different species, allowing the development of multiple publicly available resources[1](Fernandez-Pozo et al., 2015). The genomes of the cultivated tomato (Heinz 1706) and its wild ancestor (*S. pimpinellifolium*) were published in 2012 (Tomato Genome Consortium, 2012). The genome sequence of *S. pennellii*, a wild tomato species used as gene donor for the cultivated *S. lycopersicum* because of its extreme tolerance to abiotic stresses, was released in 2014 (Bolger et al., 2014). Its sequencing is enhancing the use of the already available introgression lines (ILs), in which genomic regions of *S. lycopersicum* are replaced with the corresponding segments of *S. pennellii*. Those publications allowed several re-sequencing projects to be developed with

TABLE 1. Availability of *in vitro* regeneration/genetic transformation protocols and of genomic resources for some vegetable crops.

Family	Cultivated species	Estimated genome size (Mb)[a]	Regeneration and transformation[b]	Genomic resources[c]
Solanaceae	*Solanum lycopersicum*	900	++/+++	+++
	Solanum melongena	1130	+++	++
	Capsicum annuum	3300	–	++/+++
Brassicaceae	*Brassica rapa*	284	++	++
	Brassica oleracea	378	++	++
	Raphanus sativus	529	–	+
Cucurbitaceae	*Cucumis sativus*	350	++	+++
	Cucumis melo	375	++	+++
	Citrullus lanatus	425	++	++
	Cucurbita pepo	538	++	+
	Cucurbita maxima	450	++	–
Asteraceae	*Lactuca sativa*	2700	+++	+
	Cichorium intybus	1300–1400	++	+
	Cynara cardunculus	1084	–	++
Chenopodiaceae	*Spinacia oleracea*	1002	++	+
Apiaceae	*Daucus carota*	473	+++	++
	Foeniculum vulgare	4450	–/+	–
	Apium graveolens	3000	–/+	–
	Petroselinum crispum	2201	–	–
Leguminosae	*Phaseolus vulgaris*	587	+	+++
	Vigna unguiculata	620	+	–
	Pisum sativum	4300	+	–
	Vicia faba	13000	+	–
Amaryllidaceae	*Allium cepa*	16000	+/++	+
	Allium sativum	15901	+	–

[a] *Solanum lycopersicum* (Tomato Genome Consortium, 2012); *S. melongena* (Hirakawa et al., 2014); *C. annuum* (Qin et al., 2014); *B. oleracea* (Liu et al., 2014); *B. rapa* (Wang X. et al., 2011); *R. sativus* (Kitashiba et al., 2014); *C. sativus* (Huang et al., 2009); *C. melo* (Garcia-Mas et al., 2012); *C. lanatus* (Guo et al., 2013); *C. pepo* (Plant DNA C-values database release 6.0, 2012 http://data.kew.org/cvalues/); *C. maxima* (Plant DNA C-values database release 6.0, 2012 http://data.kew.org/cvalues/); *L. sativa* (Truco et al., 2013); *C. intybus* (Gonthier et al., 2010); *C. cardunculus* (Scaglione et al., 2016); *S. oleracea* (Plant DNA C-values database release 6.0, 2012 http://data.kew.org/cvalues/); *D. carota* (Iorizzo et al., 2016); *F. vulgare* (Plant DNA C-values database release 6.0, 2012 http://data.kew.org/cvalues/); *A. graveolens* (Fu et al., 2013); *P. crispum* (Plant DNA C-values database release 6.0, 2012 http://data.kew.org/cvalues/); *P. vulgaris* (Schmutz et al., 2014); *V. unguiculata* (Timko et al., 2008); *P. sativum* (Macas et al., 2007); *V. faba* (Kaur et al., 2012); *A. cepa* (Finkers et al., 2015); *A. sativum* (Sun et al., 2012); *A. porrum* (Plant DNA C-values database release 6.0, 2012 http://data.kew.org/cvalues/); *A. officinalis* (Li et al., 2014).

[b] "–" Protocols not available or available with low repeatability in a few genotypes ⇒ "+++" various protocols (indirect or direct transformation methods in combination with different explant types, regeneration through DNSO or SE either from tissue explants or protoplasts) available for multiple genotypes.

[c] "–" Little information available ⇒ "+++" re-sequenced genomes from cultivated and wild species available in addition to the complete reference genome sequence; functional information available.

different scientific purposes (100 Tomato Genome Sequencing Consortium, 2014; Lin T. et al., 2014). Such recent efforts have led to the accumulation of a huge amount of valuable data on sequence diversity in the tomato clade that are of practical importance in the design of effective breeding strategies based on the generation of precise sequence changes in the target genome. In parallel, genome-wide association studies (GWAS) are revealing thousands of genetic variations associated with agronomically important traits (Ranc et al., 2012; Shirasawa et al., 2013; Ruggieri et al., 2014; Sauvage et al., 2014; Zhang J. et al., 2015), while much less numerous are the studies that rely on targeted enrichment methods for the discovery of genome-wide sequence variations from specific candidate *loci* (Terracciano et al., 2016).

Whole genome sequences of the hot (*C. annuum* cv. CM334) as well as of the cultivated pepper Zunla-1 (*C. annuum* L.) and its wild progenitor *Chiltepin* (*C. annuum* var. *glabriusculum*) have been released into the public domain together with re-sequencing data of cultivated, semi-wild/wild accessions (Kim et al., 2014; Qin et al., 2014). Genome resequencing of representative pepper accessions as well as single nucleotide polymorphism (SNP) discovery through genotyping by sequencing (Ahn et al., 2016; Taranto et al., 2016) have allowed genetic diversity in pepper to be unlocked. Nevertheless, the association between DNA variation and the observed phenotypic variability is still poor (Nimmakayala et al., 2016a).

A draft genome sequence of eggplant is also available (Hirakawa et al., 2014). An additional sequencing effort by an Italian consortium aiming at the generation of a gold-standard reference genome is ongoing (Barchi et al., 2016). Based on a fairly large number of accessions, a couple of GWAS have been published so far in eggplant allowing a series of novel marker/trait associations to be detected (Cericola et al., 2014; Portis et al., 2015).

Within the Brassicaceae, the sequencing of the *B. rapa* genome (Wang X. et al., 2011) has been followed by the release of the draft genome assembly of *B. oleracea* (Liu et al., 2014), a species characterized by a large morphological diversity and including different crops such as cauliflower, broccoli, cabbages, Brussels sprouts, kohlrabi, and kales. The investigation on the genetic variability in these species is still quite limited (Lin K. et al., 2014).

Within the Cucurbitaceae family, the complete genome of cucumber (*C. sativus* L.) and melon (*C. melo* L.) as well as the draft genome of watermelon (*C. lanatus*) and zucchini (*C. pepo*) are already available (Huang et al., 2009; Garcia-Mas et al., 2012; Guo et al., 2013)[2]. In case of *C. lanatus* the resequencing of 20 accessions has also been performed (Guo et al., 2013) providing a large source of haplotype data with great potential for next-generation breeding. In addition, medium to large SNP catalogs generated by genotyping by sequencing are available in watermelon (Nimmakayala et al., 2014) and melon (Nimmakayala et al., 2016b). The latter has been used to perform GWAS for fruit firmness. In the near future, it is also expected the publication of the pumpkin (*C. maxima* Duch.) genome sequence[3].

The genome of the globe artichoke (*C. cardunculus* var. *scolymus*), a species belonging to the Asteraceae, was released in 2016 (Scaglione et al., 2016), while the first draft of the lettuce (*Lactuca sativa*) and chicory (*Cichorium intybus*) genomes has been announced[4] (Galla et al., 2014). A transcriptome-based SNP discovery, through Illumina sequencing of 11 representative accessions of the three *C. cardunculus*taxa generated 195k variants (Scaglione et al., 2012). More recently, a re-sequencing approach, followed by a whole genome SNP mining strategy, as well as the identification of PAV (presence-absence variation), have been applied for globe artichoke (Acquadro et al., 2016).

A high quality chromosome-scale assembly of carrot (*Daucus carota* subsp. *carota* L.) genome is now available (Iorizzo et al., 2016) for the scientific community interested in the Apiaceae genetic improvement. Transcriptome-based analysis on the allelic diversity of wild and cultivated accession is also available (Iorizzo et al., 2013).

Common bean (*P. vulgaris* L.) genomes of two genotypes from Andean and Mesoamerican gene pools (Schmutz et al., 2014; Vlasova et al., 2016) have been published as part of the Leguminosae family. In addition, an International Consortium for Pea Genome Sequencing (PGS) has been formed to explore options for sequencing the pea genome that is particularly complex and large (~4.3 Gb[5]). Several GWAS have been performed in common bean to explore broader genetic diversity in order to establish marker-trait associations for future application in breeding programs (Cichy et al., 2015; Kamfwa et al., 2015; Perseguini et al., 2016).

Knowledge on Amaryllidaceae and Asparagaceae genomes is scarce compared to Solanaceae and Cucurbitaceae. Genomic resources for garlic (*Allium sativum* L.), onion (*A. cepa* L.), and asparagus (*Asparagus officinalis* L.) are limited because of their large, extremely complex, repetitive, and often polyploid genomes and long generation times. Nevertheless, the Sequon – Onion Genome Sequencing project is underway[6] with the main goal to generate a high quality sequence for the gene-rich regions of a doubled haploid (DH) onion line as a reference for the Amaryllidaceae family.

All these genome sequencing initiatives, usually paralleled by the generation of transcriptome-derived sequences in case of species of minor interest, have led to the development of clade-oriented databases dedicated to the genomics of specific crop families (Wang X. et al., 2011; Fernandez-Pozo et al., 2015)[7]. The genome size in various vegetables and the availability of genomic resources are summarized in Table 1.

Genome Editing in Vegetable Crops

Current genome editing approaches rely on the induction of cuts in double-strand DNA (DSB, double-strand breaks), which are then "repaired" through two different processes: non-homologous end joining (NHEJ) or homology-directed repair (HDR; Cardi, 2016). In the former case, in the absence of foreign donor sequences, small changes (mostly frame-shift mutations due to insertions and deletions) are induced in the original sequence during the repair process, generally resulting in the loss of function of the target gene and a mutated phenotype. On the other hand, if appropriate DNA fragments homologous to the target sequence are also inserted into the cell, they can, using the precise HDR mechanism, replace (correct) some nucleotide sequences of the gene to be modified or add new genes or regulatory elements in a predetermined position of the genome. Genome editing strategies recently applied to vegetable crops are reported in Table 2. In all but one case, they have been based on the induction of targeted mutations through NHEJ repair of DSBs determined by either transcription activator-like effector nucleases (TALEN) or clustered regularly interspaced short palindromic repeats/CRISPR-associated 9 endonuclease system (CRISPR/Cas9).

TABLE 2. Genome editing approaches applied to vegetable crops.

Species	Trait	Gene	Genetic modification/double-strand breaks (DSB) repair mechanism[a]	Technology	DNA delivery	Tissue culture system	Reference
Solanum lycopersicum (tomato)	Plant development	*PROCERA* (*PRO*)	Indel mutations (gene knockout)/NHEJ	TALEN	*Agrobacterium tumefaciens* (binary vector)	Cotyledon explants	Lor et al., 2014
	Leaf development	*ARGONAUTE7* (*SlAGO7*)	Indel mutations (gene knockout)/NHEJ	CRISPR/Cas9	*A. tumefaciens* (binary vector)	Cotyledon explants	Brooks et al., 2014
	Root development	*SHORT-ROOT* (*SHR*)	Indel mutations (gene knockout)/NHEJ	CRISPR/Cas9	*A. rhizogenes* (binary vector)	Cotyledon explants[b]	Ron et al., 2014
	Fruit ripening	*Ripening inhibitor* (*RIN*)	Indel mutations (gene knockout)/NHEJ	CRISPR/Cas9	*A. tumefaciens* (binary vector)	Cotyledon explants	Ito et al., 2015
	Anthocyanin biosynthesis	*Anthocyanin 1* (*ANT1*)	Insertion of a novel promoter/HDR	TALEN, CRISPR/Cas9	*A. tumefaciens* (geminivirus replicons)	Cotyledon explants	Čermák et al., 2015
	Plant development	*Self-pruning 5G* (*sp5g*), *Self-pruning* (*sp*)	Indel mutations (gene knockout)/NHEJ	CRISPR/Cas9	*A. tumefaciens* (binary vector)	Cotyledon explants	Soyk et al., 2017
Brassica oleracea (broccoli[c])	Plant development, fruit dehiscence	*Gibberellin 3 beta-dioxygenase 1* (*BolC.GA4.a*)	Indel mutations (gene knockout)/NHEJ	CRISPR/Cas9	*A. tumefaciens* (binary vector)	Cotyledonary petiole explants	Lawrenson et al., 2015
Lactuca sativa (lettuce)	Plant development	*BRASSINOSTEROID INSENSITIVE 2* (*BIN2*)	Indel mutations (gene knockout)/NHEJ	CRISPR/Cas9	Polyethylene glycol (PEG)[d]	Cotyledon protoplasts	Woo et al., 2015
Cucumis sativus (cucumber)	Virus resistance	*Eukaryotic translation initiation factor 4E* (*eIF4E*)	Indel mutations (gene knockout), SNPs/NHEJ	CRISPR/Cas9	*A. tumefaciens* (binary vector)	Cotyledon explants	Chandrasekaran et al., 2016

[a] *Double-Strand Breaks induced by sequence-specific nucleases were repaired either by non-homologous end joining (NHEJ) or homology-directed repair (HDR).*
[b] *Hairy roots were induced from explants after co-cultivation with A. rhizogenes.*
[c] *A doubled haploid genotype derived from a cross between Chinese broccoli (B. oleracea ssp. alboglabra) and broccoli (B. oleracea ssp. italica) was used.*
[d] *Preassembled complexes of purified Cas9 protein and guide RNA were transfected into plant protoplasts.*

In tomato, TALE nucleases under the control of an estrogen-inducible promoter have been used to knockout one DELLA gene (*PRO*) involved in negative regulation of gibberellic acid (GA) signaling. Consistently with the induction of frameshift mutations, loss of DELLA function and increased GA response, plants carrying two different mutant alleles (biallelic) or the same mutation in homozygous condition showed longer internodes than the wild type and lighter green leaves with smoother margins. Induced *pro* mutations were transmitted to the progenies according to Mendelian inheritance, while the transgene encoding the TALE nuclease segregated away from the TALEN-induced mutations and was not present in some of the plants showing the mutant phenotype (Lor et al., 2014). In the same species, a CRISPR/Cas9 construct has been designed to induce loss-of-function mutations in the *ARGONAUTE7*(*SlAGO7*) gene, which through the synthesis of short interfering RNAs causes post-transcriptional silencing of *AUXIN RESPONSE FACTOR* genes and regulates organ polarity (Brooks et al., 2014). Forty-eight percent of T_0 transformed plants showed the expected phenotype with needle-like or wiry leaves. Only one regenerant, however, contained the deletion of the expected size after the contemporary use of two single guide RNAs (sgRNAs), the remaining being homozygous, biallelic or chimeric for small insertions and deletions. Induced mutations could be transmitted through the germline and the loss of the Cas9 transgene was shown in some progenies. A similar approach has been used in the same work to carry out functional studies in three homologs of the gene *Solyc11g064850* involved in multiple aspects of tomato reproductive development. CRISPR/Cas has also been used in functional studies aiming to understand, in a hairy root tissue culture system, the role of the tomato homolog of the transcription factor *SHORT-ROOT* (*SHR*), which in Arabidopsis is known to regulate the expression of another transcription factor (*SCARECROW, SCR*) and induce a short root phenotype (Ron et al., 2014). Three regions of the tomato *RIN* gene, encoding a MADS-box transcription factor regulating fruit ripening, have also been recently targeted for inducing sequence-specific mutations by NHEJ. A range of indel mutations have been detected already in T_0 plants in homozygous, heterozygous, biallelic or chimeric condition. Homozygous plants for the desired mutation have been easily recovered in following generation and some T_1 plants did not carry any T-DNA. Induced mutations provoked either the formation of a truncated RIN protein or no protein accumulation, affecting fruit ripening in a variety of modes. Differently from the conventional *rin*mutation, however, no other genes were affected (Ito et al., 2015). Recently, the CRISPR/Cas9 system has been utilized to inactivate two genes, *sp5g* (*self-pruning 5g*) and *sp* (*self-pruning*), involved in photoperiod response, flowering and control of determinate growth. Regardless of day length, "double-determinate CR-sp/sp5g plants" showed an earlier burst of flowering and earliness for fruit set when compared to indeterminate and determinate *sp*-classic controls (Soyk et al., 2017). Alternatively to frameshift mutations determined by imprecise NHEJ repair pathway, Cermak et al. (2015) pursued in tomato the precise insertion by homologous recombination (HR) of the strong constitutive promoter 35S upstream of endogenous *ANT1*, which encodes a Myb transcription factor and induces anthocyanin accumulation in purple tissues. Besides the 35S promoter and flanking recombination sequences, however, the donor template had to include a *nptII* gene for the selection of transformed cells. In order to increase template production in plant cells and gene targeting (GT) frequency, template and nuclease sequences (either TALEN or CRISPR/Cas9) have been cloned in place of endogenous viral genes encoding coat and movement proteins in a modified *Bean Yellow Dwarf Virus* (BeYDV) genome. As in previous reports in tomato, DNA delivery in plant cells has been accomplished by Agrobacterium-mediated transformation of cotyledon explants. By employing non-integrating geminivirus replicons GT frequency achieved by TALEN and CRISPR/Cas9 systems was comparable and about one order of magnitude higher than using non-replicating T-DNA vectors. Furthermore, no off-target mutations could be detected.

In *B. oleracea*, CRISPR/Cas9 has been used to induce indel mutations in two regions of the *BolC.GA4.a*gene, which similarly to the homolog *GA4* gene in Arabidopsis, is involved in gibberellin biosynthesis (Lawrenson et al., 2015). Regenerated plants showed a range of mutations in the target gene and two of them exhibited also the expected dwarf phenotype and alterations in pod valve margins. Some off-target mutations have been detected in another gene (*BolC.GA4.b*) which, compared with the *GA4.a* gene, showed two mismatches in target region 2 (no off-targeting, however, has been detected in target region 1 with four mismatches).

An alternative innovative approach for delivering editing reagents in plant cells has been recently reported in a number of species, including lettuce (Woo et al., 2015). In the latter, the homolog of the Arabidopsis BRASSINOSTEROID INSENSITIVE 2 (BIN2) gene, encoding a negative regulator in the brassinosteroid (BR) signaling pathway, has been knocked-out after transfecting PEG-treated protoplasts with a mixture of Cas9 and a sgRNA targeting the third exon of the gene. Overall mutation frequency in protoplast-derived calli was ~46%: 5.7 and 40% of calli contained monoallelic and biallelic mutations at the target site, respectively. No off-target mutations were detected and plants regenerated *via* organogenesis from mutant calli transmitted the mutations to the progeny.

Finally, the virus resistance of cucumber plants has been investigated after mutating the Eukaryotic translation initiation factor 4E (*eIF4E*) gene in two sites: in the first case, the gene was completely knocked-down, whereas, in the second, translation of two-thirds of the protein product was still possible (Chandrasekaran et al., 2016). Chimeric T_0 plants containing indel and SNP mutations were regenerated and T_3 progenies were submitted to virus tests. Non-transgenic homozygous mutant plants showed either immunity or resistance to *Cucumber vein yellowing virus* (CVYV), *Zucchini yellow mosaic virus* (ZYMV), and *Papaya ring spot mosaic virus-W* (PRSV-W), although resistance breaking was observed in some cases. The same plants had no resistance to *Cucumber mosaic virus* (CMV) and *Cucumber green mottle mosaic virus* (CGMMV). After specific analyses, no off-target activity could be detected.

CONCLUSION AND PERSPECTIVES

Vegetables include a broad number of species with different reproductive features and genetic structures, which largely affect the access to and the exploitation of genetic variability as well as the efficiency of breeding efforts. In the past decades, genetic improvement programs relied on the selection of superior families or plants from intra-specific or inter-specific cross-derived populations, involving the combination of favorable traits from different genetic backgrounds. The low efficiency of phenotypic selection, especially for quantitatively inherited agronomic traits, has been overcome by the introduction of molecular marker technologies that have improved the efficiency of selection, allowing the detection of specific regions and/or genes to introgress *via* MAB (marker assisted breeding) programs. These approaches, however, are not free from limitations due to the large number of crosses and derived plants to manage. Artificial mutagenesis, developed in the '60, allowed to increase the range of variability available for direct or indirect use, with beneficial effects particularly in vegetatively propagated crops, but the necessity to recover the desirable trait in very large populations composed of plants carrying also many undesirable mutations limits its use.

Priorities in the development of novel vegetable cultivars include the improvement of quality traits, the introduction of resistance to a broad range of pests and pathogens, and tolerance to abiotic stresses such as drought, salinity, and low/high temperature. Most of these resistance/tolerance traits need often to be transferred from wild germplasm, taking many generations to remove the deleterious genes that go along with the desirable traits due to linkage drag (Bai and Lindhout, 2007). In some cases, genes are in secondary or tertiary gene pools requiring tissue culture and/or cytogenetic manipulations in order to obtain fertile hybrids and offsprings. Furthermore, the polygenic nature of many agricultural and biochemical characters warrants the development of extensive trial populations and appropriate experimental designs for the identification of the underlying genes.

Genome editing approaches, aiming to either functional studies or genetic improvement, help face some of the above-mentioned issues and allow to overcome some limitations of classical breeding (Abdallah et al., 2015; Nogue et al., 2016; Rani et al., 2016) (Figure 2). The possibility to easily recover in regenerated plants or first generation progenies homozygous/biallelic mutations induced by targeted mutagenesis or other editing approaches will ease the production of novel sets of parental lines to be used in F_1 hybrid production. An additional advantage consists in the possibility to produce modified plants eventually non-transgenic, by expressing editing reagents transiently or inducing sorting-out of transgenic sequences by segregation or by the use of recombinases (Wang Y. et al., 2011; Mahfouz et al., 2016). The possibility to use genome editing tools either to achieve the crop ideotype by modifying major genes underlying key vegetative and reproductive traits, or induce the *de novo* domestication of wild relatives by manipulating monogenic yield-related traits, has been recently exemplified in tomato (Zsögön et al., 2017).

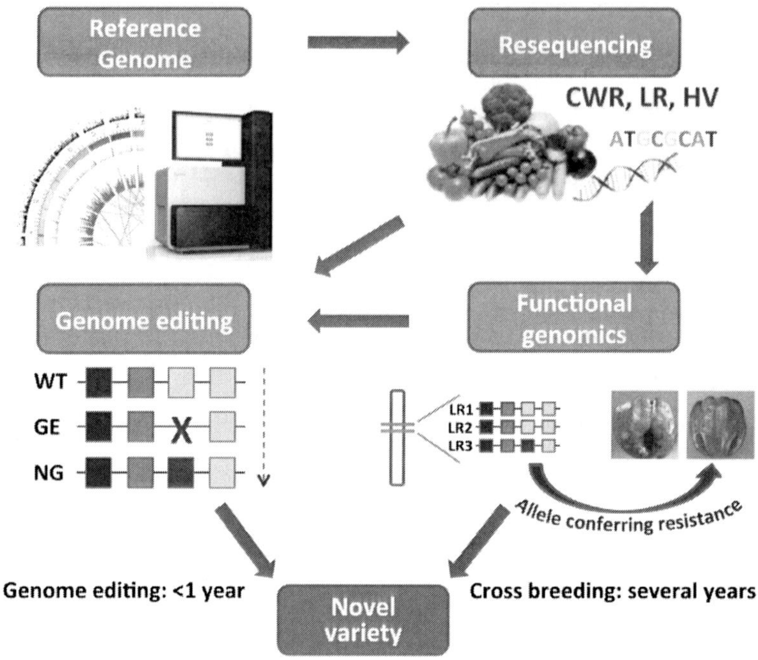

FIGURE 2. Development and exploitation of genomic resources by cross breeding and by genome editing. Resequencing of CWR, LR, and heirloom varieties (HV) in vegetables and subsequent identification of the allelic variation associated to relevant traits allow to introduce precise genome modifications accelerating the breeding of novel varieties. WT, wild type; GE, genome editing process; NG, new "genome edited" genotype; LR1, 2, 3, landraces. Alleles are indicated by colored boxes.

In the last 20 years, several efforts have been made to achieve precise targeted mutagenesis and gene insertion in higher plants, but only the development of site directed nucleases, especially the CRISPR/Cas system, has allowed a more widespread use of such approaches (Cardi and Stewart, 2016). Nevertheless, compared to cereals and other major crops (e.g., potato or oilseed rape), their application is still limited to few vegetable crops and traits. Besides regulatory and patenting issues, that will be hopefully sorted out in a near future (Egelie et al., 2016; Schinkel and Schillberg, 2016; Sprink et al., 2016), major limiting factors for a more common application of genome editing to cultivated vegetable genotypes include the availability of genomic information as well as of efficient protocols for transformation and regeneration. Furthermore, genome editing components have to be efficiently delivered into plant cells and, in case of modifications *via* HDR, frequency of HR optimized (Voytas, 2013; Altpeter et al., 2016; Steinert et al., 2016).

Whole genome sequences of many vegetable crops are already available, while allele mining efforts, based on whole genome resequencing and/or targeted resequencing of a fairly large number of accessions, are underway to search out valuable allelic variants in landraces and crop wild relatives. These efforts are generally paralleled by GWAS that are revealing a series of novel marker/trait associations. All such sequence-based knowledge is an essential prerequisite to transfer technologically advanced methods to vegetable species, allowing overcoming the barriers to a more widespread use of new technologies. Indeed, the possibility to use genome editing approaches to selectively knockout target genes or induce specific mutations is also effective to understand gene functions.

Although efficient ways to deliver editing reagents in plant cells must be still achieved in most crops (Ledford, 2016), it can be envisaged that in vegetables with substantial information on gene sequence and function as well as good regeneration/transformation potential (e.g., tomato, eggplant, lettuce), a range of editing approaches including gene knockout, gene editing or gene replacement/stacking can be applied soon to induce recessive or dominant novel traits. On the other hand, in cases where tissue culture and/or genomic resources are available but still not optimal, applications of genome editing will be likely limited to the introduction of (recessive) frameshift mutations in target genes, for instance those encoding negative regulators or factors necessary for essential pathogen functions [susceptibility (S) genes] (van Schie and Takken, 2014; Sun et al., 2016). In addition, provided that information on regulatory sequences are available, modified CRISPR/Cas9 complexes can be used to mediate the transcriptional activation/repression of the expression of endogenous genes (Piatek and Mahfouz, 2016). In case transformation and/or regeneration protocols are lacking in cultivated genotypes of interest, genome editing approaches can be applied to selected genotypes with good transformation/regeneration ability in order to introduce relevant mutations in the primary gene pool and subsequently transfer them by conventional means. Nevertheless, in some important vegetables (e.g., artichoke, pepper, legumes) it seems necessary to focus on the development of repeatable regeneration/transformation protocols.

The development of such protocols has so far largely been the result of empirical experimentation regarding gene delivery and culture parameters. As previously reported, however, the genetic control of regeneration and transformation ability has been established in various species and the transfer of responsible genes achieved in some instances. More recently, the function of a number of genes involved in genetic and epigenetic mechanisms underpinning various steps of either DNSO or SE, and their possible manipulation for improving response of recalcitrant genotypes, have been described in several reviews (Duclercq et al., 2011; Pulianmackal et al., 2014; Xu and Huang, 2014; Ikeuchi et al., 2016). Engineering the production of specific plant and bacterial proteins could also help enhance Agrobacterium-mediated transformation (Gelvin, 2003; Altpeter et al., 2016). In order to increase feasibility of massive screening of edited products, an automated platform has been recently developed for transformation and genome editing using plant protoplasts (Dlugosz et al., 2016). Alternatively, transformation protocols not relying on tissue culture and *in vitro* regeneration should be developed in recalcitrant genotypes (Altpeter et al., 2016), but although they have been reported in several crops, as described above, their reliability and general applicability is questioned (Finer, 2010).

As far as future applications are concerned, virus resistance represents a major concern in many vegetable crops. Besides recessive resistance achieved by manipulating host genes coding for factors required by invading viruses, as previously discussed for cucumber (Chandrasekaran et al., 2016), acquisition of dominant resistance has been recently reported by expressing in *Nicotiana benthamiana* sgRNAs directed toward various ssDNA Geminivirus, that replicate in plant cells through an intermediate dsDNA stage and are known to affect a large number of vegetables (Zaidi et al., 2016). Results obtained with *tomato yellow leaf curl virus* (TYLCV) and BeYDV are particularly interesting considering their possible transfer in vegetable host species (Ali et al., 2015; Baltes et al., 2015).

To date, genome editing has been mainly focused on the control of single variants underlying qualitative traits. Quantitative variation is instead mediated by several nucleotides (QTN, quantitative trait nucleotides) with large and small effects on the phenotype. Editing quantitative traits is feasible once the availability of datasets of sequences and phenotypes will enable to discover large numbers of QTNs (Jenko et al., 2015). A possible future achievement could be to perform a small number of edits on few QTNs with major effects. A further constraint in vegetable breeding is the manipulation of the reproductive system (e.g., apomixis and self-incompatibility). Such traits are under the control of several candidate genes (Hand and Koltunow, 2014; Yamamoto and Nishio, 2014) and genome editing methods could facilitate the identification of their roles, enhancing the possibility to fix desirable genotypes and accelerate the breeding rate.

FOOTNOTES

[1]https://solgenomics.net/

[2]https://cucurbigene.upv.es/genome-v3.2/

[3]http://www.icugi.org/cgi-bin/ICuGI/index.cgi

[4]https://lgr.genomecenter.ucdavis.edu/

[5]https://www.france-genomique.org/spip/spip.php?article141&lang=fr

[6]http://www.oniongenome.net/

[7]http://www.icugi.org/cgi-bin/ICuGI/index.cgi

REFERENCES

1. 100 Tomato Genome Sequencing Consortium, Aflitos, S., Schijlen, E., De Jong, H., De Ridder, D., Smit, S., et al. (2014). Exploring genetic variation in the tomato (*Solanum* section Lycopersicon) clade by whole-genome sequencing. *Plant J.* 80, 136–148. doi: 10.1111/tpj.12616
2. Abdallah, N. A., Prakash, C. S., and Mchughen, A. G. (2015). Genome editing for crop improvement: challenges and opportunities. *GM Crops Food* 6, 183–205. doi: 10.1080/21645698.2015.1129937
3. Acquadro, A., Barchi, L., Portis, E., Mangino, G., Valentino, D., Mauromicale, G., et al. (2016). "Whole genome resequencing in *Cynara cardunculus*: detection of intra-specific variability," in *Proceedings of the 20th EUCARPIA General Congress: Plant Breeding: the Art of Bringing Science to Life*, Zürich, 334.
4. Ahn, Y.-K., Karna, S., Jun, T.-H., Yang, E.-Y., Lee, H.-E., Kim, J.-H., et al. (2016). Complete genome sequencing and analysis of *Capsicum annuum* varieties. *Mol. Breed.* 36:140. doi: 10.1007/s11032-016-0557-9
5. Ahn, Y. K., Yoon, M. K., and Jeon, J. S. (2013). Development of an efficient agrobacterium-mediated transformation system and production of herbicide-resistant transgenic plants in garlic (*Allium sativum* L.). *Mol. Cells* 36, 158–162. doi: 10.1007/s10059-013-0142-6
6. Ali, Z., Abulfaraj, A., Idris, A., Ali, S., Tashkandi, M., and Mahfouz, M. M. (2015). CRISPR/Cas9-mediated viral interference in plants. *Genome Biol.* 16:238. doi: 10.1186/s13059-015-0799-6
7. Altpeter, F., Springer, N. M., Bartley, L. E., Blechl, A. E., Brutnell, T. P., Citovsky, V., et al. (2016). Advancing crop transformation in the era of genome editing. *Plant Cell* 28, 1510–1520. doi: 10.1105/tpc.16.00196
8. Anderson, P. K., Cunningham, A. A., Patel, N. G., Morales, F. J., Epstein, P. R., and Daszak, P. (2004). Emerging infectious diseases of plants: pathogen pollution, climate change and agrotechnology drivers. *Trends Ecol. Evol.* 19, 535–544. doi: 10.1016/j.tree.2004.07.021
9. Anzidei, M., Bennici, A., Schiff, S., Tani, C., and Mori, B. (2000). Organogenesis and somatic embryogenesis in *Foeniculum vulgare*: histological observations of developing embryogenic callus. *Plant Cell Tissue Organ Cult.* 61, 69–79. doi: 10.1023/A:1006454702620
10. Aragão, F. J. L., and Campos, F. A. P. (2007). "Common bean and cowpea," in *Transgenic Crops IV*, eds E.-C. Pua and M. R. Davey (Heidelberg: Springer Berlin Heidelberg), 263–276. doi: 10.1007/978-3-540-36752-9_14
11. Aswath, C. R., Mo, S. Y., Kim, D. H., and Park, S. W. (2006). *Agrobacterium* and biolistic transformation of onion using non-antibiotic selection marker phosphomannose isomerase. *Plant Cell Rep.* 25, 92–99. doi: 10.1007/s00299-005-0022-4

12. Bai, Y., and Lindhout, P. (2007). Domestication and breeding of tomatoes: what have we gained and what can we gain in the future? *Ann. Bot.* 100, 1085–1094. doi: 10.1093/aob/mcm150
13. Bakshi, S., Saha, B., Roy, N. K., Mishra, S., Panda, S. K., and Sahoo, L. (2012). Successful recovery of transgenic cowpea (*Vigna unguiculata*) using the 6-phosphomannose isomerase gene as the selectable marker. *Plant Cell Rep.* 31, 1093–1103. doi: 10.1007/s00299-012-1230-3
14. Bakshi, S., and Sahoo, L. (2013). How relevant is recalcitrance for the recovery of transgenic cowpea: implications of selection strategies. *J. Plant Growth Regul.* 32, 148–158. doi: 10.1007/s00344-012-9284-6
15. Baltes, N. J., Hummel, A. W., Konecna, E., Cegan, R., Bruns, A. N., Bisaro, D. M., et al. (2015). Conferring resistance to geminiviruses with the CRISPR–Cas prokaryotic immune system. *Nat. Plants* 1:15145. doi: 10.1038/nplants.2015.145
16. Baranski, R. (2008). Genetic transformation of carrot (*Daucus carota*) and other Apiaceae species. *Transgenic Plant J.* 2, 18–38.
17. Barchi, L., Delledonne, M., Lanteri, S., Dal Molin, A., Minio, A., Ferrarini, A., et al. (2016). "A high quality eggplant genome sequence: a new tool for the analysis of Solanaceae family evolution and for the molecular deciphering of complex traits," in *Proceedings of the 20th EUCARPIA General Congress*, Zurich.
18. Barraza, A., Cabrera-Ponce, J. L., Gamboa-Becerra, R., Luna-Martinez, F., Winkler, R., and Alvarez-Venegas, R. (2015). The *Phaseolus vulgaris* PvTRX1h gene regulates plant hormone biosynthesis in embryogenic callus from common bean. *Front. Plant Sci.* 6:577. doi: 10.3389/fpls.2015.00577
19. Baskaran, P., Soós, V., Balázs, E., and Van Staden, J. (2016). Shoot apical meristem injection: a novel and efficient method to obtain transformed cucumber plants. *S. Afr. J. Bot.* 103, 210–215. doi: 10.1016/j.sajb.2015.09.006
20. Bezirganoglu, I., Hwang, S. Y., Shaw, J. F., and Fang, T. J. (2014). Efficient production of transgenic melon via *Agrobacterium*-mediated transformation. *Genet. Mol. Res.* 13, 3218–3227. doi: 10.4238/2014.April.25.7
21. Bolger, A., Scossa, F., Bolger, M. E., Lanz, C., Maumus, F., Tohge, T., et al. (2014). The genome of the stress-tolerant wild tomato species *Solanum pennellii*. *Nat. Genet.* 46, 1034–1038. doi: 10.1038/ng.3046
22. Brooks, C., Nekrasov, V., Lippman, Z. B., and Van Eck, J. (2014). Efficient gene editing in tomato in the first generation using the clust ered regularly interspaced short palindromic repeats/CRISPR-associated9 system. *Plant Physiol.* 166, 1292–1297. doi: 10.1104/pp.114.247577
23. Cardi, T. (2016). Cisgenesis and genome editing: combining concepts and efforts for a smarter use of genetic resources in crop breeding. *Plant Breed.* 135, 139–147. doi: 10.1111/pbr.12345
24. Cardi, T., and Stewart, C. N. Jr (2016). Progress of targeted genome modification approaches in higher plants. *Plant Cell Rep.* 35, 1401–1416. doi: 10.1007/s00299-016-1975-1
25. Cericola, F., Portis, E., Lanteri, S., Toppino, L., Barchi, L., Acciarri, N., et al. (2014). Linkage disequilibrium and genome-wide association analysis for anthocyanin pigmentation and fruit color in eggplant. *BMC Genomics* 15:896. doi: 10.1186/1471-2164-15-896
26. Cermak, T., Baltes, N. J., Cegan, R., Zhang, Y., and Voytas, D. F. (2015). High-frequency, precise modification of the tomato genome. *Genome Biol.* 16, 232. doi: 10.1186/s13059-015-0796-9
27. Chandrasekaran, J., Brumin, M., Wolf, D., Leibman, D., Klap, C., Pearlsman, M., et al. (2016). Development of broad virus resistance in non-transgenic cucumber using CRISPR/Cas9 technology. *Mol. Plant Pathol.* 17, 1140–1153. doi: 10.1111/mpp.12375
28. Chen, W., Chiu, C., Liu, H., Lee, T., Cheng, J., Lin, C., et al. (1998). Gene transfer via pollen-tube pathway for anti-fusarium wilt in watermelon. *Biochem. Mol. Biol. Int.* 46, 1201–1209. doi: 10.1080/15216549800204762

29. Chin, D. P., Bao, J. H., and Mii, M. (2009). Transgenic spinach plants produced by *Agrobacterium*-mediated method based on the low temperature-dependent high plant regeneration ability of leaf explants. *Plant Biotechnol.* 26, 243–248. doi: 10.5511/plantbiotechnology.26.243
30. Chowrira, G. M., Akella, V., Fuerst, P. E., and Lurquin, P. F. (1996). Transgenic grain legumes obtained byin planta electroporation-mediated gene transfer. *Mol. Biotechnol.* 5, 85–96. doi: 10.1007/BF02789058
31. Cichy, K. A., Wiesinger, J. A., and Mendoza, F. A. (2015). Genetic diversity and genome-wide association analysis of cooking time in dry bean (*Phaseolus vulgaris* L.). *Theor. Appl. Genet.* 128, 1555–1567. doi: 10.1007/s00122-015-2531-z
32. Citadin, C. T., Ibrahim, A. B., and Aragão, F. J. L. (2011). Genetic engineering in Cowpea (*Vigna unguiculata*): history, status and prospects. *GM Crops* 2, 144–149. doi: 10.4161/gmcr.2.3.18069
33. Curtis, I. S., and Nam, H. G. (2001). Transgenic radish (*Raphanus sativus* L. longipinnatus Bailey) by floral-dip method – plant development and surfactant are important in optimizing transformation efficiency. *Transgenic Res.* 10, 363–371. doi: 10.1023/A:1016600517293
34. Davey, M. R., Anthony, P., Van Hooff, P., Power, J. B., and Lowe, K. C. (2007). "Lettuce," in *Biotechnology in Agriculture and Forestry. Transgenic Crops IV*, eds E. C. Pua and M. R. Davey (Heidelberg: Springer-Verlag), 221–249. doi: 10.1007/978-3-540-36752-9_12
35. Derks, F. H. M. (1992). *Organelle Transfer by Protoplast Fusion in Solanaceae.* Ph.D. thesis, Free University of Amsterdam, Amsterdam.
36. Desjardins, Y. (2014). "Fruit and vegetables and health: an overview," in *Horticulture: Plants for People and Places*, eds G. R. Dixon and D. E. Aldous (Dordrecht: Springer Science+Business Media), 965–1000. doi: 10.1007/978-94-017-8560-0_2
37. Dewir, Y. H., Murthy, H. N., Ammar, M. H., Alghamdi, S. S., Al-Suhaibani, N. A., Alsadon, A. A., et al. (2016). In vitro rooting of leguminous plants: difficulties, alternatives, and strategies for improvement. *Hortic. Environ. Biotechnol.* 57, 311–322. doi: 10.1007/s13580-016-0060-6
38. Dlugosz, E. M., Lenaghan, S. C., and Stewart, C. N. Jr. (2016). A robotic platform for high-throughput protoplast isolation and transformation. *J. Vis. Exp.* 2016:e54300. doi: 10.3791/54300
39. Duclercq, J., Sangwan-Norreel, B., Catterou, M., and Sangwan, R. S. (2011). De novo shoot organogenesis: from art to science. *Trends Plant Sci.* 16, 597–606. doi: 10.1016/j.tplants.2011.08.004
40. Eady, C., Davis, S., Catanach, A., Kenel, F., and Hunger, S. (2005). *Agrobacterium tumefaciens*-mediated transformation of leek (*Allium porrum*) and garlic (*Allium sativum*). *Plant Cell Rep.* 24, 209–215. doi: 10.1007/s00299-005-0926-z
41. Eady, C. C., Weld, R. J., and Lister, C. E. (2000). *Agrobacterium tumefaciens*-mediated transformation and transgenic-plant regeneration of onion (*Allium cepa* L.). *Plant Cell Rep.* 19, 376–381. doi: 10.1007/s002990050743
42. Egelie, K. J., Graff, G. D., Strand, S. P., and Johansen, B. (2016). The emerging patent landscape of CRISPR-Cas gene editing technology. *Nat. Biotechnol.* 34, 1025–1031. doi: 10.1038/nbt.3692
43. Fernandez-Pozo, N., Menda, N., Edwards, J. D., Saha, S., Tecle, I. Y., Strickler, S. R., et al. (2015). The Sol Genomics Network (SGN)–from genotype to phenotype to breeding. *Nucleic Acids Res.* 43, D1036–D1041. doi: 10.1093/nar/gku1195
44. Finer, J. J. (2010). "Plant nuclear transformation," in *Genetic Modification of Plants*, 1st Edn, eds F. Kempken and C. Jung (Heidelberg: Springer-Verlag), 499–550.
45. Finkers, R., Van Workum, W., Van Kaauwen, M., Huits, H., Jungerius, A., Vosman, B., et al. (2015). "A de novo assembly for the 16GB *Allium cepa* genome, tears of joy," in *Proceedings of the Largest Ag-Genomics Meeting in the World: Plant & Animal Genome XXIII*, San Diego, CA.

46. Fu, N., Wang, Q., and Shen, H. L. (2013). De novo assembly, gene annotation and marker development using Illumina paired-end transcriptome sequences in celery (*Apium graveolens* L.). *PLoS ONE* 8:e57686. doi: 10.1371/journal.pone.0057686
47. Galla, G., Chedina, A., Tiozzo, S., and Barcaccia, G. (2014). "Towards a first high-quality genome draft for leaf chicory, radicchio (*Cichorium intybus* L.)," in *Proceedings of the 58th Italian Society of Agricultural Genetics Annual Congress*, Alghero.
48. Gantait, S., Mandal, N., and Das, P. K. (2010). An Overview on in vitro culture of genus allium. *Am. J. Plant Physiol.* 5, 325–337. doi: 10.3923/ajpp.2010.325.337
49. Garcia-Mas, J., Benjak, A., Sanseverino, W., Bourgeois, M., Mir, G., Gonzalez, V. M., et al. (2012). The genome of melon (*Cucumis melo*L.). *Proc. Natl. Acad. Sci. U.S.A.* 109, 11872–11877. doi: 10.1073/pnas.1205415109
50. Gatti, I., Guindón, F., Bermejo, C., Espósito, A., and Cointry, E. (2016). In vitro tissue culture in breeding programs of leguminous pulses: use and current status. *Plant Cell Tissue Organ Cult.* 127, 543–559. doi: 10.1007/s11240-016-1082-6
51. Geekiyanage, S., Takase, T., Watanabe, S., Fukai, S., and Kiyosue, T. (2006). The combined effect of photoperiod, light intensity and GA_3on adventitious shoot regeneration from cotyledons of spinach (*Spinacia oleracea* L.). *Plant Biotechnol.* 23, 431–435. doi: 10.5511/plantbiotechnology.23.431
52. Gelvin, S. B. (2003). Improving plant genetic engineering by manipulating the host. *Trends Biotechnol.* 21, 95–98. doi: 10.1016/S0167-7799(03)00005-2
53. Gómez-Montes, E. O., Oliver-Salvador, C., Durán-Figueroa, N., Badillo-Corona, J. A., and Salas, C. E. (2015). Optimization of direct shoot regeneration using cotyledonary explants and true leaves from lettuce cv. Romaine (*Lactuca sativa* L.) by surface response methodology. *Plant Growth Regul.* 77, 327–334. doi: 10.1007/s10725-015-0067-5
54. Gonthier, L., Bellec, A., Blassiau, C., Prat, E., Helmstetter, N., Rambaud, C., et al. (2010). Construction and characterization of two BAC libraries representing a deep-coverage of the genome of chicory (*Cichorium intybus* L., Asteraceae). *BMC Res. Notes* 3:225. doi: 10.1186/1756-0500-3-225
55. Guo, S., Zhang, J., Sun, H., Salse, J., Lucas, W. J., Zhang, H., et al. (2013). The draft genome of watermelon (*Citrullus lanatus*) and resequencing of 20 diverse accessions. *Nat. Genet.* 45, 51–58. doi: 10.1038/ng.2470
56. Hand, M. L., and Koltunow, A. M. (2014). The genetic control of apomixis: asexual seed formation. *Genetics* 197, 441–450. doi: 10.1534/genetics.114.163105
57. Heidmann, I., De Lange, B., Lambalk, J., Angenent, G. C., and Boutilier, K. (2011). Efficient sweet pepper transformation mediated by the BABY BOOM transcription factor. *Plant Cell Rep.* 30, 1107–1115. doi: 10.1007/s00299-011-1018-x
58. Hirakawa, H., Shirasawa, K., Miyatake, K., Nunome, T., Negoro, S., Ohyama, A., et al. (2014). Draft genome sequence of eggplant (*Solanum melongena* L.): the representative *solanum* species indigenous to the old world. *DNA Res.* 21, 649–660. doi: 10.1093/dnares/dsu027
59. Hnatuszko-Konka, K., Kowalczyk, T., Gerszberg, A., Wiktorek-Smagur, A., and Kononowicz, A. K. (2014). *Phaseolus vulgaris* — Recalcitrant potential. *Biotechnol. Adv.* 32, 1205–1215. doi: 10.1016/j.biotechadv.2014.06.001
60. Huang, S., Li, R., Zhang, Z., Li, L., Gu, X., Fan, W., et al. (2009). The genome of the cucumber, *Cucumis sativus* L. *Nat. Genet.* 41, 1275–1281. doi: 10.1038/ng.475
61. Ikeuchi, M., Ogawa, Y., Iwase, A., and Sugimoto, K. (2016). Plant regeneration: cellular origins and molecular mechanisms. *Development*143, 1442–1451. doi: 10.1242/dev.134668
62. Ilori, C. O., and Pellegrineschi, A. (2011). Transgene expression in cowpea (*Vigna unguiculata* (L.) Walp.) through *Agrobacterium*transformation of pollen in flower buds. *Afr. J. Biotechnol.* 10, 11821–11828.
63. Iorizzo, M., Ellison, S., Senalik, D., Zeng, P., Satapoomin, P., Huang, J., et al. (2016). A high-quality carrot genome assembly provides new insights into carotenoid

accumulation and asterid genome evolution. *Nat. Genet.* 48, 657–666. doi: 10.1038/ng.3565

64. Iorizzo, M., Senalik, D. A., Ellison, S. L., Grzebelus, D., Cavagnaro, P. F., Allender, C., et al. (2013). Genetic structure and domestication of carrot (*Daucus carota* subsp. sativus) (Apiaceae). *Am. J. Bot.* 100, 930–938. doi: 10.3732/ajb.1300055
65. Ito, Y., Nishizawa-Yokoi, A., Endo, M., Mikami, M., and Toki, S. (2015). CRISPR/Cas9-mediated mutagenesis of the RIN locus that regulates tomato fruit ripening. *Biochem. Biophys. Res. Commun.* 467, 76–82. doi: 10.1016/j.bbrc.2015.09.117
66. Jakhar, M. L., and Choudhary, M. R. (2012). Regeneration of in vitro plantlets through organogenesis in fennel (*Foeniculum Vulgare*Mill.). *J. Plant Sci. Res.* 28, 203–209.
67. Jenko, J., Gorjanc, G., Cleveland, M. A., Varshney, R. K., Whitelaw, C. B., Woolliams, J. A., et al. (2015). Potential of promotion of alleles by genome editing to improve quantitative traits in livestock breeding programs. *Genet. Sel. Evol.* 47, 55. doi: 10.1186/s12711-015-0135-3
68. Kamfwa, K., Cichy, K. A., and Kelly, J. D. (2015). Genome-wide association analysis of symbiotic nitrogen fixation in common bean. *Theor. Appl. Genet.* 128, 1999–2017. doi: 10.1007/s00122-015-2562-5
69. Kaur, S., Pembleton, L. W., Cogan, N. O., Savin, K. W., Leonforte, T., Paull, J., et al. (2012). Transcriptome sequencing of field pea and faba bean for discovery and validation of SSR genetic markers. *BMC Genomics* 13:104. doi: 10.1186/1471-2164-13-104
70. Kenel, F., Eady, C., and Brinch, S. (2010). Efficient *Agrobacterium tumefaciens*-mediated transformation and regeneration of garlic (*Allium sativum*) immature leaf tissue. *Plant Cell Rep.* 29, 223–230. doi: 10.1007/s00299-009-0814-z
71. Kim, S., Park, M., Yeom, S. I., Kim, Y. M., Lee, J. M., Lee, H. A., et al. (2014). Genome sequence of the hot pepper provides insights into the evolution of pungency in *Capsicum* species. *Nat. Genet.* 46, 270–278. doi: 10.1038/ng.2877
72. Kitashiba, H., Li, F., Hirakawa, H., Kawanabe, T., Zou, Z., Hasegawa, Y., et al. (2014). Draft sequences of the radish (*Raphanus sativus*L.) *genome. DNA Res.* 21, 481–490. doi: 10.1093/dnares/dsu014
73. Klenotičová, H., Smykalová, I., Švábová, L., and Griga, M. (2013). Resolving browning during the establishment of explant cultures in Vicia faba L. for genetic transformation. *Acta Univ. Agric. Silvic. Mendel. Brun.* 61, 1279–1288. doi: 10.11118/actaun201361051279
74. Klocke, E., Nothnagel, T., and Schumann, G. (2010). "Vegetables," in *Genetic Modification of Plants*, 1st Edn, eds F. Kempken and C. Jung (Heidelberg: Springer-Verlag).
75. Koornneef, M., Bade, J., Hanhart, C., Horsman, K., Schel, J., Soppe, W., et al. (1993). Characterization and mapping of a gene controlling shoot regeneration in tomato. *Plant J.* 3, 131–141. doi: 10.1111/j.1365-313X.1993.tb00016.x
76. Kothari, S. L., Joshi, A., Kachhwaha, S., and Ochoa-Alejo, N. (2010). Chilli peppers–a review on tissue culture and transgenesis. *Biotechnol. Adv.* 28, 35–48. doi: 10.1016/j.biotechadv.2009.08.005
77. Kumar, P., and Srivastava, D. K. (2016). Biotechnological advancement in genetic improvement of broccoli (*Brassica oleracea* L. var. italica), an important vegetable crop. *Biotechnol. Lett.* 38, 1049–1063. doi: 10.1007/s10529-016-2080-9
78. Kurtz, S. M., and Lineberger, R. D. (1983). Genotypic differences in morphogenic capacity of cultured leaf explants of tomato [*Lycopersicon esculentum*]. *J. Am. Soc. Hortic. Sci.* 108, 710–714.
79. Lawrenson, T., Shorinola, O., Stacey, N., Li, C., Ostergaard, L., Patron, N., et al. (2015). Induction of targeted, heritable mutations in barley and *Brassica oleracea* using RNA-guided Cas9 nuclease. *Genome Biol.* 16, 258. doi: 10.1186/s13059-015-0826-7
80. Ledford, H. (2016). A better way to hack plant DNA. *Nature* 539, 16–17. doi: 10.1038/539016a

81. Li, S. F., Gao, W. J., Zhao, X. P., Dong, T. Y., Deng, C. L., and Lu, L. D. (2014). Analysis of transposable elements in the genome of *Asparagus officinalis* from high coverage sequence data. *PLoS ONE* 9:e97189. doi: 10.1371/journal.pone.0097189
82. Lin, K., Zhang, N., Severing, E. I., Nijveen, H., Cheng, F., Visser, R. G., et al. (2014). Beyond genomic variation–comparison and functional annotation of three *Brassica rapa* genomes: a turnip, a rapid cycling and a Chinese cabbage. *BMC Genomics* 15:250. doi: 10.1186/1471-2164-15-250
83. Lin, T., Zhu, G., Zhang, J., Xu, X., Yu, Q., Zheng, Z., et al. (2014). Genomic analyses provide insights into the history of tomato breeding. *Nat. Genet.* 46, 1220–1226. doi: 10.1038/ng.3117
84. Liu, L., Gu, Q., Ijaz, R., Zhang, J., and Ye, Z. (2016). Generation of transgenic watermelon resistance to *Cucumber mosaic virus*facilitated by an effective *Agrobacterium*-mediated transformation method. *Sci. Hortic.* 205, 32–38. doi: 10.1016/j.scienta.2016.04.013
85. Liu, S., Liu, Y., Yang, X., Tong, C., Edwards, D., Parkin, I. A., et al. (2014). The *Brassica oleracea* genome reveals the asymmetrical evolution of polyploid genomes. *Nat. Commun.* 5:3930. doi: 10.1038/ncomms4930
86. Liu, Z., Park, B.-J., Kanno, A., and Kameya, T. (2005). The novel use of a combination of sonication and vacuum infiltration in *Agrobacterium*-mediated transformation of kidney bean (*Phaseolus vulgaris* L.) with lea gene. *Mol. Breed.* 16, 189.
87. Lombardi-Crestana, S., Azevedo, M. D., Silva, G. F. F. E., Pino, L. E., Appezzato-Da-Gloria, B., Figueira, A., et al. (2012). The tomato (*Solanum lycopersicum* cv. Micro-Tom) natural genetic variation Rg1 and the DELLA mutant procera control the competence necessary to form adventitious roots and shoots. *J. Exp. Bot.* 63, 5689–5703. doi: 10.1093/jxb/ers221
88. Lor, V. S., Starker, C. G., Voytas, D. F., Weiss, D., and Olszewski, N. E. (2014). Targeted mutagenesis of the tomato PROCERA gene using transcription activator-like effector nucleases. *Plant Physiol.* 166, 1288–1291. doi: 10.1104/pp.114.247593
89. Macas, J., Neumann, P., and Navratilova, A. (2007). Repetitive DNA in the pea (*Pisum sativum* L.) genome: comprehensive characterization using 454 sequencing and comparison to soybean and *Medicago truncatula*. *BMC Genomics* 8:427. doi: 10.1186/1471-2164-8-427
90. Maggioni, L. (2004). "Conservation and use of vegetable genetic resources: a european perspective," in *Proceedings of the International Society for Horticultural Science (ISHS)*, Leuven, 13–30. doi: 10.17660/actahortic.2004.637.1
91. Mahfouz, M. M., Cardi, T., and Neal Stewart, C. Jr (2016). Next-generation precision genome engineering and plant biotechnology. *Plant Cell Rep.* 35, 1397–1399. doi: 10.1007/s00299-016-2009-8
92. Manamohan, M., Prakash, N., and Sharath Chandra, G. (2011). "Cucurbits," in *Advances in Horticulture Biotechnology*, eds K. Nirmal Babu, H. P. Singh, and V. A. Parthasarathy (New Delhi: Westville Publishing House), 227–252.
93. Matvieieva, N. A. (2015). *Agrobacterium*-mediated transformation of compositae plants. I. Construction of transgenic plants and "hairy" roots with new properties. *Biotechnol. Acta* 8, 19–31. doi: 10.15407/biotech8.01.019
94. Menin, B., Moglia, A., Comino, C., Lanteri, S., Van Herpen, T. W. J. M., and Beekwilder, J. (2012). "In vitro callogenesis and *Agrobacterium*-mediated transformation of globe artichoke," in *Proceedings of the International Society for Horticultural Science (ISHS)*, Leuven, 267–271. doi: 10.17660/actahortic.2012.961.34
95. Michael, T. P., and VanBuren, R. (2015). Progress, challenges and the future of crop genomes. *Curr. Opin. Plant Biol.* 24, 71–81. doi: 10.1016/j.pbi.2015.02.002
96. Mikschofsky, H., and Broer, I. (2012). Feasibility of *Pisum sativum* as an expression system for pharmaceuticals. *Transgenic Res.* 21, 715–724. doi: 10.1007/s11248-011-9573-z
97. Milojević, J., Tubić, L., Nolić, V., Mitić, N., Ćalić-Dragosavac, D., Vinterhalter, B., et al. (2012). Hygromycin promotes somatic embryogenesis in spinach. *Plant Cell Tissue Organ Cult.* 109, 573–579. doi: 10.1007/s11248-011-9573-z

98. Nadolska-Orczyk, A., and Malepszy, S. (1989). In vitro culture of *Cucumis sativus* L. *Theor. Appl. Genet.* 78, 836–840. doi: 10.1007/BF00266667

99. Nanasato, Y., Okuzaki, A., and Tabei, Y. (2013). Improving the transformation efficiency of *Cucurbita* species: factors and strategy for practical application. *Plant Biotechnol.* 30, 287–294. doi: 10.5511/plantbiotechnology.13.0331a

100. Nguyen, A. H., Hodgson, L. M., Erskine, W., and Barker, S. J. (2016). An approach to overcoming regeneration recalcitrance in genetic transformation of lupins and other legumes. *Plant Cell Tissue Organ Cult.* 127, 623–635. doi: 10.1007/s11240-016-1087-1

101. Nguyen, Q. V., Boo, K. H., Sun, H. J., Cao, D. V., Lee, D., Ko, S. H., et al. (2013a). Evaluation of factors influencing *Agrobacterium*-mediated spinach transformation and transformant selection by EGFP fluorescence under low-selective pressure. *In vitro Cell. Dev. Biol. Plant* 49, 498–509. doi: 10.1007/s11627-013-9534-8

102. Nguyen, Q. V., Sun, H. J., Boo, K. H., Lee, D., Lee, J.-H., Lim, P. O., et al. (2013b). Effect of plant growth regulator combination and culture period on in vitro regeneration of spinach (*Spinacia oleracea* L.). *Plant Biotechnol. Rep.* 7, 99–108. doi: 10.1007/s11816-012-0242-3

103. Nimmakayala, P., Abburi, V. L., Saminathan, T., Almeida, A., Davenport, B., Davidson, J., et al. (2016a). Genome-wide divergence and linkage disequilibrium analyses for *Capsicum baccatum* revealed by genome-anchored single nucleotide polymorphisms. *Front. Plant Sci.* 7:1646. doi: 10.3389/fpls.2016.01646

104. Nimmakayala, P., Levi, A., Abburi, L., Abburi, V. L., Tomason, Y. R., Saminathan, T., et al. (2014). Single nucleotide polymorphisms generated by genotyping by sequencing to characterize genome-wide diversity, linkage disequilibrium, and selective sweeps in cultivated watermelon. *BMC Genomics* 15:767. doi: 10.1186/1471-2164-15-767

105. Nimmakayala, P., Tomason, Y. R., Abburi, V. L., Alvarado, A., Saminathan, T., Vajja, V. G., et al. (2016b). Genome-wide differentiation of various melon horticultural groups for use in GWAS for fruit firmness and construction of a high resolution genetic map. *Front. Plant Sci.* 7:1437.

106. Nogue, F., Mara, K., Collonnier, C., and Casacuberta, J. M. (2016). Genome engineering and plant breeding: impact on trait discovery and development. *Plant Cell Rep.* 35, 1475–1486. doi: 10.1007/s00299-016-1993-z

107. O'Sullivan, D. M., and Angra, D. (2016). Advances in faba bean genetics and genomics. *Front. Genet.* 7:150. doi: 10.3389/fgene.2016.00150

108. Park, J.-I., Ahmed, N. U., Kim, H.-R., and Nou, I.-S. (2012). Advances in in vitro culture of the Brassicaceae crop plants. *J. Plant Biotechnol.* 39, 13–22. doi: 10.1007/978-1-4939-1695-5_23

109. Perseguini, J. M., Oblessuc, P. R., Rosa, J. R., Gomes, K. A., Chiorato, A. F., Carbonell, S. A., et al. (2016). Genome-wide association studies of anthracnose and angular leaf spot resistance in common bean (*Phaseolus vulgaris* L.). *PLoS ONE* 11:e0150506. doi: 10.1371/journal.pone.0150506

110. Piatek, A., and Mahfouz, M. M. (2016). Targeted genome regulation via synthetic programmable transcriptional regulators. *Crit. Rev. Biotechnol.* doi: 10.3109/07388551.2016.1165180 [Epub ahead of print].

111. Pino, L. E., Lombardi-Crestana, S., Azevedo, M. S., Scotton, D. C., Borgo, L., Quecini, V., et al. (2010). The Rg1 allele as a valuable tool for genetic transformation of the tomato 'Micro-Tom' model system. *Plant Methods* 6:23. doi: 10.1186/1746-4811-6-23

112. Portis, E., Cericola, F., Barchi, L., Toppino, L., Acciarri, N., Pulcini, L., et al. (2015). Association mapping for fruit, plant and leaf morphology traits in eggplant. *PLoS ONE* 10:e0135200. doi: 10.1371/journal.pone.0135200

113. Pulianmackal, A. J., Kareem, A. V., Durgaprasad, K., Trivedi, Z. B., and Prasad, K. (2014). Competence and regulatory interactions during regeneration in plants. *Front. Plant Sci.* 5:142. doi: 10.3389/fpls.2014.00142

114. Punja, Z. K., Jayaraj, J., and Wally, O. (2007). "Carrot," in *Biotechnology in Agriculture and Forestry. Transgenic Crops IV*, eds E. C. Pua and M. R. Davey (Heidelberg: Springer-Verlag), 277–294. doi: 10.1007/978-3-540-36752-9_15

115. Qin, C., Yu, C., Shen, Y., Fang, X., Chen, L., Min, J., et al. (2014). Whole-genome sequencing of cultivated and wild peppers provides insights into *Capsicum* domestication and specialization. *Proc. Natl. Acad. Sci. U.S.A.* 111, 5135–5140. doi: 10.1073/pnas.1400975111

116. Ranc, N., Munos, S., Xu, J., Le Paslier, M. C., Chauveau, A., Bounon, R., et al. (2012). Genome-wide association mapping in tomato (*Solanum lycopersicum*) is possible using genome admixture of *Solanum lycopersicum* var. *cerasiforme. G3 (Bethesda)* 2, 853–864. doi: 10.1534/g3.112.002667

117. Rani, R., Yadav, P., Barbadikar, K. M., Baliyan, N., Malhotra, E. V., Singh, B. K., et al. (2016). CRISPR/Cas9: a promising way to exploit genetic variation in plants. *Biotechnol. Lett.* 38, 1991–2006. doi: 10.1007/s10529-016-2195-z

118. Ron, M., Kajala, K., Pauluzzi, G., Wang, D., Reynoso, M. A., Zumstein, K., et al. (2014). Hairy root transformation using *Agrobacterium*rhizogenes as a tool for exploring cell type-specific gene expression and function using tomato as a model. *Plant Physiol.* 166, 455–469. doi: 10.1104/pp.114.239392

119. Rotino, G. L., Sala, T., and Toppino, L. (2014). "Eggplant," in *Alien Gene Transfer in Crop Plants*, 1st Edn, Vol. 2, eds A. Pratap and J. Kumar (New York, NY: Springer-Verlag).

120. Ruggieri, V., Francese, G., Sacco, A., D'alessandro, A., Rigano, M. M., Parisi, M., et al. (2014). An association mapping approach to identify favourable alleles for tomato fruit quality breeding. *BMC Plant Biol.* 14:337. doi: 10.1186/s12870-014-0337-9

121. Sauvage, C., Segura, V., Bauchet, G., Stevens, R., Do, P. T., Nikoloski, Z., et al. (2014). Genome-wide association in tomato reveals 44 candidate loci for fruit metabolic traits. *Plant Physiol.* 165, 1120–1132. doi: 10.1104/pp.114.241521

122. Saxena, S. N., Kothari, P., Rathore, S. S., Ishan-Ulla-Khan, and Saxena, R. (2012). Organogenesis in fennel (*Foeniculum vulgare* Mill.). *Int. J. Seed Spices* 2, 1–4.

123. Scaglione, D., Lanteri, S., Acquadro, A., Lai, Z., Knapp, S. J., Rieseberg, L., et al. (2012). Large-scale transcriptome characterization and mass discovery of SNPs in globe artichoke and its related taxa. *Plant Biotechnol. J.* 10, 956–969. doi: 10.1111/j.1467-7652.2012.00725.x

124. Scaglione, D., Reyes-Chin-Wo, S., Acquadro, A., Froenicke, L., Portis, E., Beitel, C., et al. (2016). The genome sequence of the outbreeding globe artichoke constructed de novo incorporating a phase-aware low-pass sequencing strategy of F1 progeny. *Sci. Rep.*6:19427. doi: 10.1038/srep19427

125. Schinkel, H., and Schillberg, S. (2016). Genome editing: intellectual property and product development in plant biotechnology. *Plant Cell Rep.* 35, 1487–1491. doi: 10.1007/s00299-016-1988-9

126. Schmutz, J., Mcclean, P. E., Mamidi, S., Wu, G. A., Cannon, S. B., Grimwood, J., et al. (2014). A reference genome for common bean and genome-wide analysis of dual domestications. *Nat. Genet.* 46, 707–713. doi: 10.1038/ng.3008

127. Shirasawa, K., Fukuoka, H., Matsunaga, H., Kobayashi, Y., Kobayashi, I., Hirakawa, H., et al. (2013). Genome-wide association studies using single nucleotide polymorphism markers developed by re-sequencing of the genomes of cultivated tomato. *DNA Res.* 20, 593–603. doi: 10.1093/dnares/dst033

128. Silva Dias, J. (2014). Guiding strategies for breeding vegetable cultivars. *Agric. Sci.* 5, 9–32. doi: 10.4236/as.2014.51002

129. Solís-Ramos, L. Y., González-Estrada, T., Nahuath-Dzib, S., Zapata-Rodriguez, L. C., and Castaño, E. (2009). Overexpression of WUSCHEL in *C. chinense* causes ectopic morphogenesis. *Plant Cell Tissue Organ Cult.* 96, 279–287. doi: 10.1007/s11240-008-9485-7

130. Somers, D. A., Samac, D. A., and Olhoft, P. M. (2003). Recent advances in legume transformation. *Plant Physiol.* 131, 892–899. doi: 10.1104/pp.102.017681

131. Song, D., Han, Q., Dong, Z., and He, Z. (2014). Genetic transformation of lettuce (*Lactuca sativa*): a review. *Afr. J. Biotechnol.* 13, 1686–1693. doi: 10.5897/AJB2014.13651
132. Song, G.-Q., Loskutov, A. V., and Sink, K. C. (2007). Highly efficient *Agrobacterium tumefaciens*-mediated transformation of celery (*Apium graveolens* L.) through somatic embryogenesis. *Plant Cell Tissue Organ Cult.* 88, 193–200. doi: 10.1007/s11240-006-9190-3
133. Soyk, S., Muller, N. A., Park, S. J., Schmalenbach, I., Jiang, K., Hayama, R., et al. (2017). Variation in the flowering gene SELF PRUNING 5G promotes day-neutrality and early yield in tomato. *Nat. Genet.* 49, 162–168. doi: 10.1038/ng.3733
134. Sparrow, P. A. C., Dale, P. J., and Irwin, J. A. (2004). The use of phenotypic markers to identify *Brassica oleracea* genotypes for routine high-throughput *Agrobacterium*-mediated transformation. *Plant Cell Rep.* 23, 64–70. doi: 10.1007/s00299-004-0818-7
135. Sparrow, P. A. C., Goldsack, C. M. P., and Østergaard, L. (2011). "Transformation technology in the Brassicaceae," in *Genetics and Genomics of the Brassicaceae*, eds R. Schmidt and I. Bancroft (New York, NY: Springer-Verlag), 505–525. doi: 10.1007/978-1-4419-7118-0_18
136. Sprink, T., Eriksson, D., Schiemann, J., and Hartung, F. (2016). Regulatory hurdles for genome editing: process- vs. product-based approaches in different regulatory contexts. *Plant Cell Rep.* 35, 1493–1506. doi: 10.1007/s00299-016-1990-2
137. Steinert, J., Schiml, S., and Puchta, H. (2016). Homology-based double-strand break-induced genome engineering in plants. *Plant Cell Rep.* 35, 1429–1438. doi: 10.1007/s00299-016-1981-3
138. Sun, K., Wolters, A. M., Vossen, J. H., Rouwet, M. E., Loonen, A. E., Jacobsen, E., et al. (2016). Silencing of six susceptibility genes results in potato late blight resistance. *Transgenic Res.* 25, 731–742. doi: 10.1007/s11248-016-9964-2
139. Sun, X., Zhou, S., Meng, F., and Liu, S. (2012). De novo assembly and characterization of the garlic (*Allium sativum*) bud transcriptome by Illumina sequencing. *Plant Cell Rep.* 31, 1823–1828. doi: 10.1007/s00299-012-1295-z
140. Svabova, L., Smykal, P., Griga, M., and Ondrej, V. (2005). *Agrobacterium*-mediated transformation of *Pisum sativum* in vitro and in vivo. *Biol. Plant.* 49, 361–370. doi: 10.1007/s10535-005-0009-6
141. Taranto, F., D'agostino, N., Greco, B., Cardi, T., and Tripodi, P. (2016). Genome-wide SNP discovery and population structure analysis in pepper (*Capsicum annuum*) using genotype-by-sequencing. *BMC Genomics* 17:943. doi: 10.1186/s12864-016-3297-7
142. Terracciano, I., Cantarella, C., and D'agostino, N. (2016). "Hybridization-based enrichment and next generation sequencing to explore genetic diversity in plants," in *Dynamics of Mathematical Models in Biology*, eds A. Rogato, V. Zazzu, and M. R. Guarracino (Cham: Springer International Publishing).
143. Timko, M. P., Rushton, P. J., Laudeman, T. W., Bokowiec, M. T., Chipumuro, E., Cheung, F., et al. (2008). Sequencing and analysis of the gene-rich space of cowpea. *BMC Genomics* 9:103. doi: 10.1186/1471-2164-9-103
144. Tomato Genome Consortium (2012). The tomato genome sequence provides insights into fleshy fruit evolution. *Nature* 485, 635–641. doi: 10.1038/nature11119
145. Truco, M. J., Ashrafi, H., Kozik, A., Van Leeuwen, H., Bowers, J., Reyes Chin Wo, S., et al. (2013). An ultra high-density, transcript-based, genetic map of lettuce. *G3 (Bethesda)* 3, 617–631. doi: 10.1534/g3.112.004929
146. van Schie, C. C., and Takken, F. L. (2014). Susceptibility genes 101: how to be a good host. *Annu. Rev. Phytopathol.* 52, 551–581. doi: 10.1146/annurev-phyto-102313-045854
147. Veltcheva, M., Svetleva, D., Petkova, S., and Perl, A. (2005). In vitro regeneration and genetic transformation of common bean (*Phaseolus vulgaris* L.)—Problems and progress. *Sci. Hortic.* 107, 2–10. doi: 10.1016/j.scienta.2005.07.005

148. Vinoth, A., and Ravindhran, R. (2015). Efficient plant regeneration of watermelon (*Citrullus lanatus* Thunb.) via somatic embryogenesis and assessment of genetic fidelity using ISSR markers. *In Vitro Cell. Dev. Biol. Plant* 52, 107–115. doi: 10.1007/s11627-015-9731-8
149. Vinterhalter, D., Sretenovič-Rajičič, T., Vinterhalter, B., and Ninkovic, S. (2007). Genetic transformation of *Brassica oleracea* vegetables. *Transgenic Plant J.* 1, 340–355.
150. Vlasova, A., Capella-Gutierrez, S., Rendon-Anaya, M., Hernandez-Onate, M., Minoche, A. E., Erb, I., et al. (2016). Genome and transcriptome analysis of the Mesoamerican common bean and the role of gene duplications in establishing tissue and temporal specialization of genes. *Genome Biol.* 17, 32. doi: 10.1186/s13059-016-0883-6
151. Voytas, D. F. (2013). Plant genome engineering with sequence-specific nucleases. *Annu. Rev. Plant Biol.* 64, 327–350. doi: 10.1146/annurev-arplant-042811-105552
152. Wang, S.-L., Ku, S. S., Ye, X.-G., He, C.-F., Kwon, S. Y., and Choi, P. S. (2015). Current status of genetic transformation technology developed in cucumber (*Cucumis sativus* L.). *J. Integr. Agric.* 14, 469–482. doi: 10.1016/S2095-3119(14)60899-6
153. Wang, X., Wang, H., Wang, J., Sun, R., Wu, J., Liu, S., et al. (2011). The genome of the mesopolyploid crop species *Brassica rapa*. *Nat. Genet.* 43, 1035–1039. doi: 10.1038/ng.919
154. Wang, Y., Yau, Y. Y., Perkins-Balding, D., and Thomson, J. G. (2011). Recombinase technology: applications and possibilities. *Plant Cell Rep.* 30, 267–285. doi: 10.1007/s00299-010-0938-1
155. Wędzony, M., Szechyńska-Hebda, M., Żur, I., Dubas, E., and Krzewska, M. (2014). "Tissue culture and regeneration: a prerequisite for alien gene transfer," in *Alien Gene Transfer in Crop Plants: Innovations, Methods and Risk Assessment*, Vol. 1, eds A. Pratap and J. Kumar (New York, NY: Springer), 43–75.
156. Woo, J. W., Kim, J., Kwon, S. I., Corvalan, C., Cho, S. W., Kim, H., et al. (2015). DNA-free genome editing in plants with preassembled CRISPR-Cas9 ribonucleoproteins. *Nat. Biotechnol.* 33, 1162–1164. doi: 10.1038/nbt.3389
157. Xu, L., and Huang, H. (2014). "Chapter One – Genetic and epigenetic controls of plant regeneration," in *Current Topics in Developmental Biology*, ed. G. Brigitte (Cambridge, MA: Academic Press), 1–33.
158. Yamamoto, M., and Nishio, T. (2014). Commonalities and differences between *Brassica* and *Arabidopsis* self-incompatibility. *Hortic. Res.* 1, 14054. doi: 10.1038/hortres.2014.54
159. Zaidi, S., Tashkandi, M., Mansoor, S., and Mahfouz, M. (2016). Engineering plant immunity: using CRISPR/Cas9 to generate virus resistance. *Front. Plant Sci.* 7:1673. doi: 10.3389/fpls.2016.01673
160. Zhang, H., Gao, P., Wang, X., and Luan, F. (2014). An improved method of *Agrobacterium tumefaciens*-mediated genetic transformation system of melon (*Cucumis melo* L.). *J. Plant Biochem. Biotechnol.* 23, 278–283. doi: 10.1007/s13562-013-0211-0
161. Zhang, J., Zhao, J., Xu, Y., Liang, J., Chang, P., Yan, F., et al. (2015). Genome-wide association mapping for tomato volatiles positively contributing to tomato flavor. *Front. Plant Sci.* 6:1042. doi: 10.3389/fpls.2015.01042
162. Zhang, N., Huang, X., Bao, Y., Wang, B., Liu, L., Dai, L., et al. (2015). Genome-wide identification and expression profiling of WUSCHEL-related homeobox (WOX) genes during adventitious shoot regeneration of watermelon (*Citrullus lanatus*). *Acta Physiol. Plant.* 37, 224. doi: 10.1007/s11738-015-1964-y
163. Zheng, S.-J., Henken, B., Kyun Ahn, Y., Krens, F. A., and Kik, C. (2004). The development of a reproducible *Agrobacterium tumefaciens*transformation system for garlic (*Allium sativum* L.) and the production of transgenic garlic resistant to beet armyworm (*Spodoptera exigua* Hübner). *Mol. Breed.* 14, 293–307. doi: 10.1023/B:MOLB.0000047775.83715.b5

164. Zheng, S.-J., Khrustaleva, L., Henken, B., Sofiari, E., Jacobsen, E., Kik, C., et al. (2001). *Agrobacterium tumefaciens*-mediated transformation of *Allium cepa* L.: the production of transgenic onions and shallots. *Mol. Breed.* 7, 101–115. doi: 10.1023/A:1011348229189

165. Zsögön, A., Cermak, T., Voytas, D., and Peres, L. E. P. (2017). Genome editing as a tool to achieve the crop ideotype and de novo domestication of wild relatives: case study in tomato. *Plant Sci.* 256, 120–130. doi: 10.1016/j.plantsci.2016.12.012

Index